# Wireless Communications, Networking and Signal Processing

# Wireless Communications, Networking and Signal Processing

Editor: Grant Harper

NY RESEARCH PRESS

New York

Published by NY Research Press
118-35 Queens Blvd., Suite 400,
Forest Hills, NY 11375, USA
www.nyresearchpress.com

Wireless Communications, Networking and Signal Processing
Edited by Grant Harper



International Standard Book Number: 978-1-63238-648-9 (Hardback)



**Cataloging-in-Publication Data**

Wireless communications, networking and signal processing / edited by Grant Harper.
p. cm.
Includes bibliographical references and index.
ISBN 978-1-63238-648-9
1. Wireless communication systems. 2. Signal processing.
3. Communication--Network analysis. I. Harper, Grant.
TK5103.2 .W57 2019
621.384--dc23

Printed in China.

# Contents

# Preface

Every book is initially just a concept; it takes months of research and hard work to give it the final shape in which the readers receive it. In its early stages, this book also went through rigorous reviewing. The notable contributions made by experts from across the globe were first molded into patterned chapters and then arranged in a sensibly sequential manner to bring out the best results.

Wireless communication enables the transmission of data between two or more points without the use of wires or any other electrical conductor. Wireless technology is a modern networking technology that is both cost and time effective. It is characterized by the use of radio wave communication and signal processing. This technology is central to modern mobile technology and telecommunication industry and is undergoing rapid advancement. Some advanced models employ sound or electromagnetic radiations like light for short and long distance communication. This book is compiled to provide information about the key principles and techniques related to wireless communications. It aims to shed light on some of the unexplored aspects of this field and provide insights into the emerging applications in various other areas of science. Those who want to broaden their expanse of knowledge will find this book immensely helpful.

It has been my immense pleasure to be a part of this project and to contribute my years of learning in such a meaningful form. I would like to take this opportunity to thank all the people who have been associated with the completion of this book at any step.

**Editor**

# 1

# An inter-frame dynamic double-threshold energy detection for spectrum sensing in cognitive radios

Caimei Fu[1], Youming Li[1*], Yucheng He[2], Ming Jin[1], Gang Wang[1] and Peng Lei[1]

**Abstract**

This paper proposes an inter-frame dynamic double threshold (IF-DDT) spectrum sensing algorithm in order to improve the sensing performance based on energy detection (ED) in cognitive radios (CRs). Based on both the activity model of the primary user (PU) and the sensing mechanism of the secondary user (SU), the proposed algorithm exploits the relationship between two adjacent sensing frames and designs dynamic double thresholds for each sensing frame to enhance spectrum sensing performance when the received energy cannot give a reliable local decision. The detection probability and false alarm probability of the proposed sensing scheme are analyzed, and an algorithm for searching the optimal dynamic double thresholds is derived with very low complexity according to the Neyman-Pearson (NP) test criterion. Theoretical analysis and simulation results show that the proposed algorithm outperforms the ED algorithm.

**Keywords:** Cognitive radios, Dynamic double thresholds, Energy detection, Inter-frame spectrum sensing

## 1 Introduction

Nowadays, wireless communications have been experiencing an increasing tension of the frequency resource which, however, is far from fully utilized under the static spectrum assignment policy [1]. Cognitive radio (CR) technology provides a promising solution to the conflict by allowing the second user (SU) to opportunistically access the licensed spectrum band when it is temporarily not occupied by the primary user (PU) [2].

Spectrum sensing is a critical aspect of CR systems that aims to identify the working state of the PU (i.e., ON or OFF, indicating whether the licensed spectrum is occupied or not, respectively) before allowing the SU temporarily access the channel without causing harmful interference to the PU. Typically, the SU operates spectrum sensing frame by frame, and each frame is divided into two parts: the sensing duration and the data duration [3–5]. The SU detects whether the licensed spectrum is accessed within the sensing duration, and then transmits data within the data duration if the sensing result is OFF (which identifies that the PU's state is OFF); otherwise, the SU keeps silent.

The accuracy of spectrum sensing is measured by two probabilities: the false alarm probability (denoted as $P_f$) and the detection probability (denoted as $P_d$). The false alarm probability is defined as the probability that the PU's state is identified as ON when its real state is OFF, while the detection probability is defined as the probability that the PU's state is identified as ON when its real state is also ON. Clearly, a low false alarm probability improves the efficiency of the unused spectrum, whereas a high detection probability reduces the resulting interference in the licensed users [4, 6].

A number of spectrum sensing methods have been proposed for CR systems, such as energy detection (ED), cyclostationarity feature detection, second-order statistics detection, matched-filtering detection, compressive sensing detection [7, 8], and multiple antenna detection [9]. The energy detection has been preferred due to its feasible applicability and low implementation complexity. The cyclostationarity feature detection differentiates noise from PU signals by exploiting the cyclostationarity features of the received signals. However, it has high

*Correspondence: liyouming@nbu.edu.cn
[1]College of Information Science and Engineering, Ningbo University, Ningbo, China
Full list of author information is available at the end of the article

computational complexity and requires the knowledge of the cyclic frequencies which is difficult to be obtained. The second-order statistics-based detection is subject to the modulations of primary signals. The matched-filtering detection is the optimum method only when perfect information about the waveforms of the PU is given. The compressive sensing detection makes use of the sparse structure of primary signals. The multiple antenna detection effectively overcomes the noise uncertainty and improves the detection performance based on the spatial correlation in the multi-antenna receiver. Nevertheless, it requires the SU equipped with multi-antenna receivers.

In order to avoid error detection caused by dramatic energy decrease, the average signal power of several past sensing frames is used in [10] for the ED schemes. However, conventional ED-based spectrum sensing schemes usually employ only a single fixed threshold. When the received energy of the SU nears the threshold, the decision for the PU's state is subject to a high probability of misjudgment, thus yielding a loss in sensing performance. In order to alleviate this problem, the double-threshold energy detection method is discussed in [11]. This method can decrease the interference of the SU to the PU. However, how to set the two thresholds and deal with the received energy falling between the two thresholds are not discussed. Double-threshold energy detection method is also discussed in cooperative spectrum sensing environment [12]. In this method, each SU sends its observed energy to the fusion center (FC) when the energy falls between the two thresholds; otherwise, the SU sends its decision results. The final decision is made based on the soft combination of the received energy and hard combination of the received decision results in the FC. Therefore, the performance improvements of cooperative spectrum sensing schemes are obtained at the cost of extra control messages and increased computation complexities [13].

An alternative way of improving the ED schemes is to take into account the PU's activity model during the sensing process. A model of two-state Markov chain is used as an adequate mean accurately describing spectrum occupancy in the time domain in [14–17]. On the one hand, the durations of the PU's states (ON and OFF) are respectively exponentially distributed [18, 19]. On the other hand, the frame length of the SU is much shorter than both the average durations of the PU's ON and OFF states [20, 21], which implies that the PU's activities over successive sensing frames are not independent, especially between adjacent frames. Furthermore, the PU stays at the same state with a high probability during the whole sensing frame [22]. These properties can be exploited to improve the sensing performance of the ED schemes.

In [22], the received energy sequence over successive sensing frames is modeled as a continuous hidden Markov chain, and the final decision is made according to the combined observations from all previous sensing frames. However, this scheme needs to perform simultaneously spectrum sensing and data transmission and hence suffers from implementation difficulties due to self-interference [23]. In [24] and [25], the PU activity is modeled as a Markov chain, and the final decision is also based on history observations but using different combining rules from that of [22]. However, in the schemes in [24] and [25], each sensing frame is further divided into many short slots for either dynamic spectrum sensing or data transmission. This frequent sensing-transmitting alternating strategy may cause synchronization difficulty on the receiving terminal.

In this paper, using a two-state Markov chain model of the PU's activity, an inter-frame dynamic double threshold (IF-DDT) ED scheme is proposed. Firstly, the relationship between two adjacent sensing frames is analyzed, and the IF-DDT scheme for each sensing frame is designed based on dynamic double thresholds. Then, according to the Neyman-Pearson (NP) test criterion, an optimization problem is formulated to acquire the optimal dynamic double thresholds which maximize the detection probability while maintaining a maximum tolerable false alarm probability. Finally, the optimization problem is transformed into an easily solvable problem and an algorithm for searching the optimal dynamic double thresholds are derived with very low complexity. Theoretical analysis and computer simulations show that the proposed IF-DDT scheme can achieve performance improvement due to dynamic double thresholds even when the received energy cannot give a reliable local decision. Compared to reference [11], theoretical analysis and the implementation method are provided for the proposed double-threshold energy detection scheme.

The remainder of this paper is organized as follows. In Section 2, the system model is presented. In Section 3, the IF-DDT scheme is proposed, and the optimal dynamic double thresholds are derived. Section 4 evaluates the proposed scheme for spectrum sensing through simulations. Finally, Section 5 concludes the paper.

## 2 System model

We consider a CR system with one PU, one SU, and a single channel licensed to the PU. The PU operates between ON and OFF states alternately, while the SU executes spectrum sensing to the licensed channel and opportunistically transmits data in a frame-wise manner. In the sequel, the PU's ON and OFF states are represented by integers "1" and "0," respectively. The system model is presented in Fig. 1 , which shows the PU's activity model and the SU's sensing mechanism.

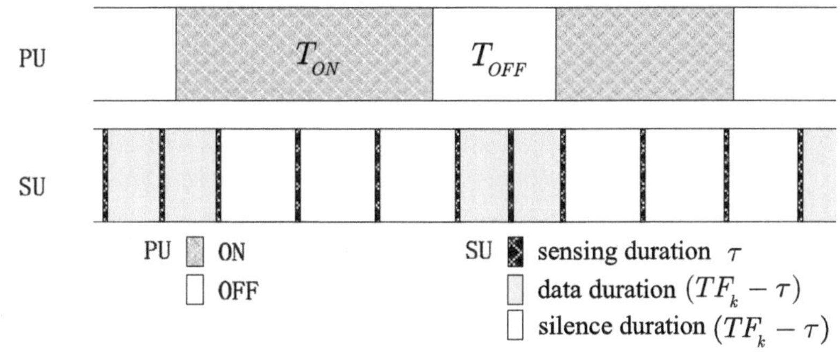


**Fig. 1** The PU's activity and the SU's sensing mechanism

### 2.1 The PU's activity model and the SU's sensing mechanism

Let $T_1$ and $T_0$ denote the durations that the PU stays respectively in ON and OFF states as shown in Fig. 1. It is well accepted that the durations are exponentially distributed with expectations $\overline{T}_1 = \alpha$ and $\overline{T}_0 = \beta$ [25]. The PU's average activity period is defined as $\overline{T}_{PU} = \overline{T}_1 + \overline{T}_0$, and the PU's traffic load is defined as $\eta_{\mathrm{PU}} = \overline{T}_1/\overline{T}_{\mathrm{PU}}$.

We assume that the SU allocates each frame $k$ with a dynamic length $TF_k$, consisting of fixed sensing duration $\tau$ and varying data transmission duration $TF_k - \tau$ for $k = 1, 2, \ldots$. Furthermore, we assume that the lengths satisfy the conditions that $\tau \ll TF_k$, $TF_k \ll \overline{T}_1$, and $TF_k \ll \overline{T}_0$.

Let $z_k \in \{0, 1\}$ denote the real state of the PU within the sensing duration of any specific frame $k$ at the SU. On the above assumptions, it is clear that there is a high probability that $z_k$ remains unchanged within a whole frame and that $z_k$ is closely related to the past state $z_{k-1}$ with respect to the preceding frame $k - 1$ for $k > 1$ [18, 20, 26].

The PU's activity can be modeled as a two-state discrete-time Markov chain with the stationary distribution

$$P\{z_k = 0\} = \frac{\alpha}{\alpha+\beta}, \quad P\{z_k = 1\} = \frac{\beta}{\alpha+\beta}, \tag{1}$$

and the transition matrix

$$\mathbf{P}_{k-1} = \begin{bmatrix} p_{00,k-1} & p_{01,k-1} \\ p_{10,k-1} & p_{11,k-1} \end{bmatrix} = \frac{1}{\alpha+\beta}\begin{bmatrix} \alpha + \beta e^{-(\alpha+\beta)TF_{k-1}} & \beta - \beta e^{-(\alpha+\beta)TF_{k-1}} \\ \alpha - \alpha e^{-(\alpha+\beta)TF_{k-1}} & \beta + \alpha e^{-(\alpha+\beta)TF_{k-1}} \end{bmatrix} \tag{2}$$

where $p_{ij,k-1} \triangleq P\{z_k = j | z_{k-1} = i\}$ denotes the state transition probability over two adjacent frames $k - 1$ and $k$ for $i, j \in \{0, 1\}$; it is easy to identify that $p_{i0,k-1} + p_{i1,k-1} = 1$ for $i = 0, 1$.

Although the frame length $TF_k$ varies with $k$, it holds that $(\alpha + \beta)TF_{k-1} \ll 1$ on the above assumptions. It follows in general that $p_{00,k-1} > p_{01,k-1}$ and $p_{11,k-1} > p_{10,k-1}$. This states an inertia property of the PU activity that $z_k$ is more likely to remain the same as $z_{k-1}$ between any two adjacent frames $k - 1$ and $k$. In the CR system model, the SU is assumed to have full knowledge of parameters $\alpha$ and $\beta$ [20, 22, 25].

### 2.2 Statistics for ED

We consider the additive white Gaussian noise (AWGN) for the licensed channel. Let $N$ denote the number of signal samples of the SU. For the frame $k$, the $n_{\mathrm{th}}$ baseband signal sample is expressed as

$$x_k(n) = \begin{cases} w_k(n), & z_k = 0 \\ s_k(n) + w_k(n), & z_k = 1 \end{cases} \tag{3}$$

where $n = 1, 2, \ldots, N$; $w_k(n)$ denotes the independent and identically distributed (i.i.d.) AWGN samples with zero-mean and normalized unit variance; and $s_k(n)$ denotes the PU's signal samples with constant transmit power. Hence, the average received signal-to-noise ratio (SNR) per sample at the SU is expressed as $\gamma = \mathbb{E}\left(|s_k(n)|^2\right)$.

As the test statistic of spectrum sensing, the received energy in frame $k$ is given by

$$U_k = \sum_{n=1}^{N} |x_k(n)|^2 \tag{4}$$

Then, the test statistic $U_k$ follows the chi-square distribution with $N$ degrees of freedom. By the central limit theorem, if the sample length $N$ is large enough ($N \gg 1$), $U_k$ approximately follows the Gaussian distributions

$$U_k = \begin{cases} \mathbb{N}(N, 2N), & z_k = 0 \\ \mathbb{N}(N(1+\gamma), 2N(1+2\gamma)), & z_k = 1 \end{cases} \tag{5}$$

where $\mathbb{N}(\mu, \sigma^2)$ is the notation of the Gaussian distribution with mean $\mu$ and variance $\sigma^2$.

### 2.3 The conventional ED rule

In most conventional ED schemes, the PU's real states are assumed independent during different frames. Let $d_k$ denote the sensing result of the PU's real state $z_k$ in frame

$k$. The decision rule is expressed as a simple comparison between the received energy $U_K$ and a threshold $\lambda_{ED,k}$ as follows

$$d_k = \begin{cases} 1, & U_k \geq \lambda_{ED,k} \\ 0, & U_k < \lambda_{ED,k} \end{cases} \tag{6}$$

The false alarm probability $P_{f,k}^{ED}$ and the detection probability $P_{d,k}^{ED}$ are obtained as [19]

$$P_{f,k}^{ED} \triangleq P\{d_k = 1 | z_k = 0\} = \mathbb{Q}\left(\frac{\lambda_{ED,k} - N}{\sqrt{2N}}\right) \tag{7}$$

$$P_{d,k}^{ED} \triangleq P\{d_k = 1 | z_k = 1\} = \mathbb{Q}\left(\frac{\lambda_{ED,k} - N(1+\gamma)}{\sqrt{2N(1+2\gamma)}}\right) \tag{8}$$

where $\gamma$ is the average received SNR and

$$\mathbb{Q}(x) = \frac{1}{\sqrt{2\pi}} \int_x^{\infty} e^{-\frac{t^2}{2}} dt$$

In conventional ED schemes, it is usual to preset a fixed threshold $\lambda_{ED,k} = \lambda_{ED}$ for all $k$. Thus, the probability measures $P_{f,k}^{ED}$ and $P_{d,k}^{ED}$ are independent of $k$. It has been shown that the single fixed threshold scheme yields a high probability of erroneous decision when the received energy $U_k$ nears the threshold $\lambda_{ED,k}$ [11].

## 3 The IF-DDT scheme

In this section, we propose an IF-DDT spectrum sensing scheme by exploiting the inertia property of the PU's activity described in Section 2.1. Following the presentation of the decision rule, we formulate an optimization problem to acquire the optimal dynamic double thresholds. Furthermore, we finally transform the optimization problem into an easily solvable problem and derive the optimal dynamic double thresholds with very low complexity.

### 3.1 The decision rule using IF-DDT

To improve the probability of correct decision in the ED schemes, we consider double dynamic thresholds, by which we may elaborate the received energies into three decision ranges in order to exploit the inertia property of the PU's activity. The decision rule using IF-DDT is expressed as follows

$$d_k = \begin{cases} 1, & U_k \geq \lambda_{1,k} \\ 0, & U_k < \lambda_{0,k} \\ d_{k-1}, & otherwise \end{cases} \tag{9}$$

where $\lambda_{1,k}$ and $\lambda_{0,k}$ represent the double dynamic thresholds for frame $k$ at the SU, and $\lambda_{0,k} \leqslant \lambda_{1,k}$.

For $k = 1$, the two thresholds should be identically initialized as $\lambda_{0,1} = \lambda_{1,1}$ . For $k > 1$, if the received energy $U_k$ lies in the open range $(\lambda_{0,k}, \lambda_{1,k})$, the sensing result $d_k$ for frame $k$ retains the same as the sensing result $d_{k-1}$ for the preceding frame $k - 1$.

It is easy to show that the Markov chain for the PU's activity is reversible according to the steady-state distribution in Eq. (1), the transition matrix in Eq. (2), and hence the detailed balance equation [24, 27]

$$P\{z_{k-1} = i, z_k = j\} = P\{z_k = i, z_{k-1} = j\}$$

Define the reverse transition probability as

$$p_{ij,k}^{*} \triangleq P\{z_{k-1} = j | z_k = i\} \tag{10}$$

where $i, j \in \{0, 1\}$. Then, $p_{ij,k}^{*} = p_{ij,k-1}$ is derived.

According to the decision rule using IF-DDT in Eq. (9), the probabilities $P_{f,k}$ and $P_{d,k}$ are obtained as follows (see Appendix 1)

$$P_{f,k} = \mathbb{Q}\left(\frac{\lambda_{0,k} - N}{\sqrt{2N}}\right) P_{0,k} + \mathbb{Q}\left(\frac{\lambda_{1,k} - N}{\sqrt{2N}}\right)(1 - P_{0,k}) \tag{11}$$

$$P_{d,k} = \mathbb{Q}\left(\frac{\lambda_{0,k} - N(1+\gamma)}{\sqrt{2N(1+2\gamma)}}\right) P_{1,k} + \mathbb{Q}\left(\frac{\lambda_{1,k} - N(1+\gamma)}{\sqrt{2N(1+2\gamma)}}\right)(1 - P_{1,k}) \tag{12}$$

where

$$P_{0,k} \triangleq P\{d_{k-1} = 1 | z_k = 0\} = P_{f,k-1} p_{00,k}^{*} + P_{d,k-1} p_{01,k}^{*} \tag{13}$$

$$P_{1,k} \triangleq P\{d_{k-1} = 1 | z_k = 1\} = P_{f,k-1} p_{10,k}^{*} + P_{d,k-1} p_{11,k}^{*} \tag{14}$$

For ease of understanding the IF-DDT scheme, Fig. 2 shows the definitions for multiple events and associated probabilistic measures on the reversible Markov chain.

### 3.2 Derivation of thresholds

To derive the double thresholds, there exists two hypothesis testing criterions: the Neyman-Pearson (NP) test [8] and the Bayesian test [28]. We consider only the NP test criterion, which aims at maximizing $P_{d,k}$ with the constraint $P_{f,k} \leq P_{f,\text{target}}$, or alternatively minimizing $P_{f,k}$ with the constraint $P_{d,k} \geq P_{d,target}$, where $P_{f,\text{target}}$ and $P_{d,\text{target}}$ represent the tolerable maximum false alarm probability and minimum detection probability, respectively.

For any sensing frame $k > 1$, the probabilities $P_{0,k}$ and $P_{1,k}$ are considered constant since $p_{ij,k}^{*}$ are determined by Eq. (2), and $P_{f,k-1}$ and $P_{d,k-1}$ are obtained at frame $k - 1$. Thus, by Eqs. (11) and (12), it is clear that $P_{f,k}$ and $P_{d,k}$ are strictly decreasing with $\lambda_{0,k}$ and $\lambda_{1,k}$, respectively. According to the NP test criterion, the search for

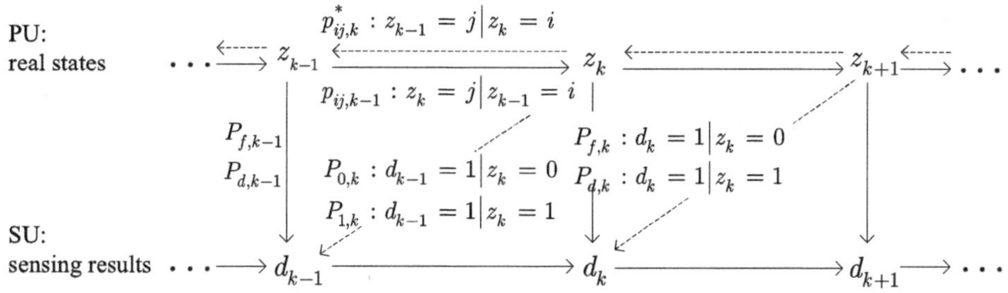


**Fig. 2** Probabilities and transitions associated with multiple events in the reversible Markov chain model

the optimal thresholds $\lambda^*_{0,k}$ and $\lambda^*_{1,k}$ with respect to each frame $k$ is formulated as the optimization problem below

$$\begin{aligned}&\max_{\lambda_{0,k},\lambda_{1,k}} \; P_{d,k}(\lambda_{0,k},\lambda_{1,k})\\ &s.t. \quad \lambda_{1,k} \geq \lambda_{0,k} \geq 0\\ &\qquad P_{f,k}(\lambda_{0,k},\lambda_{1,k}) \leq P_{f,\text{target}}\end{aligned} \tag{15}$$

where $P_{d,k}$ and $P_{f,k}$ are written as bivariate functions of thresholds.

**Theorem 1** *$P_{d,k}$ reaches the maximum value $P^*_{d,k}$ only when $P_{f,k} = P_{f,target}$, i.e., the optimization problem in Eq.* (15) *can be rewritten as (proved in Appendix* 2)

$$\begin{aligned}&\max_{\lambda_{0,k},\lambda_{1,k}} \; P_{d,k}(\lambda_{0,k},\lambda_{1,k})\\ &s.t. \quad \lambda_{1,k} \geq \lambda_{0,k} \geq 0\\ &\qquad P_{f,k}(\lambda_{0,k},\lambda_{1,k}) = P_{f,target}\end{aligned} \tag{16}$$

Based on the second constraint in Eq. (16), we define the Lagrange function

$$\begin{aligned}f(\lambda_{0,k},\lambda_{1,k},c) &= P_{d,k}(\lambda_{0,k},\lambda_{1,k})\\ &\quad + c\left(P_{f,target} - P_{f,k}(\lambda_{0,k},\lambda_{1,k})\right)\end{aligned} \tag{17}$$

where $c$ is the Lagrange multiplier.

Taking the first-order partial derivative of Eq. (17) with respect to $\lambda_{0,k}$ and $\lambda_{1,k}$, respectively

$$\frac{\partial f(\lambda_{0,k},\lambda_{1,k},c)}{\partial \lambda_{0,k}} = 0, \qquad \frac{\partial f(\lambda_{0,k},\lambda_{1,k},c)}{\partial \lambda_{1,k}} = 0 \tag{18}$$

then

$$\begin{cases}\frac{1}{\sqrt{1+2\gamma}} e^{\frac{(\lambda_{0,k}-N)^2}{4N} - \frac{[\lambda_{0,k}-N(1+\gamma)]^2}{4N(1+2\gamma)}} = c\frac{P_{0,k}}{P_{1,k}}\\ \frac{1}{\sqrt{1+2\gamma}} e^{\frac{(\lambda_{1,k}-N)^2}{4N} - \frac{[\lambda_{1,k}-N(1+\gamma)]^2}{4N(1+2\gamma)}} = c\frac{1-P_{0,k}}{1-P_{1,k}}\end{cases} \tag{19}$$

Eliminating $c$ in Eq. (19) by simple manipulations, we get

$$(\lambda_{1,k} - a)^2 - (\lambda_{0,k} - a)^2 = b \tag{20}$$

where

$$a = \frac{N}{2}, \quad b = \frac{2N(1+2\gamma)}{\gamma} \ln \frac{P_{1,k}(1-P_{0,k})}{P_{0,k}(1-P_{1,k})} \tag{21}$$

Then, recalling the second constraint in Eq. (16), we obtain

$$\mathbb{Q}\left(\frac{\lambda_{0,k}-N}{\sqrt{2N}}\right)P_{0,k} + \mathbb{Q}\left(\frac{\lambda_{1,k}-N}{\sqrt{2N}}\right)(1-P_{0,k}) = P_{f,target} \tag{22}$$

So we convert the optimization problem in Eq. (16) into a nonlinear programming problem consisting of Eqs. (20) and (22).

Clearly, the existence of the solution to the optimization problem depends upon the system parameters, including the sample length $N$ , the average $SNR$ value $\gamma$ , the frame lengths $TF_k$, the expectation of ON and OFF state period $\overline{T}_1$, $\overline{T}_0$, and the target false alarm probability $P_{f,\text{target}}$. On the other hand, the solution to the nonlinear programming problem cannot be expressed in a closed form. Therefore, in order to facilitate a heuristic approach for the solution, we derive the range for the above nonlinear programming problem firstly.

Notice that the function $\mathbb{Q}(x)$ is a strictly decreasing function. Under the initial constraint that $\lambda_{1,k} \geq \lambda_{0,k} \geq 0$ and the basic probability property that $0 < P_{0,k}, 1 - P_{0,k} < 1$, we can derive a necessary condition on the threshold $\lambda_{0,k}$ below

$$\min\left[\frac{P_{f,target}}{P_{0,k}}, \mathbb{Q}\left(-\sqrt{\frac{N}{2}}\right)\right] \geq \mathbb{Q}\left(\frac{\lambda_{0,k}-N}{\sqrt{2N}}\right) \geq P_{f,\text{target}} \tag{23}$$

Thus, the range of $\lambda_{0,k}$ is $\left[\lambda^*_{0,k}, \lambda^*\right]$ ,it is given by

$$\lambda^*_{0,k} \triangleq \sqrt{2N}\mathbb{Q}^{-1}\left\{\min\left[\frac{P_{f,\text{target}}}{P_{0,k}}, \mathbb{Q}\left(-\sqrt{\frac{N}{2}}\right)\right]\right\} + N \tag{24}$$

$$\lambda^* \triangleq \sqrt{2N}\mathbb{Q}^{-1}(P_{f,\text{target}}) + N \tag{25}$$

With incorporating the range of $\lambda_{0,k}$, we obtain a necessary condition on the threshold $\lambda_{1,k}$ as follows

$$P_{f,\text{target}} \geq \mathbb{Q}\left(\frac{\lambda_{1,k}-N}{\sqrt{2N}}\right) \geq \frac{P_{f,\text{target}} - \mathbb{Q}\left(\frac{\lambda_{0,k}^*-N}{\sqrt{2N}}\right)P_{0,k}}{1-P_{0,k}} \tag{26}$$

and the range $\left[\lambda^*, \lambda_{1,k}^*\right]$ of $\lambda_{1,k}$ is given by

$$\lambda_{1,k}^* \triangleq \sqrt{2N}\mathbb{Q}^{-1}\left(\frac{P_{f,\text{target}} - \mathbb{Q}\left(\frac{\lambda_{0,k}^*-N}{\sqrt{2N}}\right)P_{0,k}}{1-P_{0,k}}\right) + N \tag{27}$$

In practice, it is easy to show that $\lambda^* > a$ since $P_{f,\text{target}} < 0.5$ and $\mathbb{Q}^{-1}(P_{f,\text{target}}) > 0$. By analysing Eq. (20) and Eq. (24), we can obtain the relationship of $\lambda_{1,k}^*$, $\lambda_{0,k}^*$, and $a$ as follows:

$$\begin{cases} \lambda_{0,k}^* = 0, \lambda_{1,k}^* < +\infty & \frac{P_{f,\text{target}}}{P_{0,k}} \geq \mathbb{Q}\left(-\sqrt{\frac{N}{2}}\right) \\ 0 < \lambda_{0,k}^* \leq a, \lambda_{1,k}^* \to +\infty & \mathbb{Q}\left(-\frac{1}{2}\sqrt{\frac{N}{2}}\right) \leq \frac{P_{f,\text{target}}}{P_{0,k}} < \mathbb{Q}\left(-\sqrt{\frac{N}{2}}\right) \\ \lambda_{0,k}^* > a, \lambda_{1,k}^* \to +\infty & \frac{P_{f,\text{target}}}{P_{0,k}} < \mathbb{Q}\left(-\frac{1}{2}\sqrt{\frac{N}{2}}\right) \end{cases} \tag{28}$$

Combined with the range of $\lambda_{0,k}$ and $\lambda_{1,k}$ in three cases in Eqs.(28), (20), and (21) can be further demonstrated in the two-dimension Cartesian coordinate system as showed in Fig. 3. When $\frac{P_{f,\text{target}}}{P_{0,k}} \geq \mathbb{Q}\left(-\sqrt{\frac{N}{2}}\right)$, Eqs. (20) and (21) are demonstrated in Fig. 3a; when $\mathbb{Q}\left(-\frac{1}{2}\sqrt{\frac{N}{2}}\right) \leq \frac{P_{f,\text{target}}}{P_{0,k}} < \mathbb{Q}\left(-\sqrt{\frac{N}{2}}\right)$, Eqs. (20) and (21) are demonstrated in Fig. 3b; and when $\frac{P_{f,\text{target}}}{P_{0,k}} < \mathbb{Q}\left(-\frac{1}{2}\sqrt{\frac{N}{2}}\right)$, Eqs. (20) and (21) are demonstrated in Fig. 3c.

### 3.3 Solution analysis

Based on Eqs. (20) and (21), $\lambda_{1,k}$ can respectively be seen as the functions of $\lambda_{0,k}$ within the range obtained above. First, $g_1(\lambda_{0,k})$ and $g_2(\lambda_{0,k})$ are defined respectively as

$$g_1(\lambda_{0,k}) \triangleq \sqrt{2N}\mathbb{Q}^{-1}\left(\frac{P_{f,\text{target}} - \mathbb{Q}\left(\frac{\lambda_{0,k}-N}{\sqrt{2N}}\right)P_{0,k}}{1-P_{0,k}}\right) + N \tag{29}$$

$$g_2(\lambda_{0,k}) \triangleq \sqrt{(\lambda_{0,k}-a)^2 + b} + a \tag{30}$$

And $g(\lambda_{0,k})$ is further defined as

$$g(\lambda_{0,k}) \triangleq g_1(\lambda_{0,k}) - g_2(\lambda_{0,k}) \tag{31}$$

Obviously, the zero point of $g(\lambda_{0,k})$ is the solution of $\lambda_{0,k}$ to the nonlinear programming problem in Eqs. (20) and (21) meanwhile.

The first-order differential function of $g(\lambda_{0,k})$ is then deduced as

$$\begin{aligned} g'(\lambda_{0,k}) &\triangleq g_1'(\lambda_{0,k}) - g_2'(\lambda_{0,k}) \\ &= -\frac{P_{0,k}}{1-P_{0,k}} e^{\frac{1}{4N}\left[(\lambda_{1,k}-N)^2-(N-\lambda_{0,k})^2\right]} - \frac{\lambda_{0,k}-a}{\lambda_{1,k}-a} \end{aligned} \tag{32}$$

where $\lambda_{1,k} = \sqrt{2N}\mathbb{Q}^{-1}\left(\frac{P_{f,\text{target}} - \mathbb{Q}\left(\frac{\lambda_{0,k}-N}{\sqrt{2N}}\right)P_{0,k}}{1-P_{0,k}}\right) + N$.

Together with Eq. (28), the negativity of $g'(\lambda_{0,k})$ in the range of three cases in Eq. (28) is proved as follows.

(1) $\frac{P_{f,\text{target}}}{P_{0,k}} < \mathbb{Q}\left(-\frac{1}{2}\sqrt{\frac{N}{2}}\right)$. It can be easily obtained that $\frac{\lambda_{0,k}-a}{\lambda_{1,k}-a} > 0$, and thus, we can get $g'(\lambda_{0,k}) < 0$.

(2) $\frac{P_{f,\text{target}}}{P_{0,k}} \geq \mathbb{Q}\left(-\frac{1}{2}\sqrt{\frac{N}{2}}\right)$.

  (i) $\lambda_{0,k} \in (a, \lambda^*)$. Similar to (1), it can be obtained that $g'(\lambda_{0,k}) < 0$.

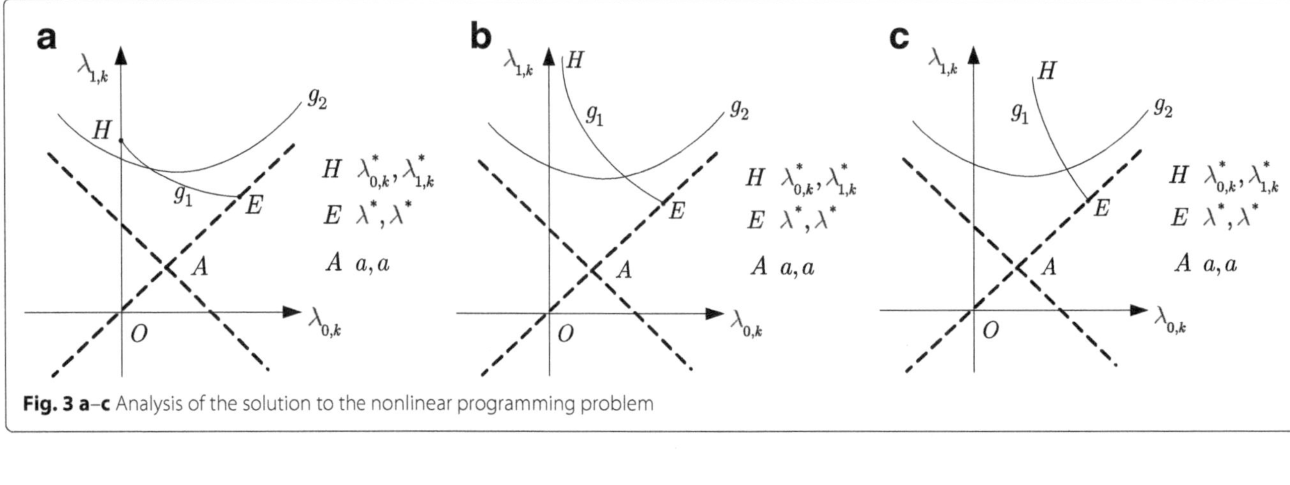


**Fig. 3 a–c** Analysis of the solution to the nonlinear programming problem

*(ii)* $\lambda_{0,k} \in (\lambda_{0,k}^{*}, a)$, Eq. (20) can be transformed into

$$\left[1-\frac{1}{2}erfc\left(\frac{N-\lambda_{0,k}}{2\sqrt{N}}\right)\right]P_{0,k} + \frac{1}{2}erfc\left(\frac{\lambda_{1,k}-N}{2\sqrt{N}}\right)(1-P_{0,k}) = P_{f,\text{target}} \tag{33}$$

where $erfc(x) = 2\mathbb{Q}(\sqrt{2}x)$.

With $\lambda_{0,k} < a$, it can be deduced that

$$\frac{N-\lambda_{0,k}}{2\sqrt{N}} > \frac{N-a}{2\sqrt{N}} = \frac{\sqrt{N}}{4} \tag{34}$$

and

$$\begin{aligned}\frac{\lambda_{1,k}-N}{2\sqrt{N}} &= \frac{1}{\sqrt{2}}\mathbb{Q}^{-1}\left(\frac{P_{f,\text{target}}-\mathbb{Q}\left(\frac{\lambda_{0,k}-N}{\sqrt{2N}}\right)P_{0,k}}{1-P_{0,k}}\right)\\ &> \frac{1}{\sqrt{2}}\mathbb{Q}^{-1}\left(\frac{P_{f,\text{target}}-\mathbb{Q}\left(-\frac{\sqrt{N}}{2\sqrt{2}}\right)P_{0,k}}{1-P_{0,k}}\right)\\ &> \frac{1}{\sqrt{2}}\mathbb{Q}^{-1}\left(\frac{P_{0,k}-\mathbb{Q}\left(-\frac{\sqrt{N}}{2\sqrt{2}}\right)P_{0,k}}{1-P_{0,k}}\right)\\ &> \frac{1}{\sqrt{2}}\mathbb{Q}^{-1}\left(1-\mathbb{Q}\left(-\frac{\sqrt{N}}{2\sqrt{2}}\right)\right)\\ &= \frac{\sqrt{N}}{4}\end{aligned} \tag{35}$$

Notice that in Eq. (35), $P_{0,k} > P_{f,\text{target}}$ since $P_{d,k-1} > P_{f,\text{target}}$ must hold to ensure a valid detection and $P_{f,k-1} = P_{f,\text{target}}$ in Eq. (13). With $\gamma \ll 1$ and $N \gg 1$, the approximation $erfc(x) \approx \frac{1}{\sqrt{\pi}x}e^{-x^2}$ can be taken in Eq. (33), and thus, we can get

$$\begin{aligned}&\frac{P_{0,k}}{1-P_{0,k}}e^{\frac{1}{4N}\left[(\lambda_{1,k}-N)^2-(N-\lambda_{0,k})^2\right]}\\ &= \frac{N-\lambda_{0,k}}{\lambda_{1,k}-N} + \frac{P_{0,k}-P_{f,\text{target}}}{1-P_{0,k}}\sqrt{\frac{\pi}{N}}(N-\lambda_{0,k})e^{\frac{1}{4N}(\lambda_{1,k}-N)^2}\end{aligned} \tag{36}$$

By taking Eq. (36) into Eq. (32), it can be proved that $g'(\lambda_{0,k}) < 0$ when $\lambda_{0,k} \in \left(\lambda_{0,k}^{*}, a\right)$ ; the nonlinear programming problem is then transformed into a problem of searching for the zero point of a strictly decreasing function, which can be easily solved by the dichotomy or Newton's method. It should be noticed that no zero point exists when $g\left(\lambda_{0,k}^{*}\right) < 0$ since $g\left(\lambda^{*}\right) < 0$.

## 4 Simulation results

In this section, numerical results are presented to evaluate the effectiveness of the IF-DDT sensing scheme. The proposed algorithm is tested under various average activity periods and traffic loads of the PU.

### 4.1 Simulation setup

Firstly, the thresholds of the IF-DDT scheme are simulated. Secondly, the actual false alarm of the new algorithm is provided to examine the correctness of the thresholds. Thirdly, the performance of the IF-DDT algorithm is quantified in different average PU active periods and PU traffic loads, respectively. Finally, the total error probability of the IF-DDT scheme is provided to verify the detection performance.

In the simulations, the sample length $N$ is set to be 1000, the sensing frame length $TF_k$ is set to be 10 ms according to the IEEE 802.22 standard, and a Markov chain with $10^6$ frames is randomly generated to model the PU state varying in each simulation. The detection performance of the proposed IF-DDT scheme is compared with that of the ED scheme and the double-threshold (DT) energy scheme in [11].

### 4.2 Numerical results

Figure 4 verifies the double thresholds of the IF-DDT algorithm changing with the frame index $k$. Figure 4a, b is plotted at low and high SNR, respectively. In both of the subfigures, $\overline{T}_0 = \overline{T}_1 = 200$ ms is set. In addition, the gray areas in the subfigures denote that the real state of the PU is ON; otherwise, it is OFF. When the received energy locates between the double thresholds and correctly detected under the IF-DDT scheme, the corresponding point is marked with a circle.

From Fig. 4a, b, we can see that the double thresholds quickly converge to stable values and actually keep invariant. It means that the calculation for the thresholds of the subsequent sensing frames can be saved as long as the system parameters remain unchanged. Furthermore, the marked points show that in lots of frames where the received energy cannot give a reliable local decision, the conventional ED algorithm seem powerless but our algorithm tends to make the right judgment. It is also indicated that the gap between the double thresholds is smaller when the SNR is high.

Figure 5 tests the actual false alarm probability of the ED scheme, IF-DDT scheme, and DT scheme versus SNR when $P_{f,\text{target}} = 0.01, 0.05, 0.1$, respectively. Other simulation settings are $\overline{T}_0 = \overline{T}_1 = 200$ ms. As shown in the figure, the actual false alarm probability of the DT scheme is slightly higher than the target false alarm probability, but the actual false alarm probability of the IF-DDT

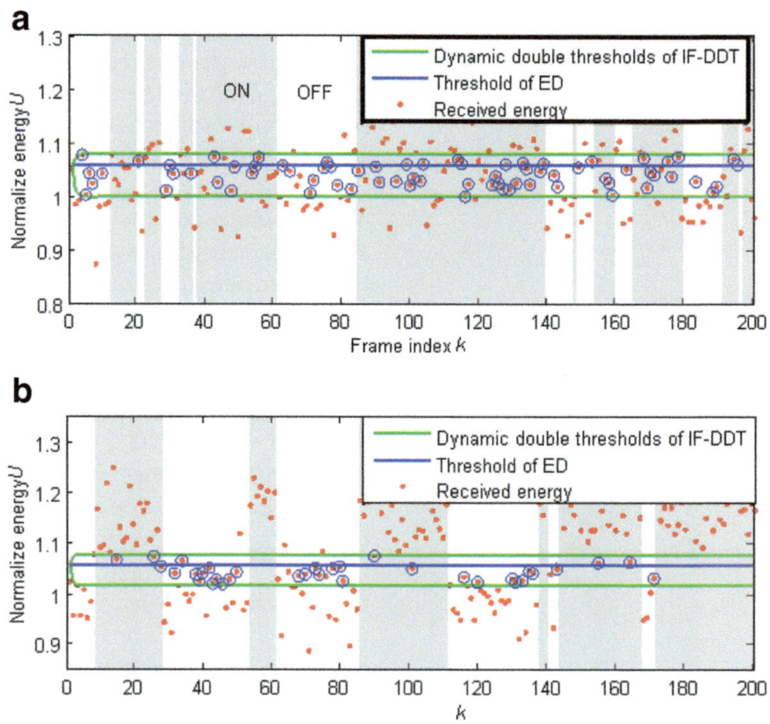


**Fig. 4** Thresholds of the IF-DDT sensing scheme. **a** Thresholds for $\bar{T}_0 = \bar{T}_1 = 200$ ms and SNR $= -14$ dB. Thresholds of the IF-DDT sensing scheme. **b** Thresholds for $\bar{T}_0 = \bar{T}_1 = 200$ ms and SNR $= -8$ dB

scheme and ED scheme is always equal to the target false alarm probability. In other words, the dynamic double thresholds derived from Eq. (15) ensure that the constraint on false alarm probability is met.

Figure 6 shows the detection probability versus SNR with different PU active periods. The average PU active parameters are set to be $\overline{T}_0 = \overline{T}_1 = 100, 200, 500$ ms, respectively. As shown in Fig. 6, the IF-DDT algorithm shows superior performance to the conventional ED and DT scheme even when the average period of the PU is short. For example, at the SNR of $-10$ dB, the detection probability of the IF-DDT algorithm reaches 0.92 when $\overline{T}_0 = \overline{T}_1 = 100$ ms, comparing with $P_d^{ED} = 0.81$ for the ED scheme and $P_d^{DT} = 0.84$ for the DT scheme.

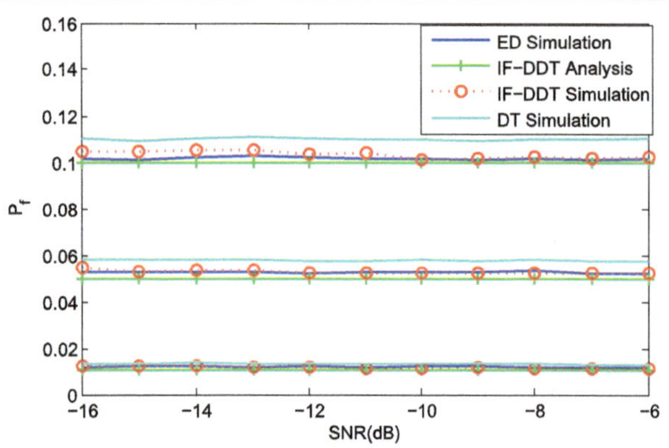


**Fig. 5** False alarm probability testing for $\overline{T}_0 = \overline{T}_1 = 200$ ms

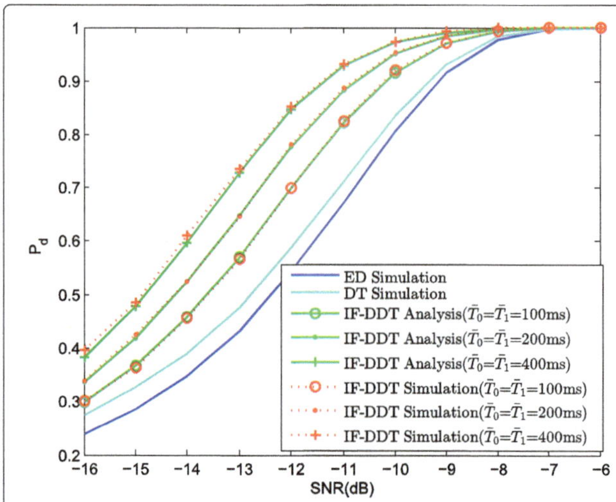


**Fig. 6** $P_d$ versus *SNR* for $P_{f,\mathrm{target}=0.1}$ and $\eta_{\mathrm{PU}} = 0.5$

This improvement benefits from the dynamic double-threshold strategy, with which the detector tends to make the correct judgment when the received energy is not reliable enough. What is more, when the PU actives with longer period and thus becomes more inertial, the IF-DDT algorithm gains higher detection probability which depicts that the PU active period has much impact on the performance of the new sensing scheme. In addition, it is clearly shown that the simulation results well coincide with the theoretical analysis in all cases.

Figure 7 tests the performance of the IF-DDT scheme when the PU's traffic load varies. And as expected, the new scheme gains obvious performance improvement compared with the convention ED and DT schemes. It is also verified that when the PU's average activity period is fixed, the detection probability of the proposed algorithm does not vary much with the PU's traffic load. For example, at the SNR of −12 dB, the detection probability is 0.78, which is slightly higher by 0.03 and 0.06, respectively, than that in the other cases. The difference is even slighter in higher SNR region.

It is important to investigate the detector's capability on timely and correctly identifying the spectrum hole when the spectrum is not occupied by the PU and vacating the spectrum when the PU starts to occupy the authorized spectrum. Therefore, to compare the performance in terms of both the detection probability and the false alarm probability, we define an erroneous decision probability $P_e$ as follows:

$$P_e = P_{\mathrm{OFF}} P_f + P_{\mathrm{ON}}(1 - P_d) \tag{37}$$

Figure 8 shows $P_e$ among IF-DDT scheme, ED scheme, and DT scheme versus SNR. The simulation environment is the same as that in Fig. 6. The figure shows that IF-DDT scheme provide the lowest $P_e$ among these schemes. In addition, the $P_e$ of the IF-DDT scheme decreases as the average period of the PU becomes longer. This is due to the fact that when the average period increases, the PU's state between contiguous frames tends to be more inertial, and then, the decision rule in Eq. (9) is more reliable; therefore, the decision based on the energy falling between the two thresholds is more reasonable. Furthermore, the performance of ED and DT schemes are much worse than those of the IF-DDT scheme.

Figures 6, 7, and 8 reveal:

1. IF-DDT sensing scheme outperforms the conventional ED scheme and DT scheme
2. The proposed method is more robust to the PU's activity period and traffic load.

## 5 Conclusions

In this paper, the PU's activity model and the SU's sensing mechanism are utilized and an inter-frame-based dynamic double-threshold spectrum sensing scheme, which exploits the relationship of two adjacent sensing frames, is proposed to enhance energy detection performance. When the received energy cannot give a reliable local decision to distinguish the ON and OFF states of the PU, the detector retains the previous sensing result. The detection probability and false alarm probability of the proposed sensing scheme are analyzed, and the optimal

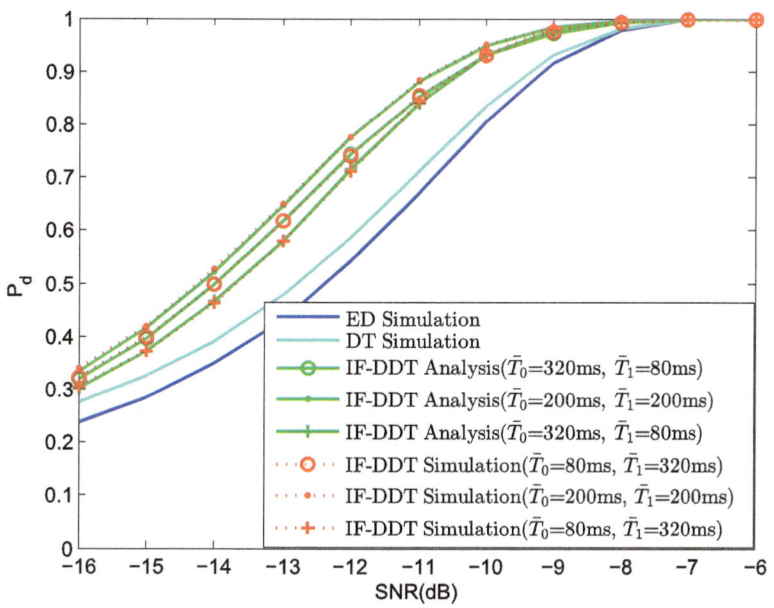


**Fig. 7** $P_d$ versus *SNR* for $P_{f,\mathrm{target}=0.1}$ and $T_{\mathrm{PU}} = 400$ ms

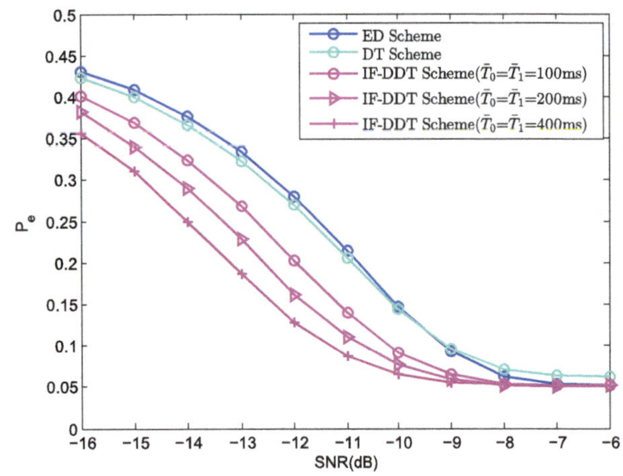

**Fig. 8** $P_e$ versus $SNR$ for $P_{f,\text{target}=0.1}$ and $\eta_{PU} = 0.5$

double thresholds are derived with very low complexity according to the Neyman-Pearson (NP) test criterion. This sensing scheme is simple but shown to outperform the ED algorithm from both theoretical analysis and simulation results.

In this paper, our algorithm is based on energy detection. However, it can be extended to other traditional spectrum sensing method, such as coherent detection and feature detection, where the PU's activity model and the SU's sensing mechanism have not been exploited. The same way can be followed to obtain the corresponding dynamic double thresholds and the detection probability together with the false alarm probability.

## Appendix 1: Derivation of Eqs. (12) and (13)

According to the definition for $P_{f,k}$ and the decision rule given by Eq. (9),

$$\begin{aligned} P_{f,k} &\triangleq P\{d_k = 1|z_k = 0\} \\ &= P\left\{U_k \geq \lambda_{1,k}|z_k = 0\right\} \\ &\quad + P\left\{\lambda_{0,k} < U_k < \lambda_{1,k}, d_{k-1} = 1|z_k = 0\right\} \\ &= P\left\{U_k \geq \lambda_{1,k}|z_k = 0\right\} \\ &\quad + P\left\{\lambda_{0,k} < U_k < \lambda_{1,k}|z_k = 0\right\} P\left\{d_{k-1} = 1|z_k = 0\right\} \\ &= \mathbb{Q}\left(\frac{\lambda_{1,k} - N}{\sqrt{2N}}\right) + \left[\mathbb{Q}\left(\frac{\lambda_{0,k} - N}{\sqrt{2N}}\right) - \mathbb{Q}\left(\frac{\lambda_{1,k} - N}{\sqrt{2N}}\right)\right] P_{0,k} \end{aligned} \tag{38}$$

where

$$\begin{aligned} P_{0,k} &\triangleq P\left\{d_{k-1} = 1|z_k = 0\right\} \\ &= P\left\{d_{k-1} = 1, z_{k-1} = 0|z_k = 0\right\} \\ &\quad + P\left\{d_{k-1} = 1, z_{k-1} = 1|z_k = 0\right\} \\ &= P\left\{d_{k-1} = 1|z_{k-1} = 0\right\} P\left\{z_{k-1} = 0|z_k = 0\right\} \\ &\quad + P\left\{d_{k-1} = 1|z_{k-1} = 1\right\} P\left\{z_{k-1} = 1|z_k = 0\right\} \\ &= P_{f,k-1}p^*_{00,k} + P_{d,k-1}p^*_{01,k} \end{aligned} \tag{39}$$

Similarly, the detection probability $P_{d,k}$ can be derived as

$$\begin{aligned} P_{d,k} &\triangleq P\{d_k = 1|z_k = 1\} \\ &= P\left\{U_k \geq \lambda_{1,k}|z_k = 1\right\} \\ &\quad + P\left\{\lambda_{0,k} < U_k < \lambda_{1,k}, d_{k-1} = 1|z_k = 1\right\} \\ &= P\left\{U_k \geq \lambda_{1,k}|z_k = 1\right\} \\ &\quad + P\left\{\lambda_{0,k} < U_k < \lambda_{1,k}|z_k = 1\right\} P\left\{d_{k-1} = 1|z_k = 1\right\} \\ &= \mathbb{Q}\left(\frac{\lambda_{1,k} - N(1+\gamma)}{\sqrt{2N(1+2\gamma)}}\right) \\ &\quad + \left[\mathbb{Q}\left(\frac{\lambda_{0,k} - N(1+\gamma)}{\sqrt{2N(1+2\gamma)}}\right) - \mathbb{Q}\left(\frac{\lambda_{1,k} - N(1+\gamma)}{\sqrt{2N(1+2\gamma)}}\right)\right] P_{1,k} \end{aligned} \tag{40}$$

and

$$P_{1,k} \triangleq P\left\{d_{k-1} = 1|z_k = 1\right\} = P_{f,k-1}p^*_{10,k} + P_{d,k-1}p^*_{11,k} \tag{41}$$

## Appendix 2: Proof of Theorem 1

Consider first $k > 1$ . In Eqs. (10) and (11), $P_{0,k}$ and $P_{1,k}$ can be viewed as constants in the open range $(0, 1)$ . Thus, the probabilities $P_{f,k}$ and $P_{d,k}$ are both strictly decreasing functions of $\lambda_{0,k}$ and $\lambda_{1,k}$ , respectively.

Assume that $\left(\lambda'_{0,k}, \lambda'_{1,k}\right)$ is the optimal solution to Eq. (16) such that $\lambda'_{0,k} < \lambda'_{1,k}$, $P_{f,k}\left(\lambda'_{0,k}, \lambda'_{1,k}\right) < P_{f,\text{target}}$, and $P_{d,k}\left(\lambda'_{0,k}, \lambda'_{1,k}\right) = P^*_{d,k}$, where $P^*_{d,k}$ is the maximum value for $P_{d,k}$ . Then, there must exist a smaller value $\lambda^*_{0,k} < \lambda'_{0,k}$ such that

$$P_{f,k}\left(\lambda'_{0,k}, \lambda'_{1,k}\right) < P_{f,k}\left(\lambda^*_{0,k}, \lambda'_{1,k}\right) = P_{f,\text{target}} \tag{42}$$

$$P_{d,k}\left(\lambda'_{0,k}, \lambda'_{1,k}\right) = P^*_{d,k} < P_{d,k}\left(\lambda^*_{0,k}, \lambda'_{1,k}\right) \tag{43}$$

which contradicts the assumption of the maximum value $P^*_{d,k}$.

On the other hand, if $\lambda'_{0,k} < \lambda'_{1,k}$ , we may also reduce $\lambda'_{1,k}$ to $\lambda^*_{1,k} \in [\lambda'_{0,k}, \lambda'_{1,k})$ with $\lambda'_{0,k}$ given, such that $P_{f,k}\left(\lambda'_{0,k}, \lambda^*_{1,k}\right) = P_{f,\text{target}}$ , followed by $P_{d,k}\left(\lambda^*_{1,k}, \lambda'_{0,k}\right) > P^*_{d,k}$.

In the special case $k = 1$ , the probabilities $P_{f,1}$ and $P_{d,1}$ become strictly decreasing functions of $\lambda_1$ as described underneath Eq. (9). Similarly, it can be proved that $P_{d,1}$ reaches $P^*_{d,1}$ only when $P_{f,1} = P_{f,\text{target}}$ using a proper threshold $\lambda^*_1$.

Therefore, it is concluded that $P_{d,k}$ reaches the maximum value $P^*_{d,k}$ only when $P_{f,k} = P_{f,\text{target}}$ with pair thresholds $\left(\lambda^*_{0,k}, \lambda^*_{1,k}\right)$ for all $k > 1$.

## Appendix 3: Proof of Eq. (23)

The proof is composed of two steps:

Firstly, $\frac{P_{f,\text{target}}}{P_{0,k}} \geq \mathbb{Q}\left(\frac{\lambda_{0,k}-N}{\sqrt{2N}}\right)$ can be derived from Eq. (20). From $\frac{\lambda_{0,k}-N}{\sqrt{2N}} = -\sqrt{\frac{N}{2}} + \frac{\lambda_{0,k}}{\sqrt{2N}} \geq -\sqrt{\frac{N}{2}}$ and the monotonic decreasing property of $\mathbb{Q}$ function, we obtain $\mathbb{Q}\left(-\sqrt{\frac{N}{2}}\right) \geq \mathbb{Q}\left(\frac{\lambda_{0,k}-N}{\sqrt{2N}}\right)$.

Therefore $\min\left[\frac{P_{f,\text{target}}}{P_{0,k}}, \mathbb{Q}\left(-\sqrt{\frac{N}{2}}\right)\right] \geq \mathbb{Q}\left(\frac{\lambda_{0,k}-N}{\sqrt{2N}}\right)$.

Secondly, from $\lambda_{0,k} \leq \lambda_{1,k}$, we have $\frac{\lambda_{0,k}-N}{\sqrt{2N}} \leq \frac{\lambda_{1,k}-N}{\sqrt{2N}}$, then $\mathbb{Q}\left(\frac{\lambda_{0,k}-N}{\sqrt{2N}}\right) \geq \mathbb{Q}\left(\frac{\lambda_{0,k}-N}{\sqrt{2N}}\right)$.

On the one hand, $\mathbb{Q}\left(\frac{\lambda_{0,k}-N}{\sqrt{2N}}\right) P_{0,k} + \mathbb{Q}\left(\frac{\lambda_{0,k}-N}{\sqrt{2N}}\right)(1-P_{0,k}) = \mathbb{Q}\left(\frac{\lambda_{0,k}-N}{\sqrt{2N}}\right)$. On the other hand, $\mathbb{Q}\left(\frac{\lambda_{0,k}-N}{\sqrt{2N}}\right) P_{0,k} + \mathbb{Q}\left(\frac{\lambda_{0,k}-N}{\sqrt{2N}}\right)(1-P_{0,k}) \geq \mathbb{Q}\left(\frac{\lambda_{0,k}-N}{\sqrt{2N}}\right) P_{0,k} + \mathbb{Q}\left(\frac{\lambda_{1,k}-N}{\sqrt{2N}}\right)(1-P_{0,k}) = P_{f,\text{target}}$. Therefore, $\mathbb{Q}\left(\frac{\lambda_{0,k}-N}{\sqrt{2N}}\right) \geq P_{f,\text{target}}$.

**Acknowledgements**

The authors would gratefully acknowledge the grants from the National Natural Science Foundation of China (61571250), the Natural Science Foundation of Ningbo City of China (2015 A610121), and the Natural Science Foundation of Fujian Province of China (2014J01243) and the K. C. Wong Magna Fund in Ningbo University.

**Authors' contributions**

PL, YH, and CF conceived and designed the study. CF and PL performed the simulation experiments. CF and PL wrote the paper. YH, YL, MJ, and GW reviewed and edited the manuscript. All authors read and approved the final manuscript.

**Competing interests**

The authors declare that they have no competing interests.

**Author details**

[1]College of Information Science and Engineering, Ningbo University, Ningbo, China. [2]School of Information Science and Engineering, National Huaqiao University, Xiamen, China.

**References**

1. G Staple, K Werbach, The end of spectrum scarcity. IEEE Spectrum. **41**(3), 48–52 (2004)
2. JI Mitola, GQ Maguire, Cognitive radio: making software radios more personal. IEEE Pers Commun. **6**(4), 13–18 (1999)
3. Y Pei, AT Hoang, Y-C Liang, in *Proc. IEEE Int. Symp. Personal, Indoor and Mobile Radio Commun. (PIMRC 2007)*. Sensing-throughput tradeoff in cognitive radio networks: how frequently should spectrum sensing be carried out?, (2007), pp. 1–5
4. S Stotas, A Nallanathan, 2012, On the throughput and spectrum sensing enhancement of opportunistic spectrum access cognitive radio networks. Wireless Commun. IEEE Trans. **11**(1), 97–107
5. J Zhang, et al, Sensing-energy efficiency tradeoff for cognitive radio networks. Iet Commun. **8**(18), 3414–3423 (2014)
6. IEEE Standard for Information Technology Telecommunications and information exchange between systems Wireless Regional Area Networks (WRAN) Specific requirements, Part 22: Cognitive Wireless RAN Medium Access Control (MAC) and Physical Layer (PHY) Specifications: Policies and Procedures for Operation in the TV Bands. IEEE std. 802.22a-2014
7. T Yücek, H Arslan, A survey of spectrum sensing algorithms for cognitive radio applications. IEEE Commun. Surv. Tutorials. **11**(1), 116–130 (2009)
8. Axell, Erik, et al, Spectrum sensing for cognitive radio:state-of-the-art and recent advances. IEEE Signal Process. Mag. **29**(3), 101–116 (2012)
9. A Taherpour, M Nasiri-Kenari, S Gazor, Multiple antenna spectrum sensing in cognitive radios. IEEE Trans. Wireless Commun. **9**(2), 814–823 (2010)
10. M López-Benítez, F Casadevall, Improved energy detection spectrum sensing for cognitive radio. Iet Commun. **6**(8), 785–796 (2012)
11. J Wu, T Luo, G Yue, in *Proc. IEEE 1st Int. Conf. Information Science and Engineering*. An energy detection algorithm based on double-threshold in cognitive radio systems, (2009), pp. 493–496
12. SQ Liu, BJ Hu, XY Wang, Hierarchical cooperative spectrum sensing based on double thresholds energy detection. IEEE Communications Letters. **16**(7), 1096–1099 (2012)
13. H Rifà-Pous, MJ Blasco, C Garrigues, Review of robust cooperative spectrum sensing techniques for cognitive radio networks. Wireless Personal Commun. **67**(2), 175–198 (2012)
14. Y Saleem, MH Rehmani, Primary radio user activity models for cognitive radio networks: a survey. J. Netw. Comput. Appl. **43**(1), 1–16 (2014)
15. SA Bayhan, Z Fatih, A Markovian approach for best-fit channel selection in cognitive radio networks. Ad Hoc Netw. **12**(1), 165–177 (2014)
16. DJ Lee, WY Yeo, Channel availability analysis of spectrum handoff in cognitive radio networks. IEEE Commun. Lett. **19**(3), 435–438 (2015)
17. M López-Benítez, F Casadevall, Empirical time-dimension model of spectrum use based on a discrete-time Markov chain with deterministic and stochastic duty cycle models. Veh. Technol. IEEE Trans. **60**(6), 2519–2533 (2011)
18. S Geirhofer, L Tong, BM Sadler, Cognitive medium access: constraining interference based on experimental models. Selected Areas Commun. IEEE J. **26**(1), 95–105 (2008)
19. MH Rehmani, et al, SURF: A distributed channel selection strategy for data dissemination in multi-hop cognitive radio networks. Comput. Commun. **36**(10), 1172–1185 (2013)
20. N Moayeri, H Guo, in *Proc. IEEE Symp. on New Frontiers in Dynamic Spectrum*. How often and how long should a cognitive radio sense the spectrum, (2010), pp. 1–10
21. Xing, Xiaoshuang, et al, Optimal spectrum sensing interval in cognitive radio networks. IEEE Trans. Parallel Distributed Syst. **25**(9), 2408–2417 (2014)
22. K Haghighi, EG Strm, E Agrell, On optimum causal cognitive spectrum reutilization strategy. IEEE J. Selected Areas Commun. **30**(10), 1911–1921 (2012)
23. J Heo, et al, Simultaneous sensing and transmission in cognitive radio. IEEE Trans. Wireless Commun. **13**(4), 1948–1959 (2014)
24. KW Choi, E Hossain, Opportunistic access to spectrum holes between packet bursts: a learning-based approach. IEEE Trans. Wireless Commun. **10**(8), 2497–2509 (2011)
25. K Haghighi, EG Strom, E Agrell, Sensing or transmission: causal cognitive radio strategies with censorship. IEEE Trans. Wireless Commun. **13**(6), 3031–3041 (2014)
26. Xu, You, et al, Joint sensing period and transmission time optimization for energy-constrained cognitive radios. EURASIP J. Wireless Commun. Netw. **2010**(1), 1–16 (2009)
27. PJ Green, Reversible jump Markov chain Monte Carlo computation and Bayesian model determination. Biometrika. **82**(4), 711–732 (1995)
28. Y Zeng, et al, A review on spectrum sensing for cognitive radio: challenges and solutions. EURASIP J. Adv. Signal Process. **2010**(1), 1–15 (2010)

2

# Energy efficient power allocation strategy for 5G carrier aggregation scenario

Weidong Gao*, Lin Ma and Gang Chuai

**Abstract**

Carrier aggregation (CA) is considered to be a potential technology in next generation wireless communications. While boosting system throughput, CA has also put forward challenges to the resource allocation problems. In this paper, we firstly construct the energy efficiency optimization problem and prove that the function is strictly quasi concave. Then we propose a binary search-based power allocation algorithm to solve the strictly quasi concave optimization problem. Simulation results show that the proposed algorithm can greatly improve the system energy efficiency while keeping low computation complexity.

**Keywords:** Carrier aggregation, Power allocation, Binary search, Strictly quasi concave, Optimization

## 1 Introduction

The next generation wireless communication system should meet the characteristic of 1G+ bits per second data rate to meet the requirements of various high-speed multimedia applications. To achieve this goal, the use of a larger bandwidth for transmission becomes the most direct way to boost transmission rate. But a large segment of continuous spectrum is not available easily for most of the wireless network operators due to the practical constraints, which makes the effective use of a plurality of non-continuous frequency spectrum a viable alternative option. The international standardization organization, for example 3GPP, has carried out research on spectrum expansion technologies, which is called carrier aggregation (CA) [1]. With CA, multiple spectrum fragments, whether continuous or not, can be aggregated together to be used by single user, which can substantially improve single user's peak data rate. Thanks to its high spectral efficiency, except for cognitive radios [2, 3], millimeter-wave communication [4], it is likely that carrier aggregation technology will also play an important role in future 5G wireless networks.

Though CA technology has significantly improved the system throughput, it has also increased the complexity of resource scheduling for the network. In addition, due to that multiple component carriers are used by the UE simultaneously, more transmission power will be consumed by the eNodeB as well as by the UE [5]. Therefore, energy-saving problem cannot be ignored for carrier aggregation and it is necessary to reduce the extra energy consumed by CA operation, thus reduce the carbon emission and contribute to green communication. Up to now, there are few researches on energy efficiency optimization algorithm focusing on power allocation in carrier aggregation systems. With respect to energy efficiency optimization for CA systems, there are mainly two kinds of algorithms. For the first one, it is assumed that the collection of resource blocks occupied by each user is known [6] and then the power allocation is performed. For the second one, the users' target data rate is constraint, and then the power and resource block are jointly allocated [7]. Because of these non-practical assumptions, these two kinds of optimization methods cannot be applied to the practical systems.

In this paper, we have proposed a binary search-based power allocation scheme for CA systems, which can greatly improves the energy efficiency of carrier aggregation system while keep low computation complexity at the base station.

## 2 System model and problem formulation

### 2.1 System model

Consider a single cell network model with a total of $K$ users and $N$ CCs. Each CC is composed of $M$ PRBs each with bandwidth $B$. Define $\mathcal{K} = \{1, 2, ..., K\}$, $\mathcal{N} = \{1, 2, ..., N\}$, and $\mathscr{M} = \{1, 2, \ldots, M\}$ as the user set, CC set, and PRB set,

* Correspondence: gaoweidong@bupt.edu.cn
Beijing University of Posts and Telecommunications, Beijing 100876, China

respectively. Indicator variable $c_{k,n,m} \in \{0, 1\}$ is used to indicate the relationships among the CCs, the users, and the PRBs, where value "1" indicates that PRB # $m$ of CC # $n$ is allocated to user # $k$ and value "0" indicates the opposite. There is a rule that each PRB can only be assigned to single user at any moment.

According to the Shannon formula, the throughput that user # $k$ can obtain from PRB # $m$ of CC # $n$ can be written as:

$$r_{k,n,m} = B\log_2\left(1 + \frac{p_{k,n,m} \times g_{k,n,m}}{BN_0}\right) \tag{1}$$

where $p_{k,n,m} \geq 0$ is the transmit power of the base station, $g_{k,n,m}$ is the channel gain, and $N_0$ is additive white Gaussian noise(AWGN) power spectral density.

Based on Eq. 1 and the definition of $c_{k,n,m}$, we can obtain the total system throughput by summing up the transmission rate of all the users as following:

$$R = \sum_{k=1}^{K}\sum_{n=1}^{N}\sum_{m=1}^{M} r_{k,n,m} \times c_{k,n,m} \tag{2}$$

The power consumption of the base station is the sum of the power consumption of each user on each resource block of each carrier, which can be expressed as follows:

$$P_{BS} = \sum_{k=1}^{K}\sum_{n=1}^{N}\sum_{m=1}^{M} p_{k,n,m} \times c_{k,n,m} \tag{3}$$

where $p_{k,n,m}$ is the base station's transmit power on PRB # $m$ of CC # $n$ for user $k$.

In addition to transmit power, there is also other extra power consumption at the base station, such as the energy consumption of air conditioning and cooling facilities. Considering that this kind of power consumption is irrelevant to the radio transmission and is steady during a long period of time, we model it as a constant $P_C$. Without losing generality, the energy efficiency is defined as the number of bits that are transmitted per joule:

$$\eta_{EE} = \frac{R}{P_{BS} + P_C} \tag{4}$$

According to Eqs. 2, 3, and 4, we can obtain the energy efficiency optimization problem for carrier aggregation system, which is modeled as problem $P$:

$$P: \max \frac{\sum_{k=1}^{K}\sum_{n=1}^{N}\sum_{m=1}^{M} r_{k,n,m} \times c_{k,n,m}}{P_{BS}+P_C}$$

st.

$$\begin{aligned}
&P_{BS} \leq P_{\max} && (C1)\\
&\sum_{k=1}^{K} c_{k,n,m} \leq 1, \forall n, m && (C2)\\
&\sum_{n=1}^{N} \tau\left(\sum_{m=1}^{M} c_{k,n,m}\right) \leq T, \forall k && (C3)\\
&c_{k,n,m} \in \{0, 1\}, p_{k,n,m} \geq 0, \forall k, n, m && (C4)
\end{aligned} \tag{5}$$

where $P_{\max}$ is the maximum transmit power of the base station. Function $\tau(x)$ is defined as following:

$$\tau(x) = \begin{cases} 1, & x > 0 \\ 0, & x = 0 \end{cases} \tag{6}$$

The constraint condition C2 limits the maximum number of CCs that each user can aggregate simultaneously. It is specified that by 3GPP that the largest number of CCs each user can aggregate should be no more than 5 [1]. Therefore, throughout this paper, the value of $T$ is set to less than or equal to 5. The constraint condition C4 guarantees the effectiveness of the indicator variable as well as the transmit power.

### 2.2 Problem analysis

Because of the existence of 0–1 indicating variables, the scope of the constrained optimization problem presented in 5 is not convex. At the same time, since the base station's transmit power, $P_{BS}$, appears both in the numerator and in the denominator of the optimization objective function, the optimization objective function is not convex. According to the above observations, the optimization problem as given in Eq. 5 is not a convex optimization problem, and it is not solved by convex optimization theory.

To solve this kind of non-convex optimization problem, reference [8] adopted a fixed value of base station transmission power and expanded the range of indicator variable $c_{k,n,m}$ from discrete value of 0 or 1 to continuous real numbers that range in [0, 1], and then convex optimization theory was used to solve the problem [9]. Since too many approximate operations are used in the above schemes, the accuracy of the solution is not fine enough. Furthermore, the total number of resource blocks is much larger for 5G carrier aggregation systems than that of the LTE systems, so the computing complexity of the scheme is too high and it cannot be effectively applied in the realistic 5G communication systems.

Assume that the total transmit power of the base station is a fixed value, which is denoted by $P_{BS}$, then the optimization problem presented by Eq. 5 is reduced to a resource allocation optimization problem that under fixed given transmit power value.

$$R_{\max}(P_{BS}) = \max\sum_{k=1}^{K}\sum_{n=1}^{N}\sum_{m=1}^{M} r_{k,n,m} \times c_{k,n,m}$$

st.

$$\sum_{k=1}^{K}\sum_{n=1}^{N}\sum_{m=1}^{M} p_{k,n,m} \times c_{k,n,m} = P_{BS} \quad (C1)$$

$$\sum_{k=1}^{K} c_{k,n,m} \le 1, \forall n, m \quad (C2)$$

$$\sum_{n=1}^{N} \tau\left(\sum_{m=1}^{M} c_{k,n,m}\right) \le T, \forall k \quad (C3)$$

$$c_{k,n,m} \in \{0, 1\}, p_{k,n,m} \ge 0, \forall k, n, m \quad (C4)$$

$$(7)$$

where $R_{\max}(P_{BS})$ represents the maximum data rate of the carrier aggregation system under determined $P_{BS}$ value. The definition of $c_{k,n,m}$, $r_{k,n,m}$, and $\tau$ function are the same with that of optimization problem (Eq. 5).

On the basis of Eq. 7, we define the optimal energy efficiency of the carrier aggregation system under determined base station transmit power $P_{BS}$, which is:

$$\eta_{EE}^{\max}(P_{BS}) \triangleq \frac{R_{\max}(P_{BS})}{P_{BS} + P_C} \quad (8)$$

Based on the above analysis, we can see that the optimization problem (Eq. 5) is equivalent to solving the following constraint optimization problem:

$$\eta_{EE}^{*} = \max_{0 \le P_{BS} \le P_{\max}} \eta_{EE}^{\max}(P_{BS}) \quad (9)$$

where the constraint conditions are the same with that in the optimization problem (Eq. 7). For the newly constructed optimization problem (Eq. 9), the most direct method is traversing all of the possible base station transmit power values to obtain the optimal power allocation scheme that has the highest energy efficiency using water-filling theorem. At last, find the power allocation scheme that has the highest energy efficiency from all possible power allocation schemes, then the optimization problem is solved.

## 3 Power allocation scheme

Considering that the base station's transmission power, $P_{BS}$, has real value, if we aim at traversing all $P_{BS}$ values at equal intervals in $[0, P_{\max}]$, the step must be set small enough to ensure sufficient accuracy. That is to say, we should solve the optimization problem for a large number of transmit power values. Obviously, the complexity of the algorithm is too high, so we need to optimize it with low complexity method.

Through in-depth analysis of the optimization problem (Eq. 9), we find that $\eta_{EE}^{\max}(P_{BS})$ is a strictly quasi concave function on $P_{BS}$ (please see Appendix for the proof). It means that $\eta_{EE}^{\max}(P_{BS})$ has only one local optimal solution, and the local optimal solution is also global optimal. For this kind of optimization problems, the literature [8] proposed an iterative algorithm based on extreme point idea, which adopted a large number of approximate operations that greatly increased the algorithm's time and space overhead. Considering this, we in this paper propose a binary search-based power allocation algorithm (BSPAA), which can provide high precision without too much iterations.

Assume that the number of users in the system is $K$ and the user set is $\mathcal{K} = \{1, 2, ..., K\}$. Since the resource allocation scheme has already been determined, it is not necessary to distinguish which component carrier each resource block belongs to, i.e., we can consider all of the resource blocks as a set, which is denoted as $\mathcal{N} = \{1, 2, ..., N\}$. Defined $p_{k,n}$ as the transmission power allocated to user #$k$ on resource block #$n$ and $c_{k,n}$ as the indication whether the resource block #$n$ is assigned to user #$k$, where value 1 represents "yes" and value "0" represents "no".

According to the Shannon formula, the transmission rate of the user #$k$ on the resource block #$n$ can be represented by the following formula:

$$r_{k,n} = B\log_2\left(1 + \frac{p_{k,n} \times g_{k,n}}{BN_0}\right) \quad (10)$$

where $g_{k,n}$ is the channel gain of user #$k$ on resource block $n$. $B$ is the bandwidth of a single resource block and $N_0$ is the Gauss white noise.

According to the Eq. 10, we can get the total downlink transmission rate of the carrier aggregation system:

$$R = \sum_{k=1}^{K}\sum_{n=1}^{N} r_{k,n} \times c_{k,n} \quad (11)$$

According to above analysis, the energy efficiency optimization problem under fixed base station's transmit power and determined resource block allocation scheme is equivalent to the optimization of the downlink transmission rate of the system, therefore we can get the following mathematical model:

$$\max\sum_{k=1}^{K}\sum_{n=1}^{N}r_{k,n}\times c_{k,n}$$
st.
$$\sum_{k=1}^{K}c_{k,n}\leq 1,\forall n \quad (C1)$$
$$\sum_{k=1}^{K}\sum_{n=1}^{N}p_{k,n}=P \quad (C2) \tag{12}$$

where $r_{k,n}$ is given by Eq. 10. Constraint C1 shows that one resource block can only be assigned to at most one user. Constraint C2 shows that the total transmit power of the base station is equal to the sum of the transmit power on all of the resource blocks.

Considering the above optimization problem, because the resource allocation scheme has been determined, the constraint C1 can be satisfied at any case and the above optimization problem can be reduced to the following form:

$$\max\sum_{k=1}^{K}\sum_{n=1}^{N}r_{k,n}\times c_{k,n}$$
st.
$$\sum_{k=1}^{K}\sum_{n=1}^{N}p_{k,n}=P \quad (C1) \tag{13}$$

It is clear that above optimization problem is a typical convex optimization problem, and the constraint condition is equality, so we can solve the problem by water -filling theorem [10]. Defined $\lambda_k$ as the Lagrange multiplier of user #$k$, then we can get the transmission power allocated to user #$k$ on resource block #$n$:

$$p_{k.n}^{*}=\left(\frac{1}{\lambda_k}-\frac{N_0}{g_{k,n}}\right)^{+} \tag{14}$$

In Eq. 14, $x^{+}$ equals to $x$ when $x$ is larger than 0, and $x^{+}$ equals to 0 when $x$ is less than or equal to 0. The Lagrange multiplier, $\lambda_k$, must satisfy the following inequality:

$$\sum_{k=1}^{K}\left(\frac{1}{\lambda_k}-\frac{N_0}{g_{k,n}}\right)^{+}=P \tag{15}$$

It can be seen from Eqs. 14 and 15 that the optimal power allocation scheme can be solved as long as the value of Lagrange multiplier is determined. The solution of the Lagrange multiplier is usually achieved by binary search method, and the specific steps can be found in [10], which is not addressed in detail here.

Binary search algorithm can determine the total transmission power of the base station while water-filling can solve the problem of power allocation under fixed transmission power and determined resource allocation scheme. Combining binary search algorithm and water-filling method, we can adopt BSPAA algorithm to achieve the optimal resource allocation results, which is as follows:

step1. Use GCSRAA algorithm to get the optimal resource allocation scheme $\mathcal{A}$.

step2. Initialize all the parameters, including the lower limit of transmission power $P_{lo}=0$, the upper limit of transmission power $P_{hi}=0$, and the transmit power adjustment step $\Delta$.

step3. Set the initial transmit power of the base station to $P_{cur}=(P_{hi}+P_{lo})/2$.

step4. Using water-filling based power allocation algorithm to calculate $\eta_{EE}(P_{cur}-\Delta)$, $\eta_{EE}(P_{cur})$, and $\eta_{EE}(P_{cur}+\Delta)$, i.e., the energy efficiency under the condition that the base station transmit power is $P_{cur}-\Delta$, $P_{cur}$, and $P_{cur}+\Delta$, respectively.

step5. If the energy efficiency values obtained in step3 satisfy the following inequality:

$$\begin{aligned}&\eta_{EE}(P_{cur}-\Delta)\leq\eta_{EE}(P_{cur})\\&\eta_{EE}(P_{cur}+\Delta)\leq\eta_{EE}(P_{cur})\end{aligned} \tag{16}$$

it means that the energy efficiency under transmit power $P_{cur}$ is higher than that under transmit power $P_{cur}-\Delta$ and $P_{cur}+\Delta$, so it can be declared that $P_{cur}$ is the global optimal solution if the adjustment step $\Delta$ is fine enough, then jump to step6.

Else if the energy efficiency values obtained in step3 satisfy the following inequality:

$$\begin{aligned}&\eta_{EE}(P_{cur}-\Delta)\leq\eta_{EE}(P_{cur})\\&\eta_{EE}(P_{cur}+\Delta)\geq\eta_{EE}(P_{cur})\end{aligned} \tag{17}$$

it means that the energy efficiency value is increasing under transmit power $P_{cur}$, i.e., the local optimal solution is larger than $P_{cur}$, so we can update the lower limit of the transmission power to $P_{cur}$ and then jump to step3.

Else if the energy efficiency values obtained in step3 satisfy the following inequality:

$$\begin{aligned}&\eta_{EE}(P_{cur}-\Delta)\geq\eta_{EE}(P_{cur})\\&\eta_{EE}(P_{cur}+\Delta)\leq\eta_{EE}(P_{cur})\end{aligned} \tag{18}$$

it means that the energy efficiency value is decreasing under transmit power $P_{cur}$, i.e., the local optimal solution is smaller than $P_{cur}$, so we can update the upper limit of the transmission power to $P_{cur}$ and then jump to step3.

step6. The algorithm ends. The current base station transmit power is the optimal value, and the corresponding power allocation scheme and resource allocation scheme are also the optimal strategy.

The algorithm is expressed in a flow chart as depicted in Fig. 1.

## 4 Numerical results

This chapter focuses on the performance and complexity analysis of the binary search-based power allocation scheme for carrier aggregation systems, to prove that the BSPAA algorithm can obtain fairly good results with lower complexity. The basic parameters used in the simulation are shown in Table 1. In this paper, two kinds of classical resource allocation algorithms for carrier aggregation system are evaluated and compared: nonlinear programming resource allocation algorithm (NPRAA) [11] and Markov-based carrier selected and resource allocation algorithm (MCSRAA). In order to simplify the simulation results, the normalized system energy efficiency is adopted in the following simulations.

Figure 2 shows the relationship between the energy efficiency of the system and the value of the base station transmission power under different constant extra power consumption values. In the simulation, the transmission power of the base station is exhaustively searched with 0.1 W search interval. As can be seen from the diagram, the energy efficiency of the system is a strictly quasi concave function of the base station transmission power. When the base station transmission power is about 10 W, the optimal system energy efficiency can be obtained. For the reason that the value of the additional power consumption is constant and independent to other system parameters, it cannot be optimized, so the

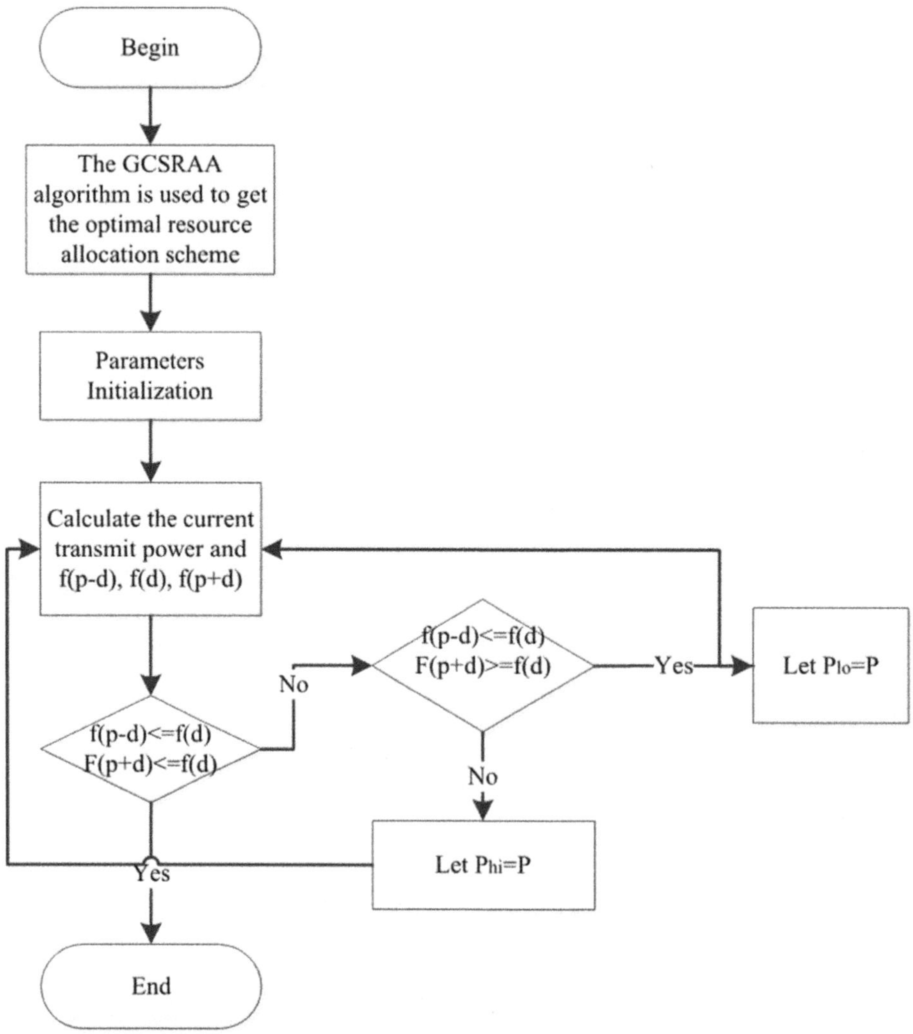


**Fig. 1** Energy efficiency optimization algorithm for carrier aggregation systems

**Table 1** Simulation parameters

| Parameters | Value |
|---|---|
| Carrier frequency | 3.5 GHz |
| Cell layout | 1 layer, 3 sectors |
| Channel model | SCME model |
| Scenario | Urban macro |
| Carrier bandwidth | 20 MHz |
| Maximal transmit power | 49 dBm |
| Minimum BS-UE distance | 35 m |
| Cell range expansion parameter | 0 dB |
| Service type | Full buffer |
| TTI length | 1 ms |
| TTI number | 5000 |

bigger its value is, the smaller the energy efficiency can be obtained by the proposed algorithm.

Figure 3 shows the relationship between the system energy efficiency and the number of component carriers that can be occupied by a single user in a carrier aggregation system with total of 5 users. For users working in carrier aggregation mode, the more component carriers a user can occupy, the more resource blocks it can be assigned, and the higher transmission rate can be obtained, which can significantly improve the energy efficiency of the whole system. Therefore, all the three curves show an upward trend for the three cases that different maximal CCs a user can occupy simultaneously. In addition, the more carrier frequencies are used in the system, the more available resource blocks there will be, so the user is more likely to be assigned to more resources and the energy efficiency value increases faster.

Figure 4 shows the relationship between the system energy efficiency and the transmission power of the base station, where the number of users is 5 and the maximum number of component carriers per user can occupy is also 5. As can be seen from the figure, the optimal performance can be achieved when the base station transmission power is about 10 W for 5, 10, and 15 carrier frequency cases, which is consistent with the observations obtained from Fig. 2. When the base station transmit power is small, increasing the transmission power of the base station can increase the power on the resource blocks assigned to the user and the transmission rate can be increased accordingly, which can improve the energy efficiency of the whole system. When the base station transmission power exceeds a certain value, the number of available resources in the system will be limited. Therefore, the energy efficiency of the system tends to be stable when increasing the transmission power of the base station. Based on the analysis of Fig. 4, it can be concluded that the higher the number of available carriers in the system, the higher the energy efficiency of the system will be.

Figure 5 shows the relationship between the system energy efficiency and the number of component carriers that single user can occupy simultaneously, where there are 15 available carrier frequencies and the base station transmission power is 10 W. For carrier aggregation users, the more component carriers a user can occupy,

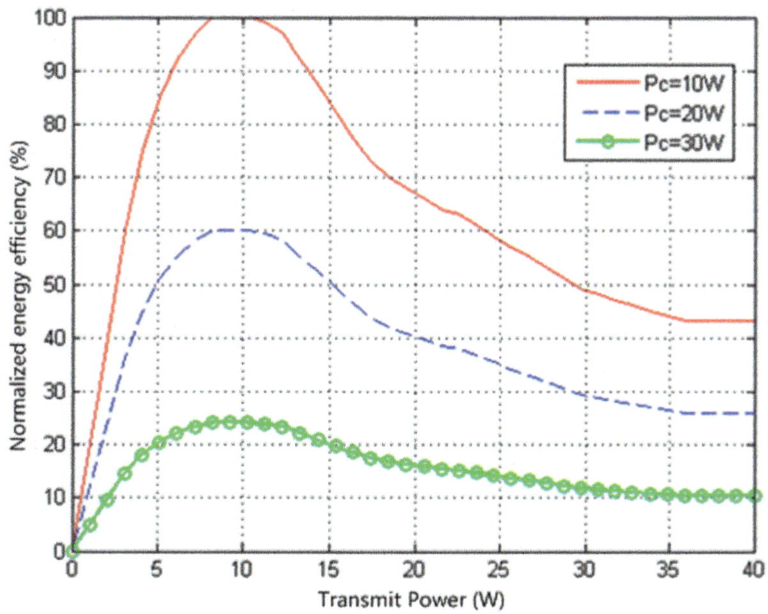


**Fig. 2** System energy efficiency v.s. transmit power

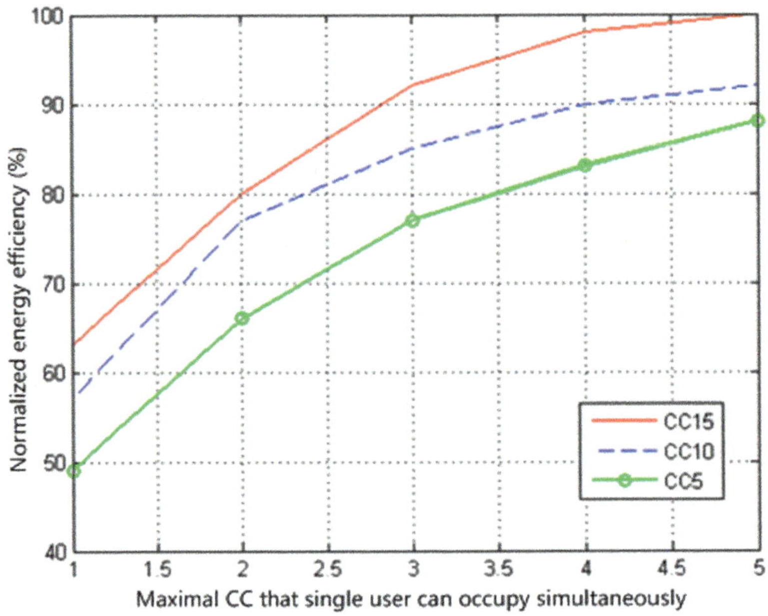


**Fig. 3** System energy efficiency v.s. maximal CC that single user can occupy simultaneously

the more resource blocks can be assigned, and the higher transmission rate can be obtained, thereby the energy efficiency of the system can be significantly improved. Therefore, all the three curves show an upward trend. In addition, it can be seen from the figure that the energy efficiency optimization algorithm BSPAA proposed in this paper has achieved an average performance improvement of 11% compared with the NLPRAA algorithm, and an average performance improvement of 25% is obtained with respect to the MCSRAA algorithm. It can be concluded that the energy efficiency optimization algorithm proposed can get better results than the existing energy efficiency optimization algorithm, and has higher energy efficiency.

Figure 6 shows the computational complexity comparison between the energy efficiency optimization

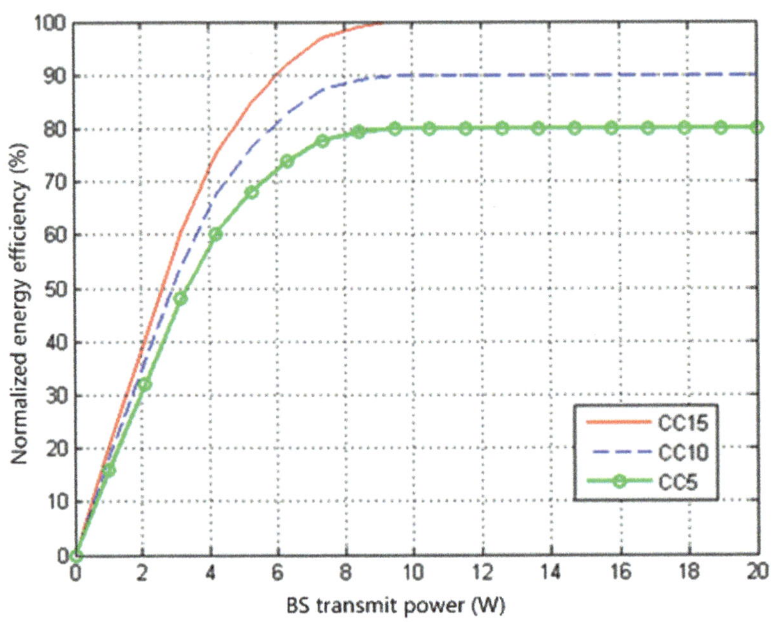


**Fig. 4** System energy efficiency v.s.BS transmit power with different number of available carrier frequencies

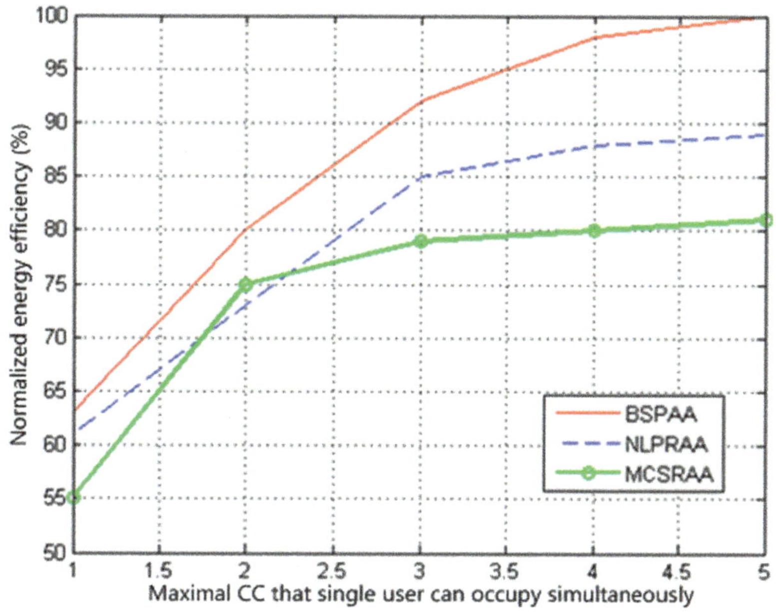


**Fig. 5** System energy efficiency v.s. maximal CC that a single user can occupy simultaneously

algorithm and the other two baseline algorithms. For the optimization algorithm proposed in this paper, because the resource allocation algorithm and the power allocation algorithm are executed independently, the total iteration number of the algorithm is the sum of that for the two sub algorithms. As can be seen from the figure, the algorithm proposed in this paper can reach 90% of the optimal performance after about 300 iterations, and the optimal performance can be obtained after about 400 iterations. However, the other two baseline algorithms can only achieve the optimal solution after 600 iterations, and the energy efficiency is far worse than the algorithm proposed in this paper. Therefore, it can be proved that the proposed algorithm can achieve better optimization results while ensuring low computation complexity.

## 5 Conclusions

This paper presents the problem of power allocation in carrier aggregation systems. Energy efficiency is adopted as the evaluation metric, and it is found that the

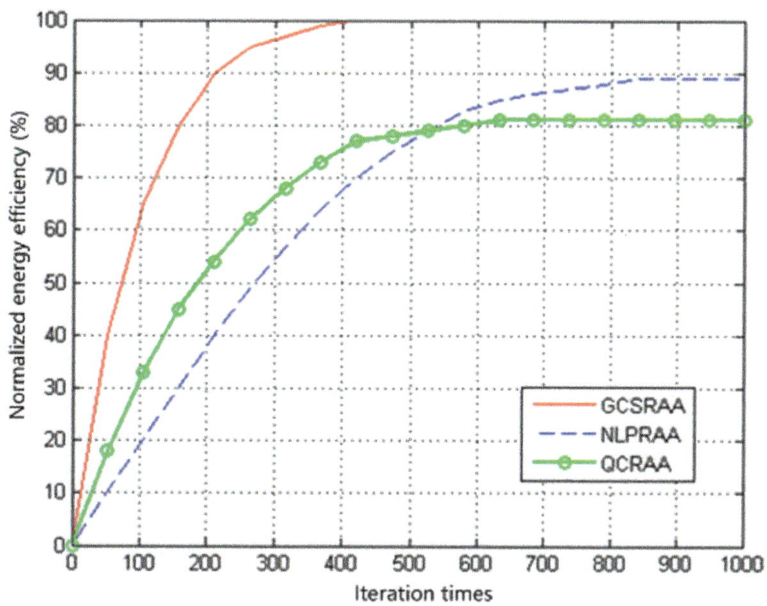


**Fig. 6** System energy efficiency v.s. iteration times

function of energy efficiency optimization is strictly quasi concave, which cannot be solved easily with traditional optimization method. We propose a binary search-based power allocation scheme, which can significantly improve the system energy efficiency while keeping low computation complexity. Future work will consider jointly optimization of component carrier selection, radio resource allocation, and power allocation to further improve the system performance.

## 6 Appendix

Firstly, we give the definition of strictly quasi concave function: for function $F(x)$and independent variable $x$, define the super set of $x$ as $\mathcal{S}_\alpha \triangleq \{x|F(x)\geq\alpha, x\geq 0\}$, where $\alpha$ is a real number greater than 0. For any $x_1\in\mathcal{S}_\alpha$, $x_2\in\mathcal{S}_\alpha$, and any real number in the interval (0, 1), if $\beta x_1+(1-\beta)x_2\in\mathcal{S}_\alpha$ holds, then the function $F(x)$ is strictly concave on $x$.

We will prove that $\eta_{EE}^{max}(P_{BS})$ is a strictly quasi concave function on $P_{BS}$ according to the definition of strictly quasi concave function. The certification process is as following:

Let the super set of independent variable $P_{BS}$ for function $\eta_{EE}^{max}(P_{BS})$ be $\mathcal{S}_\alpha \triangleq \{P_{BS}|\eta_{EE}^{max}(P_{BS})\geq\alpha, P_{BS}\geq 0\}$, where $\alpha$ is a real number greater than 0 and the definition of $P_{BS}$ is the same as above. According to the Eqs. 4, 5, 6, and 7, we substitute the expression $\eta_{EE}^{max}(P_{BS})$ into $\mathcal{S}_\alpha$ and perform a series of simplification and transposition, we can get $\mathcal{S}_\alpha=\{P_{BS}|R_{max}(P_{BS})-\alpha P_{BS}-\alpha P_C\geq 0, P_{BS}\geq 0\}$. Let $P_{BS_1}, P_{BS_2}\in\mathcal{S}_\alpha$, and $\beta$ be any real number in the interval (0, 1), next we will prove that $\beta P_{BS_1}+(1-\beta)P_{BS_2}$ also belongs to $\mathcal{S}_\alpha$.

For $\eta_{EE}^{max}(P_{BS})$, we have:

$$\eta_{EE}^{max}(\beta P_{BS_1}+(1-\beta)P_{BS_2})=R_{max}(\beta P_{BS_1}+(1-\beta)P_{BS_2}) -\alpha(\beta P_{BS_1}+(1-\beta)P_{BS_2})-\alpha P_C \tag{19}$$

We can see from Eqs. 4–1) that $R_{max}(P_{BS})$ and $P_{BS}$ are logarithmic, so $R_{max}(P_{BS})$ is a strictly convex function on $P_{BS}$, therefore we can get the following inequality:

$$R_{max}(\beta P_{BS_1}+(1-\beta)P_{BS_2})\geq R_{max}(\beta P_{BS_1})+R_{max}((1-\beta)P_{BS_2}) \tag{20}$$

Substitute Eq. 20 into the Eq. 19, we can have the following inequality:

$$R_{max}(\beta P_{BS_1}+(1-\beta)P_{BS_2})-\alpha(\beta P_{BS_1}+(1-\beta)P_{BS_2})-\alpha P_C\geq R_{max}(\beta P_{BS_1})+R_{max}((1-\beta)P_{BS_2})-\alpha(\beta P_{BS_1}+(1-\beta)P_{BS_2})-\alpha P_C \tag{21}$$

Based on the inequality (Eq. 21) and inequality (Eq. 19), the following inequalities are obtained:

$$\eta_{EE}^{max}(\beta P_{BS_1}+(1-\beta)P_{BS_2})\geq R_{max}(\beta P_{BS_1})+R_{max}((1-\beta)P_{BS_2}) -\alpha(\beta P_{BS_1}+(1-\beta)P_{BS_2})-\alpha P_C \tag{22}$$

By splitting, transposing, and merging operations to the inequality (Eq. 22), we can get the following inequalities:

$$R_{max}(\beta P_{BS_1}+(1-\beta)P_{BS_2})-\alpha(\beta P_{BS_1}+(1-\beta)P_{BS_2})-\alpha P_C\geq (R_{max}(\beta P_{BS_1})-\alpha\beta P_{BS_1}-\alpha\beta P_C)+ (R_{max}((1-\beta)P_{BS_2})-\alpha(1-\beta)P_{BS_2}-\alpha(1-\beta)P_C) \tag{23}$$

Because $R_{max}(P_{BS})$ and $P_{BS}$ are logarithmic, for any real number $0<\beta<1$, we have:

$$R_{max}(\beta P_{BS})\geq\beta R_{max}(P_{BS}) \tag{24}$$

Substitute Eq. 24 into the inequality Eq. 23, we can get the following inequality:

$$R_{max}(\beta P_{BS_1}+(1-\beta)P_{BS_2})-\alpha(\beta P_{BS_1}+(1-\beta)P_{BS_2})-\alpha P_C\geq \beta(R_{max}(P_{BS_1})-\alpha P_{BS_1}-\alpha P_C)+(1-\beta)(R_{max}(P_{BS_2})-\alpha P_{BS_2}-\alpha P_C) \tag{25}$$

Because $P_{BS_1}, P_{BS_2}\in\mathcal{S}_\alpha$, i.e., both of $P_{BS_1}$ and $P_{BS_2}$ belong to the super set $\mathcal{S}_\alpha$ defined above, the following inequality holds:

$$\begin{aligned}&\beta(R_{max}(P_{BS_1})-\alpha P_{BS_1}-\alpha P_C)\geq 0\\&(1-\beta)(R_{max}(P_{BS_2})-\alpha P_{BS_2}-\alpha P_C)\geq 0\end{aligned} \tag{26}$$

Substitute Eq. 26 into the formula Eq. 25,we can get the following inequality:

$$R_{max}(\beta P_{BS_1}+(1-\beta)P_{BS_2})-\alpha(\beta P_{BS_1}+(1-\beta)P_{BS_2})-\alpha P_C\geq 0 \tag{27}$$

So it is proved that $\beta P_{BS_1}+(1-\beta)P_{BS_2}\in\mathcal{S}_\alpha$, that is to say, $\beta P_{BS_1}+(1-\beta)P_{BS_2}$ also belongs to super set $\mathcal{S}_\alpha$, which proves the strictly quasi concave characteristic of $\eta_{EE}^{max}(P_{BS})$ on $P_{BS}$. Therefore, $\eta_{EE}^{max}(P_{BS})$ is a strictly quasi concave function on $P_{BS}$.

**Funding**

This work is supported by Major National Scientific and Technological Specific Project of China under grant number 2016ZX03001009-003.

**Authors' contributions**

WG conceived and designed the study. LM performed the experiments. GC reviewed and edited the manuscript. All authors read and agreed the manuscript.

**Competing interests**

The authors have declared that no competing interests exist.

## References


1. M Iwamura, K Etemad, Mh Fong, R Nory and R Love, Carrier aggregation framework in 3GPP LTE-advanced [WiMAX/LTE update], in IEEE Commun. Mag., vol. 48, no. 8, pp. 60–67, August 2010
2. B Li, M Sun, X Li, A Nallanathan and C Zhao, Energy detection based Spectrum sensing for cognitive radios over time-frequency doubly selective fading channels, in IEEE Trans. Signal Process., vol. 63, no. 2, pp. 402–417, Jan. 15, 2015
3. B Li, S Li, A Nallanathan and C Zhao, Deep sensing for future Spectrum and location awareness 5G communications, in IEEE J. Sel. Areas Commun., vol. 33, no. 7, pp. 1331–1344, July 2015
4. B Li, Z Zhou, H Zhang and A Nallanathan, Efficient Beamforming training for 60-GHz millimeter-wave communications: a novel numerical optimization framework," in IEEE Trans. Veh. Technol., vol. 63, no. 2, pp. 703–717, Feb. 2014
5. R Mahapatra, Y Nijsure, G Kaddoum, N Ul Hassan, C Yuen, Energy efficiency tradeoff mechanism towards wireless green communication: a survey, in IEEE Commun. Surv. Tutorials, vol. 18, no. 1, pp. 686–705, First quarter 2016
6. Z Ren, S Chen, B Hu and W Ma, Energy-efficient resource allocation in downlink OFDM wireless systems with proportional rate constraints, in IEEE Trans. Veh. Technol., vol. 63, no. 5, pp. 2139–2150, Jun 2014
7. G Yu, Q Chen, R Yin, H Zhang and GY Li, Joint downlink and uplink resource allocation for energy-efficient carrier aggregation, in IEEE Trans. Wirel. Commun., vol. 14, no. 6, pp. 3207–3218, June 2015
8. C Isheden and GP Fettweis, Energy-efficient multi-carrier link adaptation with sum rate-dependent circuit power, 2010 IEEE Global Telecommunications Conference GLOBECOM 2010, Miami, FL, 2010, pp. 1–6
9. C Schurgers, V Raghunathan, MB Srivastava, Power management for energy-aware communication systems. ACM Trans. Embed. Comput. Syst. **2**(3), 431–447 (2003)
10. W Yu, JM Cioffi, *On constant power water-filling* (Helsinki, IEEE International Conference on Communications, 2001), pp. 1665–1669
11. Y Chen, S Zhang, S Xu and GY Li, Fundamental trade-offs on green wireless networks, in IEEE Commun. Mag., vol. 49, no. 6, pp. 30–37, June 2011

# 3

# Serial distributed detection for wireless sensor networks with sensor failure

Junhai Luo* and Zuoting Liu

**Abstract**

We study serial distributed detection and fusion over noisy channels in wireless sensor networks (WSNs) with bathtub-shaped failure (BSF) rate of the sensors in this paper. In the previous work, we applied BSF rate to parallel topology and derived the Extension Log-likelihood Ration Test (ELRT) rule. Although ELRT is superior to traditional fusion rule without considering failed sensors, the detection performance decreases noticeably in the presence of a large number of failed sensors. In this paper, we construct a serial topology based on the target radiation energy attenuation model, apply BSF rate to serial topology, and derive the corresponding fusion rule. Unlike the parallel fusion, where the local sensors send their decisions to the Global Fusion Center (GFC) in the region of interest (ROI) directly, sensors in the serial topology transmit local decisions through multi-hop, short-range communications. At the same time, we extend ELRT to noisy channels. Finally, simulation results prove the effectiveness of the proposed fusion rules.

**Keywords:** Serial distributed detection, bathtub-shaped failure rate, failed sensors

## 1 Introduction

WSNs have attracted many researchers in various disciplines due to their flexibility, robustness, mobility, and cost-effectiveness. A major application of WSNs is target detection. WSN typically consists of a vast number of small, inexpensive, and low-powered sensors, which are deployed in the ROI to obtain and preprocess the received observation. The GFC is making a final decision about whether the target is present or not. There are two popular detection methods: centralized detection and distributed detection. In the centralized detection, the local sensor sends the received observation to the GFC directly without any processing. In the distributed detection, each sensor quantifies its observation into a local decision ("0" or "1") and sends it to the GFC. Although centralized detection achieves the highest performance, it is at the cost of more bandwidth and communication energy to obtain real-time results. Thus, the distributed detection is often preferable in these situations.

Distributed target detection has been extensively studied in many kinds of literature. In [1, 2], each sensor made a local decision by conducting likelihood ratio test and sent the local decision to the GFC to perform global log-likelihood ratio test, and then the GFC made a final decision. In [3], a uniformly most powerful (UMP) detector based on likelihood ratio test was developed, and an elegant test for target presence or absence was also derived. Typically, the performance of local sensors is hard to calculate. Therefore, in [4], a suboptimal fusion rule requiring less prior information was proposed, which we refer to as the counting rule (CR). CR employed the total number of decisions transmitted from local sensors for hypothesis testing at the GFC. In [5], CR was extended to the case where the total number of sensors was uncertain. Authors in [6–9] took into account imperfect communication channels between the sensors and the GFC, such as additive white Gaussian noise (AWGN) channels and fading channels. In [6], noisy communication links were considered and a Bayesian framework for distributed detection was presented, where noisy links were modeled as binary symmetric channels (BSC). In [7], distributed detection fusion in hierarchical WSNs was investigated in the presence of fading and noise, two fusion rules were derived accordingly, one utilized the complete fading channel state information (CSI), the other utilized the channel envelope statistics (CS). For resource-constrained sensor networks, a fusion rule using CSI was more preferable. Typically, local sensors communicated

*Correspondence: junhai_luo@uestc.edu.cn
School of Electronic Engineering, University of Electronic Science and Technology of China, No. 2006, Xiyuan Ave, West Hi-Tech Zone, 611731 Chengdu, China

with the GFC directly. In [10], authors considered the case where local decisions need to be relayed through the multi-hop transmission to reach the GFC and also took fading into account. In [11], based on various decisions from local sensors, a decision rule was derived in the case of unknown probability distributions. In [12, 13], decision fusion rules with unknown detection probability were investigated. Clustering-based decision fusion algorithms and fusion rules for distributed target detection in WSNs were studied in [14–16].

The structure of the WSNs can be classified into three categories: parallel topology, serial topology, and tree topology. Most researchers focus on parallel distributed detection. However, sensors are usually powered by a battery, so the energy and the communication range are limited. When sensors are far away from the GFC, the power consumption increases dramatically and the lifetime of the WSN has shortened accordingly. In [17, 18], serial distributed detection was investigated, where local decisions are transmitted to the GFC through short-range and multi-hop communication. The channel between two adjacent sensors was modeled as BSC. However, authors in [17] assumed the received energy emitted by the target at the local sensors was a deterministic value. In this case, the detection performances of the local sensors were similar.

The most common practice of traditional fusion rules, such as CR and Chair-Varshney fusion rules, is to employ all sensors in the ROI to derive a final decision. However, the signal emitted by the target often decays as the distance from the target increases. Sensors far away from the target make little contribution to the final decision at the GFC or are more likely to make a false judgment in the presence of background noise. In this paper, we simply employ sensors around the target to make the final decision.

We propose a new serial topology reconstruction method, where decisions are transmitted from sensors with lower credibility to sensors with higher credibility. Assuming the transmission channels were ideal, we applied a BSF rate of the sensor into the parallel structure and proposed ELRT in [19]. In this paper, we solve the problem when there are failed sensors in the serial structure and propose the corresponding fusion rule over noisy channels which we call serial rule (SR). In order to demonstrate the more stable detection performance of the serial structure, we also extend ELRT to noisy channels and derive the corresponding fusion rule in the same scenario which we call parallel rule (PR).

The remainder of this paper is organized as follows. In Section 2, the BSF rate of the sensor is described. In Section 3, a sensor deployment model is described. In Section 4, a detection system model is described. In Section 5, the fusion rules of the serial and parallel structure are derived. In Section 5, the performance of the proposed fusion rules is provided through simulation. In Section 6, conclusions are drawn.

## 2 BSF rate

One can characterize a lifetime distribution through three functions: reliability function, failure rate function, and mean residual life [20]. The lifetime typically means the time the product can operate regularly before it fails and can be measured in hours, miles, cycles, etc. In our original research work [19], we used failure rate function to characterize the lifetime distribution of electronic components. The failure rate function is shown in Fig. 1, which we call BSF. We can see that the failure rate function has a bathtub shape and can be visually subdivided into three stages: initial failure period, random failure period, and wear-out failure period. The initial failure period starts at zero and decreases noticeably due to early failures caused by design faults or initial implementation problems. The random failure period is relatively flat (approximately constant), which is denoted as the "useful life" phase. The failure is noticeably increasing due to material fatigue or component aging during wear-out failure period [21]. We also gave the modified failure rate function $r(t)$ as follows in [19]

$$r(t) = ab\,(at)^{b-1} + \left(\frac{a}{b}\right)(at)^{\frac{1}{b}-1} + h_0, \quad t, a, h_0 \geq 0, b > 1 \tag{1}$$

where $a$ and $b$ are related with life data of the products and $h_0$ is a worthy constant adding to the failure rate function.

Thus, sensors in this paper can be classified into four categories

- Normal sensors, sensors that can detect and transmit decision reliably

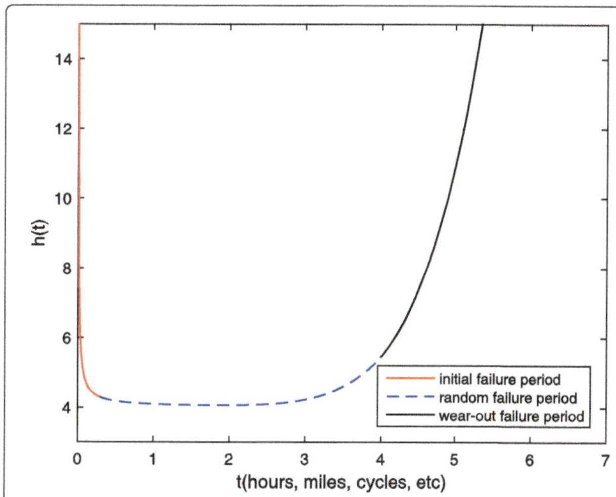


**Fig. 1** BSF rate function according to Eq. (1)

- Partially disable sensors, sensors that are still operable, but have poor detection capability
- Inoperable sensors, sensors which no longer function at all
- Failed sensors, the group of sensors consisting of the combination of the partially disable and the inoperable sensors

Let $N$ denote the initial number of the sensors, $M$ be the number of the sensors excluding the inoperable sensors, $n$ be the number of normal sensors, and $m$ be the number of failed sensors at time $t$, respectively. We can easily get that $n = N - m$. $m$ can be written as

$$m = \text{ceil}\left(\int_0^t r(t)\,dt\right) \tag{2}$$

where ceil $(\cdot)$ denotes the ceiling function.

Irrespective of the inoperable sensors at time $t$, the remaining sensor is either a normal sensor or a partially disable sensor. $R_i = 1$ which represents sensor $s_i$ is a normal sensor, and $R_i = 0$ which represents sensor $s_i$ is a partially disable sensor. The probability that sensor $s_i$ is a partially disable sensor is given as follows

$$\begin{aligned} p_i &= \Pr(R_i = 0) \\ &= \Pr(\textit{sensor } s_i \textit{ is a partially disable sensor}) \\ &= \tfrac{s}{M} \\ &= \tfrac{M-n}{M} \\ &= \tfrac{M-N+m}{M} \quad (0 \le s \le m, 0 < M \le N) \end{aligned} \tag{3}$$

where $s$ is the number of the partially disable sensors at time $t$, it is easy to note that $p_i$ is constant at time $t$ and $\Pr(R_i = 1) = 1 - p_i$. We use $p$ which denotes $\Pr(R_i = 0)$ in the latter sections.

## 3 Sensor deployment model

A sensor deployment example in the presence of failed sensors is shown in Fig. 2. $N$ sensors follow the grid deployment, and a target is deployed randomly in the ROI. The target and different types of sensors are labeled by different symbols. The received signal power emitted by the target $m_i, i = 1, 2, \ldots, N$ of sensor $s_i$ decays as the distance from the target increases. An isotropic attenuation power model adopted in this paper is defined as follows

$$m_i^2 = \frac{P_0 D_0^\gamma}{D_i^\gamma} \tag{4}$$

where $P_0$ is the signal power from the target at a reference distance $D_0$ and $D_i$ represents the Euclidean distance between the target and sensor $s_i$. The signal attenuation exponent $\gamma$ ranges from 2 to 3.

Signal power versus distance from the target is shown in Fig. 3. We can see that target radiation energy decays rapidly as the distance from the target increases. Sensors

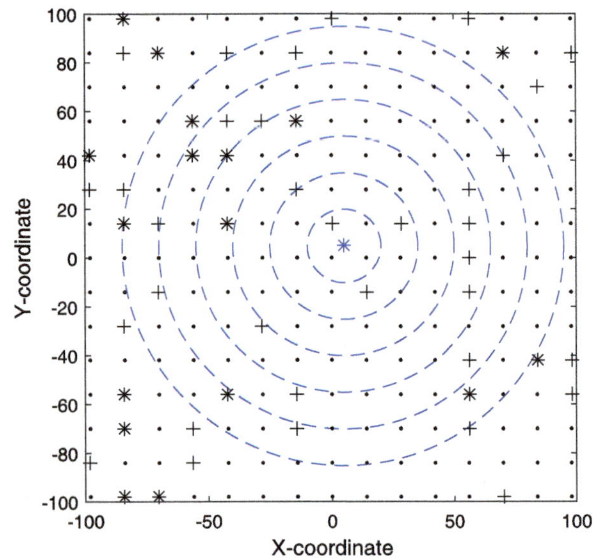


**Fig. 2** A deployment example in the presence of failed sensors. Area of the ROI:$A^2$ ; *black dot*: the normal sensors; *black star*: the inoperable sensors; *black +*: the partially disable sensors; and *blue star*: the target

far away from the target are prone to make wrong decisions influenced by background noise. Authors in [14] establish the precursor-successor relationship among the clusters based on tree topology. The cluster the target located in serves as the root cluster and the other clusters serve as the branch or leaf clusters, in this way, the root cluster can achieve the highest detection performance. Authors in [22] study a binary tree topology, where the sensor nearest the target serves as the root. Authors in [23, 24] think there are several "target spots" in the ROI where the target often appears, then analyze the area centered at the "target spot" and propose a target detection

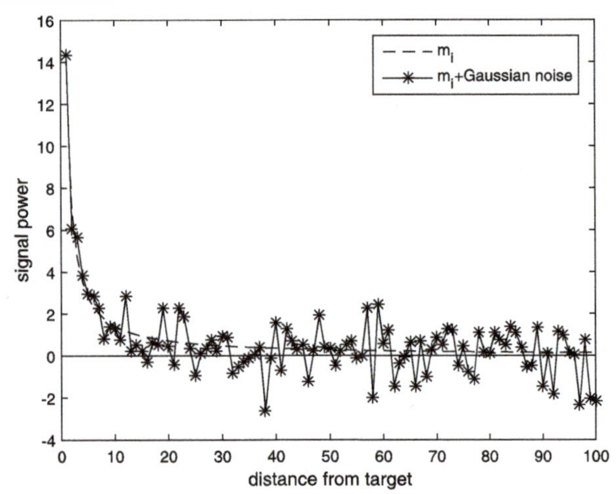


**Fig. 3** Signal power versus distance from target, where $P_0 = 180, D_0 = 1, \gamma = 2$

algorithm based on a probabilistic decision model. So, it is reasonable for us to investigate the area centered at the target in this paper.

The radiation energy of the target can be assumed as a series of concentric circles centered at the target. In this paper, we set the interval between two adjacent circles identical. Sensors lying in the same circular ring possess similar signal-to-noise ratio (SNR) and decision credibility, and the circular ring interval is 10. Due to the fast attenuation of the target radiation energy, we consider the circular area of the radius of 40 centered at the target in this paper. Firstly, sensors in the same circular ring form a serial fragment, then all fragments form the whole serial topology from the outside to the inside. We can see that decisions are transmitted from sensors with lower credibility to sensors with higher credibility in this serial structure, so the final decision at the last sensor of the serial structure possesses the highest credibility.

## 4 System model description

Figure 4 describes the serial detection system model. We leave out the inoperable sensors and form a new serial topology at time $t$. So the serial topology is dynamic. The wireless transmission channel between two adjacent sensors is modeled as BSC with transmission error probability $p_e$. The observation of sensor $s_i$ under each hypothesis is respectively given by

$$\begin{aligned} H_1 &: y_i = m_i + n_i \;;\; i = 1, \ldots, M' \\ H_0 &: y_i = n_i \;;\; i = 1, \ldots, M' \left(M' < N\right) \end{aligned} \tag{5}$$

$H_0$ and $H_1$ denote the absence and the presence of the target to be detected respectively. $M'$ is the number of sensors in the serial topology at time $t$. $y_i$ is the signal received by sensor $s_i$, and $n_i$ is the noise observed by sensor $s_i$. In this paper, we assume that noises at the local sensors are independent identically distributed (i.i.d.) and follow the standard Gaussian distribution, i.e, $n_i \sim N(0,1)$.

In Fig. 4, $v_i$ denotes the decision made by sensor $s_i$, and $u_i$ denotes the bit received by sensor $s_{i+1}$ which may differ from $v_i$. Sensor $s_1$ to $s_{M'}$ cooperatively determine whether the target is present or not in the ROI. Sensor $s_{M'}$ is the fusion center (FC) of the serial structure at time $t$. Because we apply BSF to the serial structure, $s_{M'}$ may be a partially disable sensor. If $s_{M'}$ is a partially disable sensor, the final decision is not reliable. In this paper, we let sensor $s_{M'}$ transmits its observation to the GFC, then the GFC replaces sensor $s_{M'}$ and serves as the FC of the serial structure. In this way, the $M'$th sensor of the serial structure in Fig. 4 is actually the GFC at time $t$.

In Fig. 5, a and b give the local decision-making and transmission process at sensor $s_i$ under the situation of $R_i = 1$ and $R_i = 0$, respectively. Under $H_0$, when $R_i = 1$, sensor $s_i$ makes a correct decision with probability $1 - p_{f,i}$, and an error decision with probability $p_{f,i}$, while when $R_i = 0$, sensor $s_i$ makes a correct decision with probability $1 - p'_{f,i}$, and an error decision with probability $p'_{f,i}$. During the decision transmission process, the bit $u_i$ received by sensor $s_{i+1}$ may differ from the decision $v_i$ made by sensor $s_i$ due to the noisy channel. The situation under $H_1$ is similar to that under $H_0$, except that the detection performance is denoted by $p_{d,i}$ and $p'_{d,i}$. Normally, $p'_{d,i} \le p_{d,i}, p_{f,i} \le p'_{f,i}$.

## 5 SR and PR

### 5.1 SR

Let $p(y_i|H_1)$ and $p(y_i|H_0)$ represent the conditional probability density function of the observation $y_i$ at sensor $s_i$ under $H_1$ and $H_0$, respectively. Except that sensor $s_1$ makes the local decision simply according to its observation $y_1$, the other sensors make decision according to their observations and the bit transmitted from the previous sensor. For example, sensor $s_1$ makes a decision $v_1$ according to its observation $y_1$, while sensor $s_2$ makes a decision according to its observation $y_2$ and the bit $u_1$ transmitted from sensor $s_1$, where $u_1$ may be different from $v_1$. For sensor $s_i \left(1 < i \le M' < N\right)$, the decision-making process is similar to sensor $s_2$. The likelihood ratio test at sensor $s_1$ can be expressed as

$$\Lambda_0(y_1) = \frac{p(y_1|H_1)}{p(y_1|H_0)} \underset{H_0}{\overset{H_1}{\gtrless}} \tau \tag{6}$$

Given the local threshold $\tau$ and according to Neyman-Pearson criterion, the detection probability and false

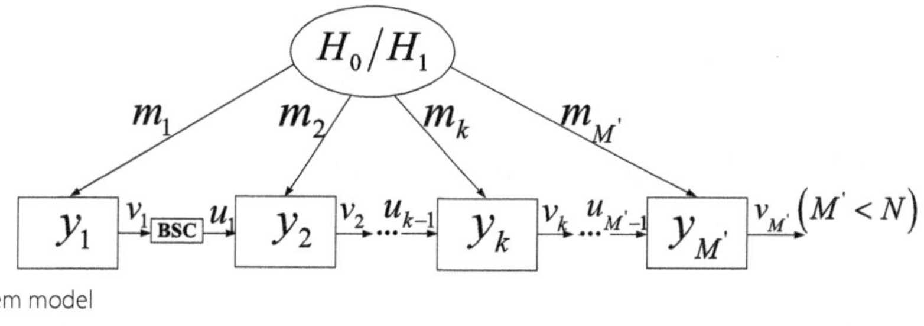


**Fig. 4** Serial detection system model

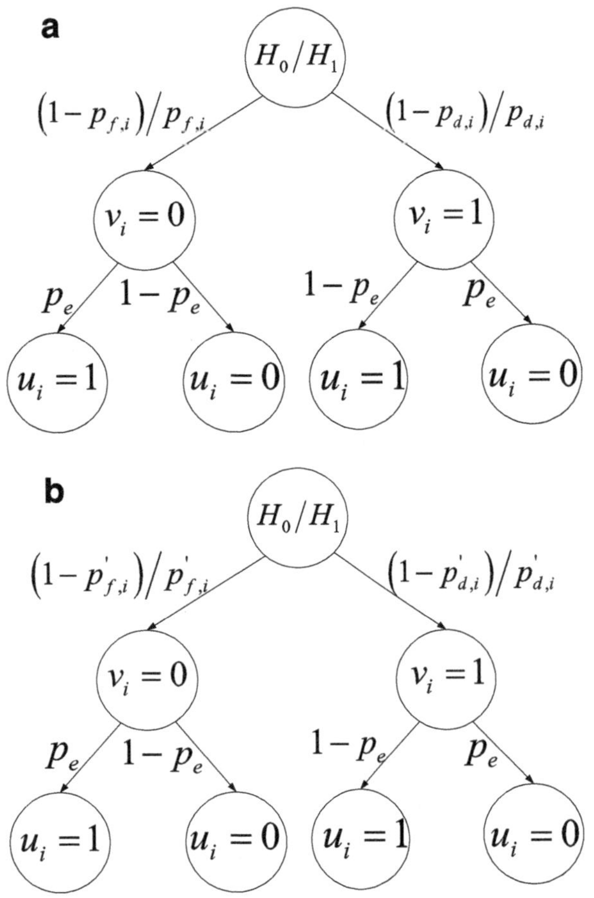


**Fig. 5** Model of serial distributed detection under noisy channel. **a** and **b** describe the situation of $R_i = 1$ and $R_i = 0$, respectively

alarm probability at sensor $s_1$ can be evaluated easily. The likelihood ratio test at sensor $s_i \left(1 < i \le M' < N\right)$ can be expressed as

$$\Lambda_0(y_i) = \frac{p(y_i, u_{i-1}|H_1)}{p(y_i, u_{i-1}|H_0)} \underset{H_0}{\overset{H_1}{\gtrless}} \tau \tag{7}$$

Because of the independence between $y_i$ and $u_{i-1}$, $\Lambda_0(y_i)$ can be further expressed as follows

$$\Lambda_0(y_i) = \frac{p(y_i|H_1)\cdot p(u_{i-1}|H_1)}{p(y_i|H_0)\cdot p(u_{i-1}|H_0)} \tag{8}$$

Let us discuss $p(u_{i-1}|H_1)$ firstly. Due to BSC, the received bit $u_{i-1}$ at sensor $s_i$ depends on the decision $v_{i-1}$ made at sensor $s_{i-1}$, thus we can get

$$\begin{aligned} p(u_{i-1}|H_1) &= \textstyle\sum_{v_{i-1}} p(u_{i-1}, v_{i-1}|H_1) \\ &= p(u_{i-1}, v_{i-1}=1|H_1) \\ &\quad + p(u_{i-1}, v_{i-1}=0|H_1) \\ &= p(u_{i-1}|v_{i-1}=1, H_1)\, p(v_{i-1}=1|H_1) \\ &\quad + p(u_{i-1}|v_{i-1}=0, H_1)\, p(v_{i-1}=0|H_1) \\ &= p(u_{i-1}|v_{i-1}=1)\, p(v_{i-1}=1|H_1) \\ &\quad + p(u_{i-1}|v_{i-1}=0)\, p(v_{i-1}=0|H_1) \end{aligned} \tag{9}$$

In the above equation, we use the Markov property that $u_{i-1}$ is only dependent on $v_{i-1}$. This property will be exploited in the following derivations. Note that the decision $v_{i-1}$ at sensor $s_{i-1}$ depends on whether $s_{i-1}$ is a partially disable sensor or not. Furthermore, we can easily get

$$\begin{aligned} a_{i-1} &= p(v_{i-1}=1|H_1) \\ &= \textstyle\sum_{R_{i-1}} p(v_{i-1}=1, R_{i-1}|H_1) \\ &= p(v_{i-1}=1, R_{i-1}=1|H_1) \\ &\quad + p(v_{i-1}=1, R_{i-1}=0|H_1) \\ &= p(v_{i-1}=1|R_{i-1}=1, H_1)\, p(R_{i-1}=1|H_1) \\ &\quad + p(v_{i-1}=1|R_{i-1}=0, H_1)\, p(R_{i-1}=0|H_1) \\ &= (1-p)\, p_{d,i-1} + p p'_{d,i-1} \end{aligned} \tag{10}$$

and

$$\begin{aligned} b_{i-1} &= p(v_{i-1}=0|H_1) \\ &= \textstyle\sum_{R_{i-1}} p(v_{i-1}=0, R_{i-1}|H_1) \\ &= p(v_{i-1}=0, R_{i-1}=1|H_1) \\ &\quad + p(v_{i-1}=0, R_{i-1}=0|H_1) \\ &= p(v_{i-1}=0|R_{i-1}=1, H_1)\, p(R_{i-1}=1|H_1) \\ &\quad + p(v_{i-1}=0|R_{i-1}=0, H_1)\, p(R_{i-1}=0|H_1) \\ &= (1-p)\left(1-p_{d,i-1}\right) + p\left(1-p'_{d,i-1}\right) \end{aligned} \tag{11}$$

when $u_{i-1} = 1$,

$$\begin{aligned} \alpha_{i-1} &= \left(\textstyle\sum_{v_{i-1}} p(u_{i-1}, v_{i-1}|H_1)\right) | (u_{i-1}=1) \\ &= (1-p_e)\, a_{i-1} + p_e b_{i-1} \end{aligned} \tag{12}$$

Similarly, when $u_{i-1} = 0$,

$$\begin{aligned} \beta_{i-1} &= \left(\textstyle\sum_{v_{i-1}} p(u_{i-1}, v_{i-1}|H_1)\right) | (u_{i-1}=0) \\ &= p_e a_{i-1} + (1-p_e)\, b_{i-1} \end{aligned} \tag{13}$$

So, $p(u_{i-1}|H_1)$ can be expressed as

$$p(u_{i-1}|H_1) = (\alpha_{i-1})^{u_{i-1}} \cdot (\beta_{i-1})^{1-u_{i-1}} \tag{14}$$

Next, let us discuss $p(u_{i-1}|H_0)$

$$\begin{aligned} p(u_{i-1}|H_0) &= \textstyle\sum_{v_{i-1}} p(u_{i-1}, v_{i-1}|H_0) \\ &= p(u_{i-1}, v_{i-1}=1|H_0) \\ &\quad + p(u_{i-1}, v_{i-1}=0|H_0) \\ &= p(u_{i-1}|v_{i-1}=1, H_0)\, p(v_{i-1}=1|H_0) \\ &\quad + p(u_{i-1}|v_{i-1}=0, H_0)\, p(v_{i-1}=0|H_0) \\ &= p(u_{i-1}|v_{i-1}=1)\, p(v_{i-1}=1|H_0) \\ &\quad + p(u_{i-1}|v_{i-1}=0)\, p(v_{i-1}=0|H_0) \end{aligned} \tag{15}$$

We can get

$$\begin{aligned} c_{i-1} &= p(v_{i-1}=1|H_0) \\ &= \sum_{R_{i-1}} p(v_{i-1}=1,R_{i-1}|H_0) \\ &= p(v_{i-1}=1,R_{i-1}=1|H_0) \\ &\quad + p(v_{i-1}=1,R_{i-1}=0|H_0) \\ &= p(v_{i-1}=1|R_{i-1}=1,H_0)\,p(R_{i-1}=1|H_0) \\ &\quad + p(v_{i-1}=1|R_{i-1}=0,H_0)\,p(R_{i-1}=0|H_0) \\ &= (1-p)\,p_{f,i-1} + p p'_{f,i-1} \end{aligned} \tag{16}$$

and

$$\begin{aligned} d_{i-1} &= p(v_{i-1}=0|H_0) \\ &= \sum_{R_{i-1}} p(v_{i-1}=0,R_{i-1}|H_0) \\ &= p(v_{i-1}=0,R_{i-1}=1|H_0) \\ &\quad + p(v_{i-1}=0,R_{i-1}=0|H_0) \\ &= p(v_{i-1}=0|R_{i-1}=1,H_0)\,p(R_{i-1}=1|H_0) \\ &\quad + p(v_{i-1}=0|R_{i-1}=0,H_0)\,p(R_{i-1}=0|H_0) \\ &= (1-p)\left(1-p_{f,i-1}\right) + p\left(1-p'_{f,i-1}\right) \end{aligned} \tag{17}$$

when $u_{i-1}=1$,

$$\begin{aligned} \gamma_{i-1} &= \left(\sum_{v_{i-1}} p(u_{i-1},v_{i-1}|H_0)\right) | (u_{i-1}=1) \\ &= (1-p_e)\,c_{i-1} + p_e d_{i-1} \end{aligned} \tag{18}$$

Similarly, when $u_{i-1}=0$,

$$\begin{aligned} \eta_{i-1} &= \left(\sum_{v_{i-1}} p(u_{i-1},v_{i-1}|H_0)\right) | (u_{i-1}=0) \\ &= p_e c_{i-1} + (1-p_e)\,d_{i-1} \end{aligned} \tag{19}$$

So, $p(u_{i-1}|H_0)$ can be expressed as

$$p(u_{i-1}|H_0) = (\gamma_{i-1})^{u_{i-1}} \cdot (\eta_{i-1})^{1-u_{i-1}} \tag{20}$$

After a series of derivation, the likelihood ratio test at sensor $s_i$ in Eq. (7) can be formulated as

$$\Lambda_0(y_i) = \frac{p(y_i|H_1)}{p(y_i|H_0)} \left(\frac{\alpha_{i-1}}{\gamma_{i-1}}\right)^{u_{i-1}} \left(\frac{\beta_{i-1}}{\eta_{i-1}}\right)^{1-u_{i-1}} \underset{H_0}{\overset{H_1}{\gtrless}} \tau \tag{21}$$

Now, the decision rule of sensor $s_i$ can be given by

$$\begin{cases} \frac{p(y_i|H_1)}{p(y_i|H_0)} \cdot \frac{\alpha_{i-1}}{\gamma_{i-1}} \underset{H_0}{\overset{H_1}{\gtrless}} \tau, & if\ u_{i-1}=1 \\ \frac{p(y_i|H_1)}{p(y_i|H_0)} \cdot \frac{\beta_{i-1}}{\eta_{i-1}} \underset{H_0}{\overset{H_1}{\gtrless}} \tau, & if\ u_{i-1}=0 \end{cases} \tag{22}$$

Furthermore, the decision fusion rule in Eq. (22) can be transformed into

$$\Lambda(y_i) = \frac{p(y_i|H_1)}{p(y_i|H_0)} \underset{H_0}{\overset{H_1}{\gtrless}} \begin{cases} \tau_{i,1}, & if\ u_{i-1}=1 \\ \tau_{i,0}, & if\ u_{i-1}=0 \end{cases} \tag{23}$$

We should note that $\Lambda(y_i)$ is another metric we defined and is different from $\Lambda_0(y_i)$, where

$$\tau_{i,1} = \tau \cdot \frac{\gamma_{i-1}}{\alpha_{i-1}},\ \tau_{i,0} = \tau \cdot \frac{\eta_{i-1}}{\beta_{i-1}} \tag{24}$$

The corresponding log-likelihood ratio form of Eq. (23) can be formulated as

$$\Lambda^*(y_i) = \ln \Lambda(y_i) \underset{H_0}{\overset{H_1}{\gtrless}} \begin{cases} \tau'_{i,1} = \ln \tau_{i,1}, & if\ u_{i-1}=1 \\ \tau'_{i,0} = \ln \tau_{i,0}, & if\ u_{i-1}=0 \end{cases} \tag{25}$$

The false alarm probability and the detection probability at sensor $s_i$ are evaluated as follows

$$\begin{aligned} p_{f,i} &= \Pr\left(\Lambda^*(y_i) \geq \tau'_{i,1}|H_0,u_{i-1}=1\right) \Pr(u_{i-1}=1|H_0) \\ &\quad + \Pr\left(\Lambda^*(y_i) \geq \tau'_{i,0}|H_0,u_{i-1}=0\right) \Pr(u_{i-1}=0|H_0) \\ &= \Pr\left(\Lambda^*(y_i) \geq \tau'_{i,1}|H_0\right)\gamma_{i-1} + \Pr\left(\Lambda^*(y_i) \geq \tau'_{i,0}|H_0\right)\eta_{i-1} \end{aligned} \tag{26}$$

$$\begin{aligned} p_{d,i} &= \Pr\left(\Lambda^*(y_i) \geq \tau'_{i,1}|H_1,u_{i-1}=1\right) \Pr(u_{i-1}=1|H_1) \\ &\quad + \Pr\left(\Lambda^*(y_i) \geq \tau'_{i,0}|H_1,u_{i-1}=0\right) \Pr(u_{i-1}=0|H_1) \\ &= \Pr\left(\Lambda^*(y_i) \geq \tau'_{i,1}|H_1\right)\alpha_{i-1} + \Pr\left(\Lambda^*(y_i) \geq \tau'_{i,0}|H_1\right)\beta_{i-1} \end{aligned} \tag{27}$$

$p_{f,i}$ and $p_{d,i}$ are recursive equations. If we get the values of $p_{f,1}$ and $p_{d,1}$, $p_{f,M'}$ and $p_{d,M'}$ can be evaluated recursively. Let $P_d$ and $P_f$ represent the global detection probability and false probability, respectively, we note that $P_d = p_{d,M'}$ and $P_f = p_{f,M'}$. Given $\tau$, $p_{f,1}$ and $p_{d,1}$ at sensor $s_1$ can be easily derived according to Eq. (6).

$$p_{f,1} = \int_\tau^\infty \frac{1}{\sqrt{2\pi}} e^{-\frac{x^2}{2}}\, dx = Q(\tau) \tag{28}$$

$$p_{d,1} = \int_\tau^\infty \frac{1}{\sqrt{2\pi}} e^{-\frac{(x-a_1)^2}{2}}\, dx = Q\left(\tau - \sqrt{\frac{P_0 d_0^r}{d_1^r}}\right) \tag{29}$$

where $Q(\cdot)$ is the complementary distribution function of the standard Gaussian, i.e,

$$Q(x) = \int_x^\infty \frac{1}{\sqrt{2\pi}} e^{-\frac{z^2}{2}}\, dz \tag{30}$$

### 5.2 PR

To demonstrate the better detection performance of SR, we extend ELRT [19] to noisy channels and propose PR in the same scenario. In the parallel structure, the local sensor $s_i$ makes a local decision $v_{pi}$ $(1 \leq i \leq M)$ and transmits it to the GFC in the ROI. We denote $u_{pi}$ as the bit received by the GFC. Due to the BSC, $u_{pi}$ may be different from $v_{pi}$

with transmission error probability $p_e$. Furthermore, we let $\overrightarrow{v_p} = (v_{p1}, v_{p2}, \ldots, v_{pM})$ and $\overrightarrow{u_p} = (u_{p1}, u_{p2}, \ldots, u_{pM})$. Due to the mutual independence among the received bits, the fusion rule at the GFC is given as follows

$$\Lambda_1 = \frac{p(\overrightarrow{u_p}|H_1)}{p(\overrightarrow{u_p}|H_0)} = \prod_{i=1}^{M} \frac{p(u_{pi}|H_1)}{p(u_{pi}|H_0)} \underset{H_0}{\overset{H_1}{\gtrless}} T \tag{31}$$

According to SR, $\Lambda_1$ can be further represented as

$$\begin{aligned}\Lambda_1 &= \prod_{i=1}^{M} \frac{\sum_{v_{pi},R_i} p(u_{pi}, v_{pi}, R_i|H_1)}{\sum_{v_{pi},R_i} p(u_{pi}, v_{pi}, R_i|H_0)} \\ &= \prod_{i=1}^{M} \left(\frac{\alpha_{pi}}{\gamma_{pi}}\right)^{u_{pi}} \left(\frac{\beta_{pi}}{\eta_{pi}}\right)^{1-u_{pi}}\end{aligned} \tag{32}$$

where

$$\begin{cases} \alpha_{pi} = (1-p_e)\,a_{pi} + p_e b_{pi} \\ \beta_{pi} = p_e a_{pi} + (1-p_e)\,b_{pi} \\ \gamma_{pi} = (1-p_e)\,c_{pi} + p_e d_{pi} \\ \eta_{pi} = p_e c_{pi} + (1-p_e)\,d_{pi} \end{cases}$$

$$\begin{cases} a_{pi} = (1-p)\,p_{p,di} + p p'_{p,di} \\ b_{pi} = (1-p)\,(1-p_{p,di}) + p\left(1-p'_{p,di}\right) \\ c_{pi} = (1-p)\,p_{p,fi} + p p'_{p,fi} \\ d_{pi} = (1-p)\,(1-p_{p,fi}) + p\left(1-p'_{p,fi}\right) \end{cases}$$

where $T$ denotes the detection threshold at the GFC, $p_e$ and $p$ are given in the previous section. $p_{p,di}$ and $p'_{p,di}$ denote the local detection probability when sensor $s_i$ is a normal sensor and a partially disable sensor, respectively. Similarly, $p_{p,fi}$ and $p'_{p,fi}$ denote the local false probability. Given the local threshold, according to Neyman-Pearson criterion, we can get the local detection probability and false probability easily. The log-likelihood ratio form is given as follows

$$\Lambda'_1 = \ln \Lambda_1 = \sum_{i=1}^{M} \left(u_{pi} \ln\left(\frac{\alpha_{pi}\eta_{pi}}{\gamma_{pi}\beta_{pi}}\right) + \ln\left(\frac{\beta_{pi}}{\eta_{pi}}\right)\right) \tag{33}$$

The final fusion rule at the GFC is as follows

$$\Lambda'_1 \underset{H_0}{\overset{H_1}{\gtrless}} T_1 \tag{34}$$

where $T_1 = \ln T$. We use $P_{pd}$ and $P_{pf}$ to represent the global detection probability and false probability of the parallel structure respectively. $P_{pd}$ and $P_{pf}$ can be evaluated as

$$\begin{aligned} P_{pd} &= \Pr\left(\Lambda'_1 > T_1|H_1\right) \\ P_{pf} &= \Pr\left(\Lambda'_1 > T_1|H_0\right) \end{aligned} \tag{35}$$

## 6 Performance analysis

We use the Matlab simulator to evaluate the performance of the proposed fusion rules in this paper, and $10^4$ Monte Carlo runs are used in each simulation. The objective is to demonstrate that serial structure has more stable detection performance than the parallel structure in the presence of a vast number of failed sensors. In [19], we studied the parallel distributed detection applying BSF rate of the sensor over ideal channels and proposed ELRT rule. Simulations demonstrated that ELRT outperformed Chair-Varshney rule in the presence of failed sensors. In this paper, we apply BSF to the serial structure over noisy channels and also extend ELRT to noisy channels, because sensors far away from the target make little contributions to the final decision at the GFC and are more likely to make wrong decisions in the presence of background noise. Thus, we consider sensors within the circular area centered at the target in this paper, the deployment in the presence of failed sensors is presented in Fig. 2. The fusion rules of the serial and parallel are given in the previous sections.

In Fig. 6, according to Eq. (2), the number of failed sensors including the partially disable sensors and the inoperable sensors is plotted as a function of $t$(hours, miles, cycles, etc) and the corresponding parameters are given at the bottom. We can see that the number of failed sensors during the initial failure period is negligible, such as $t = 0.5$. The number increases slowly at the low time, while the number grows noticeably after time 5. The curve of the number of failed sensors over time also reflects the feature of BSF.

We leave out the inoperable sensors at each time, the partially disable sensors are selected from the remaining sensors randomly. In Fig. 7, a and b show the receiver

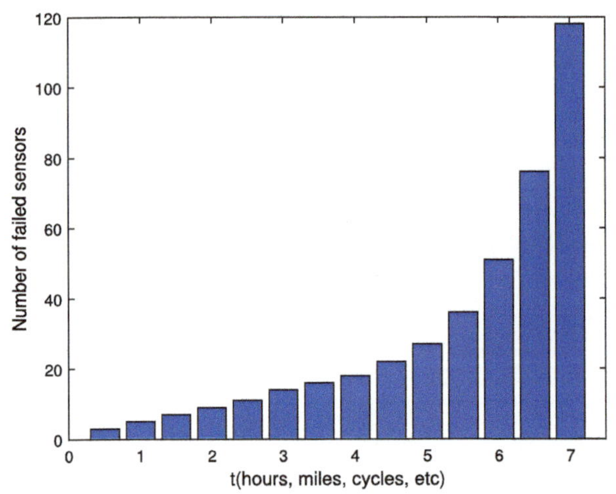


**Fig. 6** The number of failed sensors obtained by Eq. (2) at a different time: $a = 0.25, b = 8$, and $h_0 = 4$

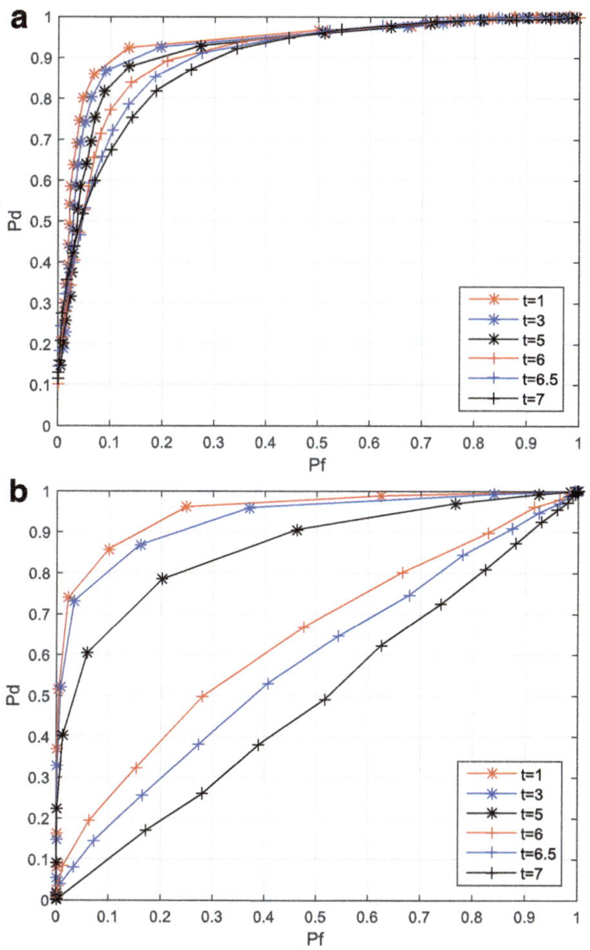


**Fig. 7** The ROC curses of SR and PR at a different time: $\gamma = 2$, $A = 200$ m, $D_0 = 1$ m, $N = 220$, $P_0 = 180$, $\frac{N-M}{m} = \frac{1}{3}$, $p'_{d,i} = \frac{p_{d,i}}{5}$, $p'_{f,i} = 0.5$, $a = 0.25$, $b = 8$, and $h_0 = 4$. **a**, **b** correspond to the ROC curses of SR and PR, respectively

operating characteristic (ROC) curves for SR and PR, respectively. The simulation parameters for these two rules are set to be identical and given at the bottom. We use $t = 1, t = 3, t = 5, t = 6, t = 6.5$, and $t = 7$ to conduct the experiment and choose $p'_{d,i} = p_{d,i}/5$ and $p'_{f,i} = 0.5$ for the partially disable sensor due to their poor performance. In Fig. 7a, we see that the performance of SR decreases slightly all the time. In Fig. 7b, we observe that the performance of PR decreases slowly before time 5 and noticeably after time 5. For PR, sensors in the parallel structure make local decisions and transmit the decisions to the GFC directly. At the low time, a small number of failed sensors have a negative but not big effect on the system performance. Meanwhile, at the high time, a large number of failed sensors have a major negative effect on the system performance. For SR, each sensor in the serial structure makes a decision according to the bit transmitted from the previous sensor and the state (partially disable or normal) of the previous sensor, and due to this cooperative detection, SR has more stable detection performance even in the presence of a large number of failed sensors at high time. We should note that BSF influences the detection performances of both serial and parallel structures.

The above analysis demonstrates the effectiveness of SR. In order to investigate the impact of SNR on the detection performance of SR, Fig. 8 shows the global detection probability $P_d$ versus SNR over time. SNR is defined as the ratio of source energy to the standard deviation of noise, namely $\mathrm{SNR} = 10\log\left(\frac{S}{N}\right)$, where $S = P_0, N = 1$ in this paper. We can see that higher SNR leads to higher global detection probability. Comparing the curves at a different time, it is easy to note that when SNR is fixed, detection probability at a lower time is superior to higher time, and this is in accordance with the results of ROC.

## 7 Conclusions

This paper investigates distributed decision fusion of serial structure in the presence of failed sensors over noisy channels. Previous literature focus on serial structure usually assumes the signal received by the local sensor is identical. While in this paper we construct a serial topology based on isotropic attenuation power model, the local decision in this topology is transmitted from sensors with lower credibility to sensors with higher credibility. We also derive the corresponding fusion rule. For comparison, we extend ELRT to noisy channels in the same scenario. Simulations show that serial is more robust than parallel in the presence of a large number of failed sensor over noisy channels. The deployment we considered in this paper is relatively ideal, we will study different deployment in the future.

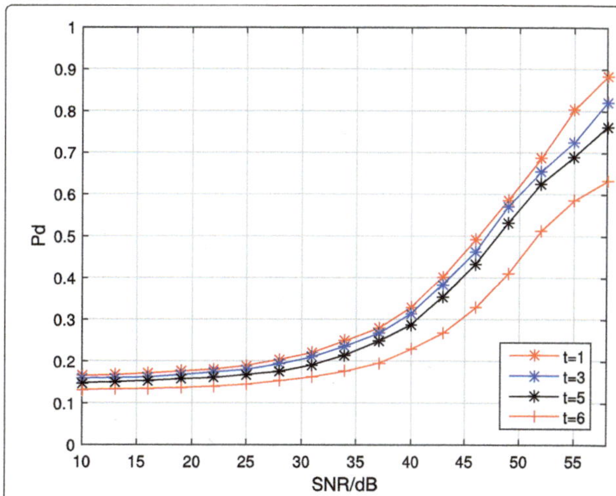


**Fig. 8** Global detection probability of SR versus SNR

**Acknowledgements**
This work was supported in part by the Overseas Academic Training Funds, University of Electronic Science and Technology of China (OATF, UESTC) (Grant No. 201506075013), and the Program for Science and Technology Support in Sichuan Province (Grant nos. 2014GZ0100 and 2016GZ0088).

**Authors' contributions**
JL conceived of and designed the research. JL and ZL performed the experiments and analyzed the result. JL and ZL wrote the paper. Both authors read and approved the final manuscript.

**Competing interests**
The authors declare that they have no competing interests.

**References**
1. Z Chair, PK Varshney, Optimal data fusion in multiple sensor detection systems. IEEE Trans. Aerosp. Electron. Syst. **22**(1), 98–101 (1986)
2. IY Hoballah, PK Varshney, Distributed Bayesian signal detection. IEEE Trans. Inf. Theory. **35**(5), 995-1000 (1989)
3. T-L Chin, YH Hu, Optimal Detector Based on Data Fusion for Wireless Sensor Networks, Global Telecommunications Conference (GLOBECOM 2011). IEEE, Kathmandu, Nepal. **6613**(1), 1–5 (2011)
4. R Niu, PK Varshney, Q Cheng, Distributed detection in a large wireless sensor network. Inf. Fusion. **7**(4), 380–394 (2006)
5. R Niu, PK Varshney, Distributed Detection and Fusion in a Large Wireless Sensor Network of Random Size. EURASIP J. Wirel. Commun. Netw. **2005**(4), 462–472 (2005)
6. Decentralized binary detection with noisy communication links. IEEE Trans. Aerosp. Electron. Syst. **42**(4), 1554–1563 (2006)
7. K Eritmen, M Keskinoz, Distributed decision fusion over fading channels in hierarchical wireless sensor networks. Wirel. Netw. **20**(5), 987–1002 (2014)
8. R Niu, B Chen, PK Varshney, Fusion of decisions transmitted over Rayleigh fading channels in wireless sensor networks. IEEE Trans. Signal Process. **54**(3), 1018–1027 (2006)
9. Y Xia, F Wang, W-S Deng, The decision fusion in the wireless network with possible transmission errors. Inf. Sci. **199**(15), 193–203 (2012)
10. Y Lin, B Chen, PK Varshney, Decision fusion rules in multi-hop wireless sensor networks. IEEE Trans. Aerosp. Electron. Syst. **41**(2), 475–488 (2005)
11. AM Aziz, A new multiple decisions fusion rule for targets detection in multiple sensors distributed detection systems with data fusion. Inf. Fusion. **18**(1), 175–186 (2014)
12. J-Y Wu, C-W Wu, T-Y Wang, et al., Channel-aware decision fusion with unknown local sensor detection probability. IEEE Trans. Signal Process. **58**(3), 1457–1463 (2010)
13. D Ciuonzo, PS Rossi, Decision fusion with unknown sensor detection probability. IEEE Signal Process. Lett. **21**(2), 208–212 (2014)
14. H Huang, L Chen, X Cao, et al., Weight-based clustering decision fusion algorithm for distributed target detection in wireless sensor networks. Int. J. Distrib. Sensor Netw. **2013**, 1-9 (2013)
15. G Ferrari, M Martalo, R Pagliari, Decentralized detection in clustered sensor networks. IEEE Trans. Aerosp. Electron. Syst. **47**(2), 959–973 (2011)
16. Q Tian, EJ Coyle, Optimal distributed detection in clustered wireless sensor networks. IEEE Trans. Signal Process. **55**(7), 3892–3904 (2007)
17. L Yin, Y Wang, D-W Yue, *Serial Distributed Detection Performance Analysis in Wireless Sensor Networks Under Noisy Channel, Wireless Communications, Networking and Mobile Computing*. (IEEE, Beijing, 2009), pp. 3092–3095
18. M Lucchi, M Chiani, *Distributed Detection of Local Phenomena with Wireless Sensor Networks, 2010 IEEE International Conference on Communications (ICC 2010)*. (IEEE, Cape Town), pp. 1–6
19. J Luo, T Li, Bathtub-shaped failure rate of sensors for distributed detection and fusion. Math. Probl. Eng. **2014**, 1-8 (2014)
20. CD Lai, M Xie, DNP Murthy, in *Advances in Reliability*. Chapter 3. Bathtubshaped failure rate life distributions, vol. 20 of Handbook of Statistics. (2001), pp. 69–104
21. M Bebbington, C-D Lai, R Zitikis, Useful periods for lifetime distributions with bathtub shaped hazard rate functions. IEEE Transactions on Reliability. **55**, 245–251 (2006)
22. J Ni, J Mei, *A Fusion Algorithm for Target Detection in Distributed Sensor Networks. Computational Intelligence and Communication Networks (CICN)*. (IEEE, Bhopal, 2014), pp. 349–353
23. T Wang, Z Peng, J Liang, et al., *Detecting Targets Based on a Realistic Detection and Decision Model in Wireless Sensor Networks. International Conference on Wireless Algorithms, Systems, and Applications WASA*. (Springer, 2015), pp. 836–844
24. T Wang, Z Peng, C Wang, et al., Extracting Target Detection Knowledge Based on Spatiotemporal Information in Wireless Sensor Networks. Int. J. Distrib. Sensor Netw. **2016**, 1–11 (2016)

# 4

# Interference management with mismatched partial channel state information

Alireza Vahid[1], Vaneet Aggarwal[2], A. Salman Avestimehr[3*] and Ashutosh Sabharwal[4]

**Abstract**

We study the fundamental limits of communications over multi-layer wireless networks where each node has only limited knowledge of the channel state information. In particular, we consider the scenario in which each source-destination pair has only enough information to perform optimally when other pairs do not interfere. Beyond that, the only other information available at each node is the global network connectivity. We propose a transmission strategy that solely relies on the available limited knowledge and combines coding with interference avoidance. We show that our proposed strategy goes well beyond the performance of interference avoidance techniques. We present an algebraic framework for the proposed transmission strategy based on which we provide a guarantee of the achievable rate. For several network topologies, we prove the optimality of our proposed strategy by providing information-theoretic outer-bounds.

## 1 Introduction

In dynamic wireless networks, optimizing system efficiency requires channel state information (CSI) in order to determine what resources are actually available. This information is acquired through feedback channel which is subject to several constraints such as delay, limited capacity, and locality. Consequently, in large-scale wireless networks, keeping track of the channel state information for making optimal decisions is typically infeasible due to the limitations of feedback channels and the significant overhead it introduces. Thus, in the absence of centralization of channel state information, nodes have limited local views of the network and make decentralized decisions based on their own *local view* of the network. The key question is how optimal decentralized decisions perform in comparison to the optimal centralized decisions.

In this paper, we consider multi-source multi-destination multi-layer wireless networks, and we seek fundamental limits of communications when sources have limited local views of the network. To model local views at wireless nodes, we consider the scenario in which each source-destination (S-D) pair has enough information to perform optimally when other pairs do not interfere. Beyond that, the only other information available at each node is the global network connectivity. We refer to this model of local network knowledge as 1-local view [1]. The motivation for this model stems from coordination protocols such as routing which are often employed in multi-hop networks to discover source-destination routes.

Our performance metric is *normalized sum capacity* defined in [2] which represents the maximum fraction of the sum capacity with full knowledge that can be always achieved when nodes have partial network knowledge. To better understand our objective, consider a multi-source multi-destination multi-layer wireless network. For each S-D pair, we define the *induced subgraph* by removing all other S-D pairs and the links that are not on a path between the chosen S-D pair. Our objective is to determine the minimum number of time slots $T$ that is required for each S-D pair to reconstruct any transmission snapshot in their induced subgraph over the original network. Normalized sum capacity is simply equal to one over $T$.

Our main contributions are as follows. We propose an algebraic framework that defines a transmission scheme that only requires 1-local view at the nodes and combines coding with interference avoidance scheduling. The scheme is a combination of three main techniques: (1) per layer interference avoidance, (2) repetition coding to allow

*Correspondence: avestimehr@ee.usc.edu
The preliminary results of this work were presented in 48th Annual Allerton Conference on Communication, Control, and Computing.
[3]Electrical Engineering Department, University of Southern California, Los Angeles 90089, CA, USA
Full list of author information is available at the end of the article

overhearing of the interference, and (3) network coding to allow interference neutralization.

We then characterize the achievable normalized sum rate of our proposed scheme and analyze its optimality for some classes of networks. We consider two-layer networks: (1) with two relays and any number of source-destination pairs, (2) with three source-destination pairs and three relays, and (3) with folded-chain structure (defined in Section 5). We also show that the gain from our proposed scheme over interference avoidance scheduling can be unbounded in $L$-nested folded-chain networks (defined in Section 5).

A related line of work was started by [3] which assumes wireless nodes are aware of the topology but not the channel state information. In this setting, all nodes have the *same* side information whereas in our work, nodes have partial and *mismatched* knowledge of the channel state information. The problem is then connected to index coding, and topological interference management scheme (TIM) is proposed. There are many results e.g., [4–9]) that follow a similar path of [3]. We generalize the problem to multi-hop networks and design our communication protocol accordingly, but TIM is designed for single-hop networks. Moreover, our formulation provides a continuous transition from no CSI to full CSI (by having more and more hops of knowledge) whereas [3] focuses only on the extreme case of no channel state information.

Many other models for imprecise network information have been considered for interference networks. These models range from having no channel state information at the sources [10–13], delayed channel state information [14–19], mismatched delayed channel state knowledge [20–22], or analog channel state feedback for fully connected interference channels [23]. Most of these works assume fully connected network or a small number of users. For networks with *arbitrary* connectivity, the first attempt to understand the role of limited network knowledge was initiated in [24, 25] in the context of single-layer networks. The authors used a message-passing abstraction of network protocols to formalize the notion of local view. The key result of [24, 25] is that local-view-based (decentralized) decisions can be either sum rate optimal or can be arbitrarily worse than the global-view (centralized) sum capacity. In this work, we focus on multi-layer setting and we show that several important additional ingredients are needed compared to the single-layer scenario.

It is worth noting that since each channel gain can range from zero to a maximum value, our formulation is similar to compound channels [26, 27] with one major difference. In the multi-terminal compound network formulations, *all* nodes are missing identical information about the channels in the network whereas in our formulation, the 1-local view results in *asymmetric* and *mismatched* information about channels at different nodes.

The rest of the paper is organized as follows. In Section 2, we introduce our network model and the new model for partial network knowledge. In Section 3 via a number of examples, we motivate our transmission strategies and the algebraic framework. In Section 4, we formally define the algebraic framework and we characterize the performance of the transmission strategy based on this framework. In Section 5, we prove the optimality of our strategies for several network topologies. Finally, Section 6 concludes the paper.

## 2 Problem formulation

In this section, we introduce our model for the wireless channel and the available network knowledge at the nodes. We also define the notion of normalized sum capacity which will be used as the performance metric for the strategies with partial network knowledge.

### 2.1 Network model and notations

We describe the two channel models we use in this paper, namely, the linear deterministic model and the Gaussian model. In both models, a network is represented by a directed graph

$$\mathcal{G} = (\mathcal{V}, \mathcal{E}, \{w_{ij}\}_{(i,j)\in\mathcal{E}}) \quad (1)$$

where $\mathcal{V}$ is the set of vertices representing nodes in the network, $\mathcal{E}$ is the set of directed edges representing links among these nodes, and $\{w_{ij}\}_{(i,j)\in\mathcal{E}}$ represents the channel gains associated with the edges.

Out of the $|\mathcal{V}|$ nodes in the network, $K$ nodes are sources and $K$ nodes are destinations. We label these source and destination nodes by $\mathsf{S}_i$s and $\mathsf{D}_i$s respectively, $i = 1, 2, \ldots, K$. The remaining $|\mathcal{V}| - 2K$ nodes are relay nodes which facilitate the communication between sources and destinations. We can simply refer to a node in $\mathcal{V}$ as $\mathsf{V}_i$, $i = 1, 2, \ldots, |\mathcal{V}|$. In this work, we focus on two-layer networks.

The two channel models used in this paper are as follows:

1. The linear deterministic model [28]: In this model, there is a non-negative integer, $w_{ij} = n_{ij}$, associated with each link $(i, j) \in \mathcal{E}$ which represents its gain. Let $q$ be the maximum of all the channel gains in this network. In the linear deterministic model, the channel input at node $\mathsf{V}_i$ at time $t$ is denoted by

$$X_{\mathsf{V}_i}[t] = [X_{\mathsf{V}_{i_1}}[t], X_{\mathsf{V}_{i_2}}[t], \ldots, X_{\mathsf{V}_{i_q}}[t]]^T \in \mathbb{F}_2^q. \quad (2)$$

The received signal at node $\mathsf{V}_j$ at time $t$ is denoted by

$$Y_{\mathsf{V}_j}[t] = [Y_{\mathsf{V}_{j_1}}[t], Y_{\mathsf{V}_{j_2}}[t], \ldots, Y_{\mathsf{V}_{j_q}}[t]]^T \in \mathbb{F}_2^q, \quad (3)$$

and is given by

$$Y_{V_j}[t] = \sum_{i:(i,j)\in\mathcal{E}} \mathbf{S}^{q-n_{ij}} X_{V_i}[t], \tag{4}$$

where $\mathbf{S}$ is the $q \times q$ shift matrix and the operations are in $\mathbb{F}_2^q$. If a link between $V_i$ and $V_j$ does not exist, we set $n_{ij}$ to be zero.

2. The Gaussian model: In this model, the channel gain $w_{ij}$ is denoted by $h_{ij} \in \mathbb{C}$. The channel input at node $V_i$ at time $t$ is denoted by $X_{V_i}[t] \in \mathbb{C}$, and the received signal at node $V_j$ at time $t$ is denoted by $Y_{V_j}[t] \in \mathbb{C}$ given by

$$Y_{V_j}[t] = \sum_i h_{ij} X_{V_i}[t] + Z_j[t], \tag{5}$$

where $Z_j[t]$ is the additive white complex Gaussian noise with unit variance. We also assume a power constraint of 1 at all nodes, i.e.,

$$\lim_{n\to\infty} \frac{1}{n}\mathbb{E}\left(\sum_{t=1}^{n} |X_{V_i}[t]|^2\right) \leq 1. \tag{6}$$

A *route* from a source $S_i$ to a destination $D_j$ is a set of nodes such that there exists an ordering of these nodes where the first one is $S_i$, the last one is $D_j$, and any two consecutive nodes in this ordering are connected by an edge in the graph.

**Definition 1** *Induced subgraph $\mathcal{G}_{ij}$ is a subgraph of $\mathcal{G}$ with its vertex set being the union of all routes (as defined above) from source $S_i$ to a destination $D_j$, and its edge set being the subset of all edges in $\mathcal{G}$ between the vertices of $\mathcal{G}_{ij}$.*

We say that S-D pair $i$ and S-D $j$ are *non-interfering* if $\mathcal{G}_{ii}$ and $\mathcal{G}_{jj}$ are two disjoint induced subgraphs of $\mathcal{G}$.

The in-degree function $d_{\text{in}}(V_i)$ is the number of incoming edges connected to node $V_i$. Similarly, the out-degree function $d_{\text{out}}(V_i)$ is the number of out-going edges connected to node $V_i$. Note that the in-degree of a source and the out-degree of a destination are both equal to 0.

The maximum degree of the nodes in $\mathcal{G}$ is defined as

$$d_{\max} = \max_{i\in\{1,\ldots,|\mathcal{V}|\}} \left(d_{\text{in}}(V_i), d_{\text{out}}(V_i)\right). \tag{7}$$

### 2.2 Partial network knowledge model

We now describe our model for partial network knowledge that we refer to as *1-local view*:

- All nodes have full knowledge of the network topology $(\mathcal{V}, \mathcal{E})$ (i.e., they know which links are in $\mathcal{G}$, but not necessarily their channel gains). The network topology knowledge is denoted by side information SI;
- Each source, $S_i$, knows the gains of all the channels that are in $\mathcal{G}_{ii}$, $i = 1, 2, \ldots, K$. This channel knowledge at a source is denoted by $L_{S_i}$.
- Each node $V_i$ (which is not a source) has the union of the information of all those sources that have a route to it, and this knowledge at node is denoted by $L_{V_i}$.

This partial network knowledge is motivated by the following argument. Suppose a network with a single S-D pair and no interference. In such a network, one expects the S-D pair to be able to communicate optimally (or close to optimal) and our model guarantees this baseline performance. In fact, our model assumes the minimum knowledge needed at different nodes such that each S-D pair can communicate optimally in the absence of interference. Moreover, our assumption is backed by message passing algorithm which is how CSI is learned in wireless networks.

For a depiction, consider the network in Fig. 1a. Source $S_1$ has the knowledge of the channel gains of the links that are denoted by solid arrows in Fig. 1a. On the other hand, relay A has the union of the information of sources $S_1$ and $S_2$. The partial knowledge of relay A is denoted by solid arrows in Fig. 1b.

**Remark 1** *Our formulation is a general version of compound channel formulation where nodes have mismatched and asymmetric lack of knowledge.*

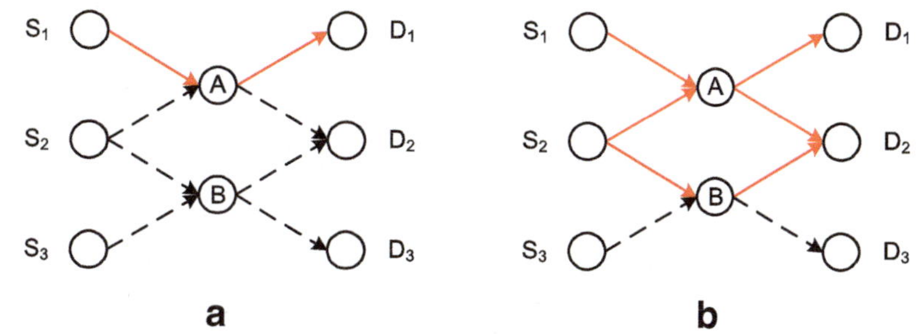


**Fig. 1 a** The partial knowledge of the channel gains available at source $S_1$ is denoted by *solid arrows* and **b** the partial knowledge of the channel gains available at relay A

### 2.3 Performance metric

Our goal is to find out the minimum number of time slots such that all induced subgraphs can be reconstructed as if there is no interference. It turns out the aforementioned objective is closely related to the notion of normalized sum capacity introduced in [2]. Here, we define the notion of normalized sum capacity which is going to be our metric for evaluating network capacity with partial network knowledge. Intuitively, normalized sum capacity represents the maximum fraction of the sum capacity with full knowledge that can be achieved when nodes have partial network knowledge.

Consider the scenario in which source $\mathsf{S}_i$ wishes to reliably communicate message $W_i \in \{1, 2, \ldots, 2^{NR_i}\}$ to destination $\mathsf{D}_i$ during $N$ uses of the channel, $i = 1, 2, \ldots, K$. We assume that the messages are independent and chosen uniformly. For each source $\mathsf{S}_i$, let message $W_i$ be encoded as $X^N_{\mathsf{S}_i}$ using the encoding function $e_i(W_i|L_{\mathsf{S}_i}, \mathsf{SI})$ which depends on the available local network knowledge, $L_{\mathsf{S}_i}$, and the global side information, SI.

Each relay in the network creates its input to the channel, $X_{\mathsf{V}_i}$, using the encoding function $f_{\mathsf{V}_i}[t]\left(Y^{(t-1)}_{\mathsf{V}_i}|L_{\mathsf{V}_i}, \mathsf{SI}\right)$ which depends on the available network knowledge, $L_{\mathsf{V}_i}$, and the side information, SI, and all the previous received signals at the relay $Y^{(t-1)}_{\mathsf{V}_i} = \left[Y_{\mathsf{V}_i}[1], Y_{\mathsf{V}_i}[2], \ldots, Y_{\mathsf{V}_i}[t-1]\right]$. A relay strategy is defined as the union of all encoding functions used by the relays, $\{f_{\mathsf{V}_i}[t]\,(Y^{(t-1)}_{\mathsf{V}_i}|L_{\mathsf{V}_i}, \mathsf{SI})\}$, $t = 1, 2, \ldots, N$ and $\mathsf{V}_i \in \bigcup_{j=1}^{L-1} \mathcal{V}_j$.

Destination $\mathsf{D}_i$ is only interested in decoding $W_i$, and it will decode the message using the decoding function $\widehat{W}_i = d_i(Y^N_{\mathsf{D}_i}|L_{\mathsf{D}_i}, \mathsf{SI})$ where $L_{\mathsf{D}_i}$ is destination $\mathsf{D}_i$'s network knowledge. We note that the local view can vary from node to node.

**Definition 2** *A Strategy $\mathcal{S}_N$ is defined as the set of: (1) all encoding functions at the sources, (2) all decoding functions at the destinations, and (3) the relay strategy for $t = 1, 2, \ldots, N$, i.e.,*

$$\mathcal{S}_N = \left\{ \begin{array}{ll} e_i(W_i|L_{\mathsf{S}_i}, \mathsf{SI}) & i = 1, 2, \ldots, K \\ f_{\mathsf{V}_i}[t]\left(Y^{(t-1)}_{\mathsf{V}_i}|L_{\mathsf{V}_i}, \mathsf{SI}\right) & t = 1, 2, \ldots, N \\ & \text{and } \mathsf{V}_i \in \bigcup_{j=1}^{L-1} \mathcal{V}_j \\ d_i(Y^N_{\mathsf{D}_i}|L_{\mathsf{D}_i}, \mathsf{SI}) & i = 1, 2, \ldots, K \end{array} \right\}. \tag{8}$$

An error occurs when $\widehat{W}_i \neq W_i$, and we define the decoding error probability, $\lambda_i$, to be equal to $P(\widehat{W}_i \neq W_i)$. A rate tuple $(R_1, R_2, \ldots, R_K)$ is said to be achievable if there exists a set of strategies $\{\mathcal{S}_j\}_{j=1}^N$ such that the decoding error probabilities $\lambda_1, \lambda_2, \ldots, \lambda_K$ go to zero as $N \to \infty$ for all network states consistent with the side information. Moreover, for S-D pair $i$, we denote the maximum achievable rate with full network knowledge by $C_i$. The sum capacity $C_{\mathsf{sum}}$ is the supremum of $\sum_{i=1}^K R_i$ over all possible encoding and decoding functions with full network knowledge.

We now define the normalized sum rate and the normalized sum capacity.

**Definition 3** ([2]) *Normalized sum rate of $\alpha$ is said to be achievable if there exists a set of strategies $\{\mathcal{S}_j\}_{j=1}^N$ such that following holds. As N goes to infinity, strategy $\mathcal{S}_N$ yields a sequence of codes having rates $R_i$ at source $\mathsf{S}_i$, $i = 1, \ldots, K$, satisfying*

$$\sum_{i=1}^{K} R_i \geq \alpha C_{sum} - \tau$$

*such that error probabilities at the destinations, $\lambda_1, \cdots \lambda_K$, go to zero for all the network states consistent with the side information and for a constant $\tau$ that is independent of the channel gains.*

**Definition 4** ([2]) *Normalized sum capacity $\alpha^*$ is defined as the supremum of all achievable normalized sum rates $\alpha$. Note that $\alpha^* \in [0, 1]$.*

**Remark 2** *While the results of this paper may appear pessimistic, they are much like any other compound channel analysis which aims to optimize the worst case scenario. We adopted the normalized capacity as our metric which opens the door for exact (or near-exact) results for several cases as we will see in this paper. We view our current formulation as a step towards the general case, much like the commonly accepted methodology where channel state information is assumed known perfectly in first steps of a new problem (MIMO, interference channels, scaling laws), even though that assumption is almost never possible.*

## 3 Motivating examples

Before diving into the main results, we use a sequence of examples to understand the mechanisms that allow us outperform interference avoidance with only 1-local view.

We first consider the single-layer network depicted in Fig. 2a. Using interference avoidance, we can create the induced subgraphs of all three S-D pairs as shown in Fig. 2b in three time slots. Thus, with interference avoidance, one can only achieve $\alpha = \frac{1}{3}$ which is the same as TDMA. However, we show that it is possible to reconstruct the three induced subgraphs in only two time slots and to achieve $\alpha = \frac{1}{2}$ with 1-local view.

Consider the linear deterministic model and the induced subgraphs of all three S-D pairs as shown in Fig. 2b. We show that any transmission strategy over these three induced subgraphs can be implemented in the

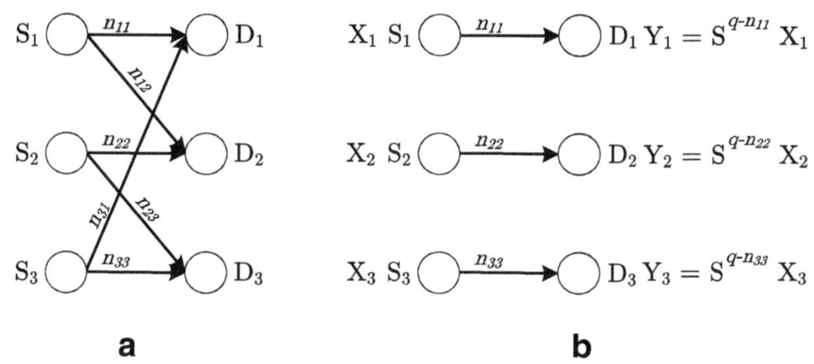


**Fig. 2 a** A network in which coding is required to achieve the normalized sum capacity and **b** the induced subgraphs

original network by using only two time slots such that all nodes receive the same signal as if they were in the induced subgraphs. This would immediately imply that a normalized sum rate of $\frac{1}{2}$ is achievable[1].

To do so, we split the communication block into two time slots of equal length. Sources $\mathsf{S}_1$ and $\mathsf{S}_2$ transmit the same codewords as if they are in the induced subgraphs over first time slot. Destination $\mathsf{D}_1$ receives the same signal as it would have in its induced subgraph, and destination $\mathsf{D}_3$ receives interference from source $\mathsf{S}_2$.

Over second time slot, $\mathsf{S}_3$ transmits the same codeword as if it was in its induced subgraph and $\mathsf{S}_2$ repeats its transmit signal from the first time slot. Destination $\mathsf{D}_2$ receives its signal interference-free. Now, if destination $\mathsf{D}_3$ adds its received signals over the two time slots, it recovers its intended signal as depicted in Fig. 3. In other words, we used interference cancellation at destination $\mathsf{D}_3$. Therefore, all S-D pairs can effectively communicate interference-free over two time slots.

We can view this discussion in an algebraic framework. To each source $\mathsf{S}_i$, we assign a transmit vector $\mathbf{T}_{\mathsf{S}_i}$ of size $2 \times 1$ where 2 corresponds to the number of time slots. If the entry at row $t$ is equal to 1, then $\mathsf{S}_i$ communicates the codeword for $\mathsf{D}_i$ during time slot $t$, $i = 1, 2, 3$, and $t = 1, 2$. In our example, we have

$$\mathbf{T}_{\mathsf{S}_1} = \begin{pmatrix} 1 \\ 0 \end{pmatrix}, \quad \mathbf{T}_{\mathsf{S}_2} = \begin{pmatrix} 1 \\ 1 \end{pmatrix}, \quad \mathbf{T}_{\mathsf{S}_3} = \begin{pmatrix} 0 \\ 1 \end{pmatrix}. \tag{9}$$

To each destination $\mathsf{D}_i$, we assign a receive matrix $\mathbf{R}_{\mathsf{D}_i}$ of size $2 \times 2$ where each row corresponds to a time slot and each column corresponds to a source that has a route to destination $\mathsf{D}_i$. More precisely, $\mathbf{R}_{\mathsf{D}_i}$ is formed by concatenating the transmit vectors of the sources that are connected to $\mathsf{D}_i$. In our example, we have

$$\mathbf{R}_{\mathsf{D}_1} = \left(\mathbf{T}_{\mathsf{S}_1} \mathbf{T}_{\mathsf{S}_3}\right) = \begin{matrix} & \mathsf{S}_1 & \mathsf{S}_3 \\ t=1 & 1 & 0 \\ t=2 & 0 & 1 \end{matrix}, \tag{10}$$

$$\mathbf{R}_{\mathsf{D}_2} = \begin{matrix} & \mathsf{S}_1 & \mathsf{S}_2 \\ t=1 & 1 & 1 \\ t=2 & 0 & 1 \end{matrix}, \quad \mathbf{R}_{\mathsf{D}_3} = \begin{matrix} & \mathsf{S}_2 & \mathsf{S}_3 \\ t=1 & 1 & 0 \\ t=2 & 1 & 1 \end{matrix}.$$

From the receive matrices in (10), we can easily check whether each destination can recover its corresponding signal or not. For instance, from $\mathbf{R}_{\mathsf{D}_1}$, we know that in time slot 1, the signal from $\mathsf{S}_1$ is received interference-free. A similar story is true for the second pair and $\mathbf{R}_{\mathsf{D}_2}$. From $\mathbf{R}_{\mathsf{D}_3}$,

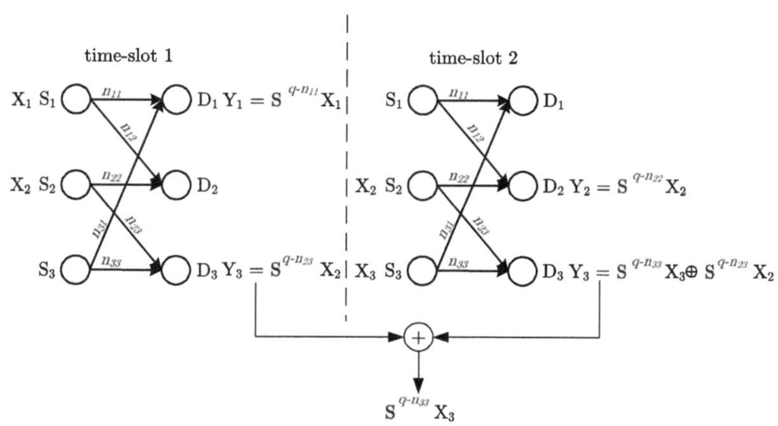


**Fig. 3** Achievability strategy for the network depicted in Fig. 2

we see that in time slot 2, the signal from $S_3$ is received with interference from $S_2$. However, with a linear row-operation, we can create a row where the only 1 appears in the column associated with $S_3$. More precisely,

$$\mathbf{R}_{D_3}(1,:) \oplus \mathbf{R}_{D_3}(2,:) = \overset{\begin{matrix} S_2 & S_3 \end{matrix}}{\begin{pmatrix} 0 & 1 \end{pmatrix}}.$$

where $\mathbf{R}_{D_3}(t,:)$ denotes row $t$ of matrix $\mathbf{R}_{D_3}$. Thus, each destination can recover its intended signal interference-free over two time slots. More precisely, we require

$$\overset{\begin{matrix} S_i & S_j \end{matrix}}{\begin{pmatrix} 1 & 0 \end{pmatrix}} \in \text{rowspan}\left(\mathbf{R}_{D_i}\right), \quad i,j \in \{1,2,3\}, \quad j \neq i. \quad (11)$$

This example illustrated that with only 1-local view, it is possible to reconstruct the three induced subgraphs in only two time slots using (repetition) coding at the sources and go beyond interference avoidance and TDMA. In the following example, we show that more sophisticated ideas are required to beat TDMA in multi-layer networks.

Consider the two-layer network of Fig. 4a. Under the linear deterministic model it is straightforward to see that by using interference avoidance or TDMA, it takes three time slots to reconstruct the induced subgraphs of Fig. 4b–d, which results in a normalized sum rate of $\alpha = \frac{1}{3}$.

We now show that using repetition coding at the sources and linear coding at the relays, it is possible to achieve $\alpha = 1/2$ and reconstruct the three induced subgraphs in only two time slots. This example was first presented in [1].

Consider any strategy for S-D pairs 1, 2, and 3 as illustrated in Fig. 4b–d. In the first layer, we implement the achievability strategy of Fig. 3 illustrated in Fig. 5. As it can be seen in this figure, at the end of the second time slot, each relay has access to the same received signal as if it was in the diamond networks of Fig. 4b–d.

In the second layer during the first time slot, relays $V_1$ and $V_2$ transmit $X_{V_1}^1$ and $X_{V_2}^1$, respectively, whereas $V_3$ transmits $X_{V_3}^2 \oplus X_{V_3}^3$ as depicted in Fig. 6. Destination $D_1$ receives the same signal as Fig. 4b. During the second time slot , relays $V_2$ and $V_3$ transmit $X_{V_2}^2$ and $X_{V_3}^2$, respectively, whereas $V_1$ transmits $X_{V_1}^1 \oplus X_{V_1}^3$. Destination $D_2$ receives the same signal as Fig. 4c. If destination $D_3$ adds its received signals over the two time slots, it recovers the same signal as Fig. 4d. Therefore, each destination receives the same signal as if it was only in its corresponding diamond network over two time slots. Hence, a normalized sum rate of $\alpha = \frac{1}{2}$ is achievable.

Again, this strategy can be viewed in an algebraic framework. We assign a transmit vector $\mathbf{T}_{S_i}$ of size $2 \times 1$ to source $S_i$:

$$\mathbf{T}_{S_1} = \begin{pmatrix} 1 \\ 0 \end{pmatrix}, \quad \mathbf{T}_{S_2} = \begin{pmatrix} 1 \\ 1 \end{pmatrix}, \quad \mathbf{T}_{S_3} = \begin{pmatrix} 0 \\ 1 \end{pmatrix}. \quad (12)$$

To each relay $V_j$ we assign a receive matrix $\mathbf{R}_{V_j}$ of size $2 \times 2$ where each row corresponds to a time slot, and each column corresponds to a source that has a route to that relay. $\mathbf{R}_{V_j}$ is formed by concatenating the transmit vectors of the sources that are connected to $V_j$. In our example, we have

$$\mathbf{R}_{V_1} = \begin{matrix} t=1 \\ t=2 \end{matrix} \overset{\begin{matrix} S_1 & S_3 \end{matrix}}{\begin{pmatrix} 1 & 0 \\ 0 & 1 \end{pmatrix}}, \quad \mathbf{R}_{V_2} = \begin{matrix} t=1 \\ t=2 \end{matrix} \overset{\begin{matrix} S_1 & S_2 \end{matrix}}{\begin{pmatrix} 1 & 1 \\ 0 & 1 \end{pmatrix}},$$

$$\mathbf{R}_{V_3} = \begin{matrix} t=1 \\ t=2 \end{matrix} \overset{\begin{matrix} S_2 & S_3 \end{matrix}}{\begin{pmatrix} 1 & 0 \\ 1 & 1 \end{pmatrix}}. \quad (13)$$

From the receive matrices in (14), we can easily check whether each relay has access to the same received signals as if it was in the diamond networks of Fig. 4b–d.

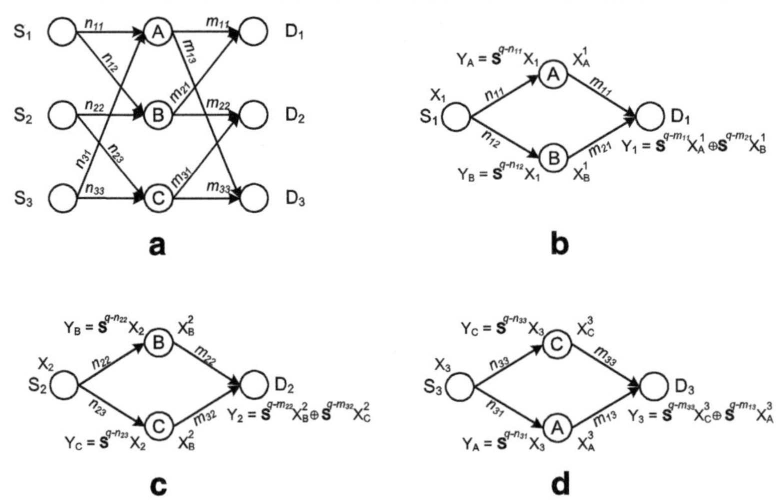


**Fig. 4 a** A two-layer network in which we need to incorporate network coding to achieve the normalized sum capacity, **b–d** the induced subgraphs of S-D pairs 1, 2, and 3, respectively

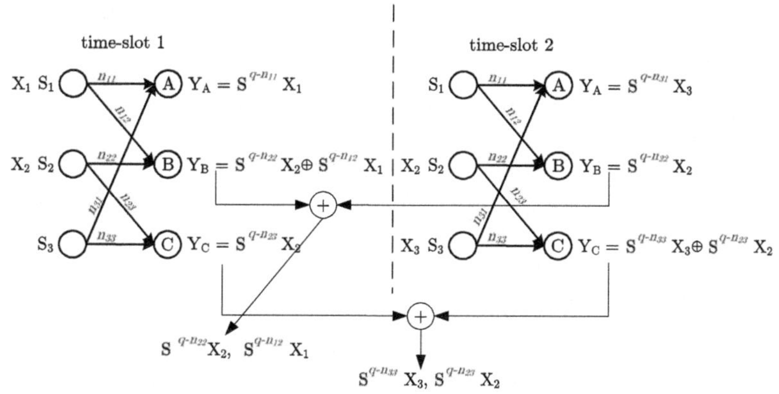


**Fig. 5** Achievability strategy for the first layer of the network in Fig. 4a

We also assign a transmit matrix $\mathbf{T}_{V_j}$ of size $2 \times 2$ to relay $V_j$ where each column corresponds to a S-D pair $i$ such that $V_j \in \mathcal{G}_{ii}$. If the entry at row $t$ and column $S_i$ of this transmit matrix $\mathbf{T}_{V_j}$ is equal to 1, then the relay communicates the signal it has for S-D pair $i$ at time $t$. In our example, we have

$$\mathbf{T}_{V_1} = \begin{array}{c} \\ t=1 \\ t=2 \end{array}\!\!\begin{array}{c} \begin{array}{cc} S_1 & S_3 \end{array} \\ \left(\begin{array}{cc} 1 & 0 \\ 1 & 1 \end{array}\right) \end{array}, \quad \mathbf{T}_{V_2} = \begin{array}{c} \\ t=1 \\ t=2 \end{array}\!\!\begin{array}{c} \begin{array}{cc} S_1 & S_2 \end{array} \\ \left(\begin{array}{cc} 1 & 0 \\ 0 & 1 \end{array}\right) \end{array},$$
$$\mathbf{T}_{V_3} = \begin{array}{c} \\ t=1 \\ t=2 \end{array}\!\!\begin{array}{c} \begin{array}{cc} S_2 & S_3 \end{array} \\ \left(\begin{array}{cc} 1 & 1 \\ 1 & 0 \end{array}\right) \end{array}. \qquad (14)$$

If in row $t$ more than one 1 appears, then the relay creates the linear combination of the signals it has for the S-D pairs that have a 1 in their column and transmits it. For instance, relay $V_1$ during time slot 2 transmits $X^1_{V_1} \oplus X^3_{V_1}$.

To each destination $D_i$, we assign a received matrix $\mathbf{R}_{D_i}$ of size $2 \times 4$ where each row corresponds to a time slot, and each column corresponds to a route from a source through a specific relay (e.g., $S_1 : V_1$). In fact, $\mathbf{R}_{D_i}$ is formed by concatenating (and reordering columns of) the transmit matrices of the relays that are connected to $D_i$. In our example, we have

$$\mathbf{R}_{D_1} = \begin{array}{c} \\ t=1 \\ t=2 \end{array}\!\!\begin{array}{c} \begin{array}{cccc} S_1{:}V_1 & S_1{:}V_2 & S_2{:}V_2 & S_3{:}V_1 \end{array} \\ \left(\begin{array}{cccc} 1 & 1 & 0 & 0 \\ 1 & 0 & 1 & 1 \end{array}\right) \end{array},$$
$$\mathbf{R}_{D_2} = \begin{array}{c} \\ t=1 \\ t=2 \end{array}\!\!\begin{array}{c} \begin{array}{cccc} S_1{:}V_2 & S_2{:}V_2 & S_2{:}V_3 & S_3{:}V_3 \end{array} \\ \left(\begin{array}{cccc} 1 & 0 & 1 & 1 \\ 0 & 1 & 1 & 0 \end{array}\right) \end{array},$$
$$\mathbf{R}_{D_3} = \begin{array}{c} \\ t=1 \\ t=2 \end{array}\!\!\begin{array}{c} \begin{array}{cccc} S_1{:}V_2 & S_2{:}V_2 & S_2{:}V_3 & S_3{:}V_3 \end{array} \\ \left(\begin{array}{cccc} 1 & 1 & 0 & 1 \\ 1 & 1 & 1 & 0 \end{array}\right) \end{array}. \qquad (15)$$

From the receive matrices in (15), we can verify whether each destination can recover its corresponding signals or not. For instance, from $\mathbf{R}_{D_1}$, we know that in time slot 1, the signals from $S_1$ through relays $V_1$ and $V_2$ are received interference-free. From $\mathbf{R}_{D_3}$, we see that at each time slot, one of the two signals that $D_3$ is interested in is received. However, with a linear row operation, we can create a row where 1s only appear in the columns associated with $S_3$.

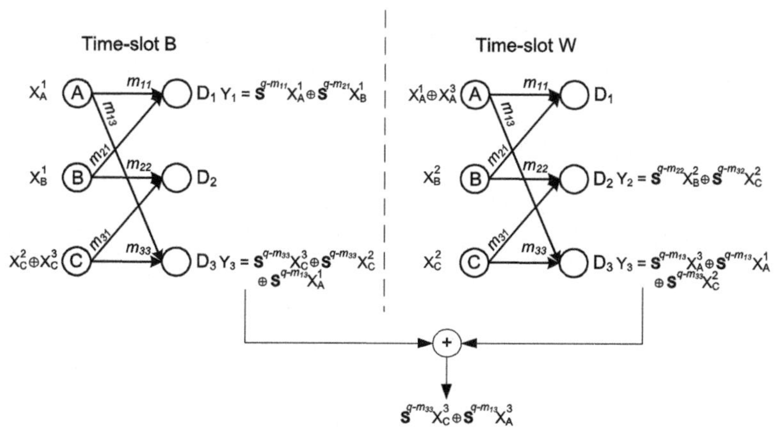


**Fig. 6** Achievability strategy for the second layer of the network in Fig. 4a

More precisely,

$$\mathbf{R}_{\mathsf{D}_3}(1,:) \oplus \mathbf{R}_{\mathsf{D}_3}(2,:) \tag{16}$$
$$= \begin{pmatrix} \overset{\mathsf{S}_1:\mathsf{V}_1}{0} & \overset{\mathsf{S}_2:\mathsf{V}_3}{0} & \overset{\mathsf{S}_3:\mathsf{V}_1}{1} & \overset{\mathsf{S}_3:\mathsf{V}_3}{1} \end{pmatrix},$$

or equivalently

$$\begin{pmatrix} \overset{\mathsf{S}_1:\mathsf{V}_1}{0} & \overset{\mathsf{S}_2:\mathsf{V}_3}{0} & \overset{\mathsf{S}_3:\mathsf{V}_1}{1} & \overset{\mathsf{S}_3:\mathsf{V}_3}{1} \end{pmatrix} \in \text{rowspan}\left(\mathbf{R}_{\mathsf{D}_3}\right).$$

Thus, each destination can recover its intended signal interference-free and a normalized sum rate of $\alpha = \frac{1}{2}$ is achievable.

## 4 An algebraic framework

In this section, given $T \in \mathbb{N}$, we define a set of conditions that if satisfied, all induced subgraphs can be reconstructed in $T$ time slots. Consider a two-layer wireless network $\mathcal{G} = (\mathcal{V}, \mathcal{E}, \{w_{ij}\}_{(i,j)\in\mathcal{E}})$ with $K$ source-destination pairs and $|\mathcal{V}| - 2K$ relay nodes. We label the sources as $\mathsf{S}_1, \ldots, \mathsf{S}_K$, the destinations as $\mathsf{D}_1, \ldots, \mathsf{D}_K$, and the relays as $\mathsf{V}_1, \ldots, \mathsf{V}_{|\mathcal{V}|-2K}$. We also need the following definitions:

**Definition 5** *We assign a number $N_{\mathsf{V}_j}$ to relay $\mathsf{V}_j$ defined as*

$$N_{\mathsf{V}_j} \triangleq \left|\left\{i|\mathsf{V}_j \in \mathcal{G}_{ii}\right\}\right|, \tag{17}$$

*and we assign a number $N_{\mathsf{D}_j}$ to destination $\mathsf{D}_j$ defined as*

$$N_{\mathsf{D}_j} \triangleq \sum_{i=1}^{K} \left(\textit{number of routes from } \mathsf{S}_i \textit{ to } \mathsf{D}_j\right). \tag{18}$$

In this section, we assign matrices to each node in $\mathcal{G}$ and the only difference between the linear deterministic model and the Gaussian model is the entries to these matrices:

1. For the linear deterministic model, all entries to these matrices are in binary field (i.e., either 0 or 1);
2. For the Gaussian model, all entries to these matrices are in $\{-1, 0, 1\}$.

We now describe the assignment of the matrices. Fix $T \in \mathbb{N}$.

- We assign a transmit vector $\mathbf{T}_{\mathsf{S}_i}$ of size $T \times 1$ to source $\mathsf{S}_i$ where each row corresponds to a time slot.
- To each relay $\mathsf{V}_j$, we assign a receive matrix $\mathbf{R}_{\mathsf{V}_j}$ of size $T \times d_{\text{in}}(\mathsf{V}_j)$, and we label each column with a source that has a route to that relay. The entry at row $t$ and column labeled by $\mathsf{S}_i$ in $\mathbf{R}_{\mathsf{V}_j}$ is equal to the entry at row $t$ of $\mathbf{T}_{\mathsf{S}_i}$.
- We assign a transmit matrix $\mathbf{T}_{\mathsf{V}_j}$ of size $T \times N_{\mathsf{V}_j}$ to each relay where each column corresponds to a S-D pair $i$ where the relay belongs to $\mathcal{G}_{ii}$.
- Finally, to each destination $\mathsf{D}_i$, we assign a receive matrix $\mathbf{R}_{\mathsf{D}_i}$ of size $T \times N_{\mathsf{D}_i}$ where each column corresponds to a route from a source $\mathsf{S}_j$ through a specific relay $\mathsf{V}_{j'}$ and is labeled as $\mathsf{S}_j : \mathsf{V}_{j'}$. The entry at row $t$ and column labeled by $\mathsf{S}_j : \mathsf{V}_{j'}$ in $\mathbf{R}_{\mathsf{D}_i}$ is equal to the entry at row $t$ and column labeled by $\mathsf{S}_j$ of $\mathbf{T}_{\mathsf{V}_{j'}}$.

An assignment of transmit and receive matrices to the nodes in the network is *valid* if the following conditions are satisfied:

**C.1** For $i = 1, 2, \ldots, K$:

$$\text{Rank}\left(\mathbf{T}_{\mathsf{S}_i}\right) = 1. \tag{19}$$

**C.2** For any relay $\mathsf{V}_j$, using linear row operations $\mathbf{R}_{\mathsf{V}_j}$ can be transformed into a matrix in which if $\mathsf{V}_j \in \mathcal{G}_{ii}$, then $\exists\, \ell$ such that all entries in row $\ell$ are zeros except for the column corresponding to source $\mathsf{S}_i$.

**C.3** For destination $\mathsf{D}_i$, using linear row operations $\mathbf{R}_{\mathsf{D}_i}$ can be transformed into a matrix in which $\exists\, \ell$ such that the $\ell^{th}$ row has only 1s in the columns corresponding to source $\mathsf{S}_i$, $i = 1, 2, \ldots, K$.

**Remark 3** *It is straightforward to show that a necessary condition for* **C.3** *is for each $\mathbf{T}_{\mathsf{V}_j}$ to have full column rank. This gives us a lower bound on $T$ which is $\max_{\mathsf{V}_j} \left|\left\{i|\mathsf{V}_j \in \mathcal{G}_{ii}\right\}\right|$. Moreover, TDMA is always a lower bound on the performance of any strategy and thus provides us with an upper bound on T. In other words, we have*

$$\max_{\mathsf{V}_j} \left|\left\{i|\mathsf{V}_j \in \mathcal{G}_{ii}\right\}\right| \leq T < K. \tag{20}$$

**Theorem 1** *For a K-user two-layer network $\mathcal{G} = (\mathcal{V}, \mathcal{E}, \{w_{ij}\}_{(i,j)\in\mathcal{E}})$ with 1-local view if there exists a valid assignment of transmit and receive matrices, then all induced subgraphs can be reconstructed in T time slots and a normalized sum rate of $\alpha = \frac{1}{T}$ is achievable.*

*Proof* We first prove the theorem for the linear deterministic model, and then, we provide the proof for the Gaussian model.

Network $\mathcal{G}$ has $K$ S-D pairs. Consider the $K$-induced subgraphs of all S-D pairs, i.e., $\mathcal{G}_{jj}$, $j = 1, 2, \ldots, K$. We show that any transmission snapshot over these induced subgraphs can be implemented in $\mathcal{G}$ over $T$ time slots such that all nodes receive the same signal as if they were in their corresponding induced subgraphs.

**Linear deterministic model.** Consider a transmission snapshot in the $K$-induced subgraphs in which:

- Node $\mathsf{V}_i$ (similarly a source) in the induced subgraph $\mathcal{G}_{jj}$ transmits $X^j_{\mathsf{V}_i}$,
- Node $\mathsf{V}_i$ (similarly a destination) in the induced subgraph $\mathcal{G}_{jj}$ receives

$$Y^j_{\mathsf{V}_i} = \sum_{i':(i',i)\in\mathcal{E}} \mathbf{S}^{q-n_{i'i}} X^j_{\mathsf{V}_{i'}}. \tag{21}$$

**Transmission strategy.** At any time slot $t$:

- Source $\mathsf{S}_i$ transmits
$$X_{\mathsf{S}_i}[t] = \mathbf{T}_{\mathsf{S}_i}(t) X^i_{\mathsf{S}_i}, \tag{22}$$
- Each relay $\mathsf{V}_j$ transmits
$$X_{\mathsf{V}_j}[t] = \sum_{i=1}^{K} \mathbf{T}_{\mathsf{V}_j}(t, \mathsf{S}_i) X^i_{\mathsf{V}_j}, \tag{23}$$
where $t = 1, \ldots, T$, and $\mathbf{T}_{\mathsf{V}_j}(t, \mathsf{S}_i)$ is the entry at row $t$ and the column corresponding to $\mathsf{S}_i$ of $\mathbf{T}_{\mathsf{V}_j}$.

Note that condition **C.1** guarantees that each source transmits its signal at least once. At any time instant $t$ relay $\mathsf{V}_i$ (similarly a destination) receives

$$Y_{\mathsf{V}_i}[t] = \sum_{i':(i',i)\in\mathcal{E}} \mathbf{S}^{q-n_{i'i}} X_{\mathsf{V}_{i'}}[t], \quad t = 1, \ldots, T, \tag{24}$$

where summation is performed in $\mathbb{F}_2^q$.

**Reconstructing the received signals.** Based on the transmission strategy described above, we need to show that at any node $\mathsf{V}_i$ received signal $Y^j_{\mathsf{V}_i}$ can be obtained.

Condition **C.2** guarantees that using linear row operations, $\mathbf{R}_{\mathsf{V}_j}$ can be transformed into a matrix in which if $\mathsf{V}_i \in \mathcal{G}_{jj}$, then $\exists\, \ell$ such that all entries in row $\ell$ are zeros except for the column corresponding to source $\mathsf{S}_j$. Note that linear operation in the linear deterministic model is simply the XOR operation. We note that two 1s in the same column of a receive matrix represent the same signals and the same channel gains. Thus, $\mathsf{V}_i$ is able to cancel out all interference by appropriately summing the received signals at different time slots.

Similar argument holds for any destination $\mathsf{D}_i$. However, since it is possible that there exist multiple routes from $\mathsf{S}_i$ to $\mathsf{D}_i$, more than a single 1 might be required after the row operations. In fact, condition **C.3** guarantees that using linear row operations, destination $\mathsf{D}_i$ can cancel out all interfering signals and only observe the intended signals from $\mathsf{S}_i$.

**Gaussian model.** The proof presented above holds for the Gaussian channel model with some modifications. First, the operations should be in real domain and the row operations are now either summation or subtraction. At any time slot $t$ relay $\mathsf{V}_j$ transmits:

$$X_{\mathsf{V}_j}[t] = \frac{1}{\sqrt{\beta_{t,\mathsf{V}_j}}} \sum_{i=1}^{K} \mathbf{T}_{\mathsf{V}_j}(t, \mathsf{S}_i) X^i_{\mathsf{V}_j} \tag{25}$$

where $t = 1, \ldots, T$. To satisfy the power constraint at the nodes, we need $\beta_{t,\mathsf{V}_j}$ to be the number of non-zero entries in row $t$ of $\mathbf{T}_{\mathsf{V}_j}$.

Moreover, we need to show that with the row operations performed on received and transmit signals, the capacity of the reconstructed induced subgraphs is "close" to the capacity of the induced subgraphs when there is no interference. The following lemma shows that the capacity of the reconstructed induced subgraphs is within a constant number of bits of the capacity of the induced subgraphs with no interference. This constant is independent of the channel gains and transmit power $P$. □

**Lemma 1** *Consider a multi-hop complex Gaussian relay network with one source* $\mathsf{S}$ *and one destination* $\mathsf{D}$ *represented by a directed graph*

$$\mathcal{G} = (\mathcal{V}, \mathcal{E}, \{h_{ij}\}_{(i,j)\in\mathcal{E}})$$

*where* $\{h_{ij}\}_{(i,j)\in\mathcal{E}}$ *represents the channel gains associated with the edges.*

*We assume that at each receive node the additive white complex Gaussian noise has variance* $\sigma^2$*. We also assume a power constraint of* $P$ *at all nodes, i.e.,* $\lim_{n\to\infty} \frac{1}{n}\mathbb{E}\left(\sum_{t=1}^{n} |X_{\mathsf{V}_i}[t]|^2\right) \leq P$*. Denote the capacity of this network by* $C(\sigma^2, P)$*. Then, for all* $T \geq 1$, $T \in \mathbb{R}$*, we have*

$$C(\sigma^2, P) - \tau \leq C(T\sigma^2, P/T) \leq C(\sigma^2, P), \tag{26}$$

*where* $\tau = |\mathcal{V}|\left(2\log T + 17\right)$ *is a constant independent of the channel gains,* $P$*, and* $\sigma^2$*.*

*Proof* First note that by increasing noise variances and by decreasing the power constraint, we only decrease the capacity. Hence, we have $C(T\sigma^2, P/T) \leq C(\sigma^2, P)$. To prove the other inequality, we use the results in [28]. The cut-set bound $\bar{C}$ is defined as

$$\bar{C}(\sigma^2, P) = \max_{p(\{X_j\}_{\mathsf{V}_j\in\mathcal{V}})} \min_{\Omega\in\Lambda_D} I(Y_{\Omega^c}; X_\Omega | X_{\Omega^c}), \tag{27}$$

where $\Lambda_D = \{\Omega : \mathsf{S} \in \Omega, \mathsf{D} \in \Omega^c\}$ is the set of all S-D cuts. [2] Also, $\bar{C}_{i.i.d}(\sigma^2, P) = \min_{\Omega\in\Lambda_D} \log|\mathbf{I} + \frac{P}{\sigma^2}\mathbf{G}_\Omega \mathbf{G}^*_\Omega|$ is the cut-set bound evaluated for i.i.d. $\mathcal{N}(0, P)$ input distributions, and $\mathbf{G}_\Omega$ is the transfer matrix associated with the cut $\Omega$, i.e., the matrix relating the vector of all the inputs at the nodes in $\Omega$, denoted by $\mathbf{X}_\Omega$, to the vector of all the outputs in $\Omega^c$, denoted by $\mathbf{Y}_{\Omega^c}$, as in $\mathbf{Y}_{\Omega^c} = \mathbf{G}_\Omega \mathbf{X}_\Omega + \mathbf{Z}_{\Omega^c}$ where $\mathbf{Z}_{\Omega^c}$ is the noise vector. In [28], it has been shown that

$$\bar{C}_{i.i.d}(\sigma^2, P) - 15|\mathcal{V}| \leq C(\sigma^2, P) \leq \bar{C}_{i.i.d}(\sigma^2, P) + 2|\mathcal{V}|, \tag{28}$$

where $|\mathcal{V}|$ is the total number of nodes in the network. Similarly, we have

$$\begin{aligned} &\bar{C}_{i.i.d}(T\sigma^2, P/T) - 15|\mathcal{V}| \leq C(T\sigma^2, P/T) \\ &C(T\sigma^2, P/T) \leq \bar{C}_{i.i.d}(T\sigma^2, P/T) + 2|\mathcal{V}|. \end{aligned} \tag{29}$$

Now, we will show that

$$C(\sigma^2, P) - C(T\sigma^2, P/T) \leq |\mathcal{V}|\left(2\log T + 17\right). \tag{30}$$

For any S-D cut, $\Omega \in \Lambda_D$, $\frac{P}{\sigma^2}\mathbf{G}_\Omega\mathbf{G}_\Omega^*$ is a positive semi-definite matrix. Hence, there exists a unitary matrix $\mathbf{U}$ such that $\mathbf{U}\mathbf{G}_{diag}\mathbf{U}^* = \frac{P}{\sigma^2}\mathbf{G}_\Omega\mathbf{G}_\Omega^*$ where $\mathbf{G}_{diag}$ is a diagonal matrix. Refer to the non-zero elements in $\mathbf{G}_{diag}$ as $g_{ii}$'s. We have:

$$\begin{aligned}
&\log|\mathbf{I} + \frac{P}{\sigma^2}\mathbf{G}_\Omega\mathbf{G}_\Omega^*| - \log|\mathbf{I} + \frac{P}{T^2\sigma^2}\mathbf{G}_\Omega\mathbf{G}_\Omega^*| \\
&= \log|\mathbf{I} + \mathbf{U}\mathbf{G}_{diag}\mathbf{U}^*| - \log|\mathbf{I} + \frac{1}{T^2}\mathbf{U}\mathbf{G}_{diag}\mathbf{U}^*| \\
&= \log|\mathbf{U}\mathbf{U}^* + \mathbf{U}\mathbf{G}_{diag}\mathbf{U}^*| - \log|\mathbf{U}\mathbf{U}^* + \frac{1}{T^2}\mathbf{U}\mathbf{G}_{diag}\mathbf{U}^*| \\
&= \log\left(|\mathbf{U}||\mathbf{I}+\mathbf{G}_{diag}||\mathbf{U}^*|\right) - \log\left(|\mathbf{U}||\mathbf{I}+\frac{1}{T^2}\mathbf{G}_{diag}||\mathbf{U}^*|\right) \\
&= \log|\mathbf{I} + \mathbf{G}_{diag}| - \log|\mathbf{I} + \frac{1}{T^2}\mathbf{G}_{diag}| \\
&= \operatorname{tr}\log\left(\mathbf{I} + \mathbf{G}_{diag}\right) - \operatorname{tr}\log\left(\mathbf{I} + \frac{1}{T^2}\mathbf{G}_{diag}\right) \\
&= \sum_i \log\left(1 + g_{ii}\right) - \sum_i \log\left(1 + \frac{1}{T^2}g_{ii}\right) \\
&= \sum_i \log\left(\frac{1 + g_{ii}}{1 + \frac{1}{T^2}g_{ii}}\right) \\
&\overset{(a)}{\leq} \sum_i \lim_{g_{ii}\to\infty} \log\left(\frac{1 + g_{ii}}{1 + \frac{1}{T^2}g_{ii}}\right) \\
&= \sum_i \log T^2 \leq 2|\mathcal{V}|\log T, \qquad (31)
\end{aligned}$$

where $(a)$ follows from the fact that $\frac{1+g_{ii}}{1+\frac{1}{T^2}g_{ii}}$ is monotonically increasing in $g_{ii}$.

Now suppose that $\min_{\Omega\in\Lambda_D}\log|\mathbf{I} + \frac{P}{T^2\sigma^2}\mathbf{G}_\Omega\mathbf{G}_\Omega^*| = \log|\mathbf{I} + \frac{P}{T^2\sigma^2}\mathbf{G}_{\Omega'}\mathbf{G}_{\Omega'}^*|$. Hence, from (31), we have

$$\begin{aligned}
&\min_{\Omega\in\Lambda_D}\log|\mathbf{I} + \frac{P}{\sigma^2}\mathbf{G}_\Omega\mathbf{G}_\Omega^*| - \min_{\Omega\in\Lambda_D}\log|\mathbf{I}+\frac{P}{T^2\sigma^2}\mathbf{G}_\Omega\mathbf{G}_\Omega^*| \\
&= \min_{\Omega\in\Lambda_D}\log|\mathbf{I} + \frac{P}{\sigma^2}\mathbf{G}_\Omega\mathbf{G}_\Omega^*| - \log|\mathbf{I} + \frac{P}{T^2\sigma^2}\mathbf{G}_{\Omega'}\mathbf{G}_{\Omega'}^*| \\
&\leq \log|\mathbf{I} + \frac{P}{\sigma^2}\mathbf{G}_{\Omega'}\mathbf{G}_{\Omega'}^*| - \log|\mathbf{I} + \frac{P}{T^2\sigma^2}\mathbf{G}_{\Omega'}\mathbf{G}_{\Omega'}^*| \\
&\overset{(a)}{\leq} 2|\mathcal{V}|\log T, \qquad (32)
\end{aligned}$$

where (a) follows from (31). From (28) and (29), we have

$$\begin{aligned}
C(\sigma^2, P) - C(T\sigma^2, P/T) &\leq \min_{\Omega\in\Lambda_D}\log|\mathbf{I} + \frac{P}{\sigma^2}\mathbf{G}_\Omega\mathbf{G}_\Omega^*| \\
&- \min_{\Omega\in\Lambda_D}\log|\mathbf{I}+\frac{P}{T^2\sigma^2}\mathbf{G}_\Omega\mathbf{G}_\Omega^*| + 17|\mathcal{V}| \overset{(a)}{\leq} |\mathcal{V}|\left(2\log T + 17\right), \qquad (33)
\end{aligned}$$

where $(a)$ follows from (32). Therefore, we get

$$C(\sigma^2, P) - \tau \leq C(T\sigma^2, P/T) \leq C(\sigma^2, P), \qquad (34)$$

where $\tau = |\mathcal{V}|\left(2\log T + 17\right)$ is a constant independent of the channel gains, $P$, and $\sigma^2$. □

## 5 Optimality of the strategies

In this section, we consider several classes of networks and derive the minimum number of time slots such that all induced subgraphs can be reconstructed as if there is no interference present. In other words, we characterize the normalized sum capacity of such networks.

### 5.1 Single-layer folded-chain networks

We start by considering a single-layer network motivated by the downlink cellular system similar to the one in Fig. 7.

**Definition 6** *A single-layer $(K, m)$ folded-chain ($1 \leq m \leq K$) network is a single-layer network with $K$ S-D pairs. In this network, source $\mathsf{S}_i$ is connected to destinations with ID's $1 + \left[\{(i-1)^+ + (j-1)\}\mathbf{mod}K\right]$ where $i = 1, \ldots, K$, $j = 1, \ldots, m$ and $(i-1)^+ = \max\{(i-1), 0\}$.*

Figure 8 is the single-layer $(3, 2)$ folded-chain network corresponding to the downlink cellular system of Fig. 7. The following theorem characterizes the normalized sum capacity of such networks.

**Theorem 2** *The normalized sum capacity of a single-layer $(K, m)$ folded-chain network with 1-local view is $\alpha^* = \frac{1}{m}$.*

To achieve the normalized sum capacity of a single-layer $(K, m)$ folded-chain network with 1-local, we need to incorporate repetition coding at the sources. We note that a single-layer $(K, K)$ folded-chain network is a $K$ user fully connected interference channel, and in that case, with

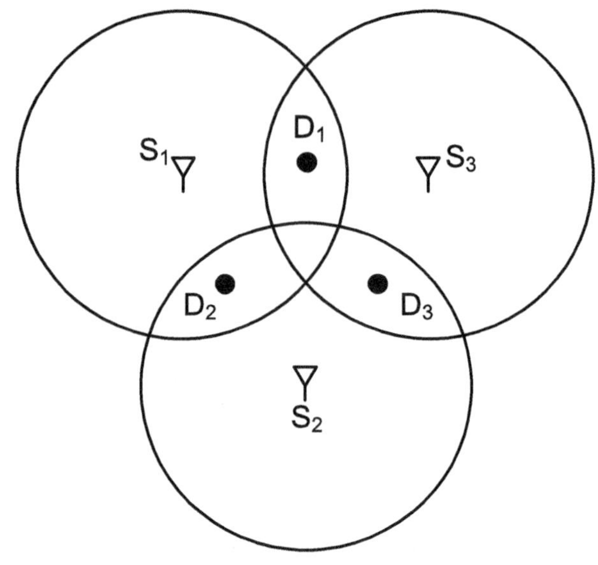


**Fig. 7** Downlink cellular network with three base stations and three destinations

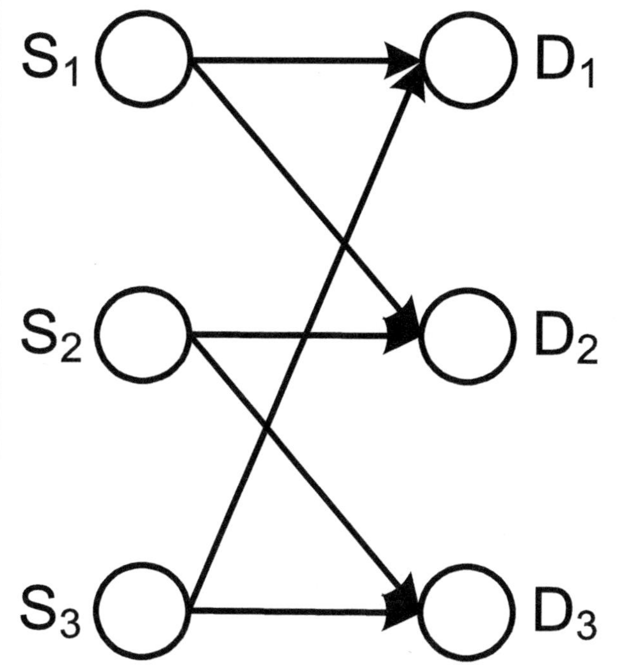


**Fig. 8** The corresponding network $\mathcal{G}$ for the downlink cellular system illustrated in Fig. 7

1-local view, interference avoidance achieves the normalized sum capacity of $1/K$.

*Proof* **Achievability**: We provide a valid assignment of transmit and receive matrices to the nodes with $T = m$. Then, by Theorem 1, we know that a normalized sum rate of $\frac{1}{m}$ is achievable.

Suppose $m < K < 2m$ (we will later generalize the achievability scheme for arbitrary $K$), let $m' = K - m$. To each source $\mathsf{S}_i$, $i = 1, \ldots, K$, we assign a transmit vector $\mathbf{T}_{\mathsf{S}_i}$ such that:

$$\mathbf{T}_{\mathsf{S}_i}(j) = 1 \Leftrightarrow j \leq i \leq j + m', \quad j = 1, \ldots, m. \tag{35}$$

□

**Remark 4** *For single-layer folded-chain networks, the assignment of the transmit vectors in (35) is the same for the linear deterministic model and the Gaussian model.*

This assignment satisfies conditions **C.1-C.3**:

- **C.1** is trivially satisfied since for any source $\mathsf{S}_i$, there exists at least one value of $j$ such that $\mathbf{T}_{\mathsf{S}_i}(j) = 1$.
- **C.2** is irrelevant since we have a single-layer network, and no relay is present.
- **C.3** is satisfied since for $1 \leq i \leq m$, $\mathbf{R}_{\mathsf{D}_i}$ has a single 1 in row $i$ and the column labeled as $\mathsf{S}_i$, and for $m < i < 2m$, the summation of rows $i - m + 1, \ldots, i - 1, i$ of $\mathbf{R}_{\mathsf{D}_i}$ has a single 1 in the column labeled as $\mathsf{S}_i$.

Since conditions **C.1-C.3** are satisfied, from Theorem 1, we know that we can achieve $\alpha = \frac{1}{m}$.

For general $K$, the achievability works as follows. Suppose, $K = c(2m-1)+r$, where $c \geq 1$ and $0 \leq r < (2m-1)$, we implement the scheme for S-D pairs $1, 2, \ldots, 2m - 1$ as if they are the only pairs in the network. The same for source-destination pairs $2m, 2m + 1, \ldots, 4m - 2$, etc. Finally, for the last $r$ S-D pairs, we implement the scheme with $m' = \max\{r - m + 1, 1\}$. This completes the proof of achievability.

**Converse**: Assume that a normalized sum rate of $\alpha$ is achievable[3]. We show that $\alpha \leq \frac{1}{m}$. Consider a single-layer $(K, m)$ folded-chain network in which the channel gain of a link from $\mathsf{S}_i$ to destinations $i, i + 1, \ldots, m$ is equal to $n \in \mathbb{N}$ (for the linear deterministic model) and $h \in \mathbb{C}$ (for the Gaussian model), $i = 1, 2, \ldots, m$, and all other channel gains are equal to zero. See Fig. 9 for a depiction.

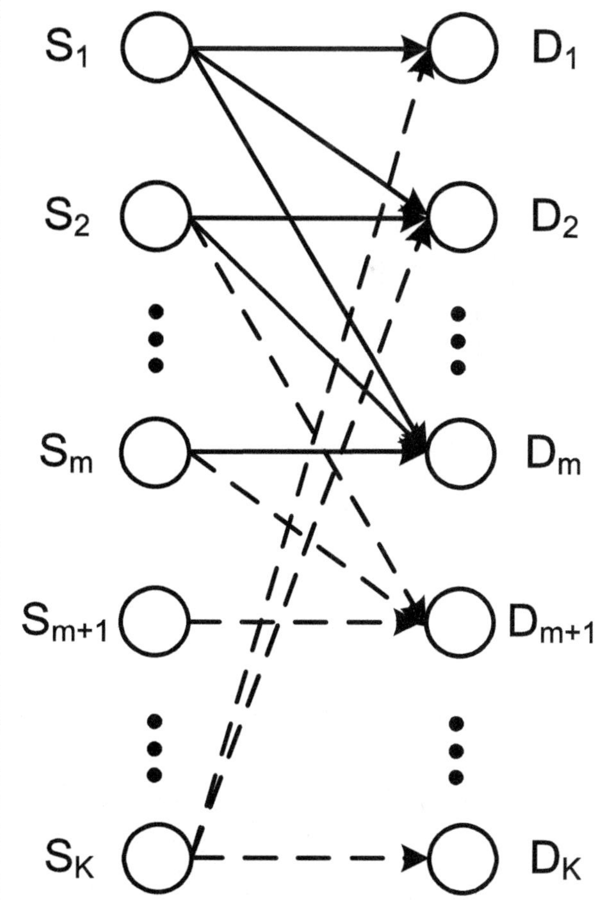


**Fig. 9** Channel gain assignment in a single-layer $(K, m)$ folded-chain network. All *solid links* have capacity $n$ (for the linear deterministic model) and $h$ (for the Gaussian model), and all *dashed links* have capacity 0

Destination $D_m$ after decoding and removing the contribution of $S_m$ has the same observation (up to the noise term) as $D_{m-1}$. Thus, $D_m$ is able to decode $W_{m-1}$. With a recursive argument, we conclude that

$$n\left(\sum_{j=1,2,\ldots,m} R_j - \epsilon_n\right) \leq H\left(Y_m^n | L_{D_m}, \mathsf{SI}\right). \tag{36}$$

The MAC capacity at destination $D_m$ in the linear deterministic model gives us

$$m(\alpha n - \tau) \leq n \Rightarrow (m\alpha - 1)n \leq d\tau. \tag{37}$$

Since this has to hold for all values of $n$, and $\alpha$ and $\tau$ are independent of $n$, we get $\alpha \leq \frac{1}{m}$. For the Gaussian model, the MAC capacity at $D_m$ gives us

$$m(\alpha \log(1 + |h|^2) - \tau) \leq \log(1 + m \times |h|^2), \tag{38}$$

which results in

$$(m\alpha - 1)\log(1 + |h|^2) \leq \log(m) + m\tau. \tag{39}$$

Since this has to hold for all values of $h$, and $\alpha$ and $\tau$ are independent of $h$, we get $\alpha \leq \frac{1}{m}$. This completes the proof of Theorem 2.

### 5.2 $K \times 2 \times K$ networks

We move to two-layer networks and start with a special class of networks defined below.

**Definition 7** *A $K \times 2 \times K$ network is a two-layer network (as defined in Section 2.1) with $|\mathcal{V}| - 2K = 2$.*

We establish the normalized sum capacity of such networks in the following theorem.

**Theorem 3** *The normalized sum capacity of a $K \times 2 \times K$ network with 1-local view is $\alpha^* = 1/d_{\max}$ where $d_{\max}$ is defined in (7).*

*Proof* The result for $K = 1$ is trivial, so we assume $K > 1$. We refer to the two relays as $A_1$ and $A_2$, see Fig. 10.

**Achievability:** We divide the S-D pair IDs into three disjoint subsets as follows:

- $\mathcal{J}_i$ is the set of all the S-D pair IDs such that the corresponding source is connected *only* to relay $A_i$, $i = 1, 2$;

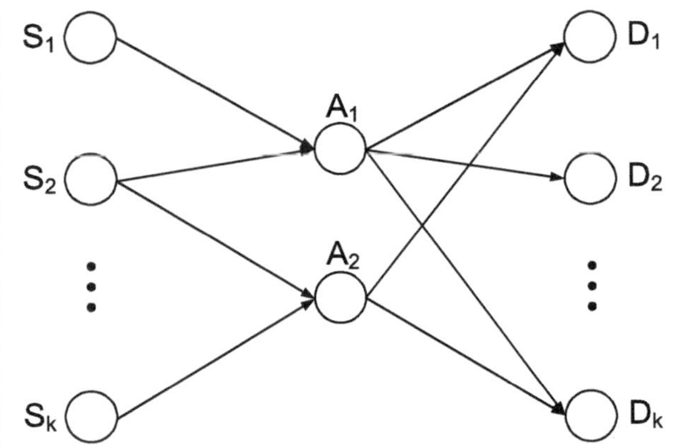


**Fig. 10** Illustration of a $K \times 2 \times K$ network

- $\mathcal{J}_{12}$ is the set of all the other S-D pair IDs. In other words, $\mathcal{J}_{12}$ is the set of all the S-D pair IDs where the corresponding source is connected to both relays.

Without loss of generality assume that $d_{\text{in}}(A_2) \geq d_{\text{in}}(A_1)$ and rearrange sources such that

$$\begin{aligned}
\mathcal{J}_1 &= \{1, 2, \ldots, d_{\text{in}}(A_1)\}, \\
\mathcal{J}_2 &= \{d_{\text{in}}(A_1) + 1, \ldots, d_{\text{in}}(A_1) + d_{\text{in}}(A_2)\}, \\
\mathcal{J}_{12} &= \{d_{\text{in}}(A_1) + d_{\text{in}}(A_2) + 1, \ldots, K\}.
\end{aligned} \tag{40}$$

We pick the smallest member of $\mathcal{J}_1$ and the smallest member of $\mathcal{J}_2$, and we set the first entry of the corresponding transmit vectors equal to 1 and all other entries equal to zero. We remove these members from $\mathcal{J}_1$ and $\mathcal{J}_2$. Then, we pick two smallest members of (updated) $\mathcal{J}_1$ and $\mathcal{J}_2$ and we set the second entry of the corresponding transmit vectors equal to 1 and all other entries equal to zero. We continue this process until $\mathcal{J}_1$ is empty. For any remaining S-D pair ID $j$, we set the $j$th entry of the corresponding transmit vector equal to 1 and all other entries equal to zero.

In the second layer, we divide S-D pair IDs based on the connection of destinations to relays, i.e., $\mathcal{J}_i'$ is the set of all the S-D pair IDs such that the corresponding destination is connected to relay $A_i$, $i = 1, 2$, and $\mathcal{J}_{12}'$ is the set of all the other S-D pair IDs. To $A_i$, we assign a transmit matrix of size $T \times \left(\mathcal{J}_i' + \mathcal{J}_{12}'\right)$, $i = 1, 2$, as described below.

Without loss of generality assume that $d_{\text{out}}(A_2) \geq d_{\text{out}}(A_1)$. We pick one member of $\mathcal{J}_1'$ and one member of $\mathcal{J}_2'$ randomly, and in the first row of $\mathbf{T}_{A_1}$ and $\mathbf{T}_{A_2}$, we set the entry at the column corresponding to the picked indices equal to 1 and all other entries in those rows equal to zero. We remove these members from $\mathcal{J}_1'$ and $\mathcal{J}_2'$. We then pick one member of (updated) $\mathcal{J}_1'$ and one member of (updated) $\mathcal{J}_2'$ randomly, and in the second row of $\mathbf{T}_{\mathsf{A}_1}$ and $\mathbf{T}_{\mathsf{A}_2}$, we set the entry at the column corresponding to the picked indices equal to 1 and all other entries in those rows equal to zero. We continue this process until $\mathcal{J}_1'$ is empty.

We then pick one of the remaining S-D pair IDs (members of $\mathcal{J}_{12}'$ and the remaining members of $\mathcal{J}_2'$), and we assign a 1 in the next available row and to the column corresponding to the picked index in the corresponding transmit matrix. We set all other entries in those rows equal to zero. We continue the process until no S-D pair ID is left.

Condition **C.1** is trivially satisfied. The corresponding transmission strategy in this case would be a "per layer" interference avoidance, i.e., if in the first hop, two sources are connected to the same relay, they do not transmit simultaneously, and if in the second hop, two destinations are connected to the same relay, they are not going to be served simultaneously. Therefore, since the scheme does not allow any interference to be created, no row operations on the receive matrix is required and conditions **C.2** and **C.3** are satisfied.

Note that according to the assignment of the vectors and matrices, we require

$$T = \max\{d_{\text{in}}(\mathsf{A}_2), d_{\text{out}}(\mathsf{A}_2)\} = d_{\max}. \tag{41}$$

Hence, from Theorem 1, we know that a normalized sum rate of $\alpha = \frac{1}{d_{\max}}$ is achievable. □

**Remark 5** *For $K \times 2 \times K$ networks, the assignment of the transmit vectors and matrices is the same for the linear deterministic model and the Gaussian model.*

**Converse:** Assume that a normalized sum rate of $\alpha$ is achievable, we show that $\alpha \leq \frac{1}{d_{\max}}$. It is sufficient to consider two cases: (1) $d_{\max} = d_{\text{in}}(\mathsf{A}_1)$ and (2) $d_{\max} = d_{\text{out}}(\mathsf{A}_1)$. Here, we provide the proof for case (1) and we postpone the proof for case (2) to Appendix B.

The proof is based on finding a worst-case scenario. Thus, to derive the upper bound, we use specific assignment of channel gains. Consider $\mathsf{D}_j$ for $j \in \mathcal{J}_1$, any such destination is either connected to relay $\mathsf{A}_1$ or to both relays. If it is connected to both, then set the channel gain from relay $\mathsf{A}_2$ equal to 0. Follow similar steps for the members of $\mathcal{J}_2$.

Now, consider $\mathsf{D}_j$ for $j \in \mathcal{J}_{12}$, such destination is either connected to only one relay or to both relays. If such destination is connected to both relays, assign the channel gain of 0 to one of the links connecting it to a relay (pick this link at random).

For all other links in the network, assign a channel gain of $n \in \mathbb{N}$ (in the linear deterministic model) and $h \in \mathbb{C}$ (in the Gaussian model). With this channel gain assignment, we have:

$$\forall j \in \mathcal{J}_1 : H\left(W_j | Y_{\mathsf{A}_1}^n, L_{\mathsf{A}_1}, \mathsf{SI}\right) \leq n\epsilon_n, \tag{42}$$

where $\epsilon_n \to 0$ as $n \to \infty$. Thus, relay $\mathsf{A}_1$ is able to decode all messages coming from sources corresponding to the members of $\mathcal{J}_1$.

A similar claim holds for relay $\mathsf{A}_2$ and all messages coming from $\mathcal{J}_2$. Relays $\mathsf{A}_1$ and $\mathsf{A}_2$ decode the messages coming from members of $\mathcal{J}_1$ and $\mathcal{J}_2$, respectively, and remove their contributions from the received signals. After removing the contributions from members of $\mathcal{J}_1$ and $\mathcal{J}_2$, relay $\mathsf{A}_1$ can decode the message of members in $\mathcal{J}_{12}$, i.e.,

$$\forall \ell \in \mathcal{J}_{12} : H\left(W_\ell | Y_{\mathsf{A}_1}^n, X_{j \in \mathcal{J}_1}^n, L_{\mathsf{A}_1}, \mathsf{SI}\right) \leq n\epsilon_n. \tag{43}$$

Thus, each relay is able to decode the rest of the messages (in the linear deterministic case, relays have the same received signals and in the Gaussian case, they receive the same codewords with different noise terms). This means that relay $\mathsf{A}_1$ is able to decode all the messages from $\mathcal{J}_1$ and $\mathcal{J}_{12}$, i.e.,

$$\sum_{j \in (\mathcal{J}_1 \cup \mathcal{J}_{12})} H\left(W_j | Y_{\mathsf{A}_1}^n, L_{\mathsf{A}_1}, \mathsf{SI}\right) \leq n\epsilon_n, \tag{44}$$

which in turn implies that

$$n\left(\sum_{j \in (\mathcal{J}_1 \cup \mathcal{J}_{12})} R_j - \epsilon_n\right) \leq H\left(Y_{\mathsf{A}_1}^n | L_{\mathsf{A}_1}, \mathsf{SI}\right). \tag{45}$$

Note that $d_{\text{in}}(\mathsf{A}_1) = |\mathcal{J}_1| + |\mathcal{J}_{12}|$.

Given the assumption of 1-local view, in order to achieve a normalized sum rate of $\alpha$, each source should transmit at a rate greater than or equal to $\alpha n - \tau$ (in linear deterministic model) for some constant $\tau$. This is due to the fact that from each source's point of view, it is possible that the other S-D pairs have capacity 0. Therefore, in order to achieve a normalized sum rate of $\alpha$, it should transmit at a rate of at least $\alpha n - \tau$. The MAC capacity at relay $\mathsf{A}_1$ gives us

$$d_{\text{in}}(\mathsf{A}_1)(\alpha n - \tau) \leq n \Rightarrow (d_{\text{in}}(\mathsf{A}_1)\alpha - 1)n \leq d_{\text{in}}(\mathsf{A}_1)\tau. \tag{46}$$

Since this has to hold for all values of $n$, and $\alpha$ and $\tau$ are independent of $n$, we get $\alpha \leq \frac{1}{d_{\text{in}}(\mathsf{A}_1)}$.

In the Gaussian case, each source should transmit at a rate greater than or equal to $\alpha \log(1 + |h|^2) - \tau$ since from each source's point of view, it is possible that the other S-D pairs have capacity 0. From the MAC capacity at relay $\mathsf{A}_1$, we get

$$d_{\text{in}}(\mathsf{A}_1)\left(\alpha \log(1 + |h|^2) - \tau\right) \leq \log\left(1 + d_{\text{in}}(\mathsf{A}_1) \times |h|^2\right), \tag{47}$$

which results in

$$d_{\text{in}}(\mathsf{A}_1)\left(\alpha \log\left(1+ |h|^2\right)-\tau\right) \leq \log(d_{\text{in}}(\mathsf{A}_1))+\log\left(1+\times|h|^2\right). \tag{48}$$

Hence, we have

$$(d_{\text{in}}(\mathsf{A}_1)\alpha - 1)\log(1 + |h|^2) \leq \log(d_{\text{in}}(\mathsf{A}_1)) + d_{\text{in}}(\mathsf{A}_1)\tau. \tag{49}$$

Since this has to hold for all values of $h$, and $\alpha$ and $\tau$ are independent of $h$, we get $\alpha \leq \frac{1}{d_{\text{in}}(\mathsf{A}_1)}$.

Combining the argument presented above with the result in Appendix B, we get

$$\alpha \leq \frac{1}{d_{\max}}. \tag{50}$$

This completes the proof of the converse.

### 5.3 3 x 3 x 3 networks

In this subsection, we consider two-layer networks with three source-destination pairs and three relays. We face networks in which we need to incorporate *network coding* techniques to achieve the normalized sum capacity with 1-local view. The coding comes in the form of repetition coding at sources and a combination of repetition and network coding at relays.

**Definition 8** *A* $3 \times 3 \times 3$ *network is a two-layer network (as defined in Section* 2.1*) with* $K = 3$ *and* $|\mathcal{V}| - 2K = 3$.

**Theorem 4** *The normalized sum capacity of a* $3 \times 3 \times 3$ *network with 1 local view,* $\alpha^*$ *is equal to*

1. *1 if and only if* $\mathcal{G}_{ii} \cap \mathcal{G}_{jj} = \emptyset$ *for* $i \neq j$.
2. *1/3 if and only if one of the graphs in Fig.* 11 *is a subgraph of the network connectivity graph* $\mathcal{G}$.
3. *1/2 otherwise.*

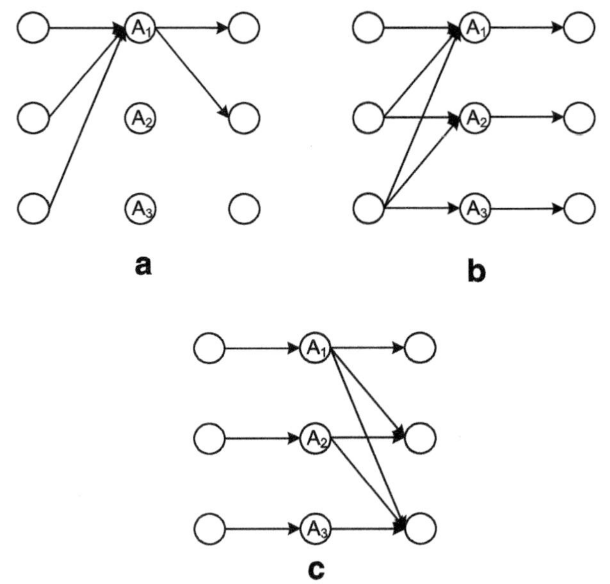


**Fig. 11 a–c** The normalized sum capacity of a 3 × 3 × 3 network with 1-local view, $\alpha^*$, is equal to 1/3 if and only if one of the graphs in this figure is a subgraph of $\mathcal{G}$

As we show in this section, the transmission strategy is a combination of three main techniques:

1. Per layer interference avoidance
2. Repetition coding to allow overhearing of the interference
3. Network coding to allow interference neutralization

**Remark 6** *From Theorems* 3 *and* 4, *we conclude that for all single-layer,* $K \times 2 \times K$, *and* $3 \times 3 \times 3$ *networks with 1-local view, the normalized sum capacity* $\alpha^* = 1/K$ *(i.e., TDMA) is optimal if and only if when all channel gains are equal and non-zero, then there exists a node that can decode all messages. We refer to such node as an "omniscient" node. We believe this observation holds for much more general network connectivities with 1-local view or in fading networks with no channel state information at the transmitters. However, this is a different line of research and it is beyond the scope of this paper. Omniscient nodes are studied in* [29, 30] *in the context of two-source two-destination multi-layer wireless networks where they dictate the asymptotic (in terms of power) behavior of these networks.*

*Proof* **Achievability:** The achievability proof for networks in category (a) is trivial as there is no interference present in such networks. For networks in category (b), a simple TDMA achieves a normalized sum rate of 1/3. Thus, we only need to prove the result for networks in category (c).

Suppose none of the graphs in Fig. 11 is a subgraph of $\mathcal{G}$ and that the network does not fall in category (a). This immediately implies $d_{\max} = 2$.

We have the following claim for such networks.

**Claim** *For a* $3 \times 3 \times 3$ *network with* $d_{\max} = 2$, *the only connectivity that results in a per layer fully connected conflict graph is the one shown in Fig. 12.*

□

If the per layer conflict graphs are not fully connected and the network does not fall in category (a), then a normalized sum rate of $\alpha = 1/2$ is easily achievable. Moreover, from claim 1, we know that with $d_{\max} = 2$, the folded-chain structure of Fig. 12 exists in at least one of the layers. In Section 3, we showed that a normalized sum rate of $\alpha = 1/2$ is achievable.

We note that these cases can be easily described within the algebraic framework of Section 4. In fact, if the folded-chain structure of Fig. 12 exists in at least one of the layers, as shown in Section 3, the transmission can be expressed as a valid assignment of transmit and receive matrices. However, if the per layer conflict graphs are not fully connected and the network does not fall in category (a), then the scheme is a per layer interference avoidance which can be easily expressed in terms of a valid assignment of transmit and receive matrices. Finally, we note that for all $3 \times 3 \times 3$ networks, the assignment of the transmit vectors and matrices is the same for the linear deterministic model and the Gaussian model.

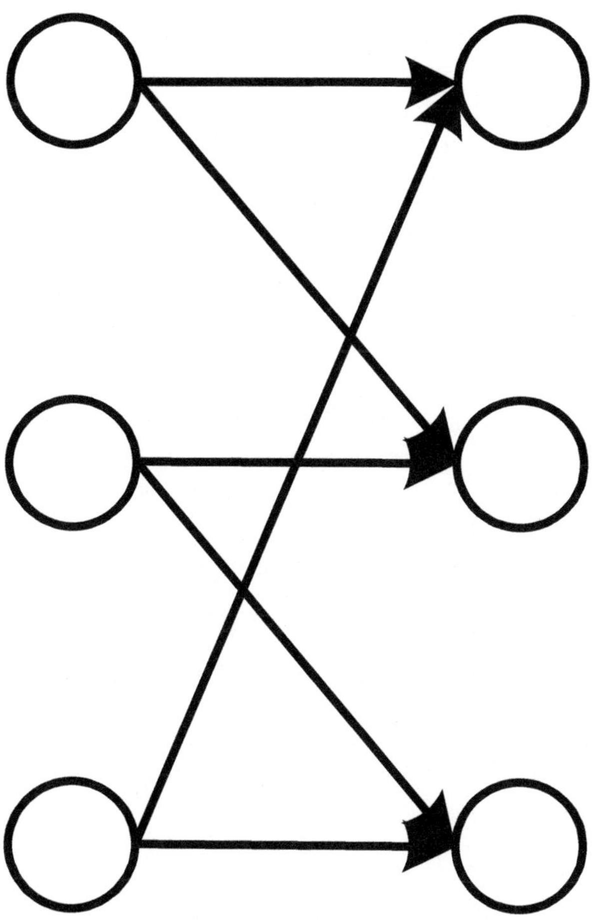

**Fig. 12** The only connectivity that results in a per layer fully connected conflict graph in a $3 \times 3 \times 3$ network with $d_{\max} = 2$

**Converse:** The forward direction of the proof for networks in category (a) is trivial as there is no interference present in such networks. For the reverse direction as shown in Lemma 3 Appendix A, for any network that does not fall into category (a), an upper bound of 1/2 on the normalized sum capacity holds. Moreover, Lemma 3 also provides the outer-bound for networks in category (c). Thus, we only need to consider $3 \times 3 \times 3$ networks in category (b).

For $3 \times 3 \times 3$ networks in category (b), we first consider the forward direction. One of the graphs in Fig. 11 is a subgraph of the network connectivity graph $\mathcal{G}$, say the graph in Fig. 11b. Assign channel gain of $n \in \mathbb{N}$ (in linear deterministic model) and $h \in \mathbb{C}$ (in Gaussian model) to the links of the subgraph and channel gain of 0 to the links that are not in the graph of Fig. 11b.

With this assignment of the channel gains, we have

$$H\left(W_i | Y_{\mathsf{A}_i}^n, L_{\mathsf{A}_i}, \mathsf{SI}\right) \leq n\epsilon_n, \qquad i = 1, 2, 3. \tag{51}$$

Basically, each destination $\mathsf{D}_i$ is *only* connected to relay $\mathsf{A}_i$, and each relay $\mathsf{A}_i$ has all the information that destination $\mathsf{D}_i$ requires in order to decode its message, $i = 1, 2, 3$. Thus, relay $\mathsf{A}_1$ can decode $\mathrm{W}_1$. After removing the contribution of $\mathsf{S}_1$, relay $\mathsf{A}_1$ is able to decode $\mathrm{W}_2$. Continuing this argument, we conclude that

$$n\left(\sum_{j=1}^{3} R_j - \epsilon_n\right) \leq H\left(Y_{\mathsf{A}_1}^n | L_{\mathsf{A}_1}, \mathsf{SI}\right). \tag{52}$$

The MAC capacity at relay $\mathsf{A}_1$, for the linear deterministic model, gives us

$$3(\alpha n - \tau) \leq n \Rightarrow (3\alpha - 1)n \leq d\tau. \tag{53}$$

Since this has to hold for all values of $n$, and $\alpha$ and $\tau$ are independent of $n$, we get $\alpha \leq \frac{1}{3}$. For the Gaussian model, the MAC capacity at relay $\mathsf{A}_1$ gives us

$$3(\alpha \log(1 + |h|^2) - \tau) \leq \log(1 + 3 \times |h|^2), \quad (54)$$

which results in

$$(3\alpha - 1) \log(1 + |h|^2) \leq \log(3) + 3\tau. \quad (55)$$

Since this has to hold for all values of $h$, and $\alpha$ and $\tau$ are independent of $h$, we get $\alpha \leq \frac{1}{3}$. The proof for the graphs in Fig. 11a, c is very similar.

For the reverse direction, if none of the graphs in Fig. 11 is a subgraph of the network connectivity graph $\mathcal{G}$, a normalized sum rate of $\alpha = \frac{1}{2}$ is achievable and optimal. This completes the converse proof.

### 5.4 Folded-chain networks

We now consider a class of networks for which we need to incorporate network coding in order to achieve the normalized sum capacity with 1-local view. This class is the generalization of the networks in Section 5.1 to two layers.

**Definition 9** *A two-layer $(K, m)$ folded-chain network $(1 \leq m \leq K)$ is a two-layer network with $K$ S-D pairs and $K$ relays in the middle. Each S-D pair $i$ has $m$ disjoint paths, through relays with indices $1 + \left[\{(i-1)^+ + (j-1)\}\mathbf{mod}K\right]$ where $i = 1, \ldots, K$, $j = 1, \ldots, m$.*

**Theorem 5** *The normalized sum capacity of a two-layer $(K, m)$ folded-chain network with 1-local view is $\alpha^* = \frac{1}{m}$.*

*Proof* **Achievability**: The result is trivial for $m = 1$. For $m = K$, the upper bound of $\alpha = \frac{1}{K}$ can be achieved simply by using TDMA. Similar to Theorem 2, suppose $m < K < 2m$. The extension to the general case would be similar to Theorem 2. Here, we describe how to construct a valid assignment of transmit and receive matrices.

The assignment of the transmit vectors in the first layer is identical to that of Theorem 2 as given in (35). We note that each relay can recover all incoming signals with this assignment. For the second layer, we have $K + 1$ steps.

• Steps 1 through $m$: Our goal is to provide destination $\mathsf{D}_i$, $1 \leq i \leq m$, during time slot $i$ with its desired signal without any interference. Therefore, row $i$ of any relay connected to $\mathsf{D}_i$ has a single 1 in the column associated with S-D pair $i$ and 0s elsewhere;

• Steps $m + 1$ through $K$: During step $i$, $m + 1 \leq i \leq K$, our goal is to provide destination $\mathsf{D}_i$ with its desired signal (interference will be handled later). To do so, consider the transmit matrix of any relay connected to $sfD_i$; and place a single 1 in the column associated with S-D pair $i$ and the row with the smallest index and least number of 1s;

• Step $K + 1$: During this step, our goal is to resolve interference and goes through the following loop:

1. Let $\mathcal{L}_j$ denote the set of row indices for which there exists at least 1 in the column associated with S-D pair $j$ of the transmit matrix of a relay connected to $\mathsf{D}_j$, $m + 1 \leq j \leq K$;
2. Set $j = m + 1$. For any $j' \neq j, j' = 1, 2, \ldots, K$, if

$$\sum_{p:\mathsf{V}_p \text{ connected to } \mathsf{D}_j} \sum_{\ell \in \mathcal{L}_j} \mathbf{T}_{\mathsf{V}_p}\left(\ell, \mathsf{S}_{j'}\right) \neq 0, \quad (57)$$

then make this summation 0 making any of the $\mathbf{T}_{\mathsf{V}_p}\left(\ell, \mathsf{S}_{j'}\right)$'s that is not previously assigned equal to $1^4$;

3. Set $j = j + 1$; If $j > K$ and during the previous loop no change has occurred, then set all entries that are not yet defined equal to zero and terminate; otherwise, go to line 2 of the loop.

We now describe the $K + 1$ steps via an example of (5, 3) two-layer folded-chain network of Fig. 13. In Fig. 14, we have demonstrated the evolution of the relays' transmit matrices at the end of steps $m$, $K$, and $K + 1$. For this example, the loop in step $K + 1$ is repeated three times.

It is easy to verify condition **C.3** for destination $\mathsf{D}_1$, $\mathsf{D}_2$, and $\mathsf{D}_3$. We have provided $\mathbf{R}_{\mathsf{D}_4}$ in (56), and as we can see, by adding the first and the second row, we can have a row that has only 1s in the columns corresponding to source $\mathsf{S}_4$. Similarly, we can show that the condition holds for $\mathsf{D}_5$.

**Converse**: Assume that a normalized sum rate of $\alpha$ is achievable, i.e., there exists a transmission strategy with 1-local view, such that for all channel realizations, it achieves a sum rate satisfying

$$\mathbf{R}_{\mathsf{D}_1} = \begin{array}{c} \\ t=1 \\ t=2 \\ t=2 \end{array} \begin{array}{c} \begin{matrix} \mathsf{S}_1:\mathsf{V}_1 & \mathsf{S}_2:\mathsf{V}_4 & \mathsf{S}_3:\mathsf{V}_4 & \mathsf{S}_3:\mathsf{V}_5 & \mathsf{S}_4:\mathsf{V}_1 & \mathsf{S}_4:\mathsf{V}_4 & \mathsf{S}_4:\mathsf{V}_5 & \mathsf{S}_5:\mathsf{V}_1 & \mathsf{S}_5:\mathsf{V}_5 \end{matrix} \\ \begin{pmatrix} 1 & 1 & 0 & 1 & 0 & 1 & 1 & 0 & 1 \\ 1 & 1 & 0 & 1 & 1 & 0 & 0 & 0 & 1 \\ 1 & 0 & 1 & 1 & 1 & 0 & 0 & 1 & 0 \end{pmatrix} \end{array}, \quad (56)$$

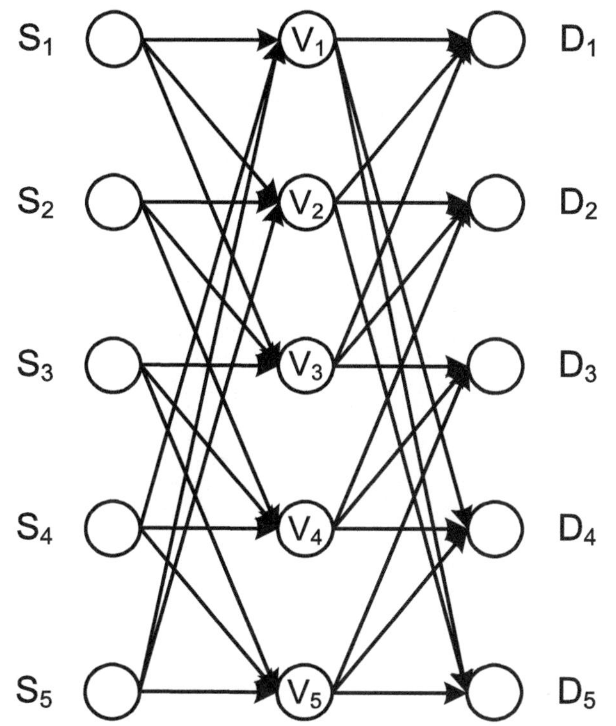


**Fig. 13** A (5, 3) two-layer folded-chain network

$$\sum_{i=1}^{K} R_i \geq \alpha C_{\text{sum}} - \tau, \tag{58}$$

with error probabilities going to zero as $N \to \infty$ and for some constant $\tau \in R$ independent of the channel gains.

Consider the first layer of the two-layer $(K, m)$ folded-chain network, where the channel gain of a link from source $i$ to relays $i, i+1, \ldots, m$ is equal to $n \in \mathbb{N}$ (for the linear deterministic model) and $h \in \mathbb{C}$ (for the Gaussian model), $i = 1, 2, \ldots, m$, and all the other channel gains are equal to zero. In the second layer, we set the channel gain from relay $i$ to destination $i$ equal to $n$ (for the linear deterministic model) or $h$ (for the Gaussian model), and all other channel gains equal to 0, $i = 1, 2, \ldots, m$. See Fig. 15 for a depiction.

With this configuration, each destination $\mathsf{D}_i$ is *only* connected to relay $\mathsf{A}_i$, and each relay $\mathsf{A}_i$ has all the information that destination $\mathsf{D}_i$ requires in order to decode its message, $i = 1, \ldots, m$. We have

$$H\left(W_i | Y_{\mathsf{A}_i}^n, L_{\mathsf{A}_i}, \mathsf{SI}\right) \leq n\epsilon_n, \qquad i = 1, \ldots, m. \tag{59}$$

At relay $\mathsf{A}_m$ after decoding and removing the contribution of $\mathsf{S}_m$, relay $\mathsf{A}_m$ is able to decode $\mathsf{W}_{m-1}$. With a

$$\begin{array}{lccc}
 & m & K & K+1 \\
\mathbf{T}_{\mathsf{V}_1}: & \begin{array}{c} \begin{matrix} \mathsf{S}_1 & \mathsf{S}_4 & \mathsf{S}_5 \end{matrix} \\ \begin{pmatrix} 1 & 0 & 0 \\ - & - & - \\ - & - & - \end{pmatrix} \end{array} & \begin{array}{c} \begin{matrix} \mathsf{S}_1 & \mathsf{S}_4 & \mathsf{S}_5 \end{matrix} \\ \begin{pmatrix} 1 & 0 & 0 \\ - & 1 & - \\ - & - & 1 \end{pmatrix} \end{array} & \begin{array}{c} \begin{matrix} \mathsf{S}_1 & \mathsf{S}_4 & \mathsf{S}_5 \end{matrix} \\ \begin{pmatrix} 1 & 0 & 0 \\ 1 & 1 & - \\ 1 & 1 & 1 \end{pmatrix} \end{array} \\
\mathbf{T}_{\mathsf{V}_2}: & \begin{array}{c} \begin{matrix} \mathsf{S}_1 & \mathsf{S}_2 & \mathsf{S}_5 \end{matrix} \\ \begin{pmatrix} 1 & 0 & 0 \\ 0 & 1 & 0 \\ - & - & - \end{pmatrix} \end{array} & \begin{array}{c} \begin{matrix} \mathsf{S}_1 & \mathsf{S}_2 & \mathsf{S}_5 \end{matrix} \\ \begin{pmatrix} 1 & 0 & 0 \\ 0 & 1 & 0 \\ 0 & 1 & 1 \end{pmatrix} \end{array} & \begin{array}{c} \begin{matrix} \mathsf{S}_1 & \mathsf{S}_2 & \mathsf{S}_5 \end{matrix} \\ \begin{pmatrix} 1 & 0 & 0 \\ 0 & 1 & 0 \\ - & - & 1 \end{pmatrix} \end{array} \\
\mathbf{T}_{\mathsf{V}_3}: & \begin{array}{c} \begin{matrix} \mathsf{S}_1 & \mathsf{S}_2 & \mathsf{S}_3 \end{matrix} \\ \begin{pmatrix} 1 & 0 & 0 \\ 0 & 1 & 0 \\ 0 & 0 & 1 \end{pmatrix} \end{array} & \begin{array}{c} \begin{matrix} \mathsf{S}_1 & \mathsf{S}_2 & \mathsf{S}_3 \end{matrix} \\ \begin{pmatrix} 1 & 0 & 0 \\ 0 & 1 & 0 \\ 0 & 0 & 1 \end{pmatrix} \end{array} & \begin{array}{c} \begin{matrix} \mathsf{S}_1 & \mathsf{S}_2 & \mathsf{S}_3 \end{matrix} \\ \begin{pmatrix} 1 & 0 & 0 \\ 0 & 1 & 0 \\ 0 & 0 & 1 \end{pmatrix} \end{array} \\
\mathbf{T}_{\mathsf{V}_4}: & \begin{array}{c} \begin{matrix} \mathsf{S}_2 & \mathsf{S}_3 & \mathsf{S}_4 \end{matrix} \\ \begin{pmatrix} - & - & - \\ 1 & 0 & 0 \\ 0 & 1 & 0 \end{pmatrix} \end{array} & \begin{array}{c} \begin{matrix} \mathsf{S}_2 & \mathsf{S}_3 & \mathsf{S}_4 \end{matrix} \\ \begin{pmatrix} - & - & 1 \\ 1 & 0 & 0 \\ 0 & 1 & 0 \end{pmatrix} \end{array} & \begin{array}{c} \begin{matrix} \mathsf{S}_2 & \mathsf{S}_3 & \mathsf{S}_4 \end{matrix} \\ \begin{pmatrix} 1 & 0 & 1 \\ 1 & 0 & 0 \\ 0 & 1 & 0 \end{pmatrix} \end{array} \\
\mathbf{T}_{\mathsf{V}_5}: & \begin{array}{c} \begin{matrix} \mathsf{S}_3 & \mathsf{S}_4 & \mathsf{S}_5 \end{matrix} \\ \begin{pmatrix} - & - & - \\ - & - & - \\ 1 & 0 & 0 \end{pmatrix} \end{array} & \begin{array}{c} \begin{matrix} \mathsf{S}_3 & \mathsf{S}_4 & \mathsf{S}_5 \end{matrix} \\ \begin{pmatrix} - & 1 & - \\ - & - & 1 \\ 1 & 0 & 0 \end{pmatrix} \end{array} & \begin{array}{c} \begin{matrix} \mathsf{S}_3 & \mathsf{S}_4 & \mathsf{S}_5 \end{matrix} \\ \begin{pmatrix} 1 & 1 & 1 \\ 1 & 0 & 1 \\ 1 & 0 & 0 \end{pmatrix} \end{array}
\end{array}$$

**Fig. 14** The evolution of the relays' transmit matrices at the end of steps $m$, $K$, and $K+1$. For this example, the loop in step $K+1$ is repeated three times

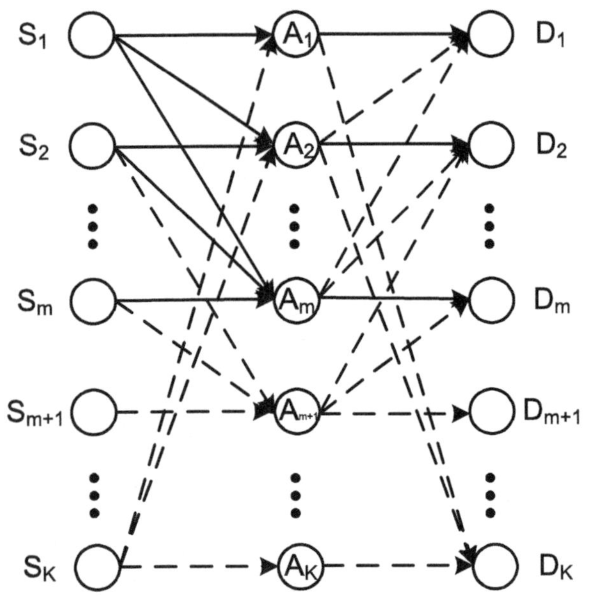


**Fig. 15** Channel gain assignment in a two-layer $(K, m)$ folded-chain network. All *solid links* have capacity $n$ (for the linear deterministic model) and $h$ (for the Gaussian model), and all *dashed links* have capacity 0

recursive argument, we conclude that

$$n\left(\sum_{j=1,2,\ldots,m} R_j - \epsilon_n\right) \leq H\left(Y^n_{\mathsf{A}_m} | L_{\mathsf{A}_m}, \mathsf{SI}\right). \tag{60}$$

The MAC capacity at relay $\mathsf{A}_m$ for the linear deterministic model gives us

$$m(\alpha n - \tau) \leq n \Rightarrow (m\alpha - 1)n \leq d\tau. \tag{61}$$

Since this has to hold for all values of $n$, and $\alpha$ and $\tau$ are independent of $n$, we get $\alpha \leq \frac{1}{m}$. For the Gaussian model, the MAC capacity at relay $\mathsf{A}_m$ gives us

$$m(\alpha \log(1 + |h|^2) - \tau) \leq \log(1 + m \times |h|^2), \tag{62}$$

which results in

$$(m\alpha - 1) \log(1 + |h|^2) \leq \log(m) + m\tau. \tag{63}$$

Since this has to hold for all values of $h$, and $\alpha$ and $\tau$ are independent of $h$, we get $\alpha \leq \frac{1}{m}$. □

### 5.5 Gain of coding over interference avoidance: nested folded-chain networks

In this subsection, we show that the gain from using coding over interference avoidance techniques can be unbounded. To do so, we first define the following class of networks.

**Definition 10** *An L-nested folded-chain network is a single-layer network with $K = 3^L$ S-D pairs, $\{\mathsf{S}_1, \ldots, \mathsf{S}_{3^L}\}$ and $\{\mathsf{D}_1, \ldots, \mathsf{D}_{3^L}\}$. For $L = 1$, an L-nested folded-chain network is the same as a single-layer $(3, 2)$ folded-chain network. For $L > 1$, an L-nested folded-chain network is formed by first creating three copies of an $(L-1)$-nested folded-chain network. Then,*

- *The i-th source in the first copy is connected to the i-th destination in the second copy, $i = 1, \ldots, 3^{L-1}$,*
- *The i-th source in the second copy is connected to the i-th destination in the third copy, $i = 1, \ldots, 3^{L-1}$,*

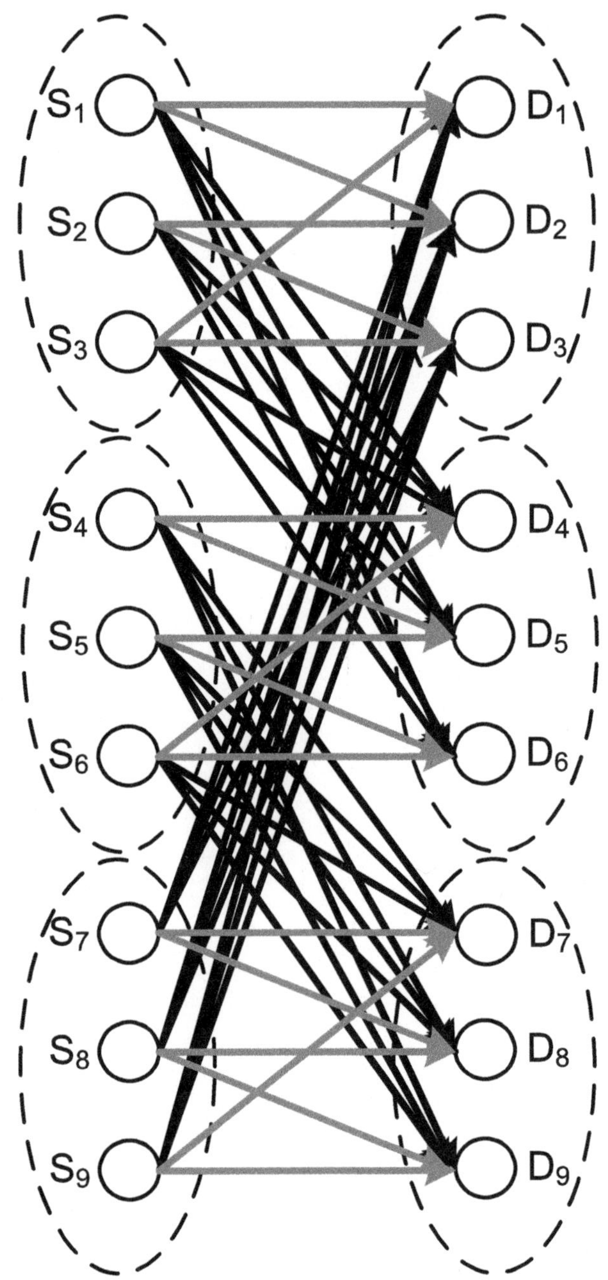


**Fig. 16** A 2-nested folded-chain network

- *The i-th source in the third copy is connected to the i-th destination in the first copy, $i = 1, \ldots, 3^{L-1}$.*

*Figure 16 illustrates a 2-nested folded-chain network.*

Consider an $L$-nested folded-chain network. The conflict graph of this network is fully connected, and as a result, interference avoidance techniques can only achieve a normalized sum rate of $\left(\frac{1}{3}\right)^L$. However, we know that for a single-layer $(3, 2)$ folded-chain network, a normalized sum rate of $\frac{1}{2}$ is achievable.

Hence, applying our scheme to an $L$-nested folded-chain network, a normalized sum rate of $\left(\frac{1}{2}\right)^L$ is achievable. For instance, consider the 2-nested folded-chain network in Fig. 16. We show that any transmission strategy over the induced subgraphs can be implemented in the original network by using only four time slots such that all nodes receive the same signal as if they were in the induced subgraphs.

To achieve a normalized sum rate of $\alpha = \left(\frac{1}{2}\right)^2$, we split the communication block into four time slots of equal length. During time slot 1, sources 1, 2, 4, and 5 transmit the same codewords as if they are in the induced subgraphs. During time slot 2, sources 3 and 6 transmit the same codewords as if they are in the induced subgraphs, and sources 2 and 5 repeat their transmit signal from the first time slot. During time slot 3, sources 7 and 8 transmit the same codewords as if they are in the induced subgraphs, and sources 4 and 5 repeat their transmit signal from the first time slot. During time slot 4, source 9 transmits the same codewords as if it is in the induced subgraph, and sources 5, 6, and 8 repeat their transmit signal.

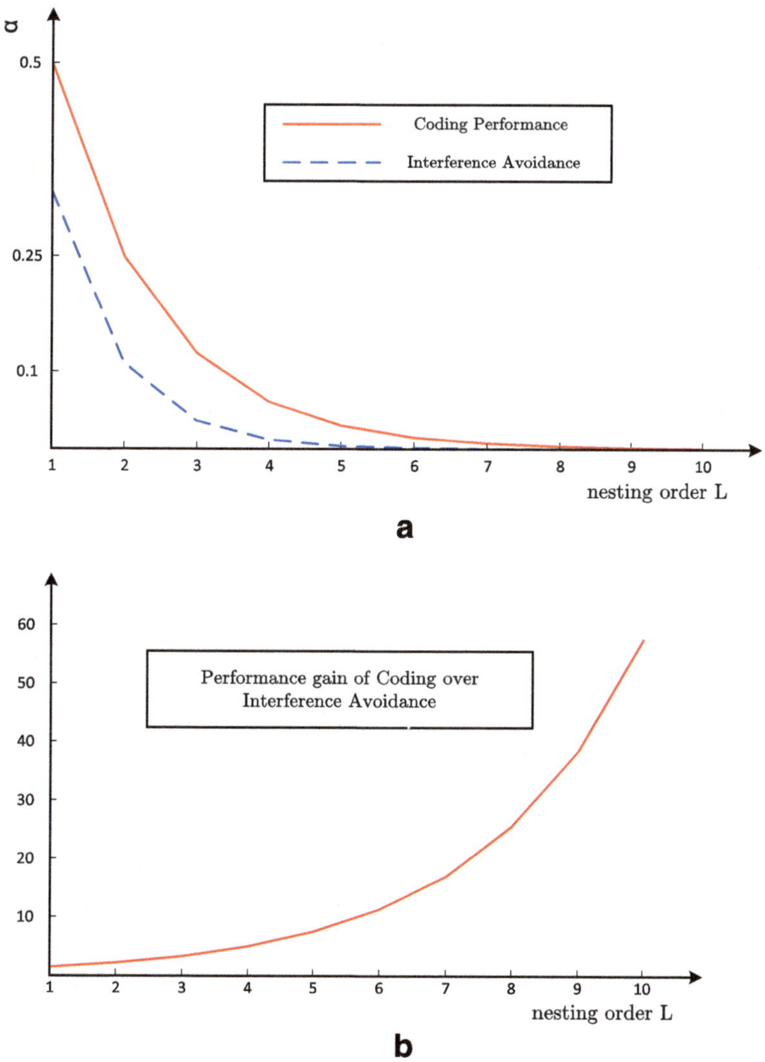


**Fig. 17 a** Achievable normalized sum rate of coding and interference avoidance in an $L$-nested folded-chain network and **b** the performance gain of coding scheme over interference avoidance

It is straightforward to verify that with this scheme, all destinations receive the same signal as if they were in the induced subgraphs. Hence, a normalized sum rate of $\alpha = \left(\frac{1}{2}\right)^2$ is achievable for the network in Fig. 16. Therefore, the gain of using coding over interference avoidance is $\left(\frac{3}{2}\right)^L$ which goes to infinity as $L \to \infty$. See Fig.17 for a depiction. As a result, we can state the following lemma.

**Lemma 2** *Consider an L-nested folded-chain network. The gain of using MCL scheduling over MIL scheduling is* $\left(\frac{3}{2}\right)^L$ *which goes to infinity as* $L \to \infty$.

The scheme required to achieve a normalized sum rate of $\alpha = \left(\frac{1}{2}\right)^L$ for an $L$-nested folded-chain network can be viewed as a simple extension of the results presented in Section 4. In a sense instead of reconstructing a single snapshot, we reconstruct $L$ snapshots of the network. The following discussion is just for the completion of the results.

To each source $\mathsf{S}_i$, we assign a transmit vector $\mathbf{T}_{\mathsf{S}_i}$ of size $(LT) \times 1$ where each row corresponds to a time slot. If we denote the transmit signal of node $\mathsf{S}_i$ in the $\ell$th snapshot by $X^{\ell}_{\mathsf{S}_i}$, then if $\mathbf{T}_{\mathsf{S}_i}(j) = 1$ and $cT + 1 \leq j < (c+1)T$ for $c = 0, 1, \ldots, L-1$, then $\mathsf{S}_i$ communicates $X^{c+1}_{\mathsf{S}_i}$. The other transmit and receive matrices can be described similarly. Conditions **C.2** and **C.3** have to be satisfied for submatrices of $\mathbf{R}_{\mathsf{V}_j}$ corresponding to rows $cT+1, cT+2, \ldots, (c+1)T-1$ for $c = 0, 1, \ldots, L-1$.

## 6 Conclusions

In this paper, we studied the fundamental limits of communications over multi-layer wireless networks where each node has limited channel state information. We developed a new transmission strategy for multi-layer wireless networks with partial channel state information (i.e., 1-local view) that combines multiple ideas including interference avoidance and network coding. We established the optimality of our proposed strategy for several classes of networks in terms of achieving the normalized sum capacity. We also demonstrated several connections between network topology, normalized sum capacity, and the achievability strategies.

So far, we have only studied cases with 1-local view. One major direction is to characterize the increase in normalized sum capacity as nodes learn more and more about the channel state information. We also focused on the case in which wireless nodes know the network connectivity globally, but the actual values of the channel gains are known for a subset of flows. Another important direction would be to understand the impact of local connectivity knowledge on the capacity and to develop distributed strategies to optimally route information with partial connectivity knowledge.

## Endnotes

[1] Since in two time slots, any transmission strategy for the diamond networks can be implemented in the original network, we can implement the strategies that achieve the capacity for any S-D pair $i$ with full network knowledge, i.e., $C_i$, over two time slots as well. Hence, we can achieve $\frac{1}{2}\left(C_1 + C_2 + C_3\right)$. On the other hand, we have $C_{\text{sum}} \leq C_1 + C_2 + C_3$. As a result, we can achieve a set of rates such that $\sum_{i=1}^{3} R_i \geq \frac{1}{2} C_{\text{sum}}$, and by the definition of normalized sum rate, we achieve $\alpha = \frac{1}{2}$.

[2] A cut $\Omega$ is a subset of $\mathcal{V}$ such that $\mathsf{S} \in \Omega, \mathsf{D} \notin \Omega$, and $\Omega^c = \mathcal{V} \setminus \Omega$.

[3] The result for $m = 1$ is trivial since we basically have an interference-free network.

[4] If $\mathbf{T}_{\mathsf{V}_p}\left(\ell, \mathsf{S}_{j'}\right)$ is not yet assigned a 0 or 1, treat it as 0 in the summation.

## Appendix A: Outer-bound of 1/2 for networks with interference

**Lemma 3** *In a K-user multi-layer network (linear deterministic or Gaussian) with 1-local view if there exists a path from* $\mathsf{S}_i$ *to* $\mathsf{D}_j$*, for some* $i \neq j$*, then the normalized sum capacity is upper-bounded by* $\alpha = 1/2$.

*Proof* Consider a path from source $\mathsf{S}_i$ to destination $\mathsf{D}_j$, $i \neq j$, $i, j \in \{1, \ldots, K\}$. Assign channel gain of $n \in \mathbb{N}$ (for the linear deterministic model) and $h \in \mathbb{C}$ (for the Gaussian model) to all edges in this path. For each one of the two S-D pairs $i$ and $j$, pick exactly one path from the source to the destination and assign channel gain of $n$ (for the linear deterministic model) or $h$ (for the Gaussian model) to all edges in these paths. Assign channel gain of 0 to all remaining edges in the network $\mathcal{G}$. See Fig. 18 for an illustration.

In order to guarantee a normalized sum rate of $\alpha$ with 1-local view, each source has to transmit at a rate greater than or equal to $\alpha n - \tau$ (for the linear deterministic model) or $\alpha \log(1 + |h|^2) - \tau$ (for the Gaussian model). This is due to the fact that from each source's point of view, it is possible that the other S-D pairs have capacity 0.

Suppose this rate is feasible, i.e., destinations $\mathsf{D}_i$ and $\mathsf{D}_j$ are able to decode $\mathrm{W}_i$ and $\mathrm{W}_j$, respectively. Since there exists a path from source $\mathsf{S}_i$ to destination $\mathsf{D}_j$, $i \neq j$, $i, j \in \{1, \ldots, K\}$, we can find a node $\mathsf{V}^* \in \mathcal{V}$ such that $\mathsf{V}^* \in \mathcal{G}_{ij}$ and $\mathsf{V}^* \in \mathcal{G}_{jj}$, see Fig. 18. Node $\mathsf{V}^*$ is able to

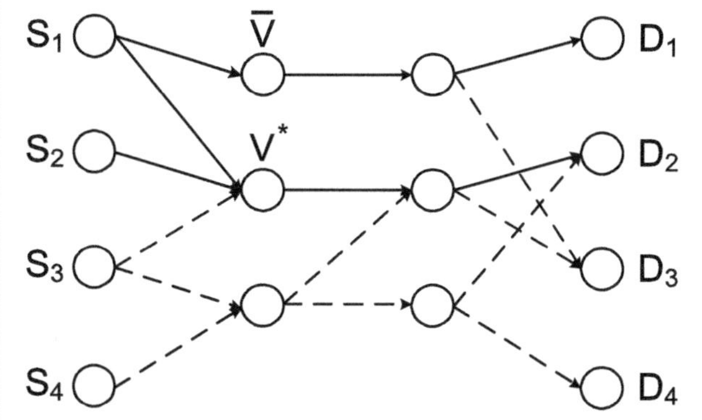


**Fig. 18** A path exists from source $S_1$ to destination $D_2$; all solid edges have capacity $n$ (for linear deterministic model) or $h$ (for the Gaussian model) and the rest have capacity 0

decode $W_j$ since it has more information than destination $D_j$. Node $V^*$ decodes $W_j$ and removes it from the received signal. Now, for the Gaussian model, it has statistically the same received signal as node $\bar{V}$ (a node in $\mathcal{G}_{ii}$ and in the same layer as $V^*$) and for the linear deterministic model, it has the same received signal as node $\bar{V}$. Node $\bar{V}$ is able to decode $W_i$ since it has more information than destination $D_i$., as a result, $V^*$ is also able to decode $W_i$. This means that there exists a node $V^* \in \mathcal{V}$ that can decode both $W_i$ and $W_j$.

Hence, the MAC capacity at $V^*$ in the linear deterministic model gives us

$$2\alpha n - 2\tau \leq n \Rightarrow (2\alpha - 1)n \leq 2\tau. \qquad (64)$$

Since this has to hold for all values of $n$, and $\alpha$ and $\tau$ are independent of $n$, we get $\alpha \leq \frac{1}{2}$.

For the Gaussian model, the MAC capacity at $V^*$ gives us

$$2(\alpha \log(1 + |h|^2) - \tau) \leq \log(1 + 2|h|^2), \qquad (65)$$

which results in

$$(2\alpha - 1) \log\left(1 + |h|^2\right) \leq 1 + 2\tau. \qquad (66)$$

Since this has to hold for all values of $h$, and $\alpha$ and $\tau$ are independent of $h$, we get $\alpha \leq \frac{1}{2}$. □

## Appendix B: Converse proof for case (2) of Theorem 3

**Converse proof for case (2)**: If a destination is only connected to relay $A_2$, assign channel gain of 0 to the link from $A_2$ to such destination. Set all the other channel gains equal to $n \in \mathbb{N}$ (in the linear deterministic model) and equal to $h \in \mathbb{C}$ (in the Gaussian model). We claim that with this channel gain assignment, a destination connected to both relays should be able to decode all messages (note that with our choice of channel gains, there is no message for destinations that are only connected to relay $A_2$).

Destinations that are connected to both relays receive the exact same signal (in the linear deterministic model) and the same codewords plus different noise terms (in the Gaussian model). Therefore, since each one of them is able to decode its message, then it should be able to decode the rest of the messages intended for destinations that are connected to both relays. They decode and remove such messages from the received signal. The remaining signal is the same codeword (plus different noise term in Gaussian model) received at the destinations that are only connected to relay $A_1$. Therefore, those messages are also decodable at a destination that is connected to both relays.

We assume 1-local view at the sources; therefore, to achieve a normalized sum rate of $\alpha$, each source should transmit at a rate greater than or equal to $\alpha n - \tau$ (in linear deterministic model). This is due to the fact that from each source's point of view, it is possible that the other S-D pairs have capacity 0; therefore, in order to achieve a normalized sum rate of $\alpha$, it should transmit at a rate of at least $\alpha n - \tau$. The above argument alongside the MAC capacity at a destination connected to both relays, results in

$$d_{\text{out}}(A_1)(\alpha n - \tau) \leq n \Rightarrow (d_{\text{out}}(A_1)\alpha - 1)n \leq d_{\text{out}}(A_1)\tau. \qquad (67)$$

Since this has to hold for all values of $n$, and $\alpha$ and $\tau$ are independent of $n$, we get $\alpha \leq \frac{1}{d_{\text{out}}(A_1)}$.

In the Gaussian case, each source should transmit at a rate greater than or equal to $\alpha \log(1+|h|^2) - \tau$, since from each source's point of view, it is possible that the other S-D pairs have capacity 0. Similar to the linear deterministic case, we get

$$d_{\text{out}}(A_1)\left(\alpha \log(1 + |h|^2) - \tau\right) \leq \log\left(1 + d_{\text{out}}(A_1) \times |h|^2\right), \qquad (68)$$

or equivalently

$$(d_{\text{out}}(A_1)\alpha - 1) \log\left(1 + |h|^2\right) \leq \log(d_{\text{out}}(A_1)) + d_{\text{out}}(A_1)\tau. \qquad (69)$$

Since this has to hold for all values of $h$, and $\alpha$ and $\tau$ are independent of $h$, we get $\alpha \leq \frac{1}{d_{\text{out}}(A_1)}$.

**Acknowledgements**
The authors would like to thank Kanes Sutuntivorakoon of Rice University for helpful discussions related to Section V. C of this paper.

**Funding**
The work of A. Vahid is in part supported by Qualcomm Innovation Fellowship. The work of A. S. Avestimehr is in part supported by NSF grant CCF-1419632 and ONR award N000141612189.

**Authors' contributions**
This work was done by AV as part of his PhD studies. He was advised by Professors VA, ASA, and AS. All authors read and approved the final manuscript.

**Competing interests**
The authors declare that they have no competing interests.

**Author details**
[1]Electrical Engineering Department, University of Colorado, Denver, Denver 80204, CO, USA. [2]School of Industrial Engineering, Purdue University, West Lafayette 47907, IN, USA. [3]Electrical Engineering Department, University of Southern California, Los Angeles 90089, CA, USA. [4]Electrical and Computer Engineering Department, Rice University, Houston 77251, TX, USA.

**References**
1. A Vahid, V Aggarwal, AS Avestimehr, A Sabharwal, in *48th Annual Allerton Conference on Communication, Control, and Computing*. On the capacity of multi-hop wireless networks with partial network knowledge (IEEE, Allerton Park, Monticello, 2010), pp. 1030–1037
2. V Aggarwal, AS Avestimehr, A Sabharwal, On achieving local view capacity via maximal independent graph scheduling. IEEE Trans. Inf. Theory. **57**(5), 2711–2729 (2011)
3. SA Jafar, Topological interference management through index coding. IEEE Trans. Inf. Theory. **60**(1), 529–568 (2014)
4. H Maleki, VR Cadambe, SA Jafar, Index coding—an interference alignment perspective. IEEE Trans. Inf. Theory. **60**(9), 5402–5432 (2014)
5. M Ji, G Caire, AF Molisch, Fundamental limits of caching in wireless D2D networks. IEEE Trans. Inf. Theory. **62**(2), 849–869 (2016)
6. N Naderializadeh, AS Avestimehr, Interference networks with no CSIT: impact of topology. IEEE Trans. Inf. Theory. **61**(2), 917–938 (2015)
7. H Sun, C Geng, SA Jafar, in *IEEE International Symposium on Information Theory Proceedings (ISIT)*. Topological interference management with alternating connectivity (IEEE, Istanbul, 2013), pp. 399–403
8. X Yi, D Gesbert, Topological interference management with transmitter cooperation. IEEE Trans. Inf. Theory. **61**(11), 6107–6130 (2015)
9. N Naderializadeh, AS Avestimehr, ITLinQ: A new approach for spectrum sharing in device-to-device communication systems. IEEE J. Selected Areas Commun. **32**(6), 1139–1151 (2014)
10. V Aggarwal, L Sankar, AR Calderbank, HV Poor. Ergodic layered erasure one-sided interference channels (IEEE, Taormina, 2009), pp. 574–578
11. C Wang, T Gou, SA Jafar, Aiming perfectly in the dark-blind interference alignment through staggered antenna switching. IEEE Trans. Signal Process. **59**(6), 2734–2744 (2011)
12. SA Jafar, in *IEEE Global Telecommunications Conference (GLOBECOM 2010)*. Exploiting channel correlations-simple interference alignment schemes with no CSIT (IEEE, Miami, 2010), pp. 1–5
13. A Vahid, MA Maddah-Ali, AS Avestimehr, Y Zhu, Binary fading interference channel with no CSIT. IEEE Trans. Inf. Theory. **63**(6), 3565–3578 (2017)
14. H Maleki, SA Jafar, S Shamai, Retrospective interference alignment over interference networks. IEEE J. Selected Topics Signal Process. **6**(3), 228–240 (2012)
15. CS Vaze, MK Varanasi, The degrees of freedom region and interference alignment for the MIMO interference channel with delayed CSIT. IEEE Trans. Inf. Theory. **58**(7), 4396–4417 (2012)
16. A Vahid, MA Maddah-Ali, AS Avestimehr, Capacity results for binary fading interference channels with delayed CSIT. IEEE Trans. Inf. Theory. **60**(10), 6093–6130 (2014)
17. MA Maddah-Ali, D Tse, Completely stale transmitter channel state information is still very useful. IEEE Trans. Inf. Theory. **58**(7), 4418–4431 (2012)
18. A Vahid, MA Maddah-Ali, AS Avestimehr, Approximate capacity region of the MISO broadcast channels with delayed CSIT. IEEE Trans. Commun. **64**(7), 2913–2924 (2016)
19. A Vahid, MA Maddah-Ali, AS Avestimehr, *Communication through collisions: opportunistic utilization of past receptions*. (IEEE International Conference on Computer Communications (INFOCOM), Toronto, 2014), pp. 2553–2561
20. A Vahid, R Calderbank, in *2015 IEEE International Symposium on Information Theory (ISIT)*. Impact of local delayed CSIT on the capacity region of the two-user interference channel (IEEE, Hong Kong, 2015), pp. 2421–2425
21. A Vahid, R Calderbank, in *2016 IEEE International Symposium on Information Theory (ISIT)*. When does spatial correlation add value to delayed channel state information? (IEEE, Barcelona, 2016), pp. 2624–2628
22. A Vahid, R Calderbank, Two-user erasure interference channels with local delayed CSIT. IEEE Trans. Inf. Theory. **62**(9), 4910–4923 (2016)
23. OE Ayach, RW Heath Jr, Interference alignment with analog channel state feedback. IEEE Transactions on Wireless Communications. **11**(2), 626–636 (2012)
24. V Aggarwal, Y Liu, A Sabharwal, in *Proceedings of International Symposium on Information Theory (ISIT)*. Message passing in distributed wireless networks (IEEE, Seoul, 2009), pp. 1090–1094
25. V Aggarwal, Y Liu, A Sabharwal, Sum capacity of interference channels with a local view: impact of distributed decisions. IEEE Trans. Inf. Theory. **58**(3), 1630–1659 (2012)
26. D Blackwell, L Breiman, AJ Thomasian, The capacity of a class of channels. Ann. Math. Stat. **30**, 1229–1241 (1959)
27. A Raja, VM Prabhakaran, P Viswanath, The two user Gaussian compound interference channel. IEEE Trans. Inf. Theory. **55**(11), 5100–5120 (2009)
28. AS Avestimehr, S Diggavi, D Tse, Wireless network information flow: a deterministic approach. IEEE Trans. Inf. Theory. **57**(4), 1872–1905 (2011)
29. I Shomorony, AS Avestimehr, Two-unicast wireless networks: characterizing the degrees of freedom. IEEE Trans. Inf. Theory. **59**(1), 353–383 (2013)
30. A Vahid, I Shomorony, R Calderbank, in *53rd Annual Allerton Conference on Communication, Control, and Computing*. Informational bottlenecks in two-unicast wireless networks with delayed CSIT (IEEE, Allerton Park, Monticello, 2015), pp. 1256–1263

# 5

# Hybrid connectivity restoration in wireless sensor and actor networks

Ke Yan, Guangchun Luo[*], Ling Tian, Qi Jia and Chengzong Peng

**Abstract**

Wireless sensor and actor networks are becoming more and more popular in the recent years. Each WSAN consists of numerous sensors and a few actors working collaboratively to carry out specific tasks. Unfortunately, actors are prone to failure due to harsh deployment environments and constrained power, which may break network connectivity resulting in disjoint components. Thus, maintaining the connectivity among actors is especially important. This paper proposes hybrid connectivity restoration (HCR), which integrates proactive selection and reactive motion. An actor protectively selects a backup node through its one-hop neighbor table and informs the backup node to supervise its stage. Once it fails, the backup node moves to the best position to restore the connectivity of the failed node's neighbors reactively. This triggers a local recovery process at the backup node, which is repeated until network connectivity is restored. In order to minimize travel distance, HCR selects the backup node which moves the shortest distance to restore connectivity. Furthermore, HCR opts to reduce the number of messages by just informing the failure to its backup node. The correctness and effectiveness of HCR are validated through both theoretical analysis and simulations.

**Keywords:** Hybrid connectivity restoration, Wireless senor and actor networks, Single-node failure

## 1 Introduction

Wireless sensor networks (WSNs) are indispensable components of Internet of Things [5–7, 9–11, 15, 19, 23–25, 32, 37, 38]. A wireless sensor and actor network (WSAN) is a special kind of WSN, which has motivated lots of research works [28]. In the corresponding applications such as environmental monitoring, battlefield surveillance, border protection, target searching and tracking, a number of sensors and actors work cooperatively to monitor a specific area and track a target of interest. Sensors are responsible for collecting data, and actors are responsible for processing data and bridging the sensors and the control center. An actor and the sensors connected to it form a self-organized sub-network. All the sub-networks collaborate with each other to carry out tasks. It is desired that all the actors in a WSAN are connected at any time.

Unfortunately, due to harsh deployed environments and limited battery power, actors may deplete energy fast. A sudden loss of a node may break network connectivity resulting in disjoint network components. Therefore, it is important to detect node failures and restore network connectivity as early as possible. Since WSANs are usually deployed far away from the control center and are operated autonomously and unattended, it is difficult and inefficient to control the restoring process in a centralized manner. Connectivity restoration therefore should be a distributed, localized, and self-healing process. In addition, a rapid connectivity restoration is desired in order to reduce the baneful influence of node failures. Moreover, the overhead such as the total travel distance and the total number of messages should be minimized considering the limited energy supply. The average travel distance should be considered as well because one node traveling too far will consume too much energy and may cause another network disconnection. A node failure disrupting network connectivity is called a cut vertex, which is difficult to identify in large-scale WSANs centralized and timely. Though there have been many distributed cut-vertex detection algorithms, they are time-consuming and resource-intensive. As a result, it is very challenging to restore network connectivity in a distributed, localized, and efficient manner.

*Correspondence: gcluo.uestc@gmail.com
School of Computer Science and Engineering, University of Electronic Science and Technology of China, Chengdu 611731, China

This paper proposes hybrid connectivity restoration (HCR) considering single-node failures, which integrates *proactive selection* and *reactive motion*. The selection of a backup node for a failure node is a proactive process. Each node identifies a backup node and is then monitored by its backup node. Once a node fails, its backup node moves to the best position that connects all the failed node's one-hop neighbors. It triggers the restoration of the backup node and the restoration is a reactive process. HCR is a distributed and localized scheme, where each node just maintains its one-hop neighborhood information. Since a failed node only affects its direct neighbors' connectivity, the main idea of the restoration is to move one of the failed node's neighbors to a new position so that all the failed node's direct neighbors can be re-connected. The node motion may trigger another disconnection on the moving node, so the restoration is a recursive process, the whole network is connected only when the motion node's directed neighbors are connected. HCR opts to efficiently restore network connectivity through selecting the most proper backup node and moving it to the best position instead of the failed node's position. The less distance the node moves, the less influence on the network connection. As aforementioned, only if a node failure breaks its directed neighborhood connectivity, it may further break network connectivity. The node whose failure breaks its directed neighborhood connectivity is called a critical node. On the opposite, the node is called an uncritical node. It should be noted that the uncritical node cannot be a cut vertex while the critical node may be a cut vertex. Though a critical node's failure may not necessarily break network connectivity, it may bring unnecessary restoration. It is much more efficient and cheaper to identify a critical node and restore its direct neighbors' connectivity than to identify a cut vertex and restore the network connectivity. To identify a cut vertex requires global information, which is impossible and inefficient in WSANs. At the same time, to identify a critical node just needs one-hop neighborhood information, and the identification is done on the node itself. Moreover, compared with moving a backup node to the failed node's position in DCR [17], it is better to move the backup node to the position which connects all the failed node's directed neighbors. Only when all the neighbors are on the boundary of the communication range, the backup node needs to move to the failed node's position.

Following are our main contributions:

- We proposed a hybrid connectivity restoration in WSAN, which integrates *proactive selection* and *reactive motion*. The proactive selection of a backup node can shorten a recovery process, the HCR offers effectiveness and timeliness.
- Different from moving a backup node to the position of failed node, HCR moves the backup node to the best position with the shortest travel distance to reconnect the failed node's one-hop neighbors. This can not only reduce a motion cost but also the total overhead, as the shorter a node travels, the fewer nodes are influenced.
- HCR opts to reduce the number of messages by just informing the failure to its backup node.
- The efficiency of HCR is mathematically analyzed and validated through simulations. The bounds on the incurred overhead are derived. HCR outperforms RIM and DCR in terms of the number of relocated nodes, total travel distance, average travel distance, and number of messages for both dense and sparse networks.

The remainder of the paper is organized as follows. Section 2 describes the system model and formulates the problem. Section 3 reviews the related works. Section 4 provides the detailed description of the proposed HCR, and the theoretical analysis is illustrated in Section 5. The simulation results are presented in Section 6. Finally, the conclusions and future work are presented in Section 7.

## 2 System model and problem formulation

### 2.1 System model

A WSAN consists of numerous sensors and a few actors (Fig. 1) which are usually randomly deployed in harsh environments far away from the control center. So a WSAN is a self-organized network where sensors are mainly responsible for collecting data and sending data to the nearest actors, and actors are responsible for making decisions. Actors process data, communicate with other actors, and send the results to the sink node. In this paper, it is assumed that actors are movable to connect other actors forming a connected network. In WSANs, actors play the role of gateways to the sink node. Therefore, it is important to maintain actor connectivity. Since sensors are static, they always try to communicate with actors in their communication range without identifying who they are. That means all the actors are the same to sensors. Therefore, actor connectivity decides network connectivity. Here, the coverage of the WSAN is not considered since an actor's failure will lose all the sensors connected to it. Because the sensors are randomly deployed in the sensing regions, it can be assumed that there is a equal number of sensors within an actor's transmission range. Therefore, moving actors has a little effect on the total coverage of the WSAN.

In WSANs, each actor has limited communication radius $R_c$. Actors can send and receive messages within the communication range to discover other actors.

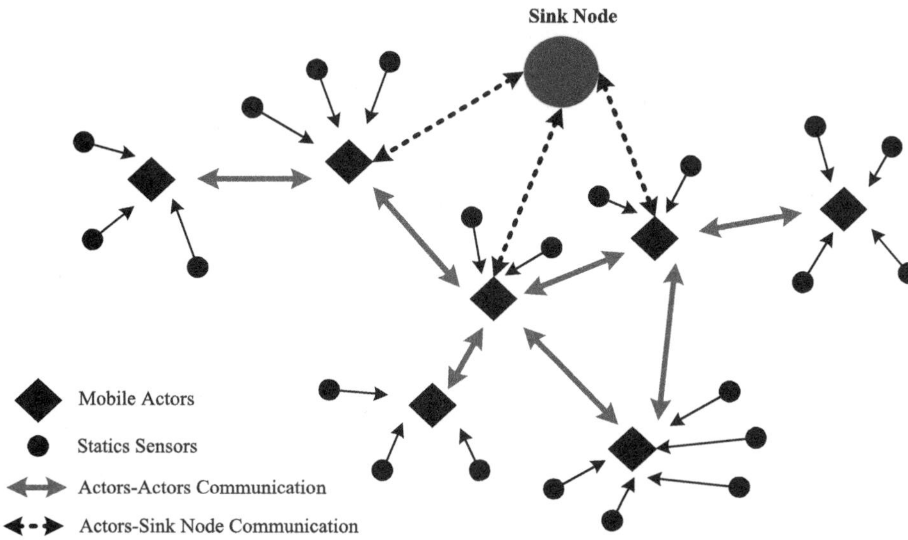


**Fig. 1** A WSAN with mobile actors

Each actor maintains a one-hop neighborhood table recording its neighbors' positions and other information. Two-hop neighborhood information or multi-hop neighborhood information can be obtained by exchanging one-hop neighborhood table with neighbors. The more information it gets, the more communication and storage it incurs. However, in most WSAN applications, actors are battery powered and have limited energy. Thus, it is more efficient to maintain less information. In this paper, each actor $i$ just maintains a one-hop neighborhood table denoted as $NT(i)$. $NT(i)$ is a two-dimensional table where each row contains one-hop neighbor information such as unique node ID (*ID*), local position (*POSITION*), and critical character (*CRITICAL*). The critical character is defined as follows.

*Critical character*: It indicates whether a node failure breaks neighbor connectivity. $CRITICAL = 1$ when the n ode failure breaks connectivity. Otherwise, $CRITICAL = 0$.

In order to restore connectivity rapidly, each actor also has backup node information. The selection of a backup node is introduced in Section 4.1. The one-hop neighborhood table and backup node information of each actor are maintained and updated during the process of connectivity restoration.

### 2.2 Problem formulation

Actors are prone to failures due to tough environments or energy depletion. The loss of an actor affects not only the sensors connected to it but also the neighboring actors. The latter case is even worse, so this paper focuses on the latter case. It is assumed that actors are movable, it is possible to restore network connectivity by relocating actors.

Generally, different node failures have different effects on network connectivity. Consider the network shown in Fig. 2. If node $A3$ fails, the network is divided into two disjoint components. If node $A8$ fails, the network is still connected and all its neighbors can communicate with each other. A leaf node's failure does not break connectivity. It is important to note that a node's failure breaking one-hop neighborhood connectivity does not necessarily mean it breaks network connectivity. For example, node $A6$ has two neighbors $A4$ and $A7$. If $A6$ fails, $A4$ and $A7$ cannot communicate with each other through $A6$, but $A5$ still connects $A4$ and $A7$. Though only a cut vertex breaks network connectivity, it needs the global network connectivity information to tell whether a node is a cut vertex which is nontrivial and inefficient in large-scale WSANs. It also incurs significant overhead in terms of computation, communication, and storage. As aforementioned, an uncritical node does not break network connectivity, and the identification of an uncritical node just needs one-hop neighborhood information. Moreover, the distance between any two sibling nodes in a WSAN is no more than $2Rc$, because both of them are in the communication

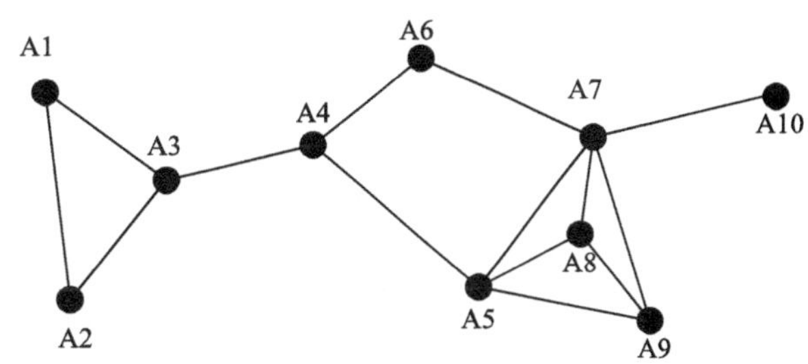


**Fig. 2** A connected network of mobile actors

range of the parent node. It takes a node to move at most *Rc* towards another to reconnect it. Therefore, compared with identifying a cut vertex, it is more efficient and significant to identify a critical node.

Though reconnecting two sibling nodes just needs to move one node for less than *Rc*, it may trigger another failure, which causes cascading failures. Even if the motion does not trigger other node failures, it may consume a lot of energy. Hence, load balancing should also be taken into consideration. In addition, a node motion may generate some other critical nodes, which increases the risk of node failure in the future. Considering all of these factors, we formulate the following problem.

*Given a connected WSAN* $G = (V, R_c)$. *V is the set of actors, and* $R_c$ *is the actors's communication radius. Each actor* $V_i$ *has ID and Positions. All the actors are homogeneous with the same* $R_c$. *Each node just knows its one-hop neighborhood information. When a single actor i fails, relocate the rest nodes so that (1) network connectivity is restored, (2) the total travel distance is minimized, (3) the average travel distance is minimized, and (4) the total number of messages is minimized.*

HCR is proposed in Section 4 to solve this problem efficiently. It is assumed that no two or more actors fail simultaneously, and no node fails during restoration.

## 3 Related works

Connectivity restoration considering single-node failures has attracted much attention, and there are a lot of surveys [22, 27, 34, 36] focusing on connectivity restoration. The existing schemes can be classified into two categories: proactive restoration and reactive restoration.

Proactive restoration schemes make use of redundant resources including nodes and paths to increase the robustness of a WSAN. When a node fails, it requires no connectivity restoration because there are redundant resources maintaining connectivity. Since the directional connection between a pair of actors is determined by the communication radius, it can only result in more redundant relay nodes to build K-disjoint paths [30]. In this case, there are K-disjoint paths between any pair of actors in a WSAN. Even K-1 paths fail, there is still a path connecting them. Consider a two-connected WSAN where there are at least two paths between any pair of nodes. CRAFT [18] establishes a bi-connected inter-partition topology while minimizing the longest path length and the number of deployed relay nodes. It strives to form the largest inner simple cycle or Backbone Polygon (BP) around the center of the damaged area where no partition lies inside, and deploys relay nodes to connect each outer partition to the BP through two non-overlapping paths. The advantage of proactive restoration schemes is that it does not disturb a network when node fails, but it requires many redundant resources. The stronger fault tolerance is, the more resources are required. Moreover, it is very difficult to place relay nodes optically, as it needs the global network information. The time complexity is very high in large-scale networks. In [20], it is proven that just listing a set of feasible sites for the relays is already at least APX-hard. Though many heuristic algorithms have been proposed, such as Genetic Algorithm [12, 21], Artificial Bee Colony Algorithm [14], and Concentric Fermat Points [31], it is still extraordinarily time-consuming.

Different from the proactive restoration schemes, the reactive restoration schemes are passive and a recovery process is triggered when a node failure is detected. They do not require reserving redundant resources. The basic idea is to reconnect the failed node's neighbors. There are two kinds of approaches: cooperative communication and relocating nodes. Cooperative communication is first proposed in [8]. It allows a node to send message beyond its communication radius with the help of its neighbors. Two nodes are able to communicate if and only if the received average signal-to-noise ratio (SNR) is no lees than the fixed threshold. Signal strength diminishes with the increase of transmission distance and overlays at the destination. CSFR [33] adopts cooperative communication to restore connectivity. Taking advantage of neighbors to transport data does not increase the neighbor's energy consumption a lot. Though it has low overhead from the current perspective, it is still a long-term process which costs a lot of energy in the long run. In addition, it is unacceptable and very time-consuming to select the help nodes.

Currently, most reactive restoration schemes reconnect a network by replacing a failure node with a proper backup node through movement which is a recursive process that may relocate the rest of the nodes. Therefore, which node moves and where to move is nontrivial. Ramezani proposed a distribute method to restore connectivity by using a centralized genetic algorithm [26] at the basic station. It strives to minimize the number of mobile nodes and the average length of all nodes' paths. It is a heuristic algorithm. As mentioned before, only a cut vertex may break network connectivity. Many approaches decide whether a node is a cut vertex firstly and deal with cut vertex failure only, such as DARA [1], PDARA [2], PCR [16], and NNN [13]. DARA identifies a cut vertex through two-hop neighborhood information. Once a failure happens, the failed node's neighbors select the most proper backup node considering node degree and distance and inform its sibling nodes. In fact, the process of identifying a cut vertex is not introduced in details in DARA. In a latter improved approach PDARA, it forms a connected dominating set (CDS). PDARA informs a particular node in advance whether a partition occurs in case of failure. They both strive to localize the scope of the recovery process and minimize the movement overhead imposed on the involved actors. In nearest non-critical neighbor (NNN), each actor periodically determines its criticality (i.e., cut vertex or not). In addition, they both maintain two-hop neighborhood information. In order to minimize the message overhead, DCR [17] identifies the critical nodes with one-hop neighborhood information, and the restoration is similar to DARA's.

Since cut vertex identification incurs significant overhead in terms of messaging and state maintenance, RIM [35] does not distinguish the importance of nodes. All the one-hop neighbors move towards the position of the failed node till the distance is "$R_c/2$". Other nodes perform a cascade inward movement to connect to the connected network when they cannot communicate with the moved nodes. RIM is simple and efficient, especially for sparse networks. But the performance degrades for dense networks. RIM involves too many unnecessary motions, especially the first step. As all the failed node's neighbors do not know each other, they all move to the position which is "$R_c/2$" far from the failed node to maintain neighbor connectivity even when they are within half of the communication radius. These may incur outward motions. In addition, the message overhead is very high.

The above methods mainly focus on the moving distance and message overhead. In fact, network lifetime is the most important factor which depends on energy efficiency and load balancing. Abdelmalek [4] proposed a two-phase restoration algorithm. It searches the redundant nodes using the cluster heads, then restores connectivity, and energy consumption is taken into consideration. CoRF [3] is another connectivity restoration algorithm that strives to increase network lifetime. It selects a backup node according to the fuzzy logic rules. In addition, there are some realistic connectivity restoration methods [29] that take obstacles or terrain elevation into consideration. The direct path movement may be impossible, or is not optimally energy-efficient. In summary, energy efficiency and load balancing are the important evaluation metrics for the connectivity restoration algorithms.

## 4 Hybrid connectivity restoration algorithm

In this section, a hybrid connectivity restoration (HCR) algorithm is proposed to restore connectivity in WSANs. HCR is a distributed, localized, and efficient approach aiming at minimizing the cost of moving nodes. Instead of identifying a cut vertex, HCR just identifies the critical nodes. Each node maintains a one-hop neighborhood table including unique node ID (*ID*), local position (*POSITION*) and critical character (*CRITICAL*). HCR combines proactive backup node selection and reactive cascade node motion.

### 4.1 Proactive backup node selection

In order to minimize the number of messages and shorten restoration time, each node will select a backup node from its neighbors before a node failure occurs. During the initialization of a one-hop neighborhood table, each node sends a broadcast message containing its (*ID*) and (*POSITION*). All the nodes in its communication range will receive the message.

After a round of information exchanging, each node will determine whether it is critical through its one-hop neighborhood table $NT$. A node is an uncritical node if and only if all its one-hop neighbors form a connected network. In Fig. 2, node $A3$ has three one-hop neighbors $A1$, $A2$, and $A4$. They form two disjoint components $\{A1, A2\}$ and $\{A4\}$. So $A3$ is a critical node. Similarly, $A4$, $A5$, $A6$, and $A7$ are critical nodes. While node $A8$ is an uncritical node for all its one-hop neighbors $A5$, $A7$, $A9$ form a connected network $\{A5, A7, A9\}$, so as node $A9$. Node $A10$ is a leaf node and just has one one-hop neighbor $A7$, so it is also an uncritical node. It is worth mentioning that a critical node's failure will not necessarily divide a network, e.g., $\{A5, A6, A7\}$. However, an uncritical node's failure must not break network connectivity. Therefore, a backup node needs to be selected only for a critical node.

Different from DARA [1] and DCR [17] which select a backup node based on distance and node degree, HCR selects a backup node based on the minimum moving cost. The minimum moving cost of all backup nodes should also be the failure cost of the failure node. For a critical node $i$, choose any node $j$ in its one-hop neighborhood table $NT(i)$, compute *BestPosition* $j'$ it should move to so as to connect all the rest nodes in $NT(i)$. As shown in Fig. 3, node $A3$ has three one-hop neighbors $A1$, $A2$, and $A4$ which form two disjoint components $\{A1, A2\}$, and $\{A4\}$. The best positions for $A1$, $A2$, and $A4$ are $A1'$, $A2'$, and $A4'$. They are all in the communication range of the rest sibling nodes. $A1'$ and $A2'$ are on the boundary of the communication range of $A4$. $A4'$ is the intersection of the communication boundary of $A1$ and $A2$. The moving cost for node $A1$, $A2$, and $A4$ are $A1A1'$, $A2A2'$, and $A4A4'$. The node with the minimum moving cost will be selected as the backup node for node $A3$. Here, node $A2$ is selected as the backup node and *BestPosition* is $A2'$. The best position for relocating the backup node is shown in Fig. 4 and the pseudo-code for backup node selection is detailed in Algorithm 2.

In order to find a backup node and *BestPosition* for node $i$, compute the best position $j'$ for each node $j \in NT(i)$ so that $d_{j,j'}$ is minimum and satisfies Eq. 1. $d_{j,j'}$ is the distance between nodes $j$ and $j'$, it is the moving cost of node $j$ for failure node $i$. For all $j \in NT(i)$, the minimum $d_{j,j'}$ should be the failure cost of node $i$, and node $j$ is elected as a backup node and $j'$ is *BestPosition*.

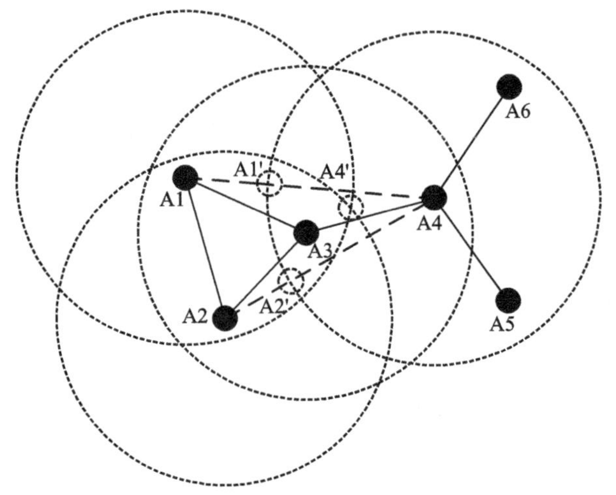


**Fig. 3** An example of backup node selection

$$\forall k \in NT(i),\ k \neq j, \quad d_{k,j'} \leq R_c \tag{1}$$

**Theorem 1** *BestPosition for node i must be the intersection of communication circle of any two nodes in NT(i) or the intersection of the line ij and communication circle of node j.*

*Proof* First, the intersection of a pair of nodes' communication boundary in $NT(i)$ must connect these pair nodes. If this intersection is within the rest sibling nodes' communication range, this node can replace the failure node to connect all the failure node's neighbors. But the intersection may not be the optimal solution. Actually, the intersection is optimal only when the moved node is beyond the communication range of the pair of nodes. As shown in Fig. 4b, node $C$ is out of the communication range of $A$ and $B$. The intersection $C_{new}$ should be the optimal position. But in Fig. 4a, node $C$ is within the communication range of $A$. Node $I$ is the intersection of $A$'s and $B$'s communication boundary. Node $I$ can connect $A$ and $B$, node $C_{new}$ can also connect $A$ and $B$, and $d_{C,C_{new}} < d_{C,I}$. $d_{C,C_{new}} + d_{C_{new},B} < d_{C,I} + d_{I,B}$ for Line Axiom, where $d_{C_{new},B} = d_{I,B} = R_c$. Thus $I$ is not the optimal position, but node $C_{new}$ is. □

**Lemma 1** *The node failure cost is no more than the nearest-neighbor distance in HCR.*

*Proof* By Theorem 1, *BestPosition* must be the intersection of any pair of nodes' communication boundary in $NT(i)$ or the intersection of the two nodes connection line and one node's communication boundary. As shown in Fig. 4, node $C_{new}$ is *BestPosition*. In Fig. 4a, $\angle CC_{new}F$ must be an obtuse angle; otherwise, $FC_{new}$ will be the tangent line of circle $B$, and node $F$ will be out of the circle. As $\angle CC_{new}F$ is an obtuse angle, $d_{C,C_{NEW}} \leq d_{C,F}$, so as Fig. 4b. Since each node's best position moving cost is less than its distance to the failure node, the node failure cost is no more than the nearest-neighbor distance. □

The previous approaches such as DARA and DCR all move the backup node to the position of the failure node, then the failure cost is equal to the nearest-neighbor distance in the best case. So HCR outperforms DARA and DCR in terms of travel distance in a motion. The shorter distance it moves, the smaller impacts on its neighbors' connectivity.

If there are two or more neighbor nodes with the same moving costs to restore the network connectivity, the one with the smallest failure cost will be selected as the backup node for the next restoration loop with the smallest moving cost. After selecting the backup node and obtaining *BestPosition*, it will send a broadcast message containing the information of the backup node and *BestPosition*. After a round of broadcasting, each node will update its backup node and *BestPosition* again and inform the changes to its new backup node only. By now, proactive backup node selection is finished. Then the node starts to

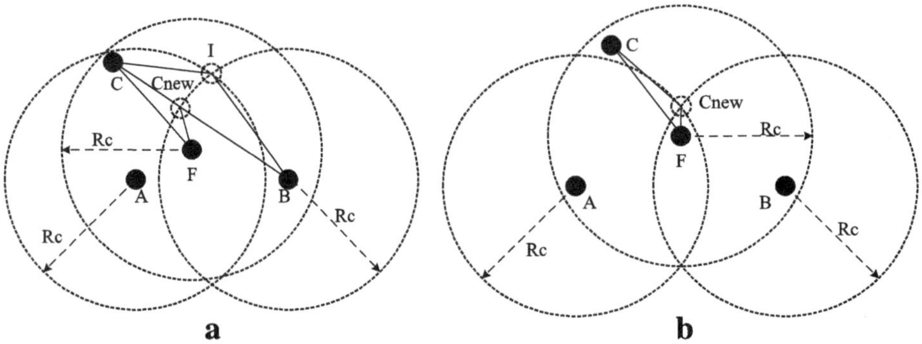


**Fig. 4** Best position for relocating the backup node. **a** Node C is out of the communication range of *A* and *B*. **b** Node *C* is within the communication range of *A* and out of *B*

send heartbeat messages to the backup node periodically to declare that it is functional. Once the backup node does not receive the heartbeat message within a period, it will start reactive cascade node motion to restore connectivity.

### 4.2 Reactive cascade node motion

As mentioned before, only a critical node will break network connectivity, so when a critical node $i$ fails, its backup node $j$ will detect the failure the first time and trigger reactive cascading node motion. Before the backup node $j$ moves to *BestPosition* $j'$, it will update its one-hop neighborhood table $NT(j)$. It replaces the failure node's position by $j'$, and checks whether its one-hop neighbors' connectivity is broken. If broken, node $j$ will be the failure node and the restoration will be triggered. Node $j$ will update its backup node $k$ and *BestPosition* $k'$. Node $k$ is selected from its rest neighbors except for the failed node $i$ in order to avoid falling into an infinite loop. The backup node $k$ will move to *BestPosition* $k'$ to connect all its neighbors in $NT(j)$ and $j'$. After selecting the backup node $k$ and *BestPosition* $k'$, node $j$ will send a message to backup node $k$ about *BestPosition* $k'$ it should move to and this will trigger a new round motion of node $k$. Node $j$ will move to $j'$ to replace the failure node $i$. This process will be repeated recursively until the failure node's neighbors are connected.

Figure 5 is an example for HCR cascading node motion. In Fig. 5a, node $A3$ fails. Its three one-hop neighbors $A1, A2$, and $A4$ become three isolated nodes and they are out of each other's communication range. For proactive backup node selection, node $A4$ is selected to move towards $A4'$ to connect $A1$ and $A2$. Due to $A4$'s movement, its neighbor $A6$ is out of the communication range, while $A5$ is still within the communication range in Fig. 5b. Before moving to $A4'$, node $A4$ selects a backup node to replace it and node $A6$ is selected. In Fig. 5c, node $A6$ moves to $A6'$ to connect node $A4'$ and node $A5$. The movement does not break its connection to node $A7$, and the whole network is connected. It is worth mentioning that $A4'$ is the intersection of $A1$'s and $A2$'s communication circles, while $A6'$ is the intersection of line $A4'A6$ and $A5$'s communication circle.

**Algorithm 1** HCR

*actor_coord*: the coordinate of actors
$R_C$: the communication range for each actor
Initialize 1-hop neighborhood table *NEIGH_TABLE* for each actor;
**for** every actor $i$ in a WSAN **do**
  **if** Is_Critical(i) **then**
    **Algorithm Backup Node Selection**
  **else**
    *back_ID* = 0; *BestPosition* =[*inf*, *inf*]; *failed_cost* = 0
  **end if**
**end for**
Each actor broadcasts a message about its failed cost *failed_cost*
**for** every actor $i$ **do**
  **if** Is_Critical($i$) and actor $i$ has two neighbors **then**
    the actor with smaller *failed_cost* is selected as a backup node for $i$
  **end if**
**end for**
Every actor sends a message to notify its backup node to monitor its status and *BestPosition*
**while** an actor node $B$ detects a failure of its neighbor $F$ OR receives a movement message **do**
  node $B$ adds *BestPosition* into its *NEIGH_TABLE*
  **while** Is_Critical($B$) **do**
    **Algorithm Backup Node Selection**
    node $B$ sends a movement message to its new backup node about *BestPosition*
    node $B$ moves to *BestPosition* to replace the failed node $F$
  **end while**
**end while**

The pseudo-code for HCR is shown in Algorithm 1. HCR is a hybrid method that selects a backup node for each actor before it fails. At the beginning of network construction, each node will broadcast a message to notify its position within its communication range, and records

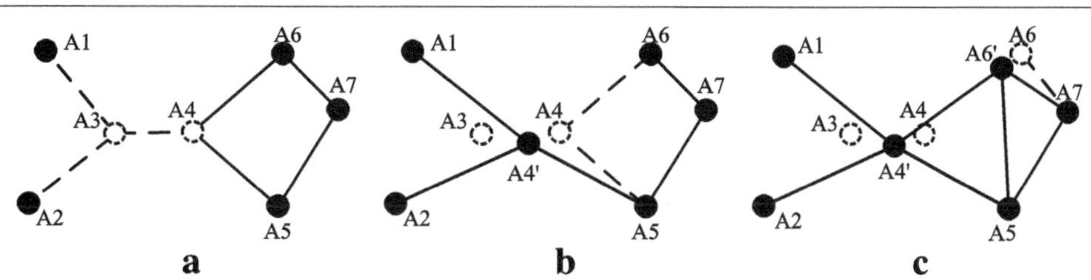


**Fig. 5** An example HCR cascading node motion. **a** Node *A3* failed and node *A4* is selected to move. **b** Node *A4* moves to *A4'*. **c** Node *A6* moves to *A6'*

its neighbors' *ID* and *Position* to build one-hop neighborhood table *NEIGH_TABLE* (line 3). Then each node will identify whether it is critical through its *NEIGH_TABLE*. For an uncritical node, its failure will not break connectivity, so *failed_cost* = 0 (line 8). While for a critical node, it will select a backup node and *BestPosition* (line 6). The algorithm for backup selection is detailed in Algorithm 2. When a node just has two neighbors, move any one towards another has the same travel distance to restore connectivity. For this case, the one with smaller *failed_cost* is chosen (line 11-16), this is to insure the next restoration has a smaller moving cost. By now, the initialization and proactive backup selection are done. Every actor sends a message to notify its backup node to monitor its status and *BestPosition*. When a node detects a failure of its neighbors, it will trigger the cascading node motion (line 18-25). The backup node will add *BestPosition* into its *NEIGH_TABLE* to tell whether it is critical. If so, it will select its new backup node before moving to replace the failed node (line 21). Otherwise, network connectivity is restored.

**Algorithm 2** Backup Node Selection

1: $R_C$: the communication range for each actor
2: *NEIGH_TABLE*: the failed node's one-hop neighborhood table
3: *F*: the failed node
4: $F'$: *BestPosition* the failed node moved to
5: **for** every actor *i* except $F'$ in *NEIGH_TABLE* **do**
6:     **for** every actor *j* in *NEIGH_TABLE* **do**
7:         Compute the intersection of circle *j* and line $L_{ij}$
8:         Add the intersection into *Candidate_Set*
9:         **for** every actor *k* in *NEIGH_TABLE* and $k \neq j$ **do**
10:             Compute the intersection of circle *j* and circle *k*, where the radius is $R_c$
11:             Add the intersection into *Candidate_Set*
12:         **end for**
13:     **end for**
14:     **for** every node *m* in *Candidate_Set* **do**
15:         **for** every actor *n* in *NEIGH_TABLE* **do**
16:             **if** $dm, n > R_c$ **then**
17:                 remove the node *m* from *Candidate_Set*
18:                 **break;**
19:             **end if**
20:         **end for**
21:     **end for**
22:     Choose the closest point in *Candidate_Set* from node *i* as its *BestPosition*
23: **end for**
24: Return the node's *ID* and *BestPosition* which has the minimum *travel_dist* in *NEIGH_TABLE*

The backup node selection is the key part of HCR and the pseudo-code is presented in Algorithm 2. It is worth mentioning that all the selection of backup node is done on the failed node or moved node and it just explores *NEIGH_TABLE*. During this process, it does not need to send message to other nodes. This reduces the message overhead. According to Theorem 1, *BestPosition* must be the intersection of any pair of nodes' communication boundary or the intersection of the two nodes connection line and one node's communication boundary. Firstly, for each node *i* in *NEIGH_TABLE*, compute its *Candidate_Set* (line 6-13). Then remove the position that is out of the rest nodes' communication range (line 14-21). Afterwards, choose the closest point in *Candidate_Set* as *BestPosition* for node *i*. Finally, return the node's *ID* and *BestPosition* which has the minimum *travel_dist* in *NEIGH_TABLE*.

## 5 Algorithm analysis

HCR combines proactive and reactive methods to handle network connectivity restoration from a single-node failure in WSANs. The selection of a backup node is proactive, while the restoration is reactive. This scheme shortens the restoration process and reduces the overhead including distance cost and message cost. Next, the performance of HCR is analyzed.

First and foremost, network connectivity after a single-node failure is restored. It is assumed that no other node fails during the restoration process and any two nodes can communicate with each other directly if they are $R_c$ apart or closer. Then network connectivity is not weakened and no new critical node is introduced during the restoration. In addition, load balancing is taken into consideration and no node travels too far while others too close. The overhead of communication and the complexity of computing are also analyzed. We introduce the following theorems.

**Theorem 2** *HCR restores network connectivity after a single-node failure.*

*Proof* Uncritical node failure will not break network connectivity since all its neighbors are connected when it is removed from the network. HCR identifies a critical node at the initialization time and selects a backup node and decides *BestPosition*. When a critical node fails, it will trigger the restoration, and its backup node moves to *BestPosition* to reconnect all its one-hop neighbors. In the following cascading node motion, the moved node will reconnect its sibling nodes until all its siblings are connected. In order to avoid an endless loop, each node can only move once during the restoration, and a moved node will not be selected as a backup node in the future. This can guarantee that HCR terminates in limited number of steps. □

**Theorem 3** *The total number of critical nodes does not increase.*

*Proof* In HCR, the selection of a backup node and *BestPosition* guarantees that the total number of critical nodes does not increase. The critical of backup node will change into that of the failed node. And the degree of each node will not decrease. The backup node will move to *BestPosition* to connect all its siblings. In Fig. 6, node $A3$ fails, and it selects a neighbor to reconnect all its neighbors. In Fig. 6a, node $A1$ is selected and moves to node $A1'$ where it is on the line $A1A4$, and $d_{A1',A4} = R_c$, $d_{A1',A2} < R_c$. Since $A1$ and $A2$ can connect before failure, we just need to guarantee that node $A1'$ is in the communication range of $A2$. In Fig. 6b, node $A4$ is selected and moves to node $A4'$ where it is the intersection of circle $A1$ and $A2$, $d_{A1,A4'} = R_c$, and $d_{A2,A4'} = R_c$. Node $A4$ is out of reach by $A1$ and $A2$ before the failure. After restoration, the unmoved critical nodes are not changed, e.g., $A2$ and $A4$ in Fig. 6a, and the moved uncritical node becomes a critical node or not. Node $A1'$ is critical, where node $A4'$ is uncritical. Before the restoration, the failed node is critical, so the total number of critical nodes does not increase. □

**Theorem 4** *The maximum distance a node travels in HCR is the communication range $R_c$.*

*Proof* In HCR, a backup node is one of the failed node's neighbors, and it moves to *BestPosition* to reconnect all its siblings. In the worst case, a backup node moves to the position of the failed node, and it must connect all its siblings. That is the relocated scheme in DCR [17]. It has been proven that the maximum distance a node travels in DCR is the communication range $R_c$. Lemma 1 proves that HCR outperforms DCR in terms of travel distance in one motion.

Take Fig. 7 as an example. Node $A3$ is the failed node, and node $A4$ moves to $A4'$ to connect $A1$ and $A2$, and $d_{A4,A4'} < d_{A4,A3} < R_c$. Due to the motion of node $A4$, node $A5$, and node $A6$ will be out of reach by $A4'$, so before moving to $A4'$, $A4$ will select a node from its one-hop neighbors $\{A5, A6\}$. Suppose $A6$ is selected and *BestPosition* is $A6'$, then $d_{A6,A6'} < d_{A6,A4} < R_c$. □

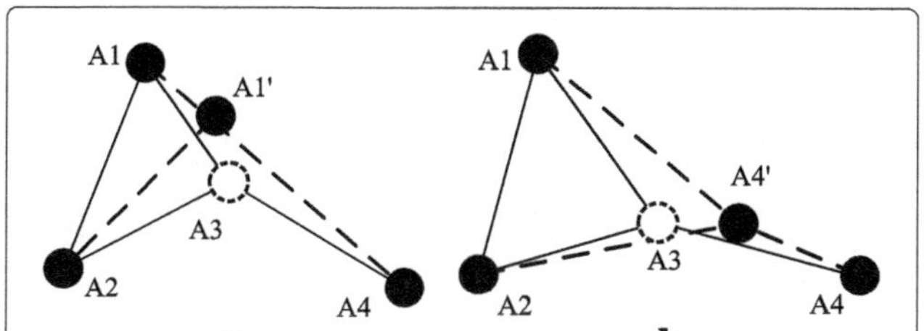

**Fig. 6** No new critical node is introduced during the restoration. **a** Node *A1* is selected to move. **b** Node *A4* is selected to move

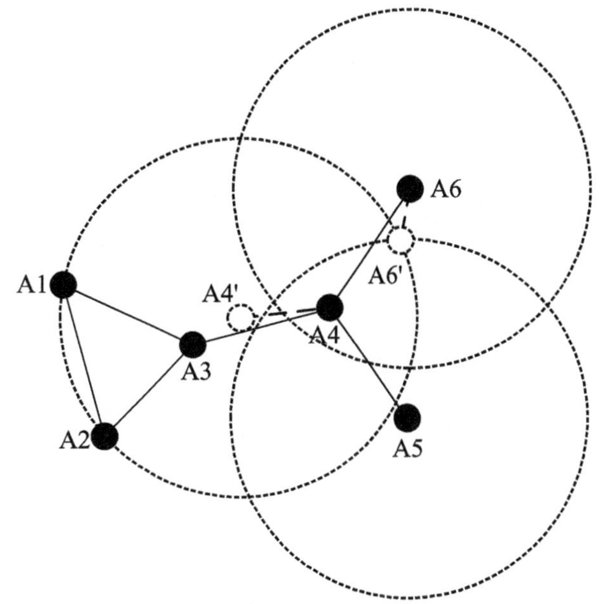

**Fig. 7** The maximum distance a node travels in HCR

**Theorem 5** *The shorter distance a node travels, the fewer nodes are affected. The probability of a node affected by its moving neighbor is* $(1 - \frac{2\theta - \sin 2\theta}{\pi})$, *where* $\theta = \arccos d/2R_c$, *$d$ is the travel distance, and $R_c$ is the communication range. It approximately equals 0.62 times of $d/R_c$.*

*Proof* Take Fig. 8a as an example. Node $A$ moves to $A'$. The nodes in the shaded area like node $B$ are within the communication range of node $A$, while out of reach by $A'$. The travel distance equals $d$. It is assumed that the nodes are deployed randomly, so a node has an equal chance to reside at any place. The probability of a node affected by its moving neighbor is the probability that it is located in the shaded area. Therefore, the probability is the shaded part area divided by the whole communication area. According to symmetry and area formula of a sector, the shaded part area equals $\pi R_c^2 - (2\theta \cdot R_c^2 - \sin 2\theta \cdot R_c^2)$, where $\theta = \arccos d/2R_c$. So the probability of a node affected by its moving neighbor is $(1 - \frac{2\theta - \sin 2\theta}{\pi})$. In Fig. 8b, the solid line reflects the probability against $d/R_c$, where the dashed line reflects the growth rate of probability. It shows that the growth rate is approximately 0.62. □

**Theorem 6** *The time complexity of the backup node and BestPosition selection is $O(n^3)$, where $n$ is the number of failed node's one-hop neighbors.*

*Proof* It has been proven that *BestPosition* must be the intersection of any pair of nodes' communication boundary in $NT(i)$ or the intersection of two nodes connection line and one node's communication boundary in

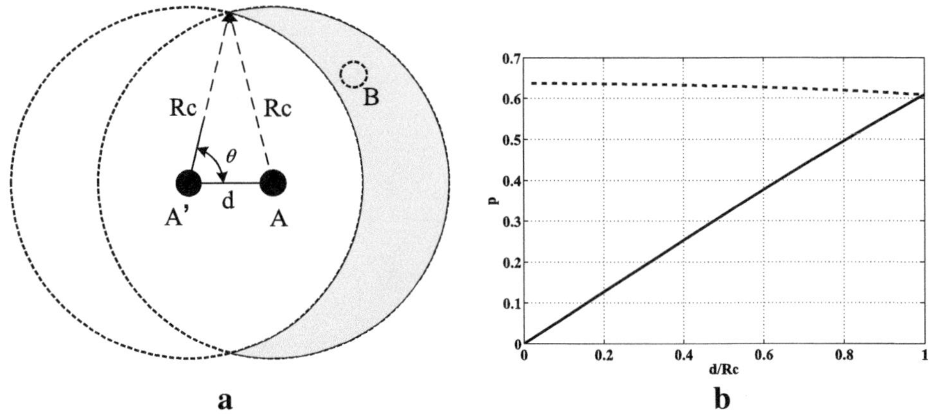


**Fig. 8** The influence of node moving distance. **a** An example of the influence of node moving. **b** The influenced probability of a node against *d/Rc*

Theorem 1. For each node, the intersection of any pair of its sibling nodes' circle is computed firstly, and the time complexity is $O(n^2)$. Then the intersection of a line that connects this node to its sibling node and the sibling node circle is obtained, and the time complexity is $O(n)$. Finally, check whether the intersection is within the communication range of all the sibling nodes and find the minimum travel distance. The time complexity is also $O\left(n^2\right)$. Therefore, the time complexity of the best position selection for each node is $O\left(n^2+n+n^2\right)$, that is $O\left(n^2\right)$. Then the node with the minimum traveling distance will be selected as the backup node, and its best position will be *BestPosition*. Hence, the time complexity of the backup node and *BestPosition* selection is $O\left(n^3\right)$. □

**Theorem 7** *The total message complexity of HCR is O(N), where N is the number of actors.*

*Proof* The selection of a backup node and *BestPosition* is done at the failed or moved nodes in HCR. It just maintains one-hop neighborhood table for each node. In addition, the failed node only sends a message to its backup node about its movement. In the worst case, there are $N-2$ nodes moving. Each moved node sends a movement message to its backup node. Therefore, the total message complexity of HCR is $O(N)$, where $N$ is the number of actors. It is worth noting that the exchange with neighbors at a new position does not count in HCR, and it is considered as a part of status update for maintaining one-hop neighborhood table. □

**Theorem 8** *The time it takes HCR to restore network connectivity is proportional to N and $R_c$, where N is the number of actors and $R_c$ is the communication range.*

*Proof* Firstly, HCR proactively selects a backup node and *BestPosition* before a node failure, so when a node fails, the backup node will move to *BestPosition*. This will trigger cascading motion. For each moved node, it will select its new backup node and *BestPosition* before its moving and send a message to notify its backup node to move to *BestPosition*. According to Theorem 7, the time complexity is $O\left(n^3\right)$ where $n$ is the number of its neighbors. Usually, $n$ is very small compared with the total number of nodes $N$, so the computing time can be ignored. In the worst case, there are $N-2$ nodes moving, and each node moves at most $R_c$, so the total time it takes HCR to restore network connectivity is $(N-2)\times R_c$, which does not exceed $(N\times R_c)$. □

## 6 Simulation results

Extensive simulations have been conducted to evaluate the performance of the proposed HCR compared with the previous algorithms RIM and DCR. The simulation settings and performance metrics are introduced in Section 6.1, and the detailed results and analysis are presented in Section 6.2.

### 6.1 Simulation settings and performance metrics

In order to evaluate the performance of HCR and compare it with the previous algorithms RIM [35] and DCR [17] fairly, we carry out all the simulations on Matlab R2012a with an Intel Core i3-3220 CPU and 8 G RAM computer. In the simulations, numerous mobile actors are deployed randomly in an area of 1000 m×1000 m. All the actors are homogeneous with the same communication range. The energy cost and lifespan are pursued during the connectivity restoration in varied applications, but the actual energy cost is difficult to model and capture during simulations, so the following four metrics are employed to evaluate the performance of HCR:

- *Number of relocated nodes*: It reports the average number of relocated nodes during a single-node failure restoration. This metric assesses the scope of connectivity restoration within a network.

- *Total distance moved*: It reports the total distance that the involved nodes move during the restoration. This metric assesses the efficiency of the restoration methods from the standpoint of a network.
- *Average travel distance*: It depicts the average distance that the involved nodes travel during the restoration. This metric assesses the efficiency standing in the perspective of a node.
- *Number of messages*: It captures the total number of messages sent among the nodes during the restoration. This metric assesses the communication overhead of the restoration methods.

In addition, to study the impact of network topology on the performance of HCR, the network topology is varied in different simulations with the following parameters:

- *Number of deployed nodes (DN)*: It reports the number of the deployed nodes in an area. Since the area is 1000 m×1000 m, this parameter actually represents the density of a WSAN. The larger the *DN*, the bigger the node density, indicating a stronger network connectivity. In a rather highly connected WSAN, a node has many one-hop neighbors, and it is easier to choose a backup node and *BestPosition*. It will increase the time of selecting at the same time.
- *Communication range* ($R_c$): It is assumed that all the actors are homogeneous with the same communication range and a pair of nodes can communicate with each other when they are within each other's communication range. $R_c$ also affects network density *DN*. Small $R_c$ will generate a rather sparse network, while large $R_c$ will increase network connectivity. This will also increase the travel distance of the involved nodes during the restoration under HCR.

### 6.2 The overall results

Both sparse and dense networks with different values of *DN* and $R_c$ are simulated to evaluate the performance of HCR compared with RIM [35] and DCR [17]. The value of *DN* is chosen from set $\{10, 20, 40, 60, 80, 100\}$ and the value of $R_c$ is chosen from set $\{20, 40, 60, 80, 100, 120\}$. For every topology, 10 WSANs are generated, and for every WSAN, 30% of the deployed nodes are randomly selected forming the failed node set. For every WSAN, RIM, DCR, and HCR are run after a node fails and the average values are computed. The overall results are shown in Table 1. We can see that HCR outperforms RIM and DCR in terms of the number of relocated nodes, total travel distance, average travel distance, and number of messages for both high-density and low-density networks. The results and analysis are detailed in the following.

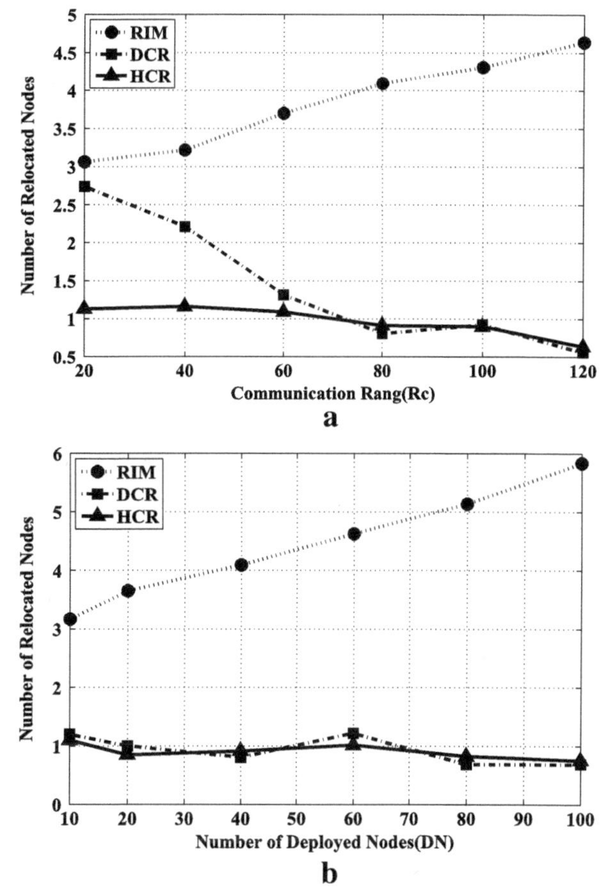


**Fig. 9** The number of relocated nodes during the restoration while varying **a** the communication range (with $DN = 40$) and **b** the network size (with $R_c = 80$)

#### 6.2.1 *Number of relocated nodes*

Figure 9a, b reports the number of the relocated nodes during the restoration under various communication range $R_c$ and network size *DN*, respectively. The plotted results are the average over multiple independent simulations. For any topology, 30% of *DN* fail randomly. The two figures indicate that DCR and HCR relocate fewer nodes than RIM. This is because RIM requires all the neighbors of the failed node to move inward. In fact, sometimes they may move outward to the position that is $R_c/2$ away from the failed node. In addition, the number of the relocated nodes grows with the increasing of $R_c$ and *DN* since the larger $R_c$ and *DN*, the more neighbors a failed node has. DCR and HCR have the same results when $R_c$ and *DN* increase, because they both identify critical nodes, and only move one of the neighbors to replace the failed node in a round of restoration. It is worth noting that the candidate selection is processed at the failed node before it fails or moves, so they just maintain one-hop neighborhood table.

**Table 1** Overall simulation results

| DN | RC | Number of relocated nodes | | | Total moved distance | | | Number of messages | | | Average travel distance | | |
|---|---|---|---|---|---|---|---|---|---|---|---|---|---|
| | | RIM | DCR | HCR | RIM | DCR | HCR | RIM | DCR | HCR | RIM | DCR | HCR |
| 10 | 20 | 2.5000 | 0.8667 | 0.7667 | 9.4481 | 11.7579 | 4.0453 | 11.2667 | 0.8000 | 0.6000 | 3.7173 | 6.4224 | 2.3910 |
| | 40 | 2.5333 | 1.1667 | 1.2000 | 22.3878 | 34.1477 | 14.3315 | 10.3333 | 1.1333 | 1.2000 | 8.5230 | 16.8824 | 6.6039 |
| | 60 | 2.7000 | 1.2667 | 1.2000 | 34.3570 | 54.9098 | 20.9049 | 11.3333 | 1.2667 | 1.1333 | 11.8734 | 27.7736 | 11.4697 |
| | 80 | 3.1667 | 1.2000 | 1.1000 | 54.5616 | 68.3552 | 25.6540 | 15.0000 | 1.2667 | 1.0667 | 16.9988 | 30.8971 | 12.3777 |
| | 100 | 2.8667 | 0.5667 | 0.8667 | 66.9697 | 41.0099 | 26.1766 | 13.0000 | 0.2000 | 0.8000 | 21.3045 | 33.2997 | 13.1408 |
| | 120 | 2.7333 | 0.9333 | 1.0333 | 71.1571 | 80.9414 | 35.6322 | 12.6000 | 0.8000 | 1.0000 | 27.5052 | 44.9467 | 17.3013 |
| 20 | 20 | 2.6667 | 1.1500 | 1.0167 | 10.1203 | 15.8547 | 5.7045 | 12.9334 | 1.1333 | 0.8667 | 3.6459 | 8.0768 | 3.2332 |
| | 40 | 2.9500 | 2.2167 | 1.2000 | 24.3734 | 64.1113 | 14.8084 | 13.3000 | 3.0000 | 0.9667 | 8.0881 | 20.7230 | 8.5739 |
| | 60 | 3.0333 | 1.8667 | 1.2000 | 38.7232 | 78.6362 | 22.8659 | 14.4000 | 2.4000 | 1.0667 | 11.9809 | 27.1874 | 12.7959 |
| | 80 | 3.6500 | 1.0000 | 0.8500 | 60.9370 | 57.9764 | 19.7107 | 20.6000 | 0.9000 | 0.6000 | 15.7861 | 31.1569 | 13.0692 |
| | 100 | 3.7833 | 0.6833 | 0.8000 | 80.4082 | 47.1660 | 22.1815 | 21.8000 | 0.4000 | 0.6333 | 21.5298 | 33.2313 | 13.5711 |
| | 120 | 3.7667 | 0.8833 | 0.8000 | 101.2945 | 72.8401 | 21.3106 | 22.7333 | 0.7333 | 0.5667 | 25.8660 | 42.7639 | 13.5360 |
| 40 | 20 | 3.0615 | 2.7385 | 1.1308 | 11.9039 | 37.9415 | 6.3112 | 15.3077 | 4.0923 | 0.8769 | 3.7377 | 9.4159 | 3.6941 |
| | 40 | 3.2154 | 2.2077 | 1.1615 | 25.6699 | 61.7150 | 11.2814 | 16.4154 | 3.0308 | 0.9385 | 7.7238 | 19.2907 | 6.6423 |
| | 60 | 3.7000 | 1.3154 | 1.0923 | 49.2846 | 58.5885 | 20.8707 | 20.4000 | 1.4154 | 0.9692 | 13.1892 | 26.9827 | 10.9955 |
| | 80 | 4.0923 | 0.8077 | 0.9154 | 69.0521 | 42.6712 | 18.4933 | 25.7077 | 0.5077 | 0.7231 | 16.7857 | 28.9037 | 10.8412 |
| | 100 | 4.3077 | 0.9231 | 0.9000 | 91.6229 | 61.6710 | 19.9848 | 27.9231 | 0.7846 | 0.7385 | 20.0957 | 33.9756 | 11.3504 |
| | 120 | 4.6385 | 0.5615 | 0.6385 | 113.5523 | 45.2215 | 20.8300 | 35.2000 | 0.3385 | 0.4923 | 24.4107 | 30.6194 | 12.0742 |
| 60 | 20 | 3.4300 | 3.4300 | 1.4600 | 13.7143 | 47.4912 | 8.3110 | 17.1500 | 5.1700 | 1.2300 | 3.9234 | 11.2197 | 4.6324 |
| | 40 | 3.7250 | 2.2900 | 1.2100 | 30.3220 | 66.0921 | 12.9398 | 20.1000 | 3.1700 | 1.0100 | 7.9336 | 20.0784 | 7.1632 |
| | 60 | 4.5600 | 1.3700 | 1.1050 | 59.5532 | 59.2711 | 17.5455 | 28.9300 | 1.5200 | 0.9900 | 12.7306 | 25.7413 | 9.1642 |
| | 80 | 4.6250 | 1.2200 | 1.0150 | 76.5166 | 67.7799 | 21.1872 | 30.6400 | 1.2200 | 0.8100 | 16.7286 | 32.0698 | 12.1174 |
| | 100 | 5.7300 | 0.5450 | 0.6150 | 117.2685 | 34.8617 | 14.5538 | 44.9600 | 0.2800 | 0.4200 | 20.5859 | 24.7201 | 9.4990 |
| | 120 | 6.8150 | 0.4950 | 0.5450 | 167.6632 | 39.0700 | 16.8272 | 67.3100 | 0.2500 | 0.3500 | 24.6528 | 28.8812 | 11.4855 |
| 80 | 20 | 3.0808 | 4.1846 | 1.3654 | 11.8660 | 59.3107 | 7.6236 | 15.0539 | 6.8077 | 1.1692 | 3.8286 | 10.6448 | 4.0656 |
| | 40 | 3.7692 | 1.3269 | 1.1769 | 31.2300 | 37.8081 | 13.7661 | 20.8461 | 1.3538 | 1.0538 | 8.0497 | 18.1350 | 7.2869 |
| | 60 | 4.3308 | 0.9231 | 0.8731 | 53.2154 | 36.9789 | 13.5749 | 29.1923 | 0.8000 | 0.7000 | 12.1273 | 20.0345 | 7.9371 |
| | 80 | 5.1346 | 0.6885 | 0.8231 | 86.9678 | 36.2184 | 15.8676 | 38.2769 | 0.3308 | 0.6000 | 17.1423 | 26.8037 | 9.7161 |
| | 100 | 5.7808 | 0.6192 | 0.7385 | 118.1022 | 41.5090 | 15.8088 | 46.8615 | 0.3385 | 0.5769 | 20.1898 | 29.5200 | 9.1799 |
| | 120 | 6.3769 | 0.5077 | 0.5846 | 155.5264 | 38.9712 | 14.9357 | 58.1000 | 0.2923 | 0.4462 | 23.9728 | 27.1466 | 9.5740 |
| 100 | 20 | 3.3333 | 2.4485 | 1.2576 | 13.0345 | 33.3660 | 6.4384 | 17.0909 | 3.3879 | 1.0061 | 3.7086 | 10.0852 | 3.6431 |
| | 40 | 3.9000 | 1.4879 | 1.0394 | 32.7427 | 41.4879 | 11.4844 | 22.4606 | 1.7030 | 0.8061 | 8.2931 | 17.3802 | 6.9119 |
| | 60 | 5.0667 | 0.8970 | 0.8152 | 63.4889 | 37.4344 | 13.0444 | 35.6909 | 0.8000 | 0.6364 | 12.2978 | 20.0826 | 7.8285 |
| | 80 | 5.8394 | 0.6909 | 0.7545 | 98.9894 | 36.4659 | 14.7784 | 48.1454 | 0.4121 | 0.5394 | 16.8274 | 24.5749 | 9.2227 |
| | 100 | 6.8515 | 0.6061 | 0.6970 | 140.9332 | 39.3920 | 15.3282 | 66.2667 | 0.3697 | 0.5515 | 20.7653 | 26.3370 | 8.7655 |
| | 120 | 8.4636 | 0.4000 | 0.5182 | 201.1593 | 29.5398 | 13.5253 | 92.2545 | 0.1333 | 0.3697 | 23.6561 | 23.8265 | 8.8305 |

Moreover, Fig. 9a shows that HCR outperforms DCR when the communication range is less than 72, and has the similar results when the communication range grows. This is because of the candidate selection. HCR will choose a node to move the least distance to *BestPosition* and to reconnect the failed node's neighbors, while DCR chooses the backup node by critical, degree, and distance. Furthermore, DCR moves the backup node to the failed node, and the failed node is not at *BestPosition* most of the time. When $R_c$ is small, the network is sparse, and the choice of a backup node is too limited, so HCD outperforms DCR. When $R_c$ is large, the network is dense, and it is likely to choose an uncritical node to restore network connectivity, so DCR can get the same result with HCR. Figure 9b shows that *DN* has little impact on HCR and DCR in terms of relocated nodes when $R_c = 80$. At the moment, the WSAN is dense for the communication range. Considering Fig. 9a, b, we can see that $R_c$ is more influential than *DN* on the number of relocated nodes.

### 6.2.2 Total moved distance

The total moved distance is an important metric for network connectivity restoration, since the actors are energy-limited while moving is an energy-consuming operation. As shown in Fig. 10a, b, HCR outperforms RIM and HCR for any $R_c$ and *DN*. As can be seen in Fig. 10a, b, RIM grows when $R_c$ or *DN* increases, and it increases linearly. This is terrible and unacceptable when the network scale is large. While the curve of DCR is unstable, it outperforms RIM when the network is dense. In the simulations, when $DN = 40$ and $R_c > 62$, DCR outperforms RIM. Similarly, when $R_c = 80$ and $DN > 20$, DCR outperforms RIM. At the same time, HCR remains stable when varying $R_c$ or *DN*.

In Fig. 10a, the curve of HCR rises slowly when $R_c < 60$, then it remains stable. It decreases with the increase of *DN* in Fig. 10b. When the communication range is small, the network is sparse and the failed node has limited neighbors to select, so HCR needs many nodes to move to restore network connectivity. Since each node just needs to move a little, the growth is slow and small. While when the communication range increases, many choices make a rapid convergence. It is the same for DCR. However, DCR moves the backup node to the failed node, while HCR moves the backup node to *BestPosition*. It has been proven in Lemma 1 and the simulation results also verify that HCR outperforms DCR in terms of total moved distance. Figure 10b shows a decreasing of the total moved distance for HCR when increasing *DN*. This is because when $R_c$ is determined, the growing of *DN* will increase the choices, and moving a shorter distance is enough to restore network connectivity.

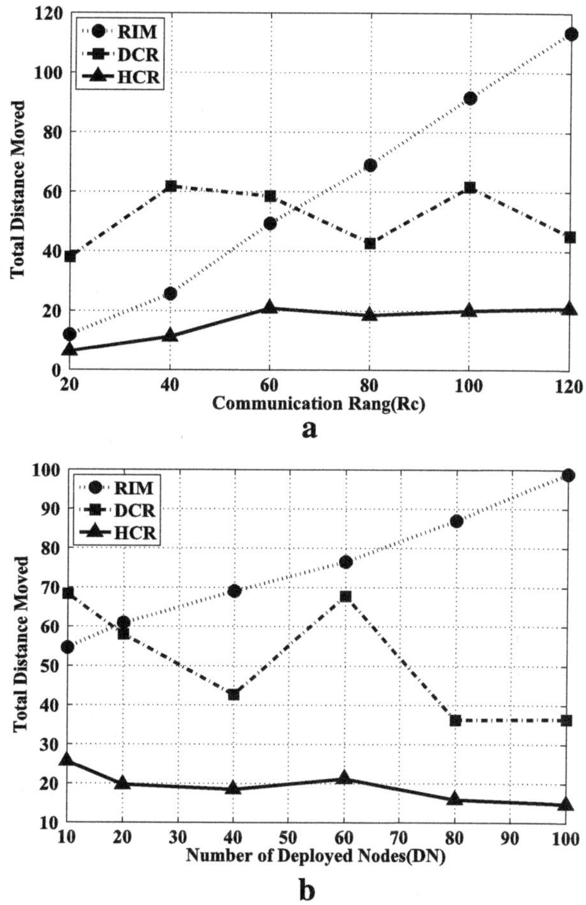


**Fig. 10** The total moved distance during the restoration while varying **a** the communication range (with $DN = 40$), and **b** the network size (with $R_c = 80$)

### 6.2.3 Average travel distance

As aforementioned, node motion is a high energy-consuming operation, and nodes are prone to be out of work due to energy depletion. A node failure will incur cascading motion and more energy consumptions creating a vicious spiral. So the average travel distance of the involved nodes is of great importance in assessing the connectivity restoration algorithms.

Figure 11a shows that the curves of DCR and RIM grow with the increase of $R_c$ while $DN = 40$. HCR also grows until $R_c = 60$, then it maintains stable. RIM outperforms DCR since it moves more nodes, and the biggest distance of RIM is limited to $R_c/2$, while DCR moves the backup node to the position of the failed node, and each involved node moves more than that in RIM. But they both increase linearly, which is unbearable in large-scale WSANs. Since the backup node moves to *BestPosition* in HCR, the involved nodes will move less than DCR, and when the network is dense, it is stable to move some distance to restore connectivity. As shown in Fig. 11a, when

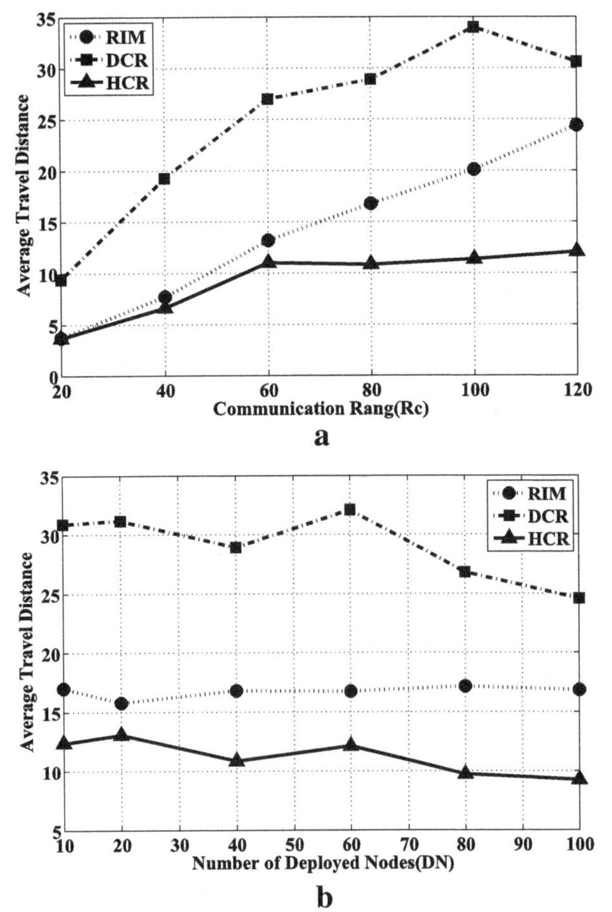


**Fig. 11** The average distance a node travels during the restoration while varying **a** the communication range (with $DN = 40$), and **b** the network size (with $R_c = 80$)

$R_c > 60$, the average distance is around 11. RIM obtains the same results as HCR when the communication range is small.

Figure 11b shows very different results from Fig. 11a. The average travel distance for each involved node almost remains unchanged with varied network size when $R_c$ is 80. This indicates that network size does not influence the average travel distance because each node moves at most $R_c/2$ in RIM, while $R_c$ is fixed to be 80. Both the curves of HCR and DCR decrease when increasing the deployed nodes with $R_c = 80$. The WSAN is dense, thus the failed node has more neighbors which increases the probability of choosing the best replacement position of nodes. Due to moving the backup node to the position of the failed node, HCR moves less than DCR.

### *6.2.4 Number of messages*

Communication cost is another important metric to assess the connectivity restoration methods because communication consumes large bandwidth resource which is very limited in WSANs. Figure 12a, b reports the total number of messages that need to be sent in restoring connectivity corporately. RIM incurs the highest message overhead since the decision of restoration is made by the failed node's neighbors in RIM. It needs to send the new position to its sibling nodes, so the communication cost is very lager. It becomes much worse when the density of a network increases because the number of neighbors grows.

Figure 12a, b indicates that HCR and DCR achieve similar message overhead when the network is dense. This is because the decision of restoration is made by the failed node. The communication is maintained just between the failed node and its backup node. Figure 12b also shows that the number of deployed nodes will influence the messages when the communication range equals 80. It is also true for other communication range, and the detailed results are shown in Table 1. It should be noted that HCR outperforms DCR when the communication range is small. In Fig. 12a, the curve of HCR is below that

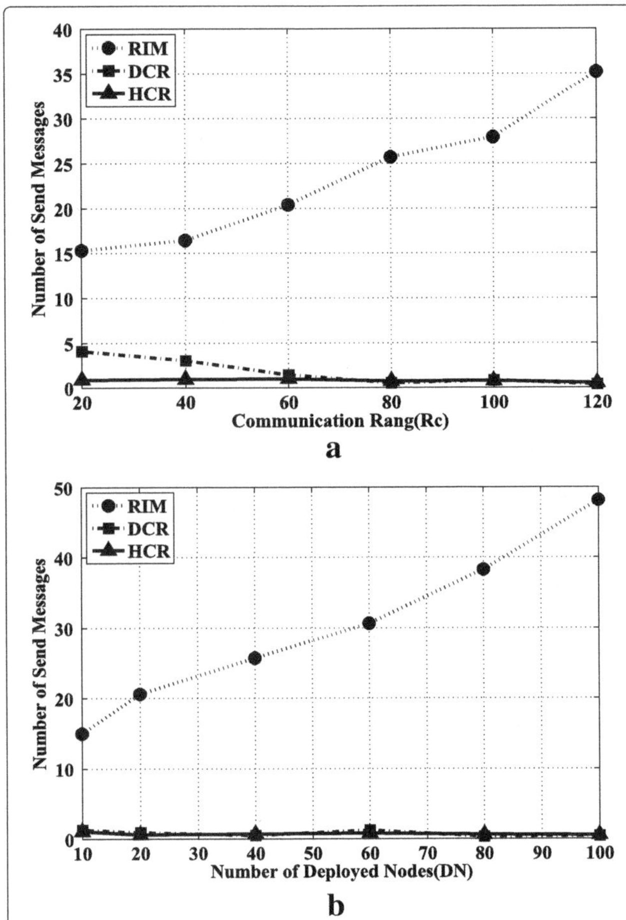


**Fig. 12** The total number of messages during the restoration while varying **a** the communication range (with $DN = 40$) and **b** the network size (with $R_c = 80$)

of DCR when the communication range is less than 60. That means HCR outperforms DCR when the network is sparse. It contributes to the motion strategy. HCR moves less distance than DCR so that it incurs less motion in cascading moving. In sparse networks, the backup node's traveling distance has a great impact on the cascading moving. Therefore, HCR outperforms DCR since it moves less during a node's motion.

In conclusion, HCR outperforms RIM and HCR in terms of the four evaluation metrics on all the aspects whether the network is sparse or dense because of its proactive backup node selection and reactive cascading node motion. Though the proactive backup node selection is time consuming, it is carried out before the node failure. This will improve the response time. During the reactive cascading node motion, the computing of *BestPosition* is complex and time-consuming. While HCR is a distributed and localized method, the number of each node's neighbors is very small in all kinds of applications, so the selection of *BestPosition* will not spend so much time, and the connectivity restoration process will be fast.

## 7 Conclusions

There is a growing interest in the applications of WSANs in the recent years. Due to the harsh employed environment and limited energy supply, WSAN is prone to be out of work, which may break network connectivity. In this paper, we investigate the problem of restoring network connectivity when a single node fails. A hybrid distributed, localized, and efficient connectivity restoration algorithm HCR is proposed to solve this problem through moving the backup node to *BestPosition*. Compared with the previous schemes, HCR performs a localized network analysis to identify critical nodes, and only a critical node's failure triggers the restoration process. It is a compromised proposal between the cut vertex identification and non-identification. It is effective and has low complexity.

The performance of HCR is analyzed mathematically and validated through simulations. The simulation results have confirmed the effectiveness of HCR in terms of all the evaluation metrics. More importantly, HCR is applicable to various network topologies, sparse or dense. The performance of HCR remains stable when varying network topology. Though a comprehensive network will increase the complexity of the selection of *BestPosition*, it is acceptable.

Though HCR is designed for restoring network connectivity after a single-node failure, that means it can only deal with a single-node failure at a time and handle the sequential node failures. It can be extended to hand multi-node failures at a time by adding one more constraint that no two nodes share the same backup node. In addition, the investigated WSANs are two-dimensional, and we plan to study three-dimensional WSANs in the future. At the same time, coverage is another factor that can be taken into consideration in connectivity restoration, which is also our future research interest.

**Acknowledgements**
This research work was partly supported by the Special Project on Youth Science and Technology Innovation Research Team of Sichuan Province, under grant No.(2015TD0002) and Science and Technology Innovation Seed Project of Sichuan Province, under grant No.(2017RZ0008). We give thanks for the insightful discussion and help of Aiguo Chen. Finally, we give great thanks to the anonymous reviewers for the their suggestions to improve the quality of this paper.

**Funding**
This research work was partly supported by the Special Project on Youth Science and Technology Innovation Research Team of Sichuan Province, under grant No. (2015TD0002) and Science and Technology Innovation Seed Project of Sichuan

**Authors' contributions**
All authors contributed equally to this work. KY and GL contributed to the conception and design of the study. KY and LT contributed to the algorithms and simulation. KY, QJ and CP contributed to the analysis and interpretation of simulation data. All authors read and approved the final manuscript.

**Authors' information**
Ke Yan received the B.S. degree and M.S. degrees from the University of Electronic Science and Technology of China, in 2012 and 2015, respectively. He is currently pursuing the Ph.D. degree in the School of Computer Science and Engineering, University of Electronic Science and Technology of China, Chengdu, China. His interests include wireless sensor network, network survivability, social network and privacy preserving.
Guangchun Luo received the B.S. degree, M.S. degrees and Ph.D. degrees in computer science from University of Electronic Science and Technology of China, Chengdu, China, in 1995, 1999 and 2004, respectively. He is currently a Professor of computer science at the University of Electronic Science and Technology of China, Chengdu, China. His research interests include computer networking, cloud computing, big data.
Ling Tian received the B.S. degree, M.S. degrees and Ph.D. degrees in computer science from University of Electronic Science and Technology of China, Chengdu, China, in 2003, 2006 and 2010, respectively. She is currently an Associate Professor with the School of Computer Science and Engineering, University of Electronic Science and Technology of China, Chengdu. She was a visiting scholar in Georgia State University in 2013. Her research interests include digital multimedia, cloud computing, big data.
Qi Jia received the B.S. degree and M.S. degrees from the University of Electronic Science and Technology of China, in 2013 and 2016, respectively. He is currently pursuing the Ph.D. degree in the School of Computer Science and Engineering, University of Electronic Science and Technology of China, Chengdu, China. His interests include wireless sensor network, big data and privacy preserving.
Chengzong Peng received the B.S. degree from University of California, Irvine in 2017. He is currently pursuing the Master degree in the School of Computer Science and Engineering, University of Electronic Science and Technology of China, Chengdu, China. His interests include mathematics, social network and privacy preserving.

**Competing interests**
The authors declare that they have no competing interests.

### References


1. AA Abbasi, M Younis, K Akkaya, Movement-assisted connectivity restoration in wireless sensor and actor networks. IEEE Trans. Parallel Distributed Syst. **20**(9), 1366–1379 (2009)
2. K Akkaya, F Senel, A Thimmapuram, S Uludag, Distributed recovery from network partitioning in movable sensor/actor networks via controlled mobility. IEEE Trans. Comput. **59**(2), 258–271 (2010)
3. U Baroudi, M Aldarwbi, in *Computer and Information Technology; Ubiquitous Computing and Communications; Dependable, Autonomic and Secure Computing; Pervasive Intelligence and Computing (CIT/IUCC/DASC/PICOM), 2015 IEEE International Conference on*. Corfl: A connectivity restoration mechanism using fuzzy logic in wireless actor and sensor networks (IEEE, Liverpool, 2015), pp. 1449–1454
4. A Boudries, M Amad, P Siarry, Novel approach for replacement of a failure node in wireless sensor network. Telecommun Syst. **65**, 341–350 (2017)
5. Z Cai, ZZ Chen, G Lin, A 3.4713-approximation algorithm for the capacitated multicast tree routing problem. Theor. Comput. Sci. **410**(52), 5415–5424 (2008)
6. Z Cai, R Goebel, G Lin, Size-constrained tree partitioning: Approximating the multicast k-tree routing problem. Theor. Comput. Sci. **412**(3), 240–245 (2011)
7. Z Cai, G Lin, G Xue, in *Proceedings of 11th Annual International Conference on Computing and Combinatorics 2005(COCOON)*. Improved approximation algorithms for the capacitated multicast routing problem (Springer, Kunming, 2005), pp. 136–145
8. M Cardei, J Wu, S Yang, Topology control in ad hoc wireless networks using cooperative communication. IEEE Trans. Mobile Comput. **5**(6), 711–724 (2006)
9. S Cheng, Z Cai, H Gao, J Li, Digging kernel component from big sensory data in wireless sensor networks. IEEE Trans. Knowl. Data Eng. **29**(4), 813–827 (2017)
10. S Cheng, Z Cai, J Li, Curve query processing in wireless sensor networks. IEEE Trans. Veh. Technol. **64**(11), 5198–5209 (2015)
11. S Cheng, Z Cai, J Li, X Fang, in *Computer Communications (INFOCOM), 2015 IEEE Conference on*. Drawing dominant dataset from big sensory data in wireless sensor networks, (Kowloon, 2015), pp. 531–539
12. SK Gupta, P Kuila, PK Jana, in *Proceedings of the Second International Conference on Computer and Communication Technologies*. Genetic algorithm for k-connected relay node placement in wireless sensor networks (Springer, New Delhi, 2016), pp. 721–729
13. N Haider, M Imran, NM Saad, MA Zakariya, in *Communications (MICC), 2013 IEEE Malaysia International Conference on*. Performance analysis of reactive connectivity restoration algorithms for wireless sensor and actor networks (IEEE, Kuala Lumpur, 2013), pp. 490–495
14. HA Hashim, B Ayinde, M Abido, Optimal placement of relay nodes in wireless sensor network using artificial bee colony algorithm. J. Netw. Comput. Appl. **64**, 239–248 (2016)
15. Z He, Z Cai, S Cheng, X Wang, Approximate aggregation for tracking quantiles and range countings in wireless sensor networks. Theor. Comput. Sci. **607**(4), 381–390 (2015)
16. M Imran, M Younis, AM Said, H Hasbullah, in *Embedded and ubiquitous computing (EUC), 2010 IEEE/IFIP 8th international conference on*. Partitioning detection and connectivity restoration algorithm for wireless sensor and actor networks (IEEE, Hong Kong, 2010), pp. 200–207
17. M Imran, M Younis, AM Said, H Hasbullah, Localized motion-based connectivity restoration algorithms for wireless sensor and actor networks. J. Netw. Comput. Appl. **35**(2), 844–856 (2012)
18. S Lee, M Younis, M Lee, Connectivity restoration in a partitioned wireless sensor network with assured fault tolerance. Ad Hoc Netw. **24**, 1–19 (2015)
19. J Li, S Cheng, Z Cai, J Yu, C Wang, Y Li, Approximate holistic aggregation in wireless sensor networks. ACM Trans. Sensor Netw. **13**(2), Article 11:1–24 (2017)
20. M Nikolov, ZJ Haas, Relay placement in wireless networks: Minimizing communication cost. IEEE Trans. Wireless Commun. **15**(5), 3587–3602 (2016)
21. K Ozera, T Oda, D Elmazi, L Barolli, in *International Conference on Broadband and Wireless Computing, Communication and Applications*. Design and implementation of a simulation system based on genetic algorithm for node placement in wireless sensor and actor networks (Springer, Cham, 2016), pp. 673–682
22. R Pallavi, GB Prakash, in *2015 International Conference on Applied and Theoretical Computing and Communication Technology (iCATccT)*. A review on network partitioning in wireless sensor and actor networks (IEEE, Davangere, 2015), pp. 771–782
23. T Qiu, N Chen, K Li, D Qiao, Z Fu, Heterogeneous ad hoc networks: Architectures, advances and challenges. Ad Hoc Netw. **55**, 143–152 (2017)
24. T Qiu, D Luo, F Xia, N Deonauth, W Si, A Tolba, A greedy model with small world for improving the robustness of heterogeneous internet of things. Comput. Netw. **101**, 127–143 (2016)
25. T Qiu, A Zhao, R Ma, V Chang, F Liu, Z Fu, A task-efficient sink node based on embedded multi-core soc for internet of things. Futur. Gener. Comput. Syst. **12**(24), 1–11 (2016)
26. T Ramezani, T Ramezani, A distributed method to reconstruct connection in wireless sensor networks by using genetic algorithm. Modern Appl. Sci. **10**(6), 50 (2016)
27. V Ranga, M Dave, AK Verma, Network partitioning recovery mechanisms in wsans: a survey. Wirel. Pers. Commun. **72**(2), 857–917 (2013)
28. P Rawat, KD Singh, H Chaouchi, JM Bonnin, Wireless sensor networks: a survey on recent developments and potential synergies. J. Supercomput. **68**(1), 1–48 (2014)
29. I Senturk, K Akkaya, S Janansefat, Towards realistic connectivity restoration in partitioned mobile sensor networks. Int. J. Commun. Syst. **29**(2), 230–250 (2016)
30. L Sitanayah, KN Brown, CJ Sreenan, A fault-tolerant relay placement algorithm for ensuring k vertex-disjoint shortest paths in wireless sensor networks. Ad Hoc Netw. **23**, 145–162 (2014)
31. R Virender, D Mayank, VA Kumar, in *Proceedings of International Conference on ICT for Sustainable Development*. Lost connectivity restoration in partitioned wireless sensor networks (Springer, Singapore, 2016), pp. 89–98
32. G Wang, J Yu, D Yu, H Yu, L Feng, P Liu, Ds-mac: An energy efficient demand sleep mac protocol with low latency for wireless sensor networks. J. Netw. Comput. Appl. **58**, 155–164 (2015)
33. H Wang, X Ding, C Huang, X Wu, Adaptive connectivity restoration from node failure (s) in wireless sensor networks. Sensors. **16**(10), 1487 (2016)
34. C Wu, H Chen, J Liu, in *Consumer Electronics, Communications and Networks (CECNet), 2013 3rd International Conference on*. A survey of connectivity restoration in wireless sensor networks (IEEE, Xianning, 2013), pp. 65–67
35. M Younis, S Lee, AA Abbasi, A localized algorithm for restoring internode connectivity in networks of moveable sensors. IEEE Trans. Comput. **59**(12), 1669–1682 (2010)
36. M Younis, IF Senturk, K Akkaya, S Lee, F Senel, Topology management techniques for tolerating node failures in wireless sensor networks: A survey. Comput. Netw. **58**, 254–283 (2014)
37. J Yu, N Wang, G Wang, D Yu, Connected dominating sets in wireless ad hoc and sensor networks a comprehensive survey. Comput. Commun. **36**(2), 121–134 (2013)
38. X Zheng, Z Cai, J Li, H Gao, A study on application-aware scheduling in wireless networks. IEEE Trans. Mobile Comput. **16**(7), 1781–1801 (2017)

# 6

# Centralized coordinated scheduling in LTE-Advanced networks

Oscar D. Ramos-Cantor[1]*, Jakob Belschner[2], Ganapati Hegde[1] and Marius Pesavento[1]

**Abstract**

This work addresses the problem associated with coordinating scheduling decisions among multiple base stations in an LTE-Advanced downlink network in order to manage inter-cell interference with a centralized controller. To solve the coordinated scheduling problem, an integer non-linear program is formulated that, unlike most existing approaches, does not rely on exact channel state information but only makes use of the specific measurement reports defined in the 3GPP standard. An equivalent integer linear reformulation of the coordinated scheduling problem is proposed, which can be efficiently solved by commercial solvers. Extensive simulations of medium to large-size networks are carried out to analyze the performance of the proposed coordinated scheduling approaches, confirming available analytical results reporting fundamental limitations in the cooperation due to out-of-cluster interference. Nevertheless, the schemes proposed in this paper show important gains in average user throughput of the cell-edge users, especially in the case of heterogeneous networks.

**Keywords:** 4G mobile communication, Scheduling algorithms, Integer linear programming

## 1 Introduction

Interference is one of the main limiting factors of today's cellular communication networks in terms of user and network throughputs, especially when operating with full frequency reuse to achieve high spectral efficiency [1–4]. In modern cellular networks, the demand for high data rates is constantly increasing [5], with the users expecting to enjoy excellent network performance irrespective of their geographic location and the load conditions of the network. Thus, new solutions are required in order to fulfill the ever increasing requirements, in particular for the users located at the cell-edge suffering from large path loss and strong inter-cell interference. Promising advances in this aspect have been made with multi-antenna technology [6–9], network densification with interference management schemes [10–12], and Coordinated Multi-Point (CoMP) transceiver techniques [13, 14].

In this work, CoMP network operation is studied, where the base stations (BSs) are prompted to cooperate with each other with the objective of improving the overall network performance, even at the expense of their individual cell or user throughputs [15]. In the literature, three main CoMP schemes are considered for the downlink scenario [16, 17]. These are (i) Joint Transmission (JT), where multiple BSs simultaneously transmit a common message to a user equipment (UE), usually located at the cell-edge, (ii) Dynamic Point Selection (DPS), where at each transmission time, the UE can be served by a different BS without triggering handover procedures, and (iii) Coordinated Scheduling (CS), where the BSs jointly make the scheduling decisions in order to manage the interference experienced by the UEs in the cooperation cluster [18]. This paper focuses on the last CoMP scheme.

The performance of the abovementioned CoMP schemes heavily depends on the channel state information (CSI) available at the transmitter. This CSI can be of different types such as instantaneous channel coefficients or user's average achievable downlink data rates, among others. In practical downlink networks, where perfect global knowledge of the instantaneous channel coefficients is not available at the BSs, CSI is typically obtained in form of achievable data rate measurement reports generated by the UEs, averaged over multiple time/frequency/space dimensions and quantized to reduce the signaling overhead. In this work, the CoMP problem formulation is

*Correspondence: oramos@nt.tu-darmstadt.de
[1]Communication Systems Group, Darmstadt University of Technology, 64283 Darmstadt, Germany
Full list of author information is available at the end of the article

based on practical considerations of the CSI, in form of periodic achievable data rate measurement reports, in the following referred to as *CSI reports*.

The network architecture, in which the CoMP schemes are implemented, also influences the performance of such schemes. Two main CoMP network architectures are typically considered, namely, centralized and decentralized [19]. In the case of centralized CoMP, a central controller is connected to multiple BSs via backhaul links. This central controller is in charge of gathering and using the CSI reports, in order to make a coordinated decision among the connected BSs. For the decentralized CoMP case, decisions are individually made by each BS based on the information exchanged with neighboring BSs. In the case of centralized CoMP, high coordination gains are achievable at the expense of high computational complexity and large signaling overhead. On the other hand, decentralized CoMP requires significantly less information exchange with lower coordination gains. This work focuses on centralized CoMP.

Over the past years, important research has been carried out regarding CoMP schemes under different network architectures and CSI assumptions. In [20] and [21], JT and DPS schemes based on the enhanced CSI reports supported by Long Term Evolution (LTE)-Advanced Release 11, in the following denoted as $\mathrm{CSI}_{\mathrm{lte}}^{\mathrm{R\text{-}11}}$ reports, have been investigated. The results therein show throughput gains for the cell-edge users mainly, and the possibility to improve mobility management by means of DPS. Barbieri et al. studied CS as a complement of enhanced inter-cell interference cancellation (eICIC) in heterogeneous networks in [22]. In their scheme, cooperation takes place in form of CS supported by beamforming in order to mitigate the interference caused by the macro BSs, to the UEs connected to the small cells. Multiple CSI reports are generated, where all possible precoders the macro BS can select from a finite precoder codebook are considered for the cooperation. The results present negligible gain for eICIC with CS, in comparison to eICIC-only. In [23], a cloud-radio access network (C-RAN) architecture is used for centralized CoMP JT in heterogeneous networks, which enables the cooperation of larger cluster sizes. In that case, gains over eICIC-only are observed, especially for large cluster sizes. Authors in [24] propose centralized and decentralized CoMP CS schemes that utilize $\mathrm{CSI}_{\mathrm{lte}}^{\mathrm{R\text{-}11}}$ reports, in which *muting* is applied to one BS at a time. A BS is called muted if it does not transmit data on a specific time/frequency resource. It has been shown that under this muting condition, both centralized and decentralized schemes achieve the same performance, favoring the decentralized scheme due to the reduced information exchange. Moreover, in [25], the authors extend the cooperation scheme of [24], to introduce muting of more than one BS per scheduling decision in a larger network. A greedy CS algorithm is presented to solve the centralized problem, which yields limited additional gain with respect to the decentralized scheme with overlapping cooperation clusters. The coordination scheme of [25] consists in a greedy optimization procedure. It is therefore suboptimal and further investigation regarding the optimally achievable performance of coordination, in the case of $\mathrm{CSI}_{\mathrm{lte}}^{\mathrm{R\text{-}11}}$ reports, has not been carried out. Additionally, the results are focused on macro-only networks, where the gains of cooperation are restricted due to similar interfering power levels experienced from multiple BSs.

Although the abovementioned works show that CoMP schemes enhance the user throughput with respect to a network operating without any cooperation, no detailed studies are carried out in order to establish the maximum achievable gains that CoMP schemes can offer in realistic network scenarios and under LTE-Advanced-specific CSI reports. In [26], it has been demonstrated from an analytical perspective that cooperative schemes have fundamentally limited gains. That is, even under the assumption of centralized coordination and ideal CSI in form of instantaneous channel coefficients, the cooperation gains are limited due to the residual interference from BSs outside of the cooperation area, the signaling overheads, and the finite nature of the time/frequency/space resources. In the paper at hand, such limits are investigated under practical conditions by means of system level evaluations. For that purpose, the CoMP CS problem is formulated and solved optimally for an LTE-Advanced downlink network with centralized architecture. The problem formulation is based on multiple CSI reports generated by the UEs and gathered by the central controller, which uses this information to determine the coordinated scheduling decisions for all connected BSs. The central controller then decides which BSs serve their connected UEs on a given time/frequency resource, and which BSs are muted in order to reduce the interference caused to the UEs served by the neighboring transmitting BSs. The main contributions of this work are summarized as follows:

- The CS problem, where BSs cooperate by muting time/frequency resources based on standardized $\mathrm{CSI}_{\mathrm{lte}}^{\mathrm{R\text{-}11}}$ reports, is formulated as an integer non-linear program (INLP).
- The non-linear CS with muting problem is reformulated into a computationally tractable *equivalent* integer linear program (ILP), which enjoys of low computational complexity and can be efficiently solved by commercial solvers.
- A configurable heuristic algorithm is proposed as an extension to the greedy algorithm in [25], which achieves an excellent trade-off between performance and computational complexity.

## 2 System model

A cellular network is considered as illustrated in Fig. 1, where a cooperation cluster consisting of $M$ BSs, operating in frequency division duplexing (FDD) mode, serves $N$ UEs in the downlink. Orthogonal frequency division multiple access (OFDMA) is assumed with frequency reuse one, where at each transmission time, all BSs can make use of the same $L$ Physical Resource Blocks (PRBs) for transmission. Thus, inter-cell interference affects the UEs, especially at the cell-edge. The operation of the cooperation cluster is managed by a central controller with backhaul connectivity to all $M$ BSs. In the following, the sets of indices $\mathcal{M} = \{1, \ldots, M\}$, $\mathcal{N} = \{1, \ldots, N\}$, and $\mathcal{L} = \{1, \ldots, L\}$ are used to address the BSs, UEs, and PRBs, respectively.

The received power at UE $n \in \mathcal{N}$, from BS $m \in \mathcal{M}$, on PRB $l \in \mathcal{L}$, is denoted as $p_{n,m,l}$. Hence, for single-input-single-output (SISO) transmission,

$$p_{n,m,l} = |g_n\, h_{n,m,l}|^2\, \phi_{m,l}, \tag{1}$$

where $\phi_{m,l}$ corresponds to the transmit power of BS $m$ on PRB $l$, the complex coefficient $h_{n,m,l}$ represents the amplitude gain of the downlink channel between BS $m$ and UE $n$ on PRB $l$, and $g_n$ is the receiver amplitude processing gain. In (1), the transmitted symbols are assumed to exhibit unit average transmit power. When summing over all PRBs, the total received power at UE $n$ from BS $m$ is obtained as

$$p_{n,m} = \sum_{l=1}^{L} p_{n,m,l}. \tag{2}$$

Generally, the serving BS for UE $n \in \mathcal{N}$ is selected as the BS from which the highest total received power is obtained, as defined in (2). In the case of heterogeneous networks, in order to achieve load balancing and cell-splitting gains, techniques like cell range expansion are applied where the UEs are instructed to add a constant off-set in the computation of the total received power of the small cells [27, 28]. In this paper, both homogeneous macro-only and heterogeneous networks are studied, where for the latter case, cell range expansion is applied. The $N \times M$ connection matrix $\mathbf{C}$ is defined, with elements

$$c_{n,m} = \begin{cases} 1 \text{ if UE } n \text{ is served by BS } m \in \mathcal{M} \\ 0 \text{ otherwise,} \end{cases} \tag{3}$$

characterizing the serving conditions between BSs and UEs, where it is assumed that only one BS serves UE $n$ over all PRBs. It is further assumed, for simplicity, that the UEs are quasi-static, such that no handover procedures are triggered between the BSs. Therefore, the connection matrix $\mathbf{C}$ is assumed to be constant during the considered operation time.

The set of indices of BSs within the cooperation cluster that interfere with UE $n \in \mathcal{N}$ is defined as $\mathcal{I}_n = \{m \mid c_{n,m} = 0,\ \forall m \in \mathcal{M}\}$, with cardinality $|\mathcal{I}_n| = M - 1$. Moreover, since UE $n$ experiences different interfering power levels from the $\mathcal{I}_n$ interfering BSs, the set $\mathcal{I}'_n \subseteq \mathcal{I}_n$ of indices of the $M'$ *strongest interfering BSs* of UE $n$, is defined such that $|\mathcal{I}'_n| = M'$. Therefore, the set $\mathcal{I}'_n$ contains the indices of the $M'$ interfering BSs with the highest total received power at UE $n$, as calculated in (2). The number of strongest interfering BSs is bounded as $0 \leq M' \leq M - 1$, where the sets $\mathcal{I}'_n$ and $\mathcal{I}_n$ are identical, if

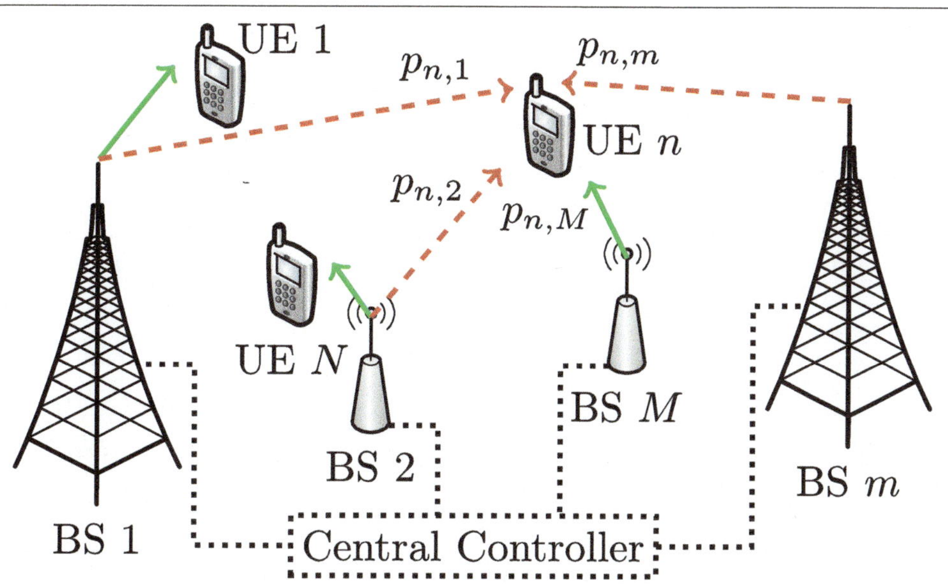


**Fig. 1** Cooperation cluster of $M$ BSs and $N$ UEs in the downlink. The BSs are connected through the backhaul to a central controller

$M' = M - 1$. The sets $\mathcal{I}_n$ and $\mathcal{I}'_n$ apply for all PRBs in the reporting period.

Within the cooperation cluster, the BSs cooperate in the form of coordinated scheduling with muting, as previously mentioned in Section 1. The central controller is then, in charge of managing the downlink transmissions of the BSs, where at each transmission time and on a per PRB basis, each BS can be requested to abstain from transmitting data. Hence, the interference caused to UEs located in neighboring BSs is reduced on the PRBs with muted BSs. Given the muting decision matrix, $\bar{\boldsymbol{\alpha}}$, of dimensions $M \times L$ and elements

$$\bar{\alpha}_{m,l} = \begin{cases} 1 \text{ if BS } m \in \mathcal{M} \text{ is muted on PRB } l \in \mathcal{L} \\ 0 \text{ otherwise,} \end{cases} \tag{4}$$

the signal-to-interference-plus-noise ratio (SINR) of UE $n \in \mathcal{N}$, served by BS $k \in \mathcal{M}$, on PRB $l$, is then defined as

$$\gamma_{n,l}(\bar{\boldsymbol{\alpha}}_l) = \frac{(1 - \bar{\alpha}_{k,l})\, p_{n,k,l}}{I^{\text{cc}}_{n,l}(\bar{\boldsymbol{\alpha}}_l) + I^{\text{oc}}_{n,l} + \sigma^2}. \tag{5}$$

The vector, $\bar{\boldsymbol{\alpha}}_l$, is equivalent to the $l$-th column of $\bar{\boldsymbol{\alpha}}$. The numerator corresponds to the average received power at UE $n$, from the serving BS $k$, on PRB $l$, as defined in (1). The first term in the denominator corresponds to the average inter-cell interference from the BSs within the cooperation cluster, with

$$I^{\text{cc}}_{n,l}(\bar{\boldsymbol{\alpha}}_l) = \sum_{m \in \mathcal{I}_n} (1 - \bar{\alpha}_{m,l})\, p_{n,m,l}, \tag{6}$$

$I^{\text{oc}}_{n,l}$ is the average out-of-cluster interference, and $\sigma^2$ is the noise power assumed, without loss of generality, to be constant for all UEs over all PRBs. It is worth noting that the out-of-cluster interference $I^{\text{oc}}_{n,l}$ is assumed to be independent of the muting decision matrix $\bar{\boldsymbol{\alpha}}$, since the central controller is not aware of the muting decisions made by the BSs outside of the cooperation cluster.

The achievable data rate of UE $n \in \mathcal{N}$ on PRB $l \in \mathcal{L}$ is modeled as a function of the UE's SINR. Hence,

$$r_{n,l} = f\left(\gamma_{n,l}\right), \tag{7}$$

where $f\left(\gamma_{n,l}\right)$ denotes a mapping from the SINR of UE $n$ on PRB $l$, to the achievable data rate.

In the literature, it is common to assume that the central controller has perfect CSI knowledge, in form of the instantaneous channel coefficients, $h_{n,m,l}$, as introduced in (1). Thus, the computation of the SINR, $\gamma_{n,l}(\bar{\boldsymbol{\alpha}}_l)$, and the achievable data rates, $r_{n,l}$, as defined in (5) and (7), respectively, is carried out in a straightforward manner for any possible muting decision $\bar{\boldsymbol{\alpha}}$. However, in practical conditions such as in LTE-Advanced networks, the CSI is typically available in form of achievable data rate measurement reports, i.e., CSI reports, which contain average information of multiple time/frequency/space resources for a subset of possible muting decisions $\bar{\boldsymbol{\alpha}}$, as defined in (4). Thus, the processing and signaling overhead is reduced, at the expense of limited CSI knowledge for the CoMP CS scheme.

## 3 CSI reporting for LTE-Advanced CoMP CS

To enable opportunistic scheduling and CoMP operation in LTE-Advanced, CSI estimation is supported by the transmission of CSI reference signals (CSI-RSs) from the BSs [29, 30]. One main feature of the CSI-RSs is the possibility to configure muted CSI-RSs, i.e., CSI-RSs with zero transmission power, enabling the UEs to estimate CSI from specific neighboring BSs without interference from the serving BS. Therefore, the UEs can generate multiple $\text{CSI}^{\text{R-11}}_{\text{lte}}$ reports that reflect different serving and interfering conditions in the network [20, 21]. These $\text{CSI}^{\text{R-11}}_{\text{lte}}$ reports are typically composed of a channel quality indicator (CQI), a precoding matrix index (PMI), and a rank indication (RI) [31]. The CQI reflects the measured/estimated SINR and the corresponding achievable data rate of a UE, when assuming a downlink transmission of rank indicated by the RI and transmit precoding vector taken from a finite-length codebook as indexed by the PMI.

For a cooperation cluster with $M$ BSs, a total of $J = 2^M - 1$ muting decisions can be made per PRB $l \in \mathcal{L}$. In this work, the practical case is considered, where the CS operation is managed by the central controller based on the $\text{CSI}^{\text{R-11}}_{\text{lte}}$ reports provided by the UEs. Since the SINRs and achievable data rates of UE $n \in \mathcal{N}$, as defined in (5) and (7), respectively, are dominated by its strongest interfering BSs [24], in the following, it is assumed that the UEs generate a total of $J' = 2^{M'}$ $\text{CSI}^{\text{R-11}}_{\text{lte}}$ reports per PRB, with $J' < J$. Then, each $\text{CSI}^{\text{R-11}}_{\text{lte}}$ report only considers the $M'$ strongest interfering BSs of UE $n$, as described by the set $\mathcal{I}'_n$, introduced in Section 2.

Each of the $J'$ $\text{CSI}^{\text{R-11}}_{\text{lte}}$ reports, which are generated by UE $n \in \mathcal{N}$ on PRB $l \in \mathcal{L}$, reflects a unique *interference scenario* for its strongest interfering BSs. More specifically, the interference scenario $j \in \mathcal{J}'$, with $\mathcal{J}' = \{1, \ldots, J'\}$, is characterized by the *muting indicator* set $\mathcal{J}_{n,j}$, which contains the indices of the (strongest) interfering BSs considered to be muted in the $j$-th $\text{CSI}^{\text{R-11}}_{\text{lte}}$ report of UE $n$. Hence, the set $\mathcal{J}_n = \mathbb{P}(\mathcal{I}'_n)$ contains all $J'$ muting indicator sets for UE $n$, with $\mathbb{P}(\cdot)$ denoting the set of all subsets of $\mathcal{I}'_n$. The set $\mathcal{J}_n$ is common to all PRBs. From the muting indicator set, $\mathcal{J}_{n,j}$, the *muting pattern* of the $j$-th $\text{CSI}^{\text{R-11}}_{\text{lte}}$ report of UE $n$, on PRB $l$, is defined as

$$\alpha_{n,m',l,j} = \begin{cases} 1 \text{ if } m' \in \mathcal{J}_{n,j} \text{ on PRB } l \\ 0 \text{ otherwise,} \end{cases} \quad \forall m' \in \mathcal{I}'_n, \tag{8}$$

i.e., $\alpha_{n,m',l,j} = 1$, if the (strongest) interfering BS $m'$, is muted on PRB $l$, under interference scenario $j$. The definition in (8) considers only the set of strongest interfering

BSs of UE $n$, i.e., $\mathcal{I}'_n$. Therefore, a constant muting state of the remaining BSs in the cooperation cluster is required, for all $\mathcal{J}'$ interference scenarios. In the following, it is assumed without loss of generality that $\alpha_{n,m,l,j} = 0$, $\forall m \notin \mathcal{I}'_n, \forall j \in \mathcal{J}'$. Although the definition of the muting pattern in (8) is similar to the definition of the muting decision in (4), the two concepts are different. The muting pattern describes the assumed muting conditions during the generation of the $\text{CSI}^{\text{R-11}}_{\text{lte}}$ reports for the different interference scenarios, while the muting decision is imposed by the central controller, to the BSs within the cooperation cluster, as the result of the implementation of the CS with muting scheme.

For the generation of the $\text{CSI}^{\text{R-11}}_{\text{lte}}$ reports, UE $n \in \mathcal{N}$ calculates the SINR and the achievable data rates, on PRB $l \in \mathcal{L}$ under interference scenario $j \in \mathcal{J}'$. Therefore, similar to (5), the SINR of UE $n$ on PRB $l$, under interference scenario $j$, is defined as

$$\gamma_{n,l,j}\left(\alpha_{n,m',l,j}\right) = \frac{p_{n,k,l}}{I^{\text{si}}_{n,l,j}\left(\alpha_{n,m',l,j}\right) + I^{\text{wi}}_{n,l} + I^{\text{oc}}_{n,l} + \sigma^2}, \quad (9)$$

where the first term in the denominator of (5) has been decomposed into two terms corresponding to the interference from the strongest interfering BSs of UE $n$, i.e., $I^{\text{si}}_{n,l,j}\left(\alpha_{n,m',l,j}\right)$, and the interference from the remaining (weakest) interfering BSs of UE $n$, denoted by $I^{\text{wi}}_{n,l}$. The interference from the strongest interfering BSs that can cooperate to improve the SINR of UE $n$ depends on interference scenario $j$, and thus, the muting pattern, as

$$I^{\text{si}}_{n,l,j}\left(\alpha_{n,m',l,j}\right) = \sum_{m' \in \mathcal{I}'_n} \left(1 - \alpha_{n,m',l,j}\right) p_{n,m',l}. \quad (10)$$

On the other hand, the interference from the weakest interfering BSs of UE $n$ is assumed to be constant and independent of the possible muting decisions, with

$$I^{\text{wi}}_{n,l} = \sum_{m \in \mathcal{I}_n \setminus \mathcal{I}'_n} p_{n,m,l}. \quad (11)$$

Furthermore, the out-of-cluster interference and the noise powers are also assumed to be constant terms among all the $J'$ interfering scenarios considered in the $\text{CSI}^{\text{R-11}}_{\text{lte}}$ reports.

To complete the information for the $\text{CSI}^{\text{R-11}}_{\text{lte}}$ reports, $r_{n,l,j}$ denotes the achievable data rate of UE $n \in \mathcal{N}$, on PRB $l \in \mathcal{L}$, under interference scenario $j \in \mathcal{J}'$. The calculation of $r_{n,l,j}$ follows the definition in (7), as a function of $\gamma_{n,l,j}$ given by (9). In this work, it is assumed that the UEs reflect their achievable data rates for the interference scenarios $j \in \mathcal{J}'$, into the CQI values to be reported in the $\text{CSI}^{\text{R-11}}_{\text{lte}}$ reports. Additionally, it is further assumed that the BSs can translate the CQI values from the $\text{CSI}^{\text{R-11}}_{\text{lte}}$ reports, into the corresponding achievable data rates in order to support the cooperative scheduling process. Therefore, in the following, the description is focused on the achievable data rates extracted from the $\text{CSI}^{\text{R-11}}_{\text{lte}}$ reports.

By inspecting (9) and (10), the SINR of UE $n \in \mathcal{N}$, on PRB $l \in \mathcal{L}$, increases when muting additional interfering BSs. Correspondingly, the achievable data rate of UE $n$, on PRB $l$, increases or remains constant with the increased SINR, if the function $f\left(\gamma_{n,l,j}\right)$ is non-decreasing.

**Example 1** *In an exemplary network with a cooperation cluster of $M = 4$ BSs and a total of $M' = 2$ strongest interfering BSs per UE, UE $n \in \mathcal{N}$ selects BS 1 and BS 2 for cooperation, such that $\mathcal{I}'_n = \{1,2\}$. Thus, UE $n$ generates $J' = 4$ $\text{CSI}^{\text{R-11}}_{\text{lte}}$ reports on PRB $l \in \mathcal{L}$, as summarized in Table 1, with $r_{n,l,2} \geq r_{n,l,3}, r_{n,l,4} \geq r_{n,l,1}$.*

## 4 CS with muting

### 4.1 Proposed INLP—problem formulation

At the central controller, the $\text{CSI}^{\text{R-11}}_{\text{lte}}$ reports, generated by the UEs and forwarded by the BSs, are used in order to compute the CS decision. The CS decision consists of two main components, namely, a scheduling decision that assigns PRBs to UEs, and a muting decision that mutes BSs on particular PRBs. The matrix variable $\bar{\mathbf{S}}$ of dimensions $N \times L$ and elements

$$\bar{s}_{n,l} = \begin{cases} 1 \text{ if PRB } l \in \mathcal{L} \text{ is assigned to UE } n \in \mathcal{N} \\ 0 \text{ otherwise,} \end{cases} \quad (12)$$

is used to denote the scheduling decision for all UEs on each PRB $l$, while the $M \times L$ matrix variable $\bar{\boldsymbol{\alpha}}$, with elements as introduced in (4), refers to the muting decision for all BSs on each PRB $l$. In the following, an INLP is proposed, to carry out joint BS muting and UE scheduling in a coordinated network.

Typically, the schedulers in mobile communications pursue a trade-off between user throughput and fairness. For that purpose, opportunistic scheduling is applied such as in the case of the Proportional Fair (PF) scheduler [32, 33]. The objective of the PF scheduler is to maximize the sum, over all UEs, of the PF metrics given by

$$\Omega_n = \frac{r_n}{R_n} \; \forall n \in \mathcal{N}, \quad (13)$$

**Table 1** $\text{CSI}^{\text{R-11}}_{\text{lte}}$ reports for UE $n$, on PRB $l$, with $M' = 2$

| Int. scenario ($j \in \mathcal{J}'$) | Mut. ind. ($\mathcal{J}_{n,j}$) | Mut. pattern ($\alpha_{n,l,j}$) | Achiev. data rate |
|---|---|---|---|
| 1 | $\{\emptyset\}$ | [0, 0, 0, 0] | $r_{n,l,1}$ |
| 2 | $\{1, 2\}$ | [1, 1, 0, 0] | $r_{n,l,2}$ |
| 3 | $\{1\}$ | [1, 0, 0, 0] | $r_{n,l,3}$ |
| 4 | $\{2\}$ | [0, 1, 0, 0] | $r_{n,l,4}$ |

where the ratio between the total instantaneous achievable data rate and the average user throughput over time, denoted by $r_n$ and $R_n$, respectively, of UE $n \in \mathcal{N}$ is considered. The total instantaneous achievable data rate of UE $n$ is calculated as

$$r_n = g\left(r_{n,l,j}, \bar{\mathbf{s}}_n, \bar{\boldsymbol{\alpha}}\right) \;\forall l \in \mathcal{L}, \forall j \in \mathcal{J}', \tag{14}$$

where $g(\cdot)$ corresponds to a function of the achievable data rates of UE $n$, denoted by $r_{n,l,j}$, over the PRBs assigned to UE $n$, as described by the $n$-th row of $\bar{\mathbf{S}}$, denoted by $\bar{\mathbf{s}}_n$, and the muting decision matrix $\bar{\boldsymbol{\alpha}}$.

The LTE-Advanced CS with muting problem can be formulated as the following INLP

$$\max_{\{\bar{\mathbf{S}},\bar{\boldsymbol{\alpha}}\}} \sum_{n\in\mathcal{N}} \Omega_n \tag{15a}$$

s.t.

$$\bar{\alpha}_{m,l} + \sum_{n\in\mathcal{N}} c_{n,m}\, \bar{s}_{n,l} \leq 1 \;\forall m \in \mathcal{M}, \forall l \in \mathcal{L}, \tag{15b}$$

$$r_n = \sum_{l\in\mathcal{L}} \rho\left(\mathbf{r}_{n,l}, \bar{\boldsymbol{\alpha}}_l, \mathcal{I}'_n\right) \bar{s}_{n,l} \;\forall n \in \mathcal{N}, \tag{15c}$$

$$\bar{s}_{n,l} \in \{0,1\} \;\forall n \in \mathcal{N}, \forall l \in \mathcal{L}, \tag{15d}$$

$$\bar{\alpha}_{m,l} \in \{0,1\} \;\forall m \in \mathcal{M}, \forall l \in \mathcal{L}, \tag{15e}$$

where the objective in (15a) is to maximize the sum of the PF metrics over all UEs, with the PF metric of UE $n \in \mathcal{N}$ calculated as in (13). The constraints in (15b) link the scheduling decision $\bar{\mathbf{S}}$ with the muting decision $\bar{\boldsymbol{\alpha}}$. If BS $m \in \mathcal{M}$ is muted on PRB $l \in \mathcal{L}$, then PRB $l$ cannot be assigned to any UE $n$ connected to BS $m$. Thus, if $\bar{\alpha}_{m,l} = 1$ in (15b), for BS $m$, the second term on the left-hand-side must be equal to zero, which is true in either of the following cases, with the connection indicator $c_{n,m}$ given by (3):

- No UEs are connected to BS $m$, i.e., $c_{n,m} = 0, \forall n \in \mathcal{N}$.
- PRB $l$ is not assigned to any UE served by BS $m$, i.e., $\bar{s}_{n,l} = 0, \forall n \in \mathcal{N}$ such that $c_{n,m} = 1$.

Furthermore, in the case that BS $m$ is not muted on PRB $l$, i.e., $\bar{\alpha}_{m,l} = 0$, the constraints in (15b) ensure that single user transmissions are carried out, where each BS is allowed to schedule a maximum of one UE per PRB. Additionally, the total instantaneous achievable data rate of UE $n$, denoted by $r_n$ as introduced in (14), is calculated in (15c), with

$$g\left(r_{n,l,j}, \bar{\mathbf{s}}_n, \bar{\boldsymbol{\alpha}}\right) = \sum_{l\in\mathcal{L}} \rho\left(\mathbf{r}_{n,l}, \bar{\boldsymbol{\alpha}}_l, \mathcal{I}'_n\right) \bar{s}_{n,l} \;\forall n \in \mathcal{N}. \tag{16}$$

In (16), $\rho\left(\mathbf{r}_{n,l}, \bar{\boldsymbol{\alpha}}_l, \mathcal{I}'_n\right)$ is a lookup table function that selects the achievable data rate of UE $n$, on PRB $l$, based on the muting decision $\bar{\boldsymbol{\alpha}}_l$ of the strongest interfering BSs of UE $n$, as indexed by $\mathcal{I}'_n$. The lookup table function $\rho(\cdot)$ selects the achievable data rate from the $J' \times 1$ vector, $\mathbf{r}_{n,l}$, with elements $r_{n,l,j}$, $\forall j \in \mathcal{J}'$, obtained from the $\text{CSI}_{\text{lte}}^{\text{R-11}}$ reports of UE $n$, on PRB $l$.

**Example 2** *Based on the Table 1 from Example 1, with $J' = 4$ $CSI_{lte}^{R\text{-}11}$ reports, the lookup table function for UE $n \in \mathcal{N}$, on PRB $l \in \mathcal{L}$, provides the results as in Table 2. Note that the value of $\rho(\cdot)$ does not depend on the muting decision of the remaining BSs.*

Due to the utilization of the $J'$ $\text{CSI}_{\text{lte}}^{\text{R-11}}$ reports, the achievable data rate of UE $n \in \mathcal{N}$, on PRB $l \in \mathcal{L}$, under interference scenario $j \in \mathcal{J}'$, is constant in the problem formulation and limited to the set of reported muting patterns. In Section 3, it has been mentioned that the achievable data rate of UE $n$, on PRB $l$, under interference scenario $j$ depends on the function $f\left(\gamma_{n,l,j}\right)$. If $f(\cdot)$ is piece-wise non-decreasing, it follows that additional (strongest) interfering BSs are only muted if the achievable data rate of UE $n$ is increased.

Moreover, the scheduling and muting matrix variables, $\bar{\mathbf{S}}$ and $\bar{\boldsymbol{\alpha}}$, are binary as described by the constraints in (15d) and (15e), respectively.

The following remarks summarize the main characteristics of the LTE-Advanced CS with muting problem formulation in (15).

- As mentioned in Section 3, given $M'$ strongest interfering BSs per UE $n \in \mathcal{N}$, a total of $J' = 2^{M'}$ interfering scenarios per UE $n$ are available. Hence, two special cases of the problem formulation are observed:
  (i) If $M' = 0$, each UE $n$ generates one $\text{CSI}_{\text{lte}}^{\text{R-11}}$ report under the assumption of no BS muting. At the central controller, the CS with muting problem formulation becomes a PF scheduler without any cooperation.
  (ii) If $M' = M - 1$, all the interfering BSs within the cooperation cluster can be muted to improve the performance of any UE, on each PRB $l \in \mathcal{L}$. If the network size is large, finding the solution while assuming cooperation of all interfering BSs for all UEs approximates an exhaustive search.
- Due to the combinatorial nature of the problem formulation, it is classified as non-deterministic polynomial-time (NP)-hard.

**Table 2** Lookup table function $\rho\left(\mathbf{r}_{n,l}, \bar{\alpha}_l, \mathcal{I}'_n\right)$ for UE $n$, on PRB $l$, with $M' = 2$

| $\bar{\alpha}_{m,l}$, $\forall m \in \mathcal{I}'_n$ | $\rho\left(\mathbf{r}_{n,l}, \bar{\alpha}_l, \mathcal{I}'_n\right)$ |
|---|---|
| [0, 0] | $r_{n,l,1}$ |
| [0, 1] | $r_{n,l,4}$ |
| [1, 0] | $r_{n,l,3}$ |
| [1, 1] | $r_{n,l,2}$ |

- The problem is non-linear because of the relation between the muting and the scheduling decision variables, $\bar{\boldsymbol{\alpha}}_l$ and $\bar{s}_{n,l}$, respectively, in the constraints in (15c).

Although the number of reported interference scenarios $J' = 2^{M'}$ can be limited by selecting a small value $M'$ of (strongest) interfering BSs per UE $n \in \mathcal{N}$, the CS with muting INLP formulation in (15) also depends on the number of UEs, i.e., $N$, and the number of PRBs, denoted by $L$. For certain network scenarios, $N$ and $L$ can be large. Therefore, given the non-linear nature of the problem in (15), finding a solution with commercial solvers may either not be possible or inefficient in terms of computation time. In the following, *separability*, *reducibility*, and *lifting* concepts are used, in order to formulate parallel ILP sub-problems that scale better with the network size.

## 4.2 Proposed ILP—parallelized sub-problem formulation

### 4.2.1 Separability

When analyzing the objective function described by (15a), the total PF metric corresponds to the sum of the individual PF metrics for all UEs. Furthermore, at each UE $n \in \mathcal{N}$, it is assumed that the total instantaneous achievable data rate is equivalent to the linear combination of the decoupled achievable data rates per scheduled PRBs, as given by (15c). Therefore, it is possible to separate the CS with muting problem in (15), into $L$ independent sub-problems, corresponding to the scheduling decision of one PRB each. By performing this parallelization, the computation time is reduced without affecting the quality of the solution, i.e., the solution of the parallelized CS with muting problem remains optimal.

### 4.2.2 Reducibility

It is expected that some of the UEs, connected to a common BS $m \in \mathcal{M}$, share one or more strongest interfering BSs. From a BS perspective, the set

$$\mathcal{J}_m = \bigcup_{n \in \mathcal{N} \mid c_{n,m}=1} \mathcal{J}_n \; \forall m \in \mathcal{M}, \tag{17}$$

contains all the unique muting indicator sets, associated to its connected UEs. Similar to the set $\mathcal{J}_n$, $\mathcal{J}_m$ is common to all PRBs in the reporting period. The number of unique muting indicator sets for BS $m$, i.e., $J'_m = |\mathcal{J}_m|$, depends on the number of UEs connected to BS $m$ and the maximum number $J'$ of reported interference scenarios per UE, as introduced in Section 3. Thus, $J' \leq J'_m \leq \sum_{n \in \mathcal{N}} c_{n,m} J'$, where the lower bound corresponds to the case when all connected UEs are interfered by the same set of strongest interfering BSs, and the upper bound represents the case with all UEs having different strongest interfering BSs. For the unique muting indicator set $\mathcal{J}_{m,j'}$, with $j' \in \mathcal{J}'_m = \{1, \ldots, J'_m\}$, the set of indices of UEs, connected to BS $m$, with equal muting indicator set is defined as

$$\mathcal{N}_{m,j'} = \{n \in \mathcal{N} \mid c_{n,m} = 1, \mathcal{J}_{m,j'} \subsetneq \mathcal{J}_n, \; \forall m \in \mathcal{M}, \forall j' \in \mathcal{J}'_m\}. \tag{18}$$

Based on the definitions in (17) and (18), and taking into account that single user transmissions take place per BS, it is sufficient that each BS $m \in \mathcal{M}$ forwards to the central controller, the $\mathrm{CSI}_{\mathrm{lte}}^{\mathrm{R\text{-}11}}$ reports related to one UE per unique muting indicator set $\mathcal{J}_{m,j'}, \forall j' \in \mathcal{J}'_m$, on PRB $l \in \mathcal{L}$, instead of the $\mathrm{CSI}_{\mathrm{lte}}^{\mathrm{R\text{-}11}}$ reports from all connected UEs. The set of indices of UEs connected to BS $m$ that maximize the PF metric, in at least one of the unique muting indicator sets indexed by $j' \in \mathcal{J}'_m$, on PRB $l$, is defined as

$$\mathcal{N}'_{m,l} = \{\hat{n} \mid \exists j : \hat{n} = \arg\max_{n \in \mathcal{N}_{m,j'}} \Omega_{n,l,j}, \; \forall j' \in \mathcal{J}'_m, \forall j \in \mathcal{J}' \mid \mathcal{J}_{n,j} = \mathcal{J}_{m,j'}\}. \tag{19}$$

The cardinality of the set $\mathcal{N}'_{m,l}$, denoted as $|\mathcal{N}'_{m,l}|$, is bounded as $1 \leq |\mathcal{N}'_{m,l}| \leq \sum_{n \in \mathcal{N}} c_{n,m}$, where the lower bound implies that only one UE provides the maximum PF metric, among all unique muting indicator sets on PRB $l$. The upper bound corresponds to the case when each UE reports different muting indicator sets with respect to the other UEs connected to BS $m$.

At the central controller, all the achievable data rates, $r_{n,l,j}, \forall n \in \mathcal{N}'_{m,l}, \forall m \in \mathcal{M}, \forall l \in \mathcal{L}, \forall j \in \mathcal{J}'$, are per definition set to zero, for the interference scenarios where UE $n$ does not provide the maximum PF metric, among the UEs connected to the same BS $m$. The set $\mathcal{N}'_l = \bigcup_{m \in \mathcal{M}} \mathcal{N}'_{m,l}$ is used to denote the indices of UEs to be considered in the reformulated ILP, on PRB $l$. The cardinality of the set $\mathcal{N}'_l$ is described as $M \leq |\mathcal{N}'_l| \leq N$. In the special case of $M' = 0$, all UEs report only one interference scenario where no cooperative interfering BS is muted, and thus, $|\mathcal{N}'_l| = M$.

### 4.2.3 Lifting

In order to linearize the constraints in (15c), a variable transformation is introduced based on the lifting technique [34]. A new coordinated decision variable is defined containing both, the scheduling and the muting decisions, as

$$s_{n,l,j} = \begin{cases} 1 \text{ if PRB } l \in \mathcal{L} \text{ is assigned to UE } n \in \mathcal{N} \\ \quad \text{under interference scenario } j \in \mathcal{J}' \\ 0 \text{ otherwise.} \end{cases} \tag{20}$$

The new decision variable, $s_{n,l,j}$, is related to the muting and scheduling decisions in (4) and (12), respectively, as

$$s_{n,l,j} = 1 \Leftrightarrow \bar{s}_{n,l} = 1 \wedge \bar{\alpha}_{m,l} = 1 \; \forall n \in \mathcal{N}, \forall l \in \mathcal{L}, \forall j \in \mathcal{J}', \forall m \in \mathcal{J}_{n,j}, \tag{21}$$

with $\wedge$ denoting the logical *and* operator. Hence, the nonlinear constraints in (15c) reduce to a linear combination of the achievable data rates, i.e., $r_{n,l,j}$, and the new decision variable, $s_{n,l,j}$.

#### 4.2.4 Problem reformulation

Using the above described concepts of separability, reducibility, and lifting, the CS with muting INLP formulation in (15) can be reformulated as an ILP, which can be efficiently solved by commercial solvers. Hence, with the set $\mathcal{N}'_l$ and defining the binary decision variable $\mathbf{S}_l$ to have dimensions $|\mathcal{N}'_l| \times J'$, the sub-problem formulation for PRB $l \in \mathcal{L}$ is

$$\max_{\{\mathbf{S}_l\}} \sum_{n \in \mathcal{N}'_l} \Omega_{n,l} \tag{22a}$$

s.t.

$$s_{n,l,j} + \sum_{k \in \mathcal{N}'_l} \sum_{i \in \mathcal{J}'} c_{k,m}\, s_{k,l,i} \leq 1 \qquad \forall n \in \mathcal{N}'_l, \forall j \in \mathcal{J}', \forall m \in \mathcal{J}_{n,j}, \tag{22b}$$

$$\sum_{n \in \mathcal{N}'_l} \sum_{j \in \mathcal{J}'} c_{n,m}\, s_{n,l,j} \leq 1 \qquad \forall m \in \mathcal{M} \setminus \cup_{n \in \mathcal{N}'_l} \mathcal{I}'_n, \tag{22c}$$

$$r_{n,l} = \sum_{j \in \mathcal{J}'} r_{n,l,j}\, s_{n,l,j} \; \forall n \in \mathcal{N}'_l, \tag{22d}$$

$$s_{n,l,j} = 0 \; \forall n \in \mathcal{N}'_l, \forall j \in \mathcal{J}' \mid r_{n,l,j} = 0, \tag{22e}$$

$$s_{n,l,j} \in \{0, 1\} \; \forall n \in \mathcal{N}'_l, \forall j \in \mathcal{J}', \tag{22f}$$

where the objective in (22a) is to maximize the sum of the PF metric over all UEs. The constraints in (22b) restrict the scheduling decisions of the strongest interfering BSs of UE $n \in \mathcal{N}'_l$, i.e., $\forall m \in \mathcal{J}_{n,j}$, in order to agree with the muting state considered in the interference scenario $j \in \mathcal{J}'$. If PRB $l$ is assigned to UE $n$, under the condition of muting the (strongest) interfering BSs indexed by the set $\mathcal{J}_{n,j} \in \mathcal{J}_n$, then no other UE connected to the muted BSs can be simultaneously scheduled on the same PRB $l$. Thus, if $s_{n,l,j} = 1$ in (22b), the second term on the left-hand-side must be equal to zero. Furthermore, in the case that $s_{n,l,j} = 0$, the constraints in (22b) ensure that single user transmissions are carried out, where each BS $m \in \mathcal{J}_{n,j}$ is allowed to schedule a maximum of one UE per PRB, over all possible interference scenarios $j \in \mathcal{J}'$. Since it is possible that specific BSs, within the cooperation cluster, do not belong to the set of strongest interfering BSs of any UE, the constraints in (22c) complement the restriction on the single user transmissions from (22b). Additionally, the total instantaneous achievable data rate of UE $n$, on PRB $l$, denoted by $r_{n,l}$, is calculated in (22d) as the achievable data rate for the selected interference scenario $j$, as defined by the coordinated decision variable $s_{n,l,j}$. It is worth noting that there is a one-to-one mapping between $r_{n,l,j}$ and $s_{n,l,j}$, thus, there is no requirement for a lookup table function as used in (16). Furthermore, the constraints in (22e) are incorporated as a preprocessing step to ensure that no PRB is scheduled to UEs for which a maximum PF metric for the corresponding interference scenario $j$ is not available. Finally, the coordinated decision variable $\mathbf{S}_l$ is binary as described by the constraints in (22f).

It can be easily proven that the problem formulations in (15) and (22) are equivalent. Furthermore, the proposed parallelized formulation in (22), reduces significantly the CS with muting problem complexity, allowing its application even for large-size networks.

### 4.3 Generalized greedy heuristic algorithm

The greedy heuristic deflation algorithm in [25] (see algorithm in Section II) iteratively solves the CS with muting problem per PRB $l \in \mathcal{L}$, where at each iteration one BS is muted, corresponding to the BS $m \in \mathcal{M}$ that, when muted, maximizes the sum of the PF metrics among all UEs on PRB $l$. The algorithm stops when muting any additional BS does not improve the sum of the PF metrics with respect to the previous iteration. There is no guarantee that the heuristic algorithm yields a globally optimal point because the quality of the scheduling decision depends directly on the gain achieved from muting one interfering BS at a time.

Given the abovementioned disadvantage of the CS with muting greedy heuristic algorithm from [25], an extension is proposed in this work, called *generalized greedy heuristic algorithm*, which trades off computational complexity with performance gains. The main difference, with respect to the algorithm in [25], is the evaluation of additional muting patterns per iteration, where for PRB $l \in \mathcal{L}$, the set of muting indicators

$$\hat{\mathcal{M}} = \bigcup_{\hat{m} \in \{1,\ldots,\tilde{m}\}} \binom{\mathcal{M}'}{\hat{m}}, \tag{23}$$

defines the muting patterns to be evaluated. In (23), the binomial coefficients of the set $\mathcal{M}'$, of possible muted BSs, are evaluated by selecting $\hat{m}$ BSs at a time. The configuration parameter $1 \leq \tilde{m} \leq M - 1$, controls the complexity of the proposed generalized greedy heuristic algorithm by determining the muting patterns to be evaluated. If $\tilde{m} = 1$, the generalized greedy heuristic algorithm reduces to the heuristic algorithm from [25]. In the case that $\tilde{m} = M - 1$, the generalized greedy heuristic algorithm performs an exhaustive search.

## 5 Simulation results

In this section, extensive simulation results are presented to evaluate the performance of the CoMP CS schemes with respect to a PF scheduler without any cooperation, referred to as "non-coop. PFS". The proposed parallelized sub-problem formulation as presented in Section 4.2, labeled as "CS-ILP", is examined together with the greedy algorithm described in [25], denoted as "CS-GA", and the proposed generalized greedy algorithm of Section 4.3, labeled as "CS-GG". In the simulations, $M' = 2$ strongest interfering BSs per UE are considered.

## 5.1 CS with muting—performance analysis

In order to study the performance of the CS with muting schemes, Monte Carlo standalone simulations have been carried out, where the $\mathrm{CSI}_{\mathrm{lte}}^{\mathrm{R\text{-}11}}$ reports are generated based on channels obtained from a 3rd Generation Partnership Project (3GPP) compliant system level simulator, as specified in [29, 35–37]. In each transmission time $t$, the average user throughput over time of UE $n \in \mathcal{N}$, used in (13), is updated based on the scheduling decisions made at the previous transmission time $t-1$, as

$$R_n(t) = \beta\, R_n(t-1) + (1-\beta)\, r_n(t-1), \tag{24}$$

with $\beta = 0.97$, denoting the forgetting factor parameter used to trade off user throughput and fairness [38]. The total instantaneous achievable data rate of UE $n$, at the previous transmission time, denoted by $r_n(t-1)$, is calculated as given by e.g., (22d).

Initially, the performance of the CS with muting algorithms in terms of average user throughput, is studied with respect to the data rates the users can achieve per symbol and to the noise power level considered in the calculation of these achievable data rates. In practical systems such as LTE-Advanced, finite modulation and coding schemes (MCSs) are used which restrict the achievable data rates per symbol to a given range [39, 40]. For the current analysis, two cases are considered with respect to the maximum achievable data rate: (i) the MCS is unbounded, denoted as "Unb. MCS", and (ii) a maximum achievable data rate of 5.4 bits/symbol is used, referred to as "B. MCS". Similarly, there are two assumptions with respect to the noise power level, where in a first case, noise free decoding is assumed, denoted as "N.-less", which considers that $\sigma^2 = \epsilon$, with $\epsilon$ arbitrarily small but larger than zero, and in a second, a typical receiver noise figure of 9 dB is considered, referred to as the "Noisy" case.

The cell-edge and the geometric mean of the user throughput are shown in Figs. 2 and 3, respectively, for a scenario with $M = 3$ BSs, $N = 30$ UEs (ten UEs per BS) and $L = 10$ PRBs. The cell-edge throughput describes the average user throughput of the cell-edge users and corresponds to the average throughput achieved by the worst 5% of the users. The usage of the geometric mean is proposed by the authors in [25] as a direct measure of the PF scheduler's objective function. The user throughputs achieved by the CS with muting schemes,

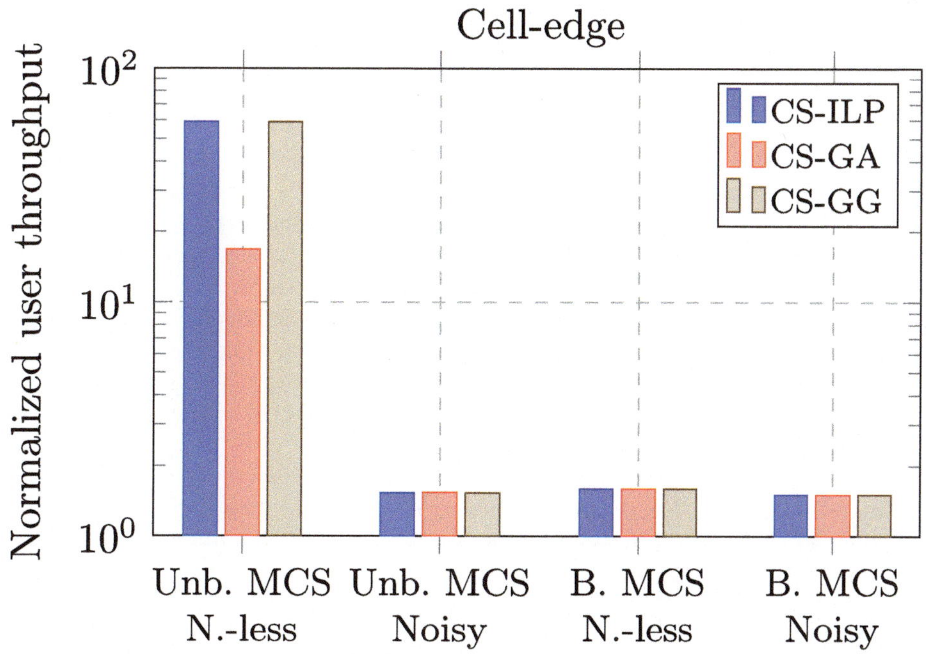


**Fig. 2** Average cell-edge user throughput for the CoMP CS schemes, normalized with respect to the non-coop. PFS. Scenario with $M = 3$ BSs, $N = 30$ UEs, $L = 10$ PRBs, and $M' = 2$ BSs. Four cases with limitations on the maximum achievable data rate and the noise power level are considered. There is no out-of-cluster interference

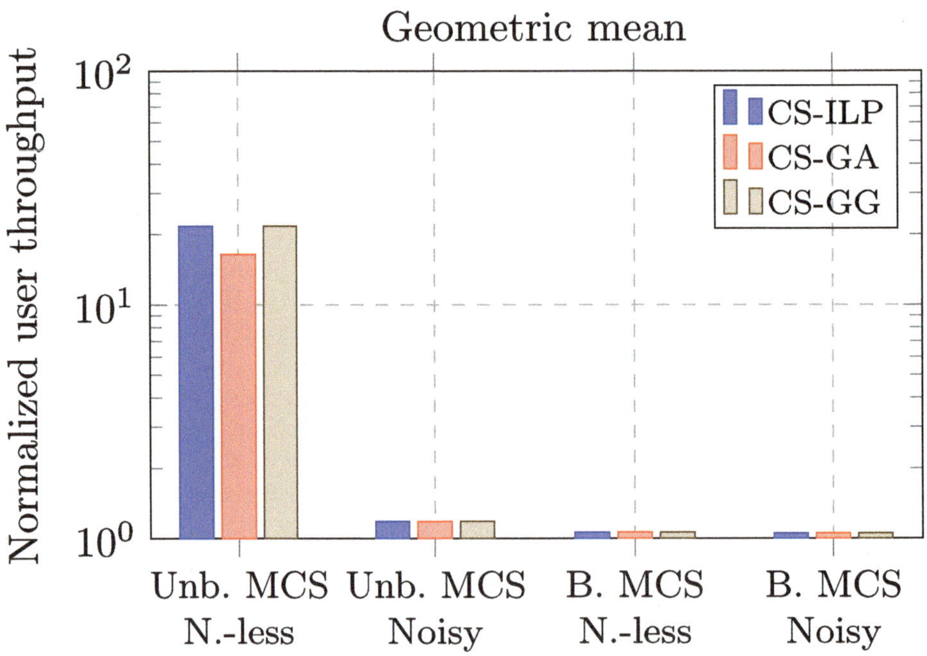


**Fig. 3** Geometric mean of the average user throughput for the CoMP CS schemes, normalized with respect to the non-coop. PFS. Scenario with $M = 3$ BSs, $N = 30$ UEs, $L = 10$ PRBs, and $M' = 2$ BSs. Four cases with limitations on the maximum achievable data rate and the noise power level are considered. There is no out-of-cluster interference

i.e., CS-ILP, CS-GA and CS-GG, are normalized by the resulting user throughput when no cooperative scheduler is applied. Four cases are considered for different combinations of maximum achievable data rate and noise power level, as specified in the horizontal axis. No additional BSs are considered in the network, hence, there is no out-of-cluster interference, such that $I_{n,l}^{\text{oc}} = 0$. It is observed that under no achievable data rate limitations and noise free receivers, i.e., Unb. MCS and N.-less, significant user throughput gains for both, the cell-edge and the geometric mean, are achieved by the cooperative schemes with respect to the non-coop. PFS. Moreover, the optimality of the proposed CS-ILP formulation is notable, with the CS-GA being unable to obtain the optimal solution as explained in Section 4.3. Due to the unboundedness of the MCS and the noise free decoder assumptions in this case, simultaneously muting the two interfering BSs can significantly increase the UE's data rate. Nevertheless, only muting one interfering BS does not yield sufficient PF metric gain, causing the CS-GA scheme to stop prematurely. Such a limitation of the CS-GA is not present in the proposed CS-GG, which achieves the same optimal performance as the CS-ILP scheme. Once limitations are assumed in either the maximum achievable data rate and/or the noise power level, the observed gains from the CS with muting schemes with respect to the non-coop. PFS approach, vanish. Due to the low number of BSs in the cooperation cluster and given the above mentioned limitations, few users benefit from the simultaneous muting of the two interfering BSs.

The average percentage of muted PRBs per BS, for the four different scheduling schemes and four combinations of maximum achievable data rate and noise power level, is presented in Table 3. The non-coop. PFS does not apply muting, therefore the table contains zero entries for all cases. For the CS with muting schemes, according to Figs. 2 and 3, the average percentage of muted resources per BS reduces when the gain of muting is restricted. It is worth noting that even when the maximum achievable data rate is assumed to be unbounded above and noiseless receivers are considered, the CS-ILP scheme mutes 2/3 of the resources per BS, which means that each BS orthogonally schedules its UEs over $1/M$-th of the available resources. Further muting resources per BS, reduces the network performance because the user throughput distribution lacks fairness among the BSs. The

**Table 3** Average percentage of muted resources per BS

| Scheduling scheme | Non-coop. PFS | CS ILP | CS GA | CS GG |
|---|---|---|---|---|
| Unb. MCS & N.-less | 0 | 0.67 | 0.53 | 0.67 |
| Unb. MCS & Noisy | 0 | 0.22 | 0.21 | 0.22 |
| B. MCS & N.-less | 0 | 0.08 | 0.08 | 0.08 |
| B. MCS & Noisy | 0 | 0.08 | 0.07 | 0.08 |

value $1/M$, represents a fundamental limit of the cooperation and agrees with analytical studies presented by Lozano et al. in [26]. Although the performance of the heuristic CS with muting schemes is near-optimal under current practical network conditions, it is envisioned that the evolution of mobile communications introduces for future networks receivers with enhanced capabilities to suppress noise and to support the usage of higher MCSs. Hence, the results in Figs. 2 and 3, provide a reference to the potential gains of these heuristic schemes with respect to the optimal performance obtained with the proposed CS-ILP.

In the following, the more practical scenario with a maximum achievable data rate of 5.4 bits/symbol and a noise figure of 9 dB, is considered in order to study the performance of the CS with muting schemes, with respect to the cooperation cluster size. For that purpose, a network of seven BSs is simulated, where a single cooperation cluster of variable size, with $M \in \{3, \ldots, 7\}$, is assumed. The BSs outside of the cooperation cluster are assumed to transmit data with maximum transmit power over the complete simulation time, such that $I_{n,l}^{oc} \geq 0$. Additionally, two alternatives for the number of strongest interfering BSs per UE are considered with $M' = M - 1$ and $M' = 2$. Each BS serves 10 UEs over $L = 10$ PRBs. The cell-edge throughput, as a function of the cooperation cluster size $M$, is shown in Fig. 4 for the UEs served by the BSs within the cooperation cluster. The presented results are normalized with respect to the user throughput achieved by the same UEs, if the non-coop. PFS is used. In accordance to the previous results, the CS with muting schemes provide gains with respect to a non-cooperative PF scheduler, with an increase in the gains for a larger cooperation cluster size. The reason for such an improvement is the opportunity of further reducing the interference, and thus enhancing the SINR, by increasing the amount of BSs involved in the coordinated scheduling procedures. It is also observable that a larger number of $M'$ strongest interfering BSs per UE improves the gains of the CS with muting schemes, at the cost of additional computational complexity and signaling overhead. In agreement with the results presented in Figs. 2 and 3, the greedy algorithm of [25] shows a near-optimal performance under practical conditions, with the proposed CS-GG algorithm performing better than the CS-GA scheme when all possible strongest interfering BSs are considered. Similar results were observed for the geometric mean of the user throughput.

### 5.2 CS with muting—potential gains

In this section, system level simulation results are presented in order to demonstrate the achievable gains of the CS with muting schemes for LTE-Advanced macro-only and heterogeneous networks in an urban deployment. In both cases, $N = 630$ UEs are served over $L = 10$ PRBs, by $M = 21$ BSs in the macro-only network and $M = 42$ BSs in the heterogeneous case where, one pico cell is located within the coverage area of a macro BS with a separation distance of 125m from the macro BS. The UEs are

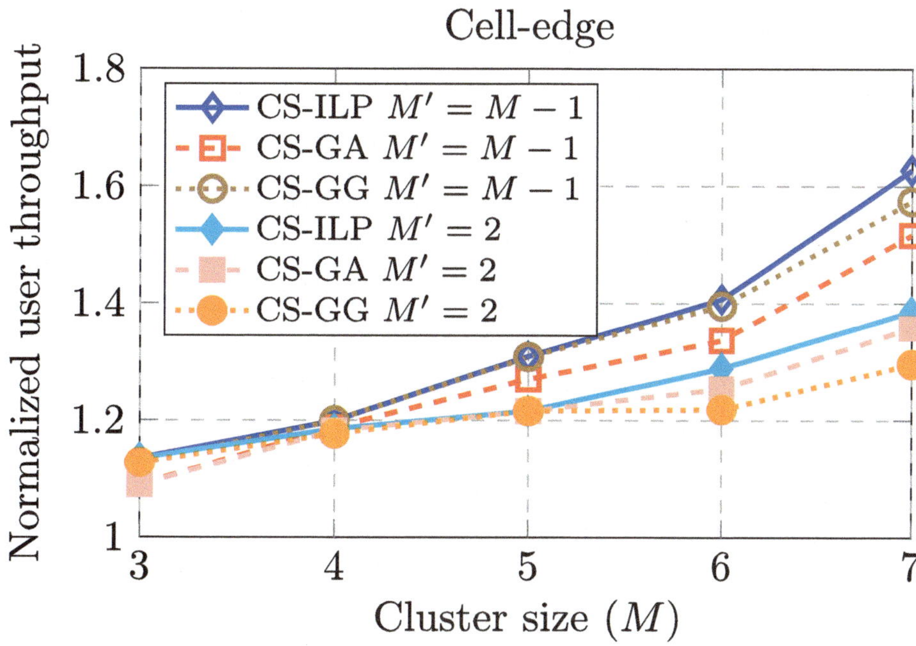


**Fig. 4** Average cell-edge user throughput, normalized with respect to the non-coop. PFS, for different cooperation cluster sizes ($M$). Scenario with $N = 10M$ UEs, $L = 10$ PRBs and $M' = \{2, M - 1\}$ BSs. There is out-of-cluster interference

uniformly distributed in the macro-only case, while in the heterogeneous network the UEs are located in a hotspot fashion, where 2/3 of the UEs are deployed in the vicinity of the pico BSs. In the heterogeneous networks cell range expansion is used with a SINR off-set of 6dB for the small cells. The out-of-cluster interference is modeled using the wrap-around technique [41], where additional BSs are deployed surrounding the $M$ BSs of interest. Additionally, $\mathrm{CSI}_{\mathrm{lte}}^{\mathrm{R\text{-}11}}$ reporting with periodicity of 5ms is applied. Full buffer conditions, ideal link adaptation and rank one transmissions are assumed, i.e., all users are always active and demand as much data as possible, there are no decoding errors and only transmit beamforming is applied, respectively. For more information on 3GPP-compliant system level simulations, including channel and path-loss models, the interested reader is referred to [29] (See 3GPP Case 1 and Case 6.2 from Section A.2.1).

The cell-edge and the geometric mean of the user throughput, normalized with respect to the non-coop, PFS, are presented in Fig. 5 for a macro-only, and in Fig. 6 for a heterogeneous network. In terms of the geometric mean, gains are limited to values around 11% for both cases, macro-only and heterogeneous networks. Additionally, the difference between the proposed schemes, i.e., CS-ILP and CS-GG, and the state-of-the-art CS-GA is negligible. For the cell-edge users, even with the limitation in the number of strongest interfering BSs, the CS with muting schemes achieve a considerable gain in performance, with gains above 40% being observable. In the case of heterogeneous networks, the cell-edge gain is even higher, due to the presence of a clear strongest interfering BS for the pico UEs, i.e., the macro BS, which is considered to cooperate within the restriction of $M' = 2$. The proposed generalized greedy algorithm CS-GG, performs better than the scheme in [25], i.e., CS-GA, which follows from the flexibility to muting additional BSs. The average percentages of muted PRBs for the CS with muting schemes in the macro-only and heterogeneous networks are presented in Table 4. One implication of the muted PRBs is the opportunity to save transmit power at the BSs, with the proposed CS-ILP and CS-GG schemes muting more PRBs than the CS-GA scheme.

It is worth to remark that the performance of equivalent decentralized CoMP CS schemes is expected to be upper bounded by their centralized counterparts, as shown in [25]. Hence, the results provided in this work give a reference of the maximum expected performance for the decentralized schemes. A decentralized CoMP CS scheme is proposed and compared with the results in the paper at hand in [42].

Finally, focusing on the proposed parallelized CS-ILP, it is recognizable that the simplifications proposed in Section 4.2, enable the implementation of such a CS with muting approach even for medium to large-size networks. Hence, instead of solving the CS with muting problem by considering the total of $N = 630$ UEs per PRB $l \in \mathcal{L}$, only $|\mathcal{N}_l'| = 136$ and $|\mathcal{N}_l'| = 213$ UEs were included in average for the macro-only and the heterogeneous network, respectively. That implies a reduction of 78 and 66% in the problem size, for each of the cases, respectively.

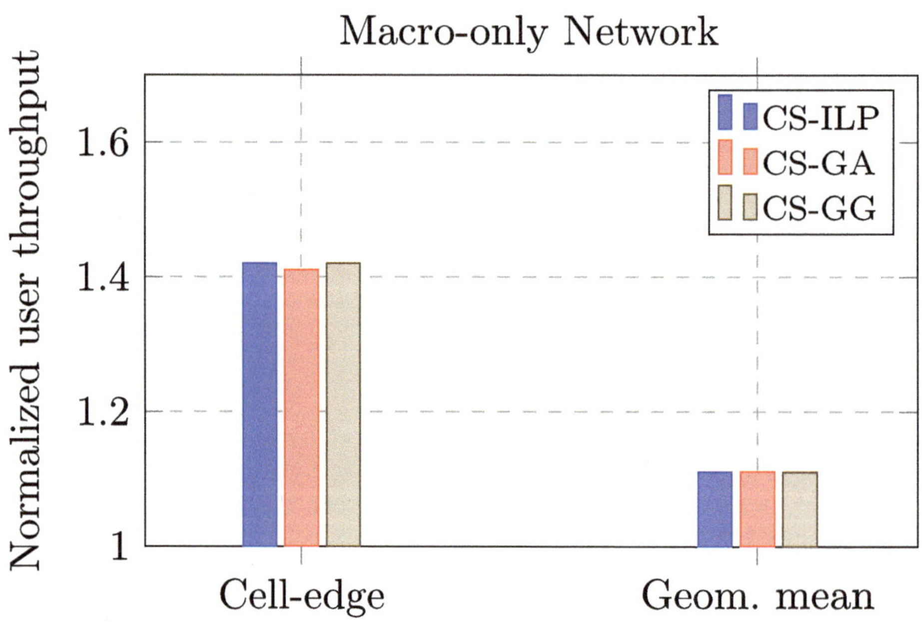


**Fig. 5** Cell-edge and geometric mean of the average user throughput, normalized with respect to the non-coop. PFS, for a scenario with $M = 21$ BSs, $N = 630$ UEs, $L = 10$ PRBs, and $M' = 2$ BSs, with wrap-around technique. Results from system level simulations of a macro-only network

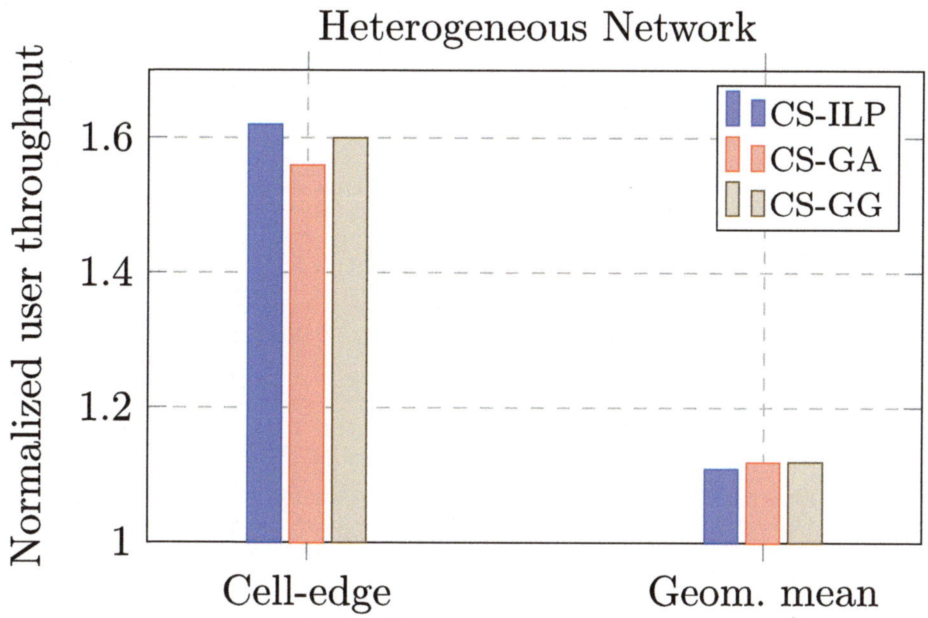


**Fig. 6** Cell-edge and geometric mean of the average user throughput, normalized with respect to the non-coop. PFS, for a scenario with $M = 42$ BSs, $N = 630$ UEs, $L = 10$ PRBs, and $M' = 2$ BSs, with wrap-around technique. Results from system level simulations of a heterogeneous network

## 6 Conclusions

In this paper, the coordinated scheduling with muting problem in the framework of LTE-Advanced networks with a centralized controller has been studied. A novel integer non-linear program formulation has been proposed to solve the problem optimally, where a computationally efficient equivalent integer linear program reformulation has been proposed to extend the applicability of the derived scheme even to large-size networks.

Extensive system level simulation results show that coordinated scheduling with muting can potentially improve the cell-edge user performance, with higher gains in heterogeneous networks. Nevertheless, these gains are limited by the remaining uncoordinated interference and the finite time/frequency/space resources to be shared in the network.

The evaluation of the proposed integer linear program formulation, as well as the state-of-the-art heuristic greedy algorithm, for alternative traffic models in the non-full buffer case, are recommended for future studies. In the case of low demand, the possibility of reducing residual interference and increasing the degrees of freedom for the cooperation can further enhance the performance gains of the mentioned coordinated scheduling schemes.

**Table 4** Average percentage of muted resources

| Network | CS-ILP | CS-GA | CS-GG |
|---|---|---|---|
| Macro-only | 0.11 | 0.10 | 0.10 |
| Heterogeneous | 0.13 | 0.08 | 0.09 |

**Acknowledgments**
We acknowledge support by the German Research Foundation and the Open Access Publishing Fund of Technische Universität Darmstadt.

**Competing interests**
The authors declare that they have no competing interests.

**Author details**
[1]Communication Systems Group, Darmstadt University of Technology, 64283 Darmstadt, Germany. [2]Technology Innovation Division, Deutsche Telekom AG, 64295 Darmstadt, Germany.

**References**

1. SG Kiani, GE Oien, D Gesbert, in *2007 IEEE Wireless Communications and Networking Conference*. Maximizing Multicell Capacity Using Distributed Power Allocation and Scheduling, (2007), pp. 1690–1694. doi:10.1109/WCNC.2007.318
2. SG Kiani, D Gesbert, Optimal and distributed scheduling for multicell capacity maximization. IEEE Trans. Wirel. Commun. **7**(1), 288–297 (2008)
3. G Boudreau, J Panicker, N Guo, R Chang, N Wang, S Vrzic, Interference coordination and cancellation for 4G networks. IEEE Commun. Mag. **47**(4), 74–81 (2009)
4. Y Yang, Q Zhang, P Shang, J Liu, in *2014 IEEE 25th Annual International Symposium on Personal, Indoor, and Mobile Radio Communication (PIMRC)*. Interference alignment based coordinated scheduling for uplink small cell enhancement, (2014), pp. 1109–1114. doi:10.1109/PIMRC.2014.7136333
5. Cisco, Cisco Visual Networking Index: Global mobile data traffic forecast update, 2015–2020 (2016). http://www.cisco.com
6. L Zheng, DNC Tse, Diversity and multiplexing: a fundamental tradeoff in multiple-antenna channels. IEEE Trans. Inf. Theory. **49**(5), 1073–1096 (2003)

7. A Paulraj, R Nabar, D Gore, *Introduction to space-time wireless communications*. (Cambridge University Press, Cambridge, 2003)
8. AB Gershman, ND Sidiropoulos, *Space-time processing for MIMO communications* (John Wiley & Sons, Chichester, 2005)
9. J Lee, J-K Han, J Zhang, MIMO Technologies in 3GPP LTE and LTE-Advanced. EURASIP J. Wirel. Commun. Netw. **2009**, 302092 (2009)
10. JG Andrews, Seven ways that HetNets are a cellular paradigm shift. IEEE Commun. Mag. **51**(3), 136–144 (2013)
11. YL Lee, TC Chuah, J Loo, A Vinel, Recent advances in radio resource management for heterogeneous LTE/LTE-A networks. IEEE Commun. Surv. Tutorials. **16**(4), 2142–2180 (2014)
12. D Lopez-Perez, I Güvenc, G de la Roche, M Kountouris, TQS Quek, J Zhang, Enhanced intercell interference coordination challenges in heterogeneous networks. IEEE Wirel. Commun. **18**(3), 22–30 (2011)
13. R Irmer, H Droste, P Marsch, M Grieger, G Fettweis, S Brueck, H-P Mayer, L Thiele, V Jungnickel, Coordinated multipoint: concepts, performance, and field trial results. IEEE Commun. Mag. **49**(2), 102–111 (2011)
14. J Lee, Y Kim, H Lee, BL Ng, D Mazzarese, J Liu, W Xiao, Y Zhou, Coordinated multipoint transmission and reception in LTE-Advanced systems. IEEE Commun. Mag. **50**(11), 44–50 (2012)
15. EG Larsson, EA Jorswieck, Competition versus cooperation on the MISO interference channel. IEEE J Sel. Areas Commun. **26**(7), 1059–1069 (2008)
16. P Baracca, F Boccardi, N Benvenuto, A dynamic clustering algorithm for downlink CoMP systems with multiple antenna UEs. EURASIP J. Wirel. Commun. Netw. **2014**, 125 (2014)
17. A Beylerian, T Ohtsuki, Multi-point fairness in resource allocation for C-RAN downlink CoMP transmission. EURASIP J. Wirel.Commun. Netw. **2016**, 12 (2016)
18. OD Ramos-Cantor, J Belschner, M Pesavento, in *2016 IEEE 17th International Workshop on Signal Processing Advances in Wireless Communications (SPAWC)*. Improved link adaptation with coordinated scheduling in non-fully loaded wireless networks, (2016), pp. 1–6. doi:10.1109/SPAWC.2016.7536902
19. L Su, C Yang, S Han, The value of channel prediction in CoMP systems with large Backhaul latency. IEEE Trans. Commun. **61**(11), 4577–4590 (2013)
20. H-L Määttänen, K Hämäläinen, J Venäläinen, K Schober, M Enescu, M Valkama, System-level performance of LTE-Advanced with joint transmission and dynamic point selection schemes. EURASIP J. Adv. Signal Process. **2012**, 247 (2012)
21. B Mondal, E Visotsky, TA Thomas, X Wang, A Ghosh, in *2012 IEEE 23rd International Symposium on Personal, Indoor and Mobile Radio Communications - (PIMRC)*. Performance of downlink comp in LTE under practical constraints, (2012), pp. 2049–2054. doi:10.1109/PIMRC.2012.6362691
22. A Barbieri, P Gaal, S Geirhofer, T Ji, D Malladi, Y Wei, F Xue, in *2012 Information Theory and Applications Workshop*. Coordinated downlink multi-point communications in heterogeneous cellular networks, (2012), pp. 7–16. doi:10.1109/ITA.2012.6181826
23. A Davydov, G Morozov, I Bolotin, A Papathanassiou, in *2013 IEEE Globecom Workshops (GC Wkshps)*. Evaluation of Joint Transmission CoMP in C-RAN based LTE-A HetNets with large coordination areas, (2013), pp. 801–806. doi:10.1109/GLOCOMW.2013.6825087
24. R Agrawal, A Bedekar, S Kalyanasundaram, N Arulselvan, T Kolding, H Kroener, in *2014 IEEE 79th Vehicular Technology Conference (VTC Spring)*. Centralized and Decentralized Coordinated Scheduling with Muting, (2014), pp. 1–5. doi:10.1109/VTCSpring.2014.7022856
25. S Gulati, S Kalyanasundaram, P Nashine, B Natarajan, R Agrawal, A Bedekar, in *2015 IEEE 82nd Vehicular Technology Conference (VTC2015-Fall)*. Performance Analysis of Distributed Multi-Cell Coordinated Scheduler, (2015), pp. 1–5. doi:10.1109/VTCFall.2015.7391069
26. A Lozano, RW Heath Jr, JG Andrews, Fundamental limits of cooperation. IEEE Trans. Inf. Theory. **59**(9), 5213–5226 (2013)
27. C Galiotto, N Marchetti, L Doyle, The role of the total transmit power on the linear area spectral efficiency gain of cell-splitting. IEEE Commun. Lett. **17**(12), 2256–2259 (2013)
28. Q Ye, B Rong, Y Chen, M Al-Shalash, C Caramanis, JG Andrews, User association for load balancing in heterogeneous cellular networks. IEEE Trans. Wirel. Commun. **12**(6), 2706–2716 (2013)
29. 3rd Generation Partnership Project, TR 36.814 v9.0.0 —Further advancements for E-UTRA physical layer aspects (Release 9) (2010)
30. E Dahlman, S Parkvall, J Sköld, *4G: LTE/LTE-Advanced for mobile broadband*, 2nd edn. (Elsevier, Oxford, 2014)
31. 3rd Generation Partnership Project, TS 36.213 v11.3.0—physical layer procedures (Release 11) (2013)
32. HJ Zhu, RHM Hafez, Scheduling schemes for multimedia service in wireless OFDM systems. IEEE Wirel. Commun. **14**(5), 99–105 (2007)
33. J Wu, NB Mehta, AF Molisch, J Zhang, Unified Spectral Efficiency analysis of cellular systems with channel-aware schedulers. IEEE Trans. Commun. **59**(12), 3463–3474 (2011)
34. E Balas, Projection, Lifting and extended formulation in integer and combinatorial optimization. Springer Ann. Oper. Res. **140**(1), 125–161 (2005)
35. 3rd Generation Partnership Project, TR 25.996 v11.0.0—Spatial channel model for Multiple Input Multiple Output (MIMO) simulations (Release 11) (2012)
36. 3rd Generation Partnership Project, TR 36.872 v12.1.0—Small cell enhancements for E-UTRA and E-UTRAN - Physical layer aspects (Release 12) (2013)
37. 3rd Generation Partnership Project, TR 36.874 v12.0.0—Coordinated multi-point operation for LTE with non-ideal backhaul (Release 12) (2013)
38. Motorola, R1-060877 – Frequency Domain Scheduling for E-UTRA (2006). http://www.3gpp.org
39. S Sesia, M Baker, I Toufik, *LTE—the UMTS Long Term Evolution: from theory to practice*, 2nd edn. (John Wiley & Sons, Chichester, 2011)
40. Y Cheng, A Philipp, M Pesavento, in *2012 Proceedings of the 20th European Signal Processing Conference (EUSIPCO)*. Dynamic rate adaptation and multiuser downlink beamforming using mixed integer conic programming, (2012), pp. 824–828
41. Y Yoon, L Duan, F Khaleghi, ACK Soong, Basic operation of the wrap-around technique for system-level simulation. 3GPP2 Technical Report (2002)
42. OD Ramos-Cantor, M Pesavento, in *2017 IEEE 18th International Workshop on Signal Processing Advances in Wireless Communications (SPAWC)*. Decentralized coordinated scheduling with muting in LTE-Advanced networks, (2017), pp. 1–5

# 7

# A spatial slotted-Aloha protocol in wireless networks for group communications

Mingyu Lee, Yunmin Kim and Tae-Jin Lee*

**Abstract**

In a large-scale group communication networks, we propose a slotted-Aloha-based access control (SA-AC) scheme with the optimal transmission probability (TP) using the stochastic geometry. Since the performance of SA-AC scheme depends on the TP of the members in a group, the analytical model presents a joint view for the downlink (DL) and the uplink (UL) with TP of the members in a group. We present simple and closed form expressions for the DL and UL joint probability by using Poisson point process (PPP). Furthermore, we analyze the dynamic TP and the optimal TP to maximize the DL/UL joint probability, for which the inclusion of a member and the transmission of the member is determined by the DL and UL thresholds. Since the requirement of the DL/UL joint probability varies with the type of services, the optimal TP has to be determined for an efficient group communication service. The performance of the SA-AC scheme with the optimal TP is demonstrated, and it is shown to be superior over the other schemes.

**Keywords:** Group communication, Spatial Aloha, Stochastic geometry

## 1 Introduction

In a large-scale group communication network, some nodes may form a cluster and clusters communicate with one another. For example, in a tactical group communication network, a commander monitors the conditions of the soldiers. A soldier equips with a monitoring system [1], and the monitored status is transmitted to the base camp through an unmanned aerial vehicle (UAV) [2]. If all the soldiers transmit their information at the same time to the UAV, the network congestion will occur increasingly as the number of soldiers increases. Thus, a hierarchical system in which relays or vehicles collect the information of the soldiers and they transmit the collected information to the UAV could be desirable. In the system, the soldiers are grouped to communicate with a relay or a vehicle, and they act as the group members and the relay or the vehicle becomes the group leader. The group leader has to cover the group members as many as possible within its communication range. The members have to choose the best group leader among the candidate group leaders. The group leader will control the transmissions of the group members within its coverage.

With a massive number of nodes, the group leaders and the members form a large-scale network and they can be modeled by the Poisson point process (PPP). PPP is a spatial point process, which is widely used to model a wireless network. However, a large-scale network with group communications in which group leaders and members send and receive information has not been studied. Since the members within the coverage of a leader are able to transmit for communications, the locations of the members are no longer independent to the location of the leader. Moreover, the interference model should be different from that in [3] because the interference from the covered area of a leader is important compared to that from the non-covered area of a leader. In addition, the covered members of a group may decide whether to transmit or not by a transmission control (TC) value, which is globally determined by the leaders in a centralized manner. Thus, a new network model to capture the behavior of hierarchical group communications needs to be setup.

Since wireless resources in a group communication network are shared in a distributed manner, the performance depends on the medium access control (MAC) scheme. In this sense, transmission probability (TP), which is decided by the TC value, should be carefully chosen. Since the TP of nodes in an Aloha-based

*Correspondence: tjlee@skku.edu
College of Information and Communication Engineering, Sungkyunkwan University, Suwon 16419, South Korea

scheme may be adaptively determined by channel variation, local topology, and target utility, a new MAC protocol can be designed by using the optimal TP which is derived from the PPP models in group communication networks.

In this paper, we first derive the dynamic TP which is determined by the statistical distributions of group leaders and the DL coverage probability. We focus on the special case to derive the closed-form expressions. The coverage probability and the average numbers of members per leader in the DL are derived for the dynamic TP. The dynamic TP has been known to be optimal in terms of the MAC layer. However, the performance using the dynamic TP has not been evaluated by considering geometrical distributions of network elements.

Next, we propose the optimal TP which maximizes the downlink (DL) and uplink (UL) joint probability. The DL/UL joint probability is that the transmission of the leader and the transmission of a member are performed within the DL coverage threshold and the UL transmission threshold. The probability is determined by the probability density function (pdf) of the distance between the leader and a member which is within the coverage of the leader. The pdf is derived and is utilized to model the UL probability and the DL/UL joint probability. Since the DL/UL joint probability is maximized for a certain distance between the leader and a member, the closed-form expressions of the optimal TP for the distance can be derived. In a practical system, the distance is usually hard to be determined. In our proposed scheme, however, a leader can determine the optimal TP by using the average number of members per leader.

Finally, we present the performance of the DL/UL joint probability and the average achievable rate for a target distance in our model. In addition, the average achievable rates for the different TPs are provided. From the analytical model, the effect of TP in a group communication network is provided. Furthermore, a new policy to determine the optimal TP is proposed by considering the geometrical effects of network elements. For the group communication network, the performance at a target distance is maximized by the proposed slotted-Aloha-based access control (SA-AC) scheme.

The contributions of our work are summarized in the following points:

- We propose a new SA-AC scheme with the optimal TP to maximize the DL/UL joint probability for group communication network.
- We define pdf of the distance between a leader and its covered member for a given DL threshold, and the closed form expression of the optimal TP to maximize the DL/UL joint probability for a target member.
- We evaluate the performance of the proposed SA-AC scheme with the traditional schemes.

The rest of this paper is organized as follows. Section 2 reviews the recent PPP studies of DL, UL, and the MAC perspectives. In Section 3, we explain the proposed SA-AC scheme and the other SA-AC schemes in group communication networks. Section 4 analyzes the dynamic TP and the optimal TP by using the PPP model and presents an access scheme with the optimal TP. Section 5 presents numerical results to demonstrate the access control schemes with different TPs. Finally, we conclude in Section 6.

## 2 Related works

Modeling of the DL of a large-scale network using stochastic geometry has been studied in [4, 5]. The performance of the DL system depends on the locations of base stations (BSs), and it has been derived by modeling BSs as a PPP [6–9]. The general expressions for the coverage probability and achievable rate at a typical user are presented in [4]. In some special cases, the closed-form expressions, which give an intuitive view for the DL system, are provided. In [5], the association probability for a certain type of BSs which are modeled as multiple PPPs with different spatial densities and system parameters is studied. The analytical model in [5] presents the outage and spectral efficiency performance for the DL system deploying different types of BSs. Most PPP models for the DL system are analyzed from the view point of a typical user with a DL communication threshold, which shows that the tractable models are fairly accurate compared to actual systems. The performance of the resource management schemes in the DL (e.g., coverage expansion [5], fractional frequency reuse (FFR) [10], and resource partitioning [11]) is analyzed by using the PPP.

The analytical model using PPP for the UL of a large-scale network has also been developed. The models in [12–14] are developed by dividing Voronoi cells from the perspective of users which are modeled as a PPP. Each user has its own BSs and UL resources for transmission. The transmission power is controlled by themselves according to its locations. For a typical transmission, the other transmissions are assumed as interference. In another approach for the UL model in [3], cells are divided with respect to BSs. The set of active users which satisfy the cutoff threshold for the UL is able to communicate with their BSs. The cutoff threshold for the UL is the average received power required at the serving BS. For a typical active user operating on an allocated channel in its serving BS, the other active users on the channel in the other BSs are assumed to be as interferences. The model provides tractability by assuming that the interfering users constitute a PPP. Since the correlations among users occur by the locations of the associated BSs and the tagged channel, the assumption is required to provide the insights for the cellular UL performance. The performance of the UL with a resource management scheme also has been analyzed with the PPP as in the DL [13].

For the clustered network scenario, [15] and [16] is researched. In these works, clusters are formed based on parent PPP points. To model the clustered networks Mattĕrn cluster process (MCP) and Thomas cluster process (TCP) is used. The interference from the inter and the intra clusters are separately analyzed and approximated using the clustered process properties. The groups or clusters are already formed as a point process without the DL/UL joint model. And the process of forming clusters is not considered when leaders and members are independently deployed.

In addition, the stochastic geometry modeling has been developed by focusing on MAC schemes, especially Aloha-based schemes [17–19]. Since a node in an ad hoc network may operate as a transmitter or a receiver, an analytical model using PPP can be classified by the model of a receiver [6]. If each transmitter node has its corresponding receiver, the model is called as "Poisson bipolar model" [20–23]. In the Poisson bipolar model, the transmitters are modeled as a PPP and are divided into two non-overlapping subsets, whether a transmitter does its role at a specific time or not by using a TP. If the transmitter and receiver are independently modeled as PPPs, the model is called as "independent receiver model." In the model, a transmitter is controlled by TP and the transmission of the transmitter can be received by all subsets of the receivers [6]. When the transmitter is modeled as a PPP and the receiver is modeled as the subset of transmitters by using a specific condition, the model is called as "mobile ad hoc network model." If the set of active transmitters is classified by TP, the set of inactive transmitters at a slot is the set of receivers [24]. Note that a source node has to send data to its far destination node in a mobile ad hoc network. Then, the intermediate nodes are required to relay data to its destination node. If some of the inactive transmitters are in the direction from a source node to a destination node, they can be the receivers of the transmitter, i.e., the source node. In [22], the optimal TPs are derived for the maximum throughput medium access, the max-min fairness medium access, and the proportional fairness medium access. In [21], the performance of the Aloha protocol which is optimized for the proportional fair medium access is analyzed. The authors in [24] propose a spatial reuse Aloha protocol for a multihop network with the optimal TP which maximizes the density of progress, the mean total distance traversed in one hop by all transmissions initialized in some unit area. However, the optimal TP for the group communications with the DL coverage and the UL transmission has not been studied.

## 3 Slotted-Aloha-based access control schemes in group communication networks

In this section, we introduce the proposed SA-AC scheme which utilizes the optimal TP maximizing the DL/UL joint probability for the target distance. The target distance can be varied by the policy of the proposed SA-AC scheme. If the policy of the proposed SA-AC scheme is to maximize the coverage, the optimal TP is determined for maximizing the performance of the outermost members. In the example, the interference for the outermost members should be decreased to satisfy the UL threshold. The distance between the leader and the outermost member is affected by the DL threshold and the interference from the other leaders. Furthermore, since the number of members in the coverage of the leader increases by extending the coverage, the interference should be controlled by TP. So the proposed SA-AC scheme computes the optimal TP which is determined by considering the DL threshold, the UL threshold, the density of the leaders, the density of the members, and the target distance.

We present a network model for group communications and SA-AC schemes with various TP. The notations are summarized in Table 1. Both the group leaders and the group members are positioned according to PPPs. The leaders are distributed according to a PPP $\Phi_l$ with $\lambda_l$, the intensity of the leaders per unit area. The members are arranged according to a PPP $\Phi_m$ with $\lambda_m$, the intensity of the members per unit area. In the network model, let us define a link from a leader to a member as DL and a link from a member to a leader as UL. In the DL, we assume that a typical member is located at the origin. The typical member in the DL is denoted as $m$, and it is associated with the nearest leader among the leaders. The leader at the origin in the UL is denoted as $l$. Since the distributions are invariant, the locations of the leaders for the typical member in the UL are the same as those in the DL. Let $l_i$ denote the leader $i$ and $m_i$ be the member $i$ where $l_i \in \Phi_l \backslash l$ and $m_i \in \Phi_m \setminus m$.

We assume an interference-limited network in which the signal-to-interference ratios (SIRs) both in the DL and the UL are considered. The transmission powers of leaders and members are $P_l$ and $P_m$, respectively. The signals in both links experience the path loss. The path loss exponent is denoted as $\alpha$. The Rayleigh fading with the unit average power is assumed between any points. Let $h_{x,y}$ be the random variable to represent the channel effect from $x$ to $y$. Thus, the SIR at the receiver with the distance $r$ in the DL from $l$ to $m$ (SIR at a typical member in the downlink) is given by

$$\mathrm{SIR}_d(r) = \frac{P_l h_{l,m} r^{-\alpha}}{I_l}, \quad (1)$$

**Table 1** Summary of notations

| Notation | Description |
|---|---|
| $l$ | Leader at the origin in the uplink |
| $l_i$ | Leader $i$ |
| $m$ | Typical member in the downlink |
| $m_i$ | Member $i$ |
| $m_c$ | Set of covered members |
| $\Phi_l$ | PPP of leaders |
| $\Phi_m$ | PPP of members |
| $\Phi_{m_c}$ | Set of locations of covered members |
| $h_{x,y}$ | Channel effect from $x$ to $y$ |
| $\lambda_l$ | Intensity of leaders |
| $\lambda_m$ | Intensity of members |
| $P_l$ | Transmission power of leaders |
| $P_m$ | Transmission power of members |
| $\alpha$ | Pathloss exponent |
| $p_{d,u}$ | Downlink, uplink joint probability |
| $p_u$ | Uplink coverage probability |
| $p_d$ | Downlink coverage probability |
| $T_d$ | Downlink threshold |
| $T_u$ | Uplink threshold |
| $p_a$ | Access control probability |
| $r$ | Distance between $m$ and its nearest leader |
| $r_{d,i}$ | Distance between a leader $i$ and a typical member |
| $r_{u,i}$ | Distance between a member $i$ and a typical leader |
| $r_{du,i}$ | Distance between a covered member $i$ and a typical leader |
| $r_{tar}$ | Target distance |
| $\bar{r}_{max}$ | Average maximum distance between $l$ and its $m_c$ |
| $I_l$ | Cumulative interference from the leaders |
| $I_m$ | Cumulative interference from the members |
| $I_{m_c}$ | Cumulative interference from the covered members |
| $\tau$ | Transmission probability (TP) |
| $\tau^{dyn}$ | TP of the dynamic framed SA-AC |
| $\tau^{fix}$ | TP of the fixed framed SA-AC |
| $K$ | Frame size |
| $K^{dyn}$ | Frame size of the dynamic framed SA-AC |
| $K^{fix}$ | Frame size of the fixed framed SA-AC |
| $SIR_d$ | SIR at a typical member in the downlink |
| $SIR_u$ | SIR at a typical leader in the uplink |
| $SIR_{d,u}$ | SIR at a typical leader in the joint DL/UL transmission |
| $\bar{N}_m$ | Average number of the covered members per leader |
| $E_{ar}$ | Average achievable rate of covered members |

where $I_l$ is the cumulative interference from the leaders except the serving leader,

$$I_l = \sum_{l_i \in \Phi_l \setminus l} P_l h_{l_i,m} {r_{d,i}}^{-\alpha}, \tag{2}$$

where $r_{d,i}$ is the distance between a leader $i$ and a typical member. Since $r$ is invariant in the DL and the UL, the SIR at a typical leader in the UL is

$$\mathrm{SIR}_u(r) = \frac{P_m h_{m,l} r^{-\alpha}}{I_m}, \tag{3}$$

where $I_m$ is the cumulative interference from the members except the member $m$. In the UL, $I_m$ is given by

$$I_m = \sum_{m_i \in \Phi_m \setminus m} P_m h_{m_i,l} {r_{u,i}}^{-\alpha}, \tag{4}$$

where $r_{u,i}$ is the distance between a member $i$ and a typical leader.

The SA-AC scheme is assumed to work in a group communication network where the time slots are synchronized. One viable solution may be equipping global positioning system (GPS) modules on the leaders and the members since the signals from the GPS satellites provide the rather accurate timing and location information. There may be also several methods which can provide the synchronization to the members without GPS modules. Once a network is grouped into several clusters, all nodes can be bounded by the parent-children relationship. All members can be synchronized to the leader by periodically exchanging synchronization and acknowledgment packets with their corresponding leaders using pair-wise synchronization method [25, 26]. The location information may be piggybacked at the synchronization and acknowledgment packet for simultaneous time synchronization and location information distribution. Depending on the velocity of the nodes, the period between the synchronization and the location information distribution can be adjusted accordingly.

The time is divided into multiple slots, and they are partitioned into control slots and data slots. The leaders and the members transmit their data with the SA-AC scheme. A leader broadcasts a control frame in a synchronized control slot. The control frame of a leader includes the information for the group of the leader, the address of the leader, and the TC value. We assume that the TC value is shared by using a specific channel and they are identical among leaders. The control frame is decodable for a member when the SIR value is greater than the DL threshold, $T_d$. The member in the coverage of the leader is denoted as the covered member. Only the covered member joins the group of the nearest leader when the SIR value for the received control frame from the nearest leader is greater than $T_d$. Once the covered member decodes the control frame successfully, it gets the information to transmit its data frame to the leader in the UL. In the UL, the covered member transmits data frame in a data slot by using a TP. The TP is determined by the TC value and the type of the TC value depends on the type of the AC scheme. If the

SA-AC scheme uses the access control probability, $p_a$, as the TC value, the TP is given by

$$\tau = p_a. \tag{5}$$

In the SA-AC scheme using $p_a$, the covered members access each slot with $p_a$. If the SA-AC scheme uses the frame size as the TC value, the TP is given by

$$\tau = \frac{1}{K}, \tag{6}$$

where $K$ is the frame size which represents the number of data slots. In the SA-AC scheme using $K$, a covered member selects and accesses a slot among the $K$ slots. In general, $1/K$ is modeled as the access probability of the contenders in a slot. The framed SA-AC scheme consists of the fixed framed SA-AC scheme and the dynamic framed SA-AC scheme. The fixed framed SA-AC scheme utilizes the fixed TP which is determined by the fixed frame size [27]. Let $\tau^{fix}$ and $K^{fix}$ denote the TP and the frame size of the fixed framed SA-AC, respectively. The fixed framed SA-AC scheme utilizes $\tau^{fix} = 1/K^{fix}$. The dynamic framed SA-AC scheme utilizes the dynamic TP determined by the frame size which is dynamically changed according to the average number of the covered members per leader. The frame size in the dynamic framed SA-AC scheme is known to be optimal when the frame size is equal to the number of contenders [18, 28]. Let $\tau^{dyn}$ and $K^{dyn}$ denote the TP and the frame size of the dynamic framed SA-AC scheme, respectively. Then the dynamic framed SA-AC scheme utilizes $\tau^{dyn} = 1/K^{dyn}$.

However, if the channel effect and pathloss are not considered, such dynamic TP may not be optimal. So we need to find a new optimal TP maximizing the performance of group communications network. The optimal TP is affected by $\lambda_l$, $\lambda_m$, $T_d$, and $T_u$, where $T_u$ is the UL threshold. In addition, it has to be determined by considering the performance of the covered members. The performance varies by the distance between a leader and the members. In Fig. 1, an example of group communications network is shown. Fig. 1a presents the DL transmissions and Fig. 1b shows the UL transmissions. In DL, the leaders broadcast the control frames to the members. $\lambda_l$ determines the distance between a member and its nearest leader, and $T_d$ decides whether the member becomes the covered member. For example, in each group, three members near the leader become the covered members. These members receive the leader's control frames without channel error. Let the member with target distance $r_{tar}$ denote the target member. The target member in group 2 receives interference signal from the leaders in groups 1 and 3, but this interference signal is tolerable for reception.

In UL, the transmission of a member occurs when the member is in the coverage of the leader and determines whether the covered member participates in the UL transmission by $\tau$ and $T_u$. The target member in group 2 transmits data to the leader. The leader in group 2 receives interference signal from the member in group 2 as well as the members in groups 1 and 3. To decode the UL transmission of the target member, the interference signals have to be controlled by the optimal TP $\tau^*$ which is determined by $\lambda_l$, $\lambda_m$, $T_d$, $T_u$, and $r_{tar}$.

## 4 Optimal transmission probability for group communications

In this section, we derive the dynamic TP and the optimal TP. To derive the dynamic TP, we need the DL coverage probability and the average number of the covered members per leader. For the optimal TP, we need to derive the UL coverage probability and the DL/UL joint probability for a target distance. The target distance is the distance between a leader and a target member for which the DL/UL joint probability is maximized. Finally, we develop an analytical model for the performance of the SA-AC schemes using the dynamic TP and the optimal TP. From the model, we derive an optimal TP of the SA-AC scheme.

In our network model, both the leaders and the members are assumed to be arranged according to PPPs. Therefore, the pdf of the distance $R$ between a member and its nearest leader [4] can be expressed as

$$f_R(r) = 2\pi r\lambda_l \exp\left(-\pi r^2\lambda_l\right), \tag{7}$$

where $r$ is the distance between a member and its nearest leader. The members are divided into the covered members and the non-covered members by $T_d$. The following lemma provides the probability that a member is in the coverage of the nearest leader.

**Lemma 1** *For the DL threshold $T_d$, the DL coverage probability that a member with the distance $r$ from its nearest leader is in the coverage of the leader is*

$$p_d(r, T_d) = e^{-\pi r^2\lambda_l\zeta_l(T_d)}, \tag{8}$$

*where* $\zeta_l(T_d) = T_d^{2/\alpha}\int_{T_d^{-2/\alpha}}^{\infty} \frac{1}{1+(u_l)^{\alpha/2}}du_l$. *Then, the DL coverage probability that a member is in the coverage of the leader is*

$$p_d(T_d) = \frac{1}{1+\zeta_l(T_d)}. \tag{9}$$

*Proof* See Appendix A.1. □

Lemma 1 shows that $p_d(r, T_d)$ decreases by increasing $r$, $T_d$, and $\lambda_l$. Since the received signal strength from the nearest leader decreases as $r$ increases, $p_d(r, T_d)$ decreases. For a certain $r$, the SIR value may not be higher than $T_d$ for large $T_d$. Since the increase of $\lambda_l$ causes the increase of interference for a certain $r$, $p_d(r, T_d)$ decreases.

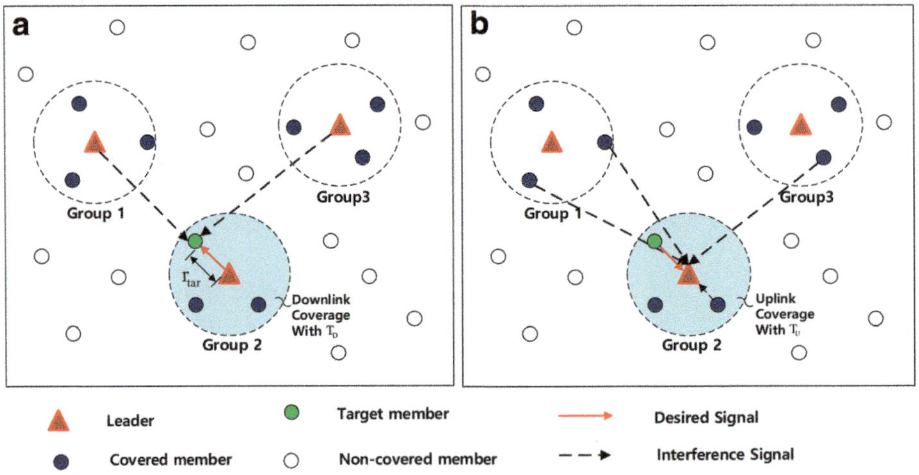


**Fig. 1** An example of group communications in a group communication network. **a** DL transmission. **b** UL transmission

Lemma 1 implies that the number of the covered members decreases by increasing $T_d$ for both $p_d(r, T_d)$ and $p_d(T_d)$. The expressions in Lemma 1 yield a closed-form when $\alpha = 4$, since $\zeta_l(T_d) = \sqrt{T_d}\left(\frac{\pi}{2} - \arctan\left(\frac{1}{\sqrt{T_d}}\right)\right)$.

From Lemma 1, we derive the dynamic TP, which determines the transmission probability of the covered members.

**Theorem 1** *For the DL threshold $T_d$, the dynamic TP that is dynamically changed by the number of the averaged covered members is*

$$\tau^{dyn} = \min\left(\frac{1}{\bar{N}_m}, 1\right) = \min\left(\frac{\lambda_l}{\lambda_m p_d(T_d)}, 1\right). \quad (10)$$

*Proof* If we assume that $S$ is the area of an entire network, $\bar{N}_m$ is obtained by dividing the average number of the covered members in $S$ by the average number of the leaders in $S$. Since the intensity of the covered members is $\lambda_m p_d(T_d)$, the average number of the covered members per leader $\bar{N}_m = \frac{\lambda_m p_d(T_d) S}{\lambda_l S} = \frac{\lambda_m p_d(T_d)}{\lambda_l}$ [5]. □

Theorem 1 shows that $\tau^{dyn}$ is adaptive to the number of the covered members per leader. We expect that $\tau^{dyn}$ decreases as $\bar{N}_m$ increases. It implies that the interference to the transmission of a typical covered member decreases as $\bar{N}_m$ increases. Since $\bar{N}_m$ is affected by $p_d(T_d)$, it increases by decreasing $T_d$. Thus, Theorem 1 indicates that $\tau^{dyn}$ decreases and the interference to the transmission of a typical covered member decreases as $T_d$ decreases in the dynamic framed SA-AC scheme.

Since the locations of the covered members are jointly changed by $T_d$ and the locations of the leaders, they are not PPP. It is challenging to model it accurately, but PPP approximation could be utilized [3]. In our model, we approximate them as a PPP. Let $\Phi_{m_c}$ and $\Phi_{m_o}$ denote the locations of the covered members and the locations of the non-covered members. We calculate the pdf of the distance between a leader and a covered member from the following theorem.

**Theorem 2** *The pdf of the distance between a leader and a covered member $m_c$ for a given $T_d$ is*

$$f_{R_{m_c}}(r) = \frac{2\pi r \lambda_l e^{-\pi r^2 \lambda_l (1+\zeta_l(T_d))}}{p_d(T_d)}. \quad (11)$$

*Proof* Since the coverage of a leader determines the distance between a leader and a covered member, the pdf is affected by the $T_d$. In our network model, the covered members are only able to transmit to their leader. Since the distribution of the distances between the leader and the members in DL is not changed in UL, the distribution of the distances between the leader and the covered members is induced by the pdf of $R$. In the UL transmission by using a TP, the interference from the covered members is obtained by normalizing the DL coverage probability at $r$ by $p_d(T_d)$. Thus, $f_{R_{m_c}}(r)$ is given by

$$f_{R_{m_c}}(r) = \frac{p_d(r, T_d) f_R(r)}{p_d(T_d)}. \quad (12)$$

Plugging $p_d(r, T_d)$ in (8) into (12), $f_{R_{m_c}}(r)$ is derived. □

Since $f_R(r)$ is conditioned by the DL coverage probability, $f_{R_{m_c}}(r)$ is affected by $T_d$. We expect that the increase of $T_d$ makes the pdf of $R_{m_c}$ within the DL coverage to increase. When $\alpha = 4$, both $p_d(r, T_d)$ and $p_d(T_d)$ have closed-form expressions, and $f_{R_{m_c}}(r)$ has also a closed-form expression. From Theorem 2, we derive the UL probability that the SIR value for the transmission of a covered member is larger than the UL threshold $T_u$. The SIR value in UL after DL is given by

$$\mathrm{SIR}_{d,u}(r) = \frac{P_m h_{m_c,l} r^{-\alpha}}{I_{m_c}}, \quad (13)$$

where $I_{m_c}$ and $SIR_{d,u}(r)$ are the cumulative interference from the covered members except for the $m_c$ at $l$ and the SIR value for the transmission of the $m_c$ at $l$ with the distance $r$ (SIR at a typical leader in the joint DL/UL transmission), respectively. In UL, $I_{m_c}$ is given by

$$I_{m_c} = \sum_{m_{c,i} \in \Phi_{m_c} \setminus m_c} P_m h_{m_{c,i},l} \left( r_{du,i}^{-\alpha} \right), \quad (14)$$

where $r_{du,i}$ is the distance between a covered member $i$ and a typical leader.

**Lemma 2** *For $\tau$, the UL coverage probability that the SIR value for the transmission of the covered member with the distance $r$ from its leader exceeds the UL threshold $T_u$ is*

$$p_u(r, \tau, T_u) = \exp\left(-\pi r^2 \lambda_m p_d(T_d) \zeta_m(T_u) \tau\right), \quad (15)$$

*where* $\zeta_m(T_u) = T_u^{2/\alpha} \frac{2\pi/\alpha}{\sin(2\pi/\alpha)}$.

*Proof* See Appendix A.2. □

Lemma 2 shows that $p_u(r, \tau, T_u)$ decreases by increasing $r$, $T_u$, and $\lambda_m$. Since the received signal strength from the covered member decreases as $r$ increases, $p_u(r, \tau, T_u)$ decreases. For a certain $r$, the SIR value for the signal may be hard to exceed $T_u$ for large $T_u$. Since the increase of $\lambda_m$ incurs the increase of the interference, $p_u(r, \tau, T_u)$ decreases. The expression in Lemma 2 is a closed-form when $\alpha = 4$, since $\zeta_m(T_u) = \frac{\pi}{2}\sqrt{T_u}$. The following lemma provides the DL/UL joint probability which quantifies the performance of the SA-AC scheme with $\tau$. We define the target member which is a member with target distance, $r_{tar}$. The DL/UL joint probability denotes the probability that a member with $r_{tar}$ from its nearest leader is in the coverage of the leader, i.e., the DL SIR exceeds the DL threshold $T_d$, and the nearest leader is in the coverage of the member transmitting the UL signal with $\tau$, i.e., the UL SIR exceeds the UL threshold $T_u$. The performance of the DL/UL joint probability for the target member shows the effect of $\tau$ for $r_{tar}$.

**Lemma 3** *For $\tau$, the DL/UL joint probability that the DL/UL SIR value for a member with $r_{tar}$ from its nearest leader exceeds $T_d$ and $T_u$ is*

$$\begin{aligned} & p_{d,u}\left(r_{tar}, \tau, T_d, T_u\right) \\ & = p_d\left(r_{tar}, T_d\right) \tau p_u\left(r_{tar}, \tau, T_u\right) \\ & = \tau e^{-\pi (r_{tar})^2 (\lambda_l \zeta_l(T_d) + \lambda_m p_d(T_d) \tau \zeta_m(T_u))}, \end{aligned} \quad (16)$$

*where* $0 \le \tau \le 1$.

*Proof* By letting $r = r_{tar}$ in (7) and (15), $p_{d,u}(r_{tar}, \tau, T_d, T_u)$ is derived. □

Since the DL/UL joint probability is affected by $T_d$, $T_u$, and $r_{tar}$, it decreases by increasing them. However, if the SA-AC scheme uses $\tau^{dyn}$ as $\tau$, $\tau^{dyn}$ increases as $T_d$ increases. Thus, the DL/UL joint probability for the SA-AC scheme is sensitive to $T_d$. If $\alpha = 4$, the closed-form expression of Theorem 3 is derived from the closed-from expressions in Lemma 1 and Lemma 2. From Lemma 3, we expect that $\tau$ varies according to $r_{tar}$. Since $p_{d,u}(r_{tar}, \tau, T_d, T_u)$ has the global extreme values for $\tau$, the optimal $\tau$ can be derived (see Appendix A.3). We now derive the optimal TP that maximizes $p_{d,u}(r_{tar}, \tau, T_d, T_u)$.

**Theorem 3** *For $T_d$ and $T_u$, the optimal TP that maximizes the DL/UL joint probability of a target member with $r_{tar}$ is*

$$\tau^* = \min\left( \frac{1}{\pi (r_{tar})^2 \lambda_m p_d(T_d) \zeta_m(T_u)}, 1 \right). \quad (17)$$

*where* $0 \le \tau^* \le 1$.

*Proof* The maximum DL/UL joint probability is found by

$$\frac{\partial p_{d,u}\left(r_{tar}, \tau, T_d, T_u\right)}{\partial \tau} = 0. \quad (18)$$

Thus the optimum $\tau$ is derived.

$$\tau^* = \frac{1}{\pi (r_{tar})^2 \lambda_m p_d(T_d) \zeta_m(T_u)}. \quad (19)$$

Since $\tau^* \le 1$ in a slot, $\tau^* = 1$ when $r_{tar} < \sqrt{1/\pi \lambda_m p_d(T_d) \zeta_m(T_u)}$. □

From Theorem 3, we derive $\tau^*$ to maximize the DL/UL joint probability for $\bar{r}_{max}$. $\bar{r}_{max}$ is the average maximum distance between a leader and its covered member. In general, since $r_{tar}$ has to be estimated by the target member and reported to the leader, $\bar{r}_{max}$ is hard to be known. However, if we assume that $\lambda_m$, $T_d$, and $T_u$ are given, $\tau^*$ can be obtained by the average number of the covered members as in the dynamic TP. The average number of the covered members can be known by using the number of associated members. Since the number of the covered members within the distance $r_{tar}$ is $\pi (r_{tar})^2 \lambda_m p_d(T_d)$, $\bar{N}_m = \pi (\bar{r}_{max})^2 \lambda_m p_d(T_d)$. So $\bar{r}_{max} = \sqrt{\frac{\bar{N}_m}{\pi \lambda_m p_d(T_d)}} = \sqrt{\frac{1}{\pi \lambda_l}}$ by plugging $\bar{N}_m$ in Theorem 1. Thus, $\tau^* = \min\left(\frac{1}{\bar{N}_m \zeta_m(T_u)}, 1\right)$. It implies that $\tau^*$ for the average number of the covered members is the same as $\tau^*$ for $\bar{r}_{max}$ in the proposed SA-AC scheme.

We now derive the average achievable rate to measure the spectral efficiency performance of the SA-AC schemes. The average achievable rate for a target member with $r_{tar}$ and the average achievable rate of the members are obtained.

**Lemma 4** *For $T_d$ and $\tau$, the average achievable rate for a target member with $r_{tar}$ is*

$$E_{ar}(r_{tar}, \tau) = \tau \mathbb{E}_{\mathrm{SIR_{d,u}(r_{tar})}} \left[ ln \left[1 + SIR_{d,u}(r_{tar})\right] | r_{tar} \right], \quad (20)$$

*where*

$$\mathbb{E}_{\mathrm{SIR_{d,u}(r_{tar})}} \left[ ln \left[1 + SIR_{d,u}\left(r_{tar}\right)\right] | r_{tar} \right] = \int_0^{\infty} \exp\left( -\pi \lambda_m p_d(T_d) \tau (r_{tar})^2 (e^t - 1)^{2/\alpha} \int_0^{\infty} \frac{1}{1 + \left(u_{ar,r_{tar}}\right)^{\alpha/2}} \mathrm{d}u_{ar,r_{tar}} \right) \mathrm{d}t. \quad (21)$$

*Furthermore, the average achievable rate of the covered members is*

$$E_{ar}(\tau) = \tau \int_0^{\infty} \frac{1}{1 + \frac{\pi}{2} \frac{\lambda_m p_d(T_d)^2 \tau}{\lambda_l} \left(e^t - 1\right)^{1/2}} dt. \approx \frac{\lambda_l}{\lambda_m p_d(T_d)^2}. \quad (22)$$

*Proof* See Appendix A.4. □

Both metrics are computed by averaging the UL coverage probability when the transmission of a covered member is governed by $\tau$. The average achievable rate of a covered member is derived from the average achievable rate for a target member. Both metrics are not closed-form expressions, and the numerical integration is required to compute them.

## 5 Numerical results

In this section, we show the numerical results for the analytical models and simulation results with DL threshold $T_d$ and UL threshold $T_u$. We use the path-loss exponent $\alpha = 4$, the transmit power of a leader $P_l = 1$ W, and the transmit power of a member $P_m = 1$ W [20]. The transmission of both the leaders and the members in a single channel is done in multiple time slots. The leader intensity $\lambda_l = 3$ leaders/km$^2$ and the member intensity $\lambda_m = 20$ members/km$^2$ [29]. Both the leaders and the members are distributed with their intensity in a 25-km$^2$ area. And we focus on the sample area in a 1-km$^2$ area to evaluate the performance. Thus, the numerical results within the sample area are obtained in the simulation. In DL, each member computes the DL SIR and checks the DL SIR if it exceeds the DL threshold $T_d$. If the DL SIR exceeds the DL threshold $T_d$, the member becomes the covered member. In UL, the covered members transmit their signals according to the TP which is determined by the traditional SA-AC schemes and the proposed SA-AC scheme. The leader computes the UL SIR for each of the covered members and checks if the UL SIR exceeds the UL threshold $T_u$. if the UL SIR exceeds the UL threshold $T_u$, the DL/UL of the covered member is successful. We compare the performance of the SA-AC scheme with $\tau^*$ to those of the SA-AC schemes with $\tau^{dyn}$ as in Theorem 1 and $\tau^{fix}$ as in [27].

In Fig. 2, we verify the proposed SA-AC scheme by comparing the performance of the analysis and the simulation.

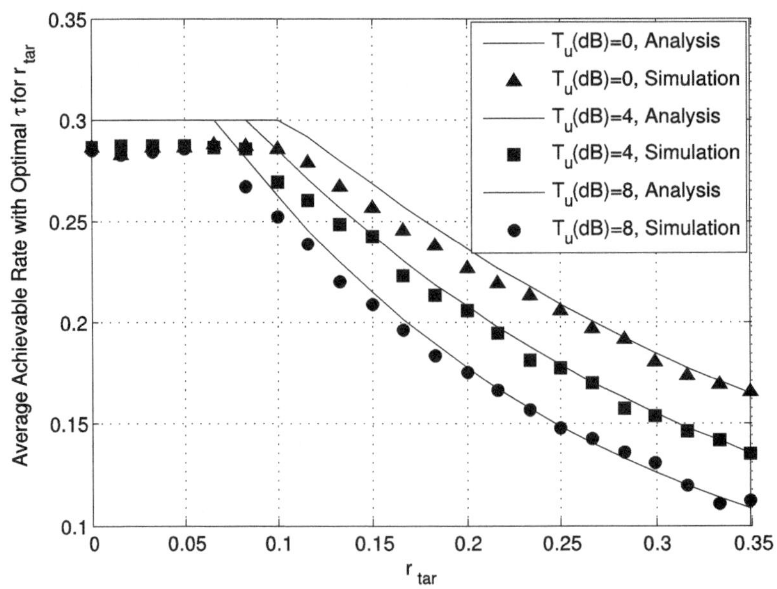


**Fig. 2** Average achievable rate with the optimal TP for varying target distance with different $T_u$ when $T_d$(dB)=−10 (marker: simulation, line: analysis)

The average achievable rate with the optimal $\tau$ is presented by varying $r_{tar}$. Since the performance is analyzed with the PPP assumption, $T_d$ is set to $-10$. The difference between the results is caused by the effect of approximation for the distribution of the covered members. The distribution of the covered members in the simulation is approximated by PPP in the analysis. The lower $T_d$ is selected, the closer the distribution of the covered members in the simulation approaches that of PPP. The exact distribution is not yet discovered [30]. The difference can be explained by the dependence between the covered members and their leaders. The dependence inherits from $T_d$. Since the locations of the covered members are jointly changed by $T_d$ and the locations of the leaders, they are not exact PPP. Since the SIR of the DL transmission is difficult to exceed $T_d$ as the distance between a covered member and its leader increases, the average maximum distance between a covered member and its leader decreases as $T_d$ increases. So the distance between the covered member and its leader is not independent. The dependence can be relaxed as $T_d$ decreases. Since the DL coverage probability increases as $T_d$ decreases, the area of a coverage area becomes larger and the distribution of the covered members approaches the PPP distribution of the members.

Since $r_{tar}$ is the distance between the target member and its leader, the signal strength of the target member decreases as $r_{tar}$ increases. Since $\tau$ is the TP of the covered members in a slot, the number of the covered members participating in the transmission in a slot increases as $\tau$ increases. The signals of the covered members except the target member become the interference to the target member. Thus, for the target member with $r_{tar}$, the SIR of the target member decreases as $r_{tar}$ increases or $\tau$ increases. Since the average achievable rate of a target member with $r_{tar}$ decreases as the SIR decreases, it also decreases as $r_{tar}$ increases or $\tau$ increases. If $r_{tar}$ decreases or $\tau$ decreases, the average achievable rate for a target member with $r_{tar}$ increases. Since the optimal $\tau$ decreases by $(r_{tar})^2$ as in Theorem 3, the interference decreases as $r_{tar}$ increases. However, the signals of the target member and its leader become weaker by $\alpha$ and the number of the transmissions decreases as the optimal $\tau$ decreases, then the average achievable rate of the covered member with the optimal $\tau$ decreases as $r_{tar}$ increases even if the interference decreases as the optimal $\tau$ decreases.

In Fig. 3, the performance of the DL/UL joint probabilities for varying $\tau$ is shown with different $r_{tar}$ when $T_d(dB) = -10$ and $T_u(dB) = 0$. The number of transmission members increases as $\tau$ increases. Since the interference increases as the number of transmission members increases, the SIR for the member with $r_{tar}$ decreases by increasing $\tau$. However, when the SIR for the member with $r_{tar}$ exceeds $T_u$ even if the interference for the member increases, $p_{d,u}(r_{tar}, \tau, T_d, T_u)$ increases as $\tau$ increases. Once the maximum $p_{d,u}(r_{tar}, \tau, T_d, T_u)$ is achieved, the performance starts to decrease by increasing $\tau$. The larger $r_{tar}$, the smaller value of the maximum $p_{d,u}(r_{tar}, \tau, T_d, T_u)$ is achieved. Since the signal becomes weaker as $r_{tar}$ increases, the SIR of the member with $r_{tar}$ is hard to exceed $T_u$ by increasing the number of the transmission members. Hence, $\tau$ has to be decreased to alleviate the interference and satisfy $T_u$ by reducing the number of the transmission members. Thus, $\tau$ to maximize $p_{d,u}(r_{tar}, \tau, T_d, T_u)$ decreases by increasing $r_{tar}$.

In Fig. 4, the performance of the optimal $\tau$ is shown for varying $r_{tar}$ with different $T_u$ when $T_d(dB) = -10$. When the signal strength of the member with $r_{tar}$ is enough to exceed $T_u$ even if all the covered members transmit their data frames, the optimal $\tau$ for $r_{tar}$ is one. The number of the inner members decreases and the signal strength increases as $r_{tar}$ decreases. So the optimal $\tau$ for $r_{tar}$ increases until one as $r_{tar}$ decreases. However, the optimal $\tau$ maximizing $p_{d,u}(r_{tar}, \tau, T_d, T_u)$ decreases by increasing $r_{tar}$ as shown in Fig. 3. The smaller $T_u$, the larger optimal $\tau$ is derived. Since the SIR for the member with $r_{tar}$ becomes easy to exceed $T_u$ as $T_u$ decreases, the optimal $\tau$ for $r_{tar}$ increases until the increase of the interference is tolerable to the member with $r_{tar}$.

In Fig. 5, the performance of the DL/UL joint probabilities is shown for varying $r_{tar}$ with the different SA-AC schemes when $T_d(dB) = -10$ and $T_u(dB) = 0$. We use the dynamic $\tau$ determined by (9) and (10), $\tau^{dyn} \approx 0.16$. Since the SIR of the member with $r_{tar}$ and the optimal $\tau$ to maximize $p_{d,u}(r_{tar}$ decreases as $r_{tar}$ increases, the performance of the SA-AC scheme with the optimal $\tau$ decreases as $r_{tar}$ increases. The performance of the SA-AC schemes with dynamic $\tau$ and fixed $\tau$ decreases as $r_{tar}$ increases. In the schemes with dynamic $\tau$ and fixed $\tau$, the values of $\tau$ do not vary for $r_{tar}$, and the SIR of the member with $r_{tar}$ may not exceed $T_u$ as $r_{tar}$ increases. When $r_{tar}$ is smaller than 0.1, the performance of the optimal $\tau$ is the same as the case with $K$ of 1. We can expect that the members within $r_{tar}$ satisfies $T_u$ even if all the members try to transmit data with $\tau = 1$. However, the DL/UL joint probability of the optimal $\tau$ is larger than that of the case with $K$ of 1 when $r_{tar} = 0.15$, since the members with $r_{tar}$ are strongly interfered by the inner members. The performance of the SA-AC scheme using the optimal $\tau$ approaches that of the SA-AC scheme using dynamic $\tau$ if $r_{tar}$ exceeds 0.1, since dynamic $\tau$ is almost the same as the optimal $\tau$ for $\bar{r}_{max}$. The optimal $\tau$ is adaptively determined by $r_{tar}$ while $\tau$ of the other schemes is not relevant to $r_{tar}$, thus $p_{d,u}(r_{tar}, \tau, T_d, T_u)$ is always maximized. Conversely, we can induce the distance whose DL/UL joint probability is maximized by $\tau$ of the other schemes.

In Fig. 6, the performance of the average achievable rate of the target member is shown for varying $r_{tar}$ with

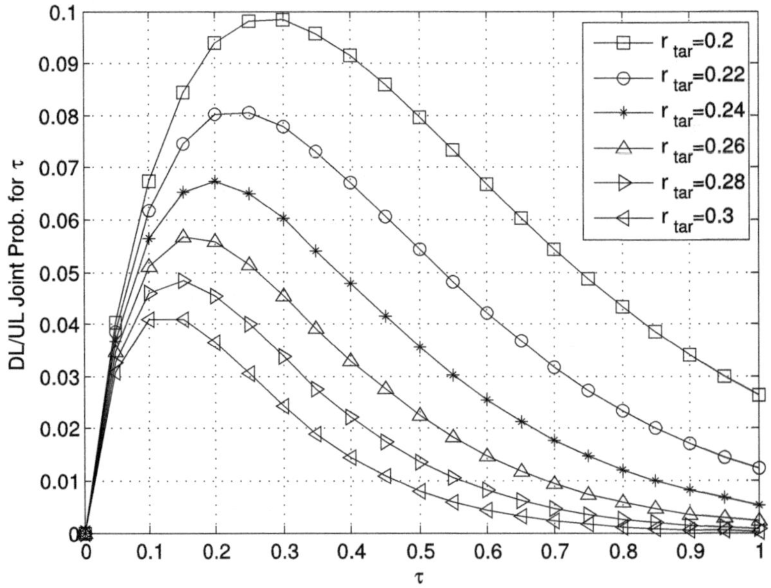


**Fig. 3** DL/UL joint probability for varying TP with different target distance when $T_d$(dB) = −10 and $T_u$(dB) = 0. It is shown that there are optimal TPs to maximize the DL/UL joint probability

the different SA-AC schemes when $T_d(dB) = -10$ and $T_u(dB) = 0$. We use the dynamic $\tau$ as in Fig. 5. Since the signal strength of the target member decreases and the optimal $\tau$ decreases to reduce the interference to the target member as $r_{tar}$ increases. So the SIR and the DL/UL joint probability for the member with $r_{tar}$ and the optimal $\tau$ decrease as $r_{tar}$ increases. In the SA-AC schemes with dynamic $\tau$ and fixed $\tau$, since the interference does not be alleviated by their TP, the performance of them decreases as $r_{tar}$ increases. However, the optimal $\tau$ case maintains the maximum performance. The optimal $\tau$ for $r_{tar}$ decreases so that the SIR for the target member with $r_{tar}$ satisfies $T_u$. However, since the case with $K$ of 1 case among the fixed $\tau$ is not adaptive to $r_{tar}$, the lowest performance is shown when $r_{tar}$ is larger than 0.2. The results show that the decrease of the DL/UL joint probability

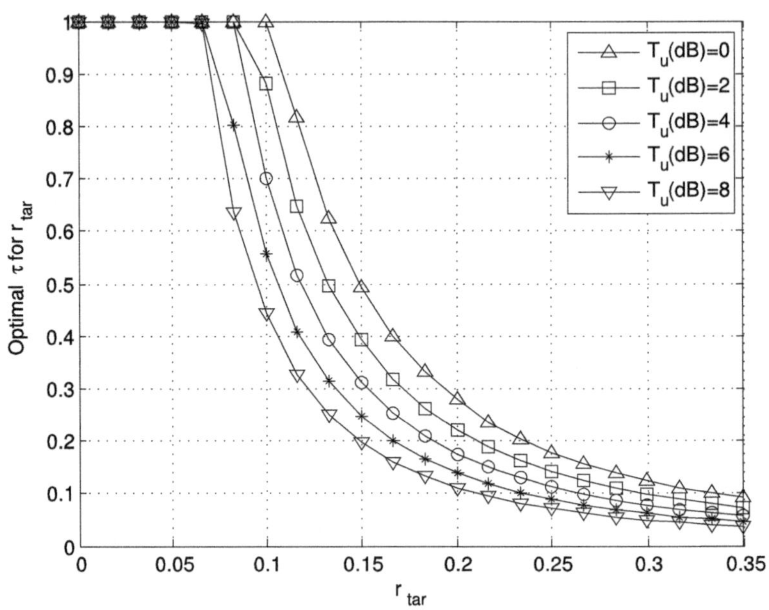


**Fig. 4** Optimal TP for varying target distance with different $T_u$ when $T_d$(dB)=-10. Since the influence of the interference is small in short target distance, the optimal TP is shown to be higher in shorter target distance

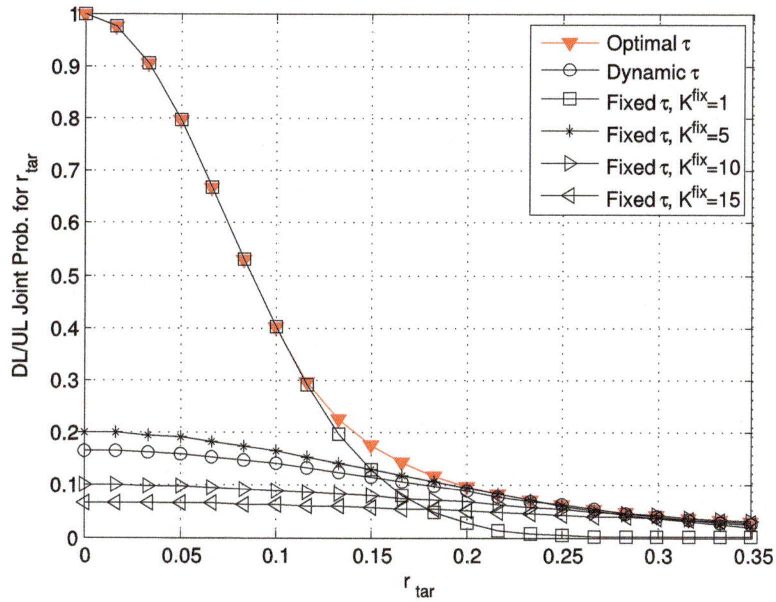


**Fig. 5** DL/UL joint probability for varying target distance with the different SA-AC schemes when $T_d(dB) = -10$ and $T_u(dB) = 0$. The proposed SA-AC scheme with optimal TP is shown to be the best by reflecting the spatial effect

induces the degradation of the average achievable rate for $r_{tar}$.

In Fig. 7, the performance of the DL/UL joint probability for the optimal $\tau$ and the performance of the UL probability to maximize the achievable rate $UL_{max}$ $\tau$ [20] are shown when $r_{tar} = 0.15$ is shown. Since the DL coverage probability decreases as $T_d$ increases, the number of the covered members decreases. Hence, the interference becomes weaker as $T_d$ decreases, and the SIR of the target member increases as $T_d$ increases. So the DL/UL joint probability increases as $T_d$ increases. Once $p_{d,u}(r_{tar}, \tau, T_d, T_u)$ is achieved, the performance starts to degrade by increasing $T_d$. For $r_{tar}$, the DL coverage probability of the target member decreases and the optimal

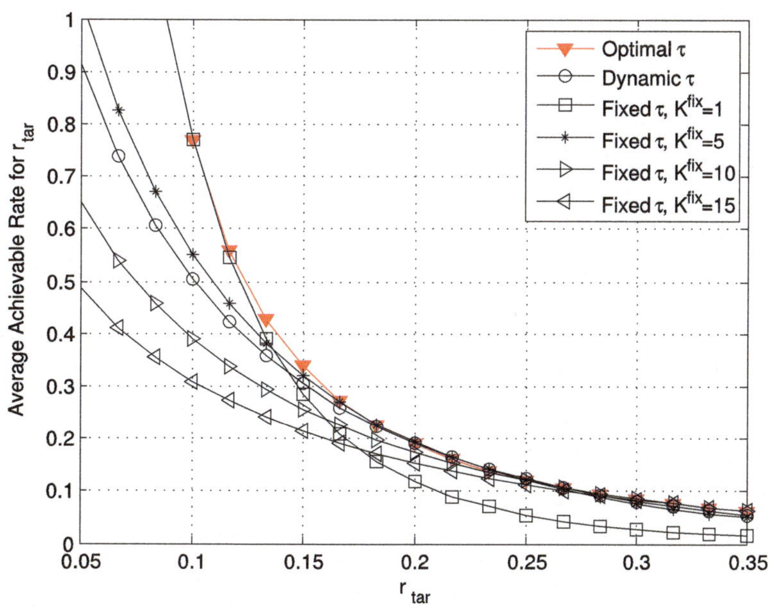


**Fig. 6** Average achievable rate for varying target distance with the different SA-AC schemes when $T_d(dB) = -10$ and $T_u(dB) = 0$. As in Fig. 5, the proposed SA-AC scheme with the optimal TP is shown to be the best

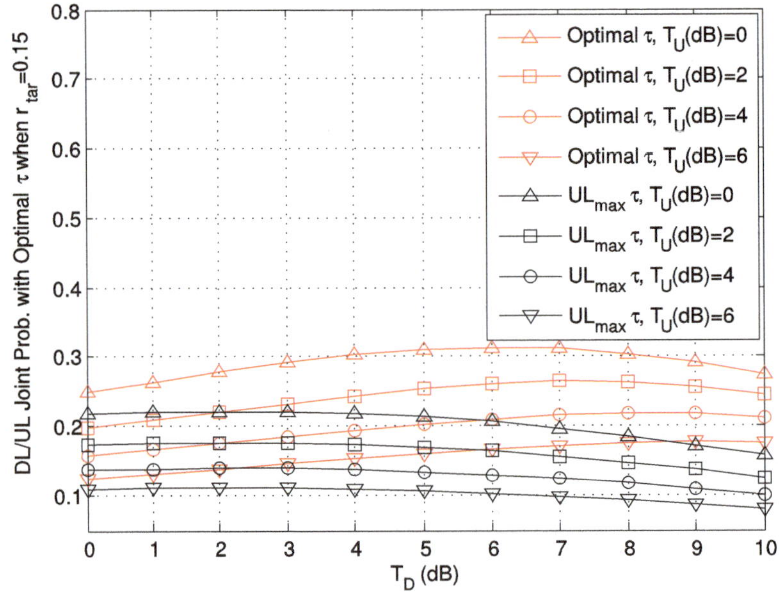


**Fig. 7** DL/UL joint probability with the optimal $\tau$ for varying $T_d$ with different $T_u$ when $r_{tar} = 0.15$

$\tau$ increases as $T_d$ increases. In addition, the DL coverage probability for a distance between a covered member and its leader increases as the distance decreases for $T_d$. Thus, the optimal $\tau$ of the covered member with smaller distance from the associated leader of the target member becomes larger. However, since the UL$_{max}$ $\tau$ is not designed to support group communications and does not consider the number of covered members by the DL threshold, i.e., UL$_{max}$ $\tau$ is determined to maximize the UL coverage probability only for the member intensity, the UL$_{max}$ $\tau$ is not changed as $T_d$ increases. So the number of transmissions of the covered members does not increase while the Interference decreases as the number of the covered members decreases. The UL SIR of the target member is hard to exceed $T_u$ after the maximum DL/UL joint probability. Since the UL coverage probability decreases as $T_u$ increases, the DL/UL joint probability decreases.

In Fig. 8, the performance of the average achievable rate with the optimal $\tau$ and the UL$_{max}$ $\tau$ for $\bar{r}_{max}$ is shown

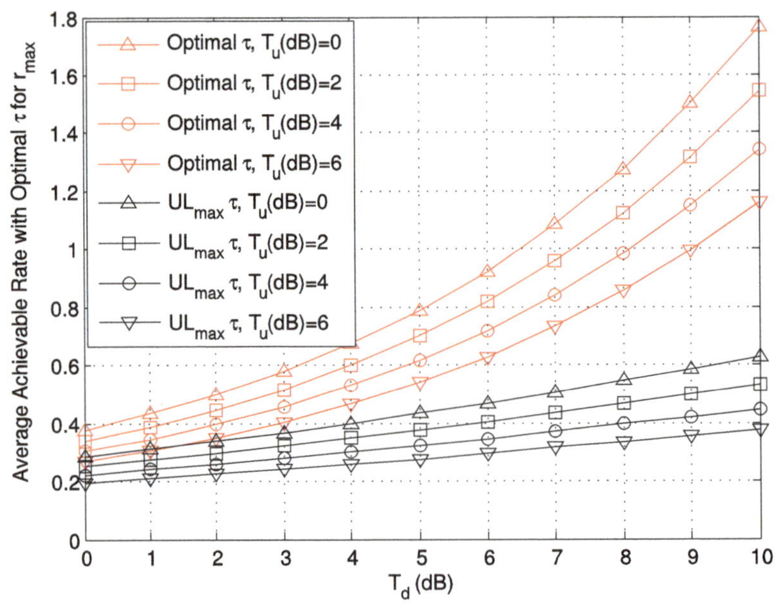


**Fig. 8** Average achievable rate with the optimal $\tau$ for varying $T_d$ with different $T_u$

for varying $T_d$ with different $T_u$. The optimal $\tau$ for $\bar{r}_{max}$ varies according to $T_d$ and $T_u$. Since the increase of $T_d$ reduces the number of the covered members, the interference for a covered member decreases as $T_d$ increases. And the signal strength of a covered member increases as the distance between the covered member and its leader decreases. Then the SIR of the member increases and it is sufficient to exceed $T_u$. Therefore, the performance of the average achievable rate increases by increasing $T_d$. The lower $T_u$, the larger the UL coverage probability becomes by increasing the optimal $\tau$. Thus, the average achievable rate with the optimal $\tau$ is maximized by increasing $T_d$ when $T_u(dB) = 0$. However, since the UL$_{max}$ $\tau$ is not changed for the coverage of the leader, it is not possible for the covered members to increase the numbers of transmissions. Although the SIR of the covered member increases by the increase of the signal strength and the decrease of the interference, the UL$_{max}$ $\tau$ might not be sufficient to increase the number of successful covered members in DL/UL.

## 6 Conclusions

In this paper, we have proposed an SA-AC scheme and developed an analytical model of the SA-AC schemes for group communications network. The proposed analytical model is affected by the intensity of leaders, the intensity of members, and the thresholds for DL and UL which are the important factors of service quality. The proposed analytical model presents the optimal TP to maximize the DL/UL joint probability at a target distance. Since the importance of the DL/UL joint probability at a target distance varies with the type of service, the optimal TP is carefully determined for providing services efficiently. The proposed analytical model has been validated via simulations, and the performance of the SA-AC schemes has been demonstrated. For group communications, the DL/UL joint probability at a target distance can be maximized by the proposed SA-AC scheme, which is superior to other schemes for the target distance. As a result, the importance of considering DL/UL joint probability in the group communication network has been proven.

## Appendix

### A.1 Proof of Lemma 1

For the DL threshold $T_d$, the DL coverage probability that a member with the distance $r$ from its nearest leader is

$$\begin{aligned} p_d(r,T_d) &= \mathbb{P}\left[\mathrm{SIR}_d(r) > T_d|r\right] \\ &= \mathbb{P}\left[h_{l,m} > P_l^{-1} r^{\alpha} I_l T_d|r\right] \\ &= \mathbb{E}\left[\exp\left(P_l^{-1} r^{\alpha} I_l T_d\right)|r\right] \\ &= \mathcal{L}_{I_l}\left(P_l^{-1} r^{\alpha} T_d\right), \end{aligned} \tag{23}$$

where $\mathcal{L}_{I_l}(s)$ is the Laplace transform of the aggregate interference received at the leader. $\mathcal{L}_{I_l}(s)$ can be expressed as

$$\begin{aligned} &\mathcal{L}_{I_l}(s) \\ &= \mathbb{E}_{I_l}\left[e^{-sI_l}\right] \\ &= \mathbb{E}_{\Phi_l,h_{l_i,m}}\left[e^{-s\left(\sum_{l_i\in\Phi_l\backslash l} P_l h_{l_i,m} r_{d,i}^{-\alpha}\right)}\right] \\ &= \mathbb{E}_{\Phi_l}\left[\prod_{l_i\in\Phi_l\backslash l} \mathbb{E}_{h_{l_i,m}}\left[e^{-h_{l_i,m}\left(sP_l r_{d,i}^{-\alpha}\right)}\right]\right] \\ &\overset{(a_1)}{=} e^{-2\pi\lambda_l\int_r^{\infty}\left(1-\mathbb{E}_{h_{l_i,m}}\left[e^{-h_{l_i,m}\left(sP_l v_l^{-\alpha}\right)}\right]\right)v_l \mathrm{d}v_l} \\ &= e^{-2\pi\lambda_l\int_r^{\infty}\left(1-\frac{1}{1+sP_l v_l^{-\alpha}}\right)v_l \mathrm{d}v_l} \\ &\overset{(a_2)}{=} e^{-2\pi\lambda_l\int_r^{\infty}\left(\frac{1}{1+\left(v_l/\left(r(T_d)^{1/\alpha}\right)\right)^{\alpha}}\right)v_l \mathrm{d}v_l} \\ &\overset{(a_3)}{=} e^{-\pi r^2\lambda_l T_d^{2/\alpha}\int_{T_d^{-2/\alpha}}^{\infty}\frac{1}{1+(u_l)^{\alpha/2}}\mathrm{d}u_l} \\ &\overset{(a_4)}{=} e^{-\pi r^2\lambda_l\zeta_l(T_d)}, \end{aligned} \tag{24}$$

where $(a_1)$ is derived from the probability generating functional (PGFL) of PPP [31], $(a_2)$ uses $s = P_l^{-1} r^{\alpha} T_d$, $(a_3)$ is derived by the change of the variable $u_l = \left(v_l/r(T_d)^{1/\alpha}\right)^2$, and $(a_4)$ utilizes the change of the variable $\zeta_l(T_d) = T_d^{2/\alpha}\int_{T_d^{-2/\alpha}}^{\infty}\frac{1}{1+(u_l)^{\alpha/2}}\mathrm{d}u_l$. Thus, plugging (24) into (23), we obtain (8). And combining (23) with (7), we obtain the DL coverage probability that a member is in the coverage of its nearest leader given by

$$p_d(T_d) = \int_0^{\infty} p_d\left(r,T_d\right) f_R(r)\mathrm{d}r. \tag{25}$$

Lemma 1 and its derivation corresponds to the interference limited network case as in [4].

### A.2 Proof of Lemma 2

For the UL threshold $T_u$ and $\tau$, the UL coverage probability that the SIR value for the transmission of the covered member with the distance $r$ from its leader exceeds $T_u$ is given by

$$\begin{aligned} p_u(r,\tau,T_u) &= \mathbb{P}\left[\mathrm{SIR}_{d,u}(r,\tau) > T_u|r\right] \\ &= \mathbb{P}\left[h_{m_c,l} > P_m^{-1} r^{\alpha} I_{m_c} T_u|r\right] \\ &= \mathbb{E}\left[\exp\left(P_m^{-1} r^{\alpha} I_{m_c} T_u\right)|r\right] \\ &= \mathcal{L}_{I_{m_c}}\left(P_m^{-1} r^{\alpha} T_u\right) \end{aligned} \tag{26}$$

where $\mathcal{L}_{I_{m_c}}(s)$ denotes the Laplace transform of the aggregate interference received at the leader. $\mathcal{L}_{I_{m_c}}(s)$ can be found in a similar way to (24). Thus,

$$\begin{aligned}
&\mathcal{L}_{I_{mc}}(s)\\
&= \mathbb{E}_{I_{mc}}\left[e^{-sI_{mc}}\right]\\
&= \mathbb{E}_{\Phi_{mc},h_{m_{c,i},l}}\left[e^{-s\left(\sum_{m_{c,i}\in\Phi_{mc}\setminus m_c} P_m h_{m_{c,i},l} r_{du,i}^{-\alpha}\right)}\right]\\
&= \mathbb{E}_{\Phi_{mc}}\left[\prod_{m_{c,i}\in\Phi_{mc}\setminus m_c}\mathbb{E}_{h_{m_{c,i},l}}\left[e^{-h_{m_{c,i},l}\left(sP_m r_{du,i}^{-\alpha}\right)}\right]\right]\\
&= \exp\Bigg(-2\pi\lambda_m p_d(T_d)\tau\\
&\int_0^\infty\left(1-\mathbb{E}_{h_{m_{c,i},l}}\left[e^{-h_{m_{c,i},l}(sP_m v_m^{-\alpha})}\right]\right)v_m\mathrm{d}v_m\Bigg)\\
&= e^{-2\pi\lambda_m p_d(T_d)\tau\int_0^\infty\left(1-\frac{1}{1+sP_m v_m^{-\alpha}}\right)v_m\mathrm{d}v_m}\\
&\overset{(a_5)}{=} e^{-2\pi\lambda_m p_d(T_d)\tau\int_0^\infty\left(\frac{1}{1+\left(v_m/(r(T_u)^{1/\alpha})\right)^\alpha}\right)v_m\mathrm{d}v_m}\\
&\overset{(a_6)}{=} e^{-\pi r^2\lambda_m p_d(T_d)\tau T_u^{2/\alpha}\int_0^\infty\frac{1}{1+(u_m)^{\alpha/2}}\mathrm{d}u_m}\\
&\overset{(a_7)}{=} e^{-\pi r^2\lambda_m p_d(T_d)\tau\zeta_m(T_u)}.
\end{aligned}\tag{27}$$

where $(a_5)$ is found from $s = P_m^{-1}r^\alpha T_u$, $(a_6)$ comes from the change of variable $u_m = \left(v_m/r(T_u)^{1/\alpha}\right)^2$, and $(a_7)$ is derived from the change of variables

$$\zeta_m(T_u) = T_u^{2/\alpha}\int_0^\infty\frac{1}{1+(u_m)^{\alpha/2}}\mathrm{d}u_m = T_u^{2/\alpha}\frac{2\pi/\alpha}{\sin(2\pi/\alpha)}. \tag{28}$$

### A.3 Proof of Optimal $\tau$

Let $f(\tau)$ be $p_{d,u}(r_{tar},\tau,T_d,T_u)$. By the extreme value theorem, if $f(\tau)$ is continuous in the interval of $\tau\in[0,1]$, then at some point of the interval, $f(\tau)$ attains a global maximum and a global minimum [32–34]. Then the point of attainment is either 1) a point where $f'(\tau)$ does not exist, 2) a point where $f'(\tau)=0$, and 3) a point at one end of the interval [33]. $f(\tau)$ can be expressed as

$$f(\tau) = \tau\exp(-a-b\tau), \tag{29}$$

where $a = \pi r_{tar}^2\lambda_l\zeta_l(T_d)$ and $b = \pi r_{tar}^2\lambda_m P_l^c(T_d)\zeta_m(T_u)$, $a\geq 0$ and $b\geq 0$. Since $f(\tau)$ is differentiable in $\tau\in[0,1]$, $f(\tau)$ is continuous and $f'(\tau)$ exists for all $\tau\in[0,1]$.

$$f'(\tau) = (1-b\tau)\exp(-a-b\tau). \tag{30}$$

When $\tau=\frac{1}{b}$, $f'(\tau)=0$. We evaluate $f(\tau)$ at the points where $f'(\tau)=0$ and at the end points as the candidates for the extreme values. The function value at the point where $f'(\tau)=0$ is

$$f\left(\frac{1}{b}\right) = \frac{1}{b}\exp(-a-1), \tag{31}$$

The function values at the end points are

$$f(0) = 0, \tag{32}$$

$$f(1) = \exp(-a-b). \tag{33}$$

Since $f(\tau)\geq 0$, $f(0)=0$ is the global minimum of $f(\tau)$.

If $0\leq b<1$, $\frac{1}{b}>1$ and $f\left(\frac{1}{b}\right)$ is not defined in the interval of $\tau\in[0,1]$. Thus, $f(1)$ is the global maximum when $0\leq b<1$. If $b\geq 1$, then $be^1\leq e^b$. Thus, $f\left(\frac{1}{b}\right)\geq f(1)$ can be verfied as follows.

$$\frac{1}{be^1}\geq\frac{1}{e^b}. \tag{34}$$

Thus, $f\left(\frac{1}{b}\right)$ is the global maximum of $f(\tau)$ when $b\geq 1$. Therefore an optimal $\tau$ exists for $f(\tau)$ in the interval of $[0,1]$.

### A.4 Proof of Lemma 4

For $T_d$ and $\tau$, the average achievable rate for a target member with $r_{tar}$ is given by

$$\begin{aligned}
&E_{ar}(r_{tar},\tau)\\
&= \tau\int_0^\infty\mathbb{P}\left[\ln\left(1+\frac{P_m h_{m_c,l} r_{tar}^{-\alpha}}{I_{m_c}}>t\right)\right]\mathrm{d}t\\
&= \tau\int_0^\infty\mathbb{P}\left[h_{m_c,l}>P_m^{-1}r_{tar}^\alpha I_{m_c}(e^t-1)\right]\mathrm{d}t\\
&= \tau\int_0^\infty\mathbb{E}\left[\exp\left(P_m^{-1}r_{tar}^\alpha I_{m_c}(e^t-1)\right)\right]\mathrm{d}t\\
&= \tau\int_0^\infty\mathcal{L}_{I_{mc}}\left(P_m^{-1}r_{tar}^\alpha(e^t-1)\right)\mathrm{d}t.
\end{aligned}\tag{35}$$

Similar to Appendix A.1 and Appendix A.2, it follows,

$$\begin{aligned}
&\mathcal{L}_{I_{mc}}\left(P_m^{-1}r_{tar}^\alpha(e^t-1)\right)\\
&= e^{-2\pi\lambda_m p_d(T_d)\tau\int_0^\infty\left(1-\frac{1}{1+(r_{tar}/v_m)^\alpha(e^t-1)}\right)v_m\mathrm{d}v_m}\\
&= e^{-2\pi\lambda_m p_d(T_d)\tau\int_0^\infty\left(\frac{1}{1+\left(v_m/\left(r(e^t-1)^{1/\alpha}\right)\right)^\alpha}\right)v_m\mathrm{d}v_m}\\
&\overset{(a_8)}{=} \exp\Bigg(-\pi\lambda_m p_d(T_d)\tau(r_{tar})^2(e^t-1)^{2/\alpha}\\
&\int_0^\infty\frac{1}{1+(u_{ar,r_{tar}})^{\alpha/2}}\mathrm{d}u_{ar,r_{tar}}\Bigg).
\end{aligned}\tag{36}$$

where $(a_8)$ uses the change variable $u_{ar,r_{tar}} = \left(\frac{v_m}{r_{tar}(e^t-1)^{1/\alpha}}\right)^2$.

When $\alpha=4$ and $r_{tar}=r$, the average achievable rate of a covered member is then given by

$$
\begin{aligned}
&E_{ar}(\tau)\\
&= \tau \int_0^\infty \int_0^\infty \mathcal{L}_{I_{mc}}\left(P_m^{-1} r^\alpha \left(e^t - 1\right)\right) f_{R_{mc}}(r) \mathrm{d}t \mathrm{d}r.\\
&\overset{(a_9)}{=} \frac{\pi \lambda_l \tau}{p_d(T_d)} \int_0^\infty \frac{2 p_d(T_d)}{2\pi\lambda_l + \pi^2 \lambda_m p_d(T_d)^2 \tau (e^t - 1)^{1/2}} \mathrm{d}t.\\
&= \tau \int_0^\infty \frac{1}{1 + \frac{\pi}{2} \frac{\lambda_m p_d(T_d)^2 \tau}{\lambda_l} (e^t - 1)^{1/2}} \mathrm{d}t.\\
&\overset{(a_{10})}{\approx} \tau \int_0^\infty \frac{1}{\frac{\pi}{2} \frac{\lambda_m p_d(T_d)^2 \tau}{\lambda_l} (e^t - 1)^{1/2}} \mathrm{d}t.\\
&\overset{(a_{11})}{=} \tau \int_0^\infty \frac{1}{u^2 + C^2} \mathrm{d}u = \tau \left[\frac{1}{C} \arctan\left(\frac{u}{C}\right)\right]_0^\infty\\
&= \tau \frac{\pi}{2C} = \frac{\lambda_l}{\lambda_m p_d(T_d)^2} \qquad (37)
\end{aligned}
$$

where $(a_9)$ utilizes the change of the variable $z = r^2$ and (9) then change the order of integrals. $(a_{10})$ comes from the approximation of $1/\left(1 + a\sqrt{e^t + 1}\right)$ to $1/\left(a\sqrt{e^t + 1}\right)$ where $a > 0$. Also, $(a_{11})$ stems from the change of variables as $C = \frac{\pi \lambda_m p_d(T_d)^2 \tau}{2\lambda_l}$ and $u = C\sqrt{e^t - 1}$.

**Acknowledgements**

This work was supported by the National Research Foundation of Korea (NRF) grant funded by the Korean government (MSIP) (2015R1A2A2A01004067), and Basic Science Research Program through the NRF funded by the Ministry of Education (NRF-2010-0020210).

**Authors' contributions**

ML's contribution is writing the paper and conducting performance analysis and simulations. YK's contribution is partly writing the paper and conducting simulations. T-JL's contribution is writing and revising the paper and guiding the direction and organization of the paper. All authors read and approved the final manuscript.

**Competing interests**

The authors declare that they have no competing interests.

**References**

1. YS Kim, HB Lee, JS Kim, HJ Baek, MS Ryu, KS Park, in *Proc. of IEEE Information Technology Applications in Biomedicine (ITAB)*. ECG, EOG detection from helmet based system, (Tokyo, 2007), pp. 191–193. http://ieeexplore.ieee.org/document/4407378/
2. A Giorgetti, M Lucchi, M Chiani, MZ Win, Throughput per pass for data aggregation from a wireless sensor network via a UAV. IEEE Trans. Aerosp. Electron. Syst. **47**(4), 2610–2626 (2011)
3. H ElSawy, E Hossain, On stochastic geometry modeling of cellular uplink transmission with truncated channel inversion power control. IEEE Trans. Wirel. Commun. **13**(8), 4454–4469 (2014)
4. JG Andrews, F Baccelli, RK Ganti, A tractable approach to coverage and rate in cellular networks. IEEE Trans. Commun. **59**(11), 3122–3134 (2011)
5. H Jo, YJ Sang, P Xia, JG Andrews, Heterogeneous cellular networks with flexible cell association: a comprehensive downlink SINR analysis. IEEE Trans. Wirel. Commun. **11**(10), 3484–3495 (2012)
6. F Baccelli, B Bàszczyszyn, *Stochastic geometry and wireless networks in foundations and trends in networking*. (INRIA & Ecole Normale Superieure, 45 rue d'Ulm, Paris, 2009). https://hal.inria.fr/inria-00403040/file/FnT2.pdf
7. M Haenggi, Outage, local throughput, and capacity of random wireless networks. IEEE Trans. Wirel. Commun. **8**(8), 4350–4359 (2009)
8. H ElSawy, E Hossain, M Haenggi, Stochastic geometry for modeling, analysis, and design of multi-tier and cognitive cellular wireless networks: a survey. IEEE Commun. Surv. Tutor. **15**(3), 996-1019 (2013). Third Quarter
9. PD Mankar, G Das, SS Pathak, Modeling and coverage analysis of BS-centric clustered users in a random wireless network. IEEE Wirel. Commun. Lett. **5**(2), 208–211 (2016)
10. H Zhuang, T Ohtsuki, A model based on Poisson point process for analyzing MIMO heterogeneous networks utilizing fractional frequency reuse. IEEE Trans. Wirel. Commun. **13**(12), 6839–6850 (2014)
11. S Singh, JG Andrews, Joint resource partitioning and offloading in heterogeneous cellular networks. IEEE Trans. Wirel. Commun. **13**(2), 888–901 (2014)
12. TD Novlan, HS Dhillon, JG Andrews, Analytical modeling of uplink cellular networks. IEEE Trans. Wirel. Commun. **12**(6), 2669–2679 (2013)
13. TD Novlan, JG Andrews, Analytical evaluation of uplink fractional frequency reuse. IEEE Trans. Commun. **61**(5), 2098–2108 (2013)
14. AH Sakr, E Hossain, Analysis of K-Tier uplink cellular networks with ambient RF energy harvesting. IEEE J. Sel. Areas Commun. **33**(10), 2226–2238 (2015)
15. M Afshang, HS Dhillon, PHJ Chong, in *Proc.of IEEE Global Communications Conference (GLOBECOM)*. Coverage and Area Spectral Efficiency of Clustered Device-to-Device Networks, (San Diego, 2015). http://ieeexplore.ieee.org/document/7417450/
16. K Han, K Huang, Wirelessly powered backscatter communication networks: modeling, coverage, and capacity. IEEE Trans. Wirel. Commun. **16**(4), 2548–2561 (2017)
17. N Abramson, in *Proc.of AFIPS*. The Aloha system-another alternative for computer communications, vol. 36, (1970), pp. 281-285
18. JF Kurose, KW Ross, *Computer networking: a top-down approach, 6th edition*. (Pearson, 2012). https://www.pearson.com/us/higher-education/product/Kurose-Computer-Networking-A-Top-Down-Approach-6th-Edition/9780132856201.html
19. RTB Ma, V Misra, D Rubenstein, An analysis of generalized slotted Aloha protocols. IEEE/ACM Trans. Networking. **17**(3), 936–949 (2009)
20. F Baccelli, B Bàszczyszyn, P Mühlethaler, Stochastic analysis of spatial and opportunistic Aloha. IEEE J. Sel. Areas Commun. **27**(7), 1105–1119 (2009)
21. F Baccelli, B Błaszczyszyn, C Singh, in *Proc. of IEEE International Conference on Computer Communications (INFOCOM)*. Analysis of a proportionally fair and locally adaptive spatial Aloha in Poisson networks, (Toronto, 2014), pp. 2544–2552. http://ieeexplore.ieee.org/document/6848201/
22. F Baccelli, C Singh, in *Proc. of IEEE International Symposium and Workshops on Modeling and Optimization in Mobile, Ad Hoc and Wireless Networks (WiOpt)*. Adaptive Spatial Aloha, Fairness and Stochastic Geometry, (Tsukuba Science City, 2013), pp. 7–14. http://ieeexplore.ieee.org/document/6576400/
23. K Huang, V Lau, Y Chen, Spectrum sharing between cellular and mobile ad hoc networks: Transmission-Capacity Tradeoff. IEEE J. Sel. Areas Commun. **27**(7), 1256–1266 (2009)
24. F Baccelli, B Bàszczyszyn, P Mühlethaler, An Aloha protocol for multihop mobile wireless networks. IEEE Trans. Inf. Theory. **52**(2), 421–436 (2006)
25. Q Li, D Rus, Global clock synchronization in sensor networks. IEEE Trans.Comput. **55**(2), 214–226 (2006)
26. AB Kulakli, K erciyes, in *Proc. of IEEE International Symposium on Computer and Information Sciences (ISCIS)*. Time synchronization algorithms based on time-sync protocol in wireless sensor networks, (Istanbul, 2008), pp. 1–5. http://ieeexplore.ieee.org/document/4717888/
27. JE Wieselthier, A Ephremides, LA Michels, An exact analysis and performance evaluation of framed ALOHA with capture. IEEE Trans. Commun. **37**(2), 125–137 (1989)
28. FC Schoute, Dynamic frame length Aloha. IEEE Trans. Commun. COM-31. **31**(4), 565–568 (1983). http://ieeexplore.ieee.org/document/1095854/
29. SM Yu, S-L Kim, in *Proc. of IEEE International Symposium on Modeling & Optimization in Mobile, Ad Hoc & Wireless Networks (WiOpt)*. Downlink Capacity and Base Station Density in Cellular Networks, (Tsukuba Science City, 2013). http://ieeexplore.ieee.org/document/6576422/

30. YJ Sang, KS Kim, Load distribution in heterogeneous cellular networks. IEEE Commun. Lett. **18**(2), 237–240 (2014)
31. D Stoyan, W Kendall, J Mecke, *Stochastic geometry and its applications, 2nd edition*. (Wiley, 1996). https://www.amazon.com/Stochastic-Geometry-Its-Applications-2nd/dp/0471950998
32. EKP Chong, SH Zak, *An introduction to optimization, 2nd edition*. (Wiley, 2001). http://as.wiley.com/WileyCDA/WileyTitle/productCd-0471654000.html
33. AE Taylor, *Advanced calculus*. (Ginn, Boston, 1955). https://www.abebooks.co.uk/servlet/BookDetailsPL?bi=22425525610&searchurl=tn%3Dadvanced%2Bcalculus%26sortby%3D17%26an%3Dtaylor%2Ba%2Be
34. FD RL Finney, BK Demana, D Waits, *Kennedy, Calculus: graphical, numerical, algebraic*. (Prentice Hall, 2003). https://www.bookdepository.com/Calculus-Graphical-Numerical-Algebraic-Student-Edition-2003c-Ross-L-Finney/9780130631312

8

# Joint beamforming and power control algorithm for cognitive MIMO broadcast channels via game theory

Shunlan Zhang[1†], Ju Ni[2*†] and Qing Peng[1]

**Abstract**

In this paper, we present a game-theoretic approach for the purpose of deriving the problem of joint beamforming and power control in cognitive radio (CR) multiple-input multiple-output (MIMO) broadcast channels (CR MIMO-BCs), where the primary users (PUs) coexist with the secondary users (SUs) and they share the same spectrum. The cognitive base station (CBS), which is equipped with multiple antennas, is capable of transmitting data to the SU's multiple-antenna receiver by employing the technology of beamforming. The proposed approach is an application of separable games, which are formally stated by the subgames of beamforming and power control. Furthermore, based on the model of noncooperative separate games, separable cost functions for the parameters of beamforming and power control are also proposed, showing that these cost functions are convex. Therefore, the convex theory of a noncooperative game can be employed to investigate the best response strategies as well as existence of Nash equilibrium solutions. Finally, we propose an iterative algorithm to achieve the optimal Nash equilibrium of the proposed joint beamforming subgame and power control subgame. Numerical results verify both the convergence and the tracking properties of the proposed algorithm for variant scenarios.

**Keywords:** Cognitive radio, MIMO broadcast channels, Beamforming, Power control, Separable games

## 1 Introduction

The radio spectrum available for wireless communication is extremely scarce due to the widely deployed wireless devices and services. On the other hand, recent studies and measurements have shown that most of the allocated bands are used inefficiently or underutilized [1–3]. In order to address the aforementioned challenges, a cognitive radio (CR) technology has recently been proposed for the purpose of dramatically improving spectrum utilization and supporting more new services [4]. In cognitive radio networks (CRNs), the licensed spectrum can be shared with secondary users (SUs), provided that they do not cause harmful interference to primary users (PUs). In order to support this spectrum reuse functionality, SUs are permitted to transmit once they detect a spectrum hole [5, 6]. Such schemes usually work when the spectrum is severely underutilized, or otherwise, SUs might not have sufficient opportunities to get channel access. Once PUs are found to be active, SUs must vacate the channels. Therefore, the secondary throughput would be significantly constrained, and the secondary system would suffer from a long latency.

As a spectral-efficient technology, multiple-input multiple-output (MIMO) is capable of providing extra spatial dimensions for signal transmission. MIMO technology can be employed by CRNs for the purpose of reducing interference at the PU and satisfying the demand of high data rate at the SU through carefully designing transmit/receive beamforming [7, 8]. As a result, SUs may access the licensed spectrum without causing harmful interference at the PUs, even if the PUs are also using the same spectrum at the same time. However, an additional spatial resource inherent in MIMO systems becomes a challenging task in the design of efficient spectrum-sharing, although this technology offers several advantages to enhance the system performances.

*Correspondence: niju800206@163.com
†Equal contributors
[2]Library of Guilin University of Electronic Technology, Jinji Road, Guilin 541004, China
Full list of author information is available at the end of the article

Furthermore, CR spectrum-sharing imposes several new challenging issues on MIMO systems [9]. First, the idea of spectrum-sharing allows simultaneous transmissions of PUs and SUs, provided that the quality of service (QoS) of PUs is guaranteed. Secondly, the primary systems cannot deliberately provide their channel estimation to the secondary systems [10]. The aforementioned challenges may impose difficulty on pre-interference cancelation at the SU's transmitter side.

CR MIMO broadcast channels (CR MIMO-BCs) have become a topic of increasing research interest for CRNs in recent years [11–13], with secondary systems coexisting with the primary systems. Unlike conventional MIMO-BCs, in CR MIMO-BCs, there exist interference between the PU link and the SU link as well as the multiple SUs' interference. Furthermore, in order to protect the primary transmission, the total-power constraint and the individual-interference-power constraint for each primary receiver are considered. Since beamforming and power control techniques play an important role in interference suppression and power constraint in CR MIMO-BCs, a joint beamforming and power control scheme over CR MIMO-BCs was proposed to minimize the transmit power while satisfying signal-to-interference-plus-noise ratio (SINR) targets for the SUs and maintaining an acceptable interference level to the PUs [14]. The development of the scheme relies crucially on the BC multiple-access channel (BC-MAC) duality result [15, 16], which is only valid for the problem with a single-sum-power constraint. In [17], the problem of joint transmit beamforming and power control was considered in CR MIMO-BCs, with the cognitive multi-antenna base station (BS) being assumed to satisfy the QoS constraints of the served SUs while protecting one primary receiver from interference. It is also assumed that the number of single-antenna users is less than that deployed at the BS. Consequently, SUs may access the primary spectrum without causing harmful interference to the PUs.

For the case of non-zero-interference-power constraint in CR MIMO-BCs, both the SUs' SNR constraints and the PUs' interference power constraints usually result in quadratically constrained quadratic programming problems, which may not be directly solved by convex tools, especially when there is a rank constraint. Semidefinite programming (SDP) relaxation can be used to convert the aforementioned problem into a convex optimization problem by dropping the rank constraint, consequently generating a local optimum [18]. It is shown in [19] that under certain conditions, a new solution can be generated from the one obtained by SDP relaxation without ruining the constraints or changing the objection function. Actually, most of the resulting problems of joint beamforming and power control are inherently non-convex, and consequently, no global optimality of an efficient solution can be guaranteed theoretically [20]. Nonetheless, in [20], sufficient conditions were presented to constrain some design parameters, making the joint beamforming and power control problem become convex. Furthermore, in [21], a semi-distributed algorithm was proposed to obtain a local optimal solution to this problem. However, in general scenarios, the obtained local optimum may not be feasible for the original problem because its rank usually does not meet the solution's requirement. As a result, approximation approaches may be used to generate a feasible solution [22–26].

In this paper, we model the problem of joint power control and beamforming in CR MIMO-BCs as a noncooperative game. Note that the noncooperative game theory for economists has been extensively investigated in terms of how rational players do not cooperate and interact in order to reach their goals. Lately, noncooperative game theory has also been widely applied to communications system. For example, in [27], joint Code Division Multiple Access (CDMA) codeword and power adaptation was formulated as a separable game in MIMO CR networks, with two corresponding subgames, i.e., the power control subgame and the codeword control subgame, being played. Furthermore, we also formulate separable games[1] to solve the problem of joint power control and beamforming. Based on the proposed noncooperative separate game model, separable cost functions for beamforming and power control parameters are also derived, showing that these cost functions are convex. Furthermore, we use the convex theory of the noncooperative game to investigate the best response strategies as well as existence of the Nash equilibrium solutions. The best response strategies of users are then obtained by constrained minimization of the user cost function subject to constraints on user SINR and beamforming vector. The corresponding algorithm is derived using a game-theoretic approach in which separable cost functions with respect to beamforming and power are defined, such that joint beamforming and power adaptation is formulated as a separable game.

The rest of this paper is organized as follows: In Section 2, we describe the system model and problem statement. In Section 3, we present joint beamforming and power control as a noncooperative game. Section 4 provides numerical simulation results and discussions. Finally, conclusions are drawn in Section 5.

Notations: Scalar is denoted by a lowercase letter. Vector and matrix are denoted by a boldface lowercase letter and a boldface uppercase letter, respectively. $\mathbf{I}_p$ denotes the $p \times p$ identity matrix. $\|\cdot\|$ represents the Euclidean norm of a vector. sgn[$\cdot$] denotes sign function. For a matrix $\mathbf{S}$, tr($\mathbf{S}$), rank($\mathbf{S}$), $\mathbf{S}^H$, and $\mathbf{S}^T$ denote its trace, rank, Hermitian matrix, and transpose matrix, respectively.

$\text{diag}(s_1, s_2, \cdots, s_n)$ denotes a diagonal matrix with diagonal elements given by $s_1, s_2, \cdots, s_n$. For a matrix $\mathbf{S}$, $\mathbf{S} \geq 0$ denotes the $\mathbf{S}$ is positively semidefined. $\{\cdot\}$ denotes the subset.

## 2 System model and problem statement

In this paper, we present an alternative approach to dealing with a variable number of active users and/or QoS requirements in a CDMA system. More specifically, we consider the system model of CR MIMO-BCs, as shown in Fig. 1, where the cognitive base station (CBS) and $K$ SUs share the spectrum with $N$ PUs, with the $N_t$-antenna CBS sending independent information signals to $K$-different SUs. Without loss of generality, each SU is assumed to have $N_r$ ($N_r \leq N_t$) antennas. Without loss of generality, both the primary base station (PBS) and PUs are assumed to be equipped with a single antenna. We assume that there is only one PU ($N = 1$) and that it is enough to demonstrate the key aspects of spectrum-sharing while avoiding unnecessary complications [28, 29], while it is pertinent to investigate the general scenario where multiple PU and SU links coexist [30].

In CR MIMO-BCs, due to the coupled structure of the transmitted signals, the BC optimization problems are usually non-convex and thus cannot be solved directly. To tackle this difficulty, the non-convex BC problem is transformed into a convex MAC problem via a so-called BC-MAC duality relationship [31]. Under a single sum-power constraint or a set of a linear power constraint, the problem of BC can be solved as a minimax optimization problem in its dual MAC setting [32]. The block diagram of BC-MAC duality relationship for the transmission process is shown in Fig. 2.

The signal $\mathbf{x} \in \mathbb{C}^{M\times 1}$ transmitted by the CBS is given by [33]

$$\mathbf{x} = \sum_{k=1}^{K} s_k \sqrt{p_k} \mathbf{u}_k = \mathbf{UPs}, \tag{1}$$

where $\mathbf{s} = [s_1, s_2, \cdots s_K]^T$, with $s_k$ denoting the data stream for the $k$th SU ($\text{SU}_\text{k}$) and satisfying $E[\,|s_k|^2] = 1$, $\mathbf{P} = \text{diag}\{\sqrt{p_1}, \sqrt{p_2}, \cdots \sqrt{p_K}\}$, with $p_k$ denoting the transmitted power allocated to $\text{SU}_\text{k}$, $\mathbf{U} = [\mathbf{u}_1, \mathbf{u}_2, \cdots \mathbf{u}_K] \in \mathbb{C}^{N_t \times K}$ is the downlink beamformer matrix, with $\mathbf{u}_k$ standing for the beamformer vector of the $\text{SU}_\text{k}$ with $\|\mathbf{u}_k\|^2 = 1$. Furthermore, we assume that a perfect CSI of primary and secondary links is available at the CBS. The received signal $\mathbf{r}_k \in \mathbb{C}^{N_r \times 1}$ at the $\text{SU}_\text{k}$ is given by

$$\mathbf{r}_k = \mathbf{H}_k \mathbf{x} + \mathbf{g}_k s_0 \sqrt{p_p} + \mathbf{n}_k$$
$$= \underbrace{\mathbf{H}_k s_k \sqrt{p_k} \mathbf{u}_k}_{\text{desired signal}} + \underbrace{\mathbf{H}_k \left( \sum_{l=1, l\neq k}^{K} s_l \sqrt{p_l} \mathbf{u}_l \right) + s_0 \sqrt{p_p} \mathbf{g}_k + \mathbf{n}_k}_{\text{interference+noise}}, \tag{2}$$

where $\mathbf{H_k} \in \mathbb{C}^{N_r \times N_t}$ denotes the channel between the CBS and $\text{SU}_\text{k}$. The vector $\mathbf{n_k} \in \mathbb{C}^{N_r \times 1}$ is the additive white Gaussian noise (AWGN), where $\mathbf{n}_k \sim CN(0, \sigma^2 \mathbf{I})$. Param-

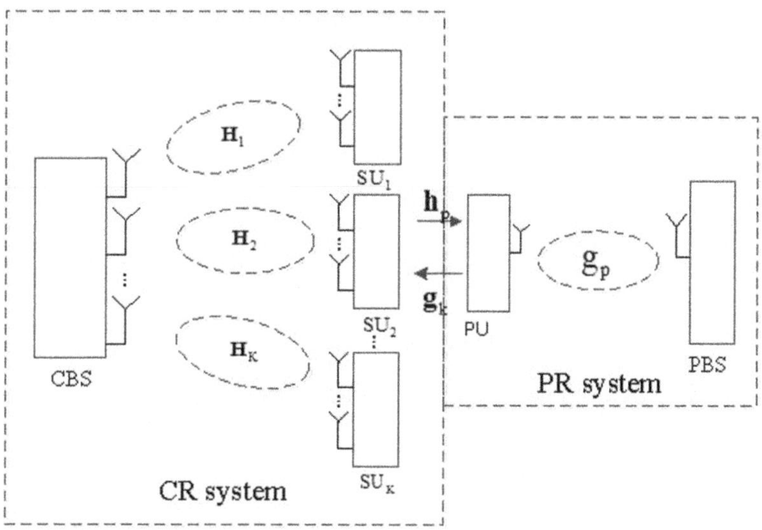


**Fig. 1** The system model for CR MIMO-BCs. The CBS has $N_t$ antennas, and each SU is equipped with $N_r$ antennas. The PBS and PU are equipped with a single antenna, respectively

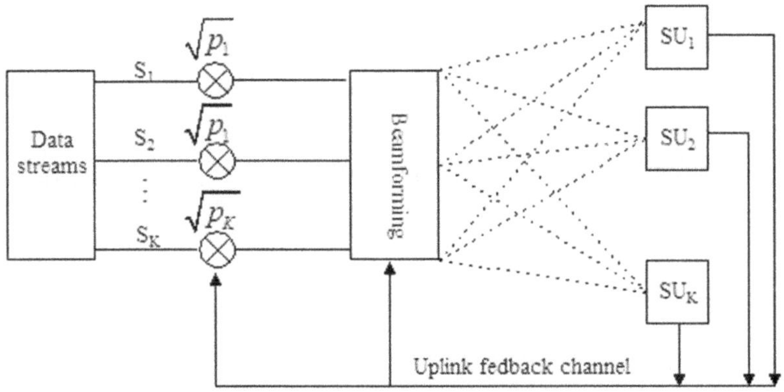


**Fig. 2** Block diagram of BC-MAC duality relationship for the transmission process

eters $p_p$ and $s_0$ denote the PBS transmission power and transmission data stream, respectively. Letting $\mathbf{g}_k$ represent the channel power gain vector from PBS to the $\mathrm{SU_k}$, the right-hand side of (2) can be decomposed into three terms: the first term is the desired signal from CBS to the $\mathrm{SU_k}$, whereas the second and third terms denote the interference plus noise from the PBS and noise, respectively. Similarly, the receive signal at the PU from PBS to PU can be written as

$$\begin{aligned} r_p &= \sqrt{p_p g_p s_0} + \mathbf{h}_p^T \mathbf{x} + n_p \\ &= \underbrace{\sqrt{p_p g_p s_0}}_{\text{desired signal}} + \underbrace{\mathbf{h}_p^T \sum_{k=1}^{K} s_k \sqrt{p_k} \mathbf{u}_k + n_p}_{\text{Interference+noise}}, \end{aligned} \tag{3}$$

where $\mathbf{h}_p$ denotes the channel power gain vector from the CBS to the PU and $n_p$ is an AWGN item.

In (2), the covariance matrix $\mathbf{Z}_k$ of the interference plus noise is given by

$$\mathbf{Z}_k = E\left[\mathbf{z}_k \mathbf{z}_k^T\right] = \sum_{l=1, l\neq k}^{K} p_l \mathbf{H}_k \mathbf{u}_l \mathbf{u}_l^T \mathbf{H}_k^T + p_p \mathbf{g}_k \mathbf{g}_k^T + \mathbf{n}_k \mathbf{n}_k^T, \tag{4}$$

where $\mathbf{z}_k = \mathbf{H}_k \left( \sum_{l=1, l\neq k}^{K} s_l \sqrt{p_l} \mathbf{u}_l \right) + s_0 \sqrt{p_p} \mathbf{g}_k + \mathbf{n}_k$. Since the combined interference and noise is not white anymore (its covariance is shown in (4)), the optimal maximum likelihood (ML) detector is equivalent to whitening the received signal, followed by applying the ML detector designed for white noise [33, 34]. We can thus whiten the received signal at the $\mathrm{SU_k}$ by multiplying $\mathbf{z}_k^{-1/2}$ with $\mathbf{r}_k$:

$$\begin{aligned} \tilde{\mathbf{r}}_k &= \mathbf{Z}_k^{-1/2} \mathbf{r}_k = \mathbf{Z}_k^{-1/2} \mathbf{H}_k b_k \sqrt{p_k} \mathbf{u}_k \\ &\quad + \mathbf{Z}_k^{-1/2} \cdot \mathbf{z}_k = \tilde{\mathbf{H}}_k b_k \sqrt{p_k} \mathbf{u}_k + \tilde{\mathbf{z}}_k, \end{aligned} \tag{5}$$

where $\tilde{\mathbf{H}}_k$ and $\tilde{\mathbf{z}}_k$ are a transformed MIMO channel matrix and Gaussian white noise after the whitening transformation, respectively. Note that the noise-whitening filter should satisfy the condition of $E\left[\tilde{\mathbf{z}}_k \tilde{\mathbf{z}}_k^T\right] = E\left[\mathbf{Z}_k^{-1/2} \mathbf{z}_k \cdot \left(\mathbf{Z}_k^{-1/2} \mathbf{z}_k\right)^T\right] = \mathbf{I}_{N_r}$. The singular value decomposition (SVD) [35] of $\tilde{\mathbf{H}}_k$ is defined as

$$\tilde{\mathbf{H}}_k = \mathbf{U}_k \Lambda_k \mathbf{V}_k^H, \tag{6}$$

where $\Lambda_k \in \mathbb{C}^{N_r \times N_t}$ is a real-valued diagonal matrix and can be partitioned into

$$\Lambda_k = \begin{bmatrix} \hat{\Lambda}_{\rho_k \times \rho_k} & \mathbf{0}_{\rho_k \times (N_t - \rho_k)} \\ \mathbf{0}_{(N_r - \rho_k) \times \rho_k} & \mathbf{0}_{(N_r - \rho_k) \times (N_t - \rho_k)} \end{bmatrix}. \tag{7}$$

Furthermore, $\rho_k = \operatorname{rank}\left(\tilde{\mathbf{H}}_k\right)$ denotes the rank of the $\mathrm{SU_k}$-transformed MIMO channel matrix, which is equivalent to the number of non-zero singular values and satisfies $\rho_k \leq \min(N_t, N_r)$. Substituting (7) into (5), we can obtain

$$\bar{\mathbf{r}}_k = \Lambda_k b_k \sqrt{p_k} \bar{\mathbf{u}}_k + \bar{\mathbf{z}}_k, \tag{8}$$

Taking the partitions in (7) into account, the last $(N_r - \rho_k)$ zero components in the received signal $\bar{\mathbf{r}}_k$ can be ignored. Furthermore, the last $(N_t - \rho_k)$ components of the transformed beamforming vector are also set to zero for the purpose of avoiding wasting transmit power on those dimensions of zero singular values.

Thus, we focus only on those dimensions corresponding to strictly positive singular values of $\tilde{\mathbf{H}}_k$. To reduce dimensionality to the rank of $\tilde{\mathbf{H}}_k$, we take the first $\rho_k$ elements in $\bar{\mathbf{r}}_k$ and obtain

$$\begin{aligned} \hat{\mathbf{r}}_k &= \left[\mathbf{I}_{\rho_k} \; \mathbf{0}_{\rho_k \times (N_r - \rho_k)}\right] \bar{\mathbf{r}}_k = \hat{\Lambda}_k s_k \sqrt{p_k} \left[\mathbf{I}_{\rho_k} \; \mathbf{0}_{\rho_k \times (N_t - \rho_k)}\right] \bar{\mathbf{u}}_k \\ &\quad + \left[\mathbf{I}_{\rho_k} \; \mathbf{0}_{\rho_k \times (N_r - \rho_k)}\right] \bar{\mathbf{z}}_k = \hat{\Lambda}_k s_k \sqrt{p_k} \hat{\mathbf{u}}_k + \hat{\mathbf{z}}_k, \end{aligned} \tag{9}$$

where $\hat{\mathbf{u}}_k = \left[\mathbf{I}_{\rho_k}\mathbf{0}_{\rho_k\times(N_t-\rho_k)}\right]\bar{\mathbf{u}}_k$ and $\hat{\mathbf{z}}_k = \left[\mathbf{I}_{\rho_k}\ \mathbf{0}_{\rho_k\times(N_r-\rho_k)}\right]\bar{\mathbf{z}}_k$. We can invert the channel in (9) to obtain the equivalent expression [36]

$$\hat{\mathbf{r}}_{k,\mathrm{inv}} = \hat{\Lambda}_k^{-1}\hat{\mathbf{r}}_k = b_k\sqrt{p_k}\hat{\mathbf{u}}_k + \hat{\Lambda}_k^{-1}\hat{\mathbf{z}}_k. \quad (10)$$

The received signal $\hat{\mathbf{r}}_{k,\mathrm{inv}}$ by matching the filtering is then given by

$$\mathbf{y}_k = \hat{\mathbf{u}}_k^T\cdot\hat{\mathbf{r}}_{k,\mathrm{inv}} = \underbrace{b_k\sqrt{p_k}}_{\text{desire dsignal}} + \underbrace{\hat{\mathbf{u}}_k^T\hat{\Lambda}_k^{-1}\hat{\mathbf{z}}_k}_{\text{interference+noise}}, \quad (11)$$

where the right-hand side of (11) comprises two terms: the first term is the desired signal and the second term denotes interference plus noise. The SINR at $\mathrm{SU_k}$ can be derived as

$$\gamma_k(\mathbf{u}_k,\mathbf{p}) = \frac{p_k}{\hat{\mathbf{u}}_k^T\hat{\Lambda}_k^{-1}\hat{\mathbf{z}}_k\hat{\mathbf{z}}_k^T\hat{\Lambda}_k^{-T}\hat{\mathbf{u}}_k} = \frac{p_k}{\hat{\mathbf{u}}_k^T\hat{\Lambda}_k^{-2}\hat{\mathbf{u}}_k} = \frac{p_k}{i_k}, \quad (12)$$

where $i_k = \hat{\mathbf{u}}_k^T\hat{\Lambda}_k^{-2}\hat{\mathbf{u}}_k$ is the interference function that depends explicitly on the $\mathrm{SU_k}$ beamformer $\hat{\mathbf{u}}_k$ as well as all the other users' beamformer $\hat{\mathbf{u}}_l$ and power $p_l$, $\forall l \neq k$, but does not depend on $\mathrm{SU_k}$ power.

In this setup, individual SUs may adjust their beamforming vector and powers in order to meet a set of specified target SINRs $\{\gamma_1^*,\gamma_2^*,\cdots,\gamma_K^*\}$ with minimum transmitted power. The target SINRs must be admissible and satisfy [37, 38]

$$\sum_{k=1}^{K}\frac{\gamma_k^*}{1+\gamma_k^*} < N_t, \quad (13)$$

where $N_t$ is the CBS antenna numbers and denotes the signal space of dimension.

Game theory provides a powerful framework for analyzing the problem of competitive utility maximization in wireless communication systems and could indicate whether a stable point, i.e., NE point, exists. A noncooperative game is[2] formally defined by using the set of players, the sets of strategies (or actions) that each player may take, and the individual player utility or cost functions.

In general, cost functions for a wireless system depend on the transmission power as well as QoS desired by a given user in the system. In CR MIMO-BCs, due to interference from other SUs and PU, the SU's transmission power needs to pay off a higher price for the purpose of achieving the abovementioned goal. On the other hand, the SU's transmission at optimal power will minimize the amount of interference. Therefore, the cost function of the $\mathrm{SU_k}$ can be defined as

$$J_k\left(p_k,\widehat{\mathbf{u}}\right) = p_k\left(i_k + \lambda\left|\mathbf{h}_p^T\mathbf{u}_k\right|^2\right), \quad (14)$$

where $\lambda$ is the pricing factor. Note that a higher $\lambda$ implies that the SUs will pay off a higher price. Due to the condition of $\mathbf{u}_k = \mathbf{V}_k\bar{\mathbf{u}}_k = \mathbf{V}_k\left[\hat{\mathbf{u}}_k\mathbf{0}_{(N_t-\rho_k)\times 1}\right]^T = \left[\hat{\mathbf{V}}_{N_t\times\rho_k}\mathbf{V}_{N_t\times(N_t-\rho_k)}\right]\left[\hat{\mathbf{u}}_k\mathbf{0}_{(N_t-\rho_k)\times 1}\right]^T = \hat{\mathbf{V}}_{N_t\times\rho_k}\hat{\mathbf{u}}_k$, the cost function of the $\mathrm{SU_k}$ in (14) can be rewritten as

$$\begin{aligned}J_k\left(p_k,\hat{\mathbf{u}}_k\right) &= p_k\cdot\left(i_k + \lambda\cdot\left|\mathbf{h}_p^T\hat{\mathbf{V}}\hat{\mathbf{u}}_k\right|^2\right)\\ &= p_k\left(\hat{\mathbf{u}}_k^T\hat{\Lambda}_k^{-2}\hat{\mathbf{u}}_k + \lambda\cdot\hat{\mathbf{u}}_k^T\hat{\mathbf{V}}^T\mathbf{h}_p\mathbf{h}_p^T\hat{\mathbf{V}}\hat{\mathbf{u}}_k\right)\\ &= p_k\hat{\mathbf{u}}_k^T\left(\hat{\Lambda}_k^{-2} + \lambda\cdot\hat{\mathbf{V}}^T\mathbf{h}_p\mathbf{h}_p^T\hat{\mathbf{V}}\right)\hat{\mathbf{u}}_k = \underbrace{p_k}_{f_k(p_k)}\ \underbrace{\hat{\mathbf{u}}_k^T\mathbf{S}_k\hat{\mathbf{u}}_k}_{g_k(\hat{\mathbf{u}}_k)},\end{aligned} \quad (15)$$

where $\mathbf{S}_k = \hat{\Lambda}_k^{-2} + \lambda\cdot\hat{\mathbf{V}}^T\mathbf{h}_p\mathbf{h}_p^T\hat{\mathbf{V}}$ is the correlation matrix of the interference plus noise.

## 3 Joint beamforming and power control as a noncooperative separable game

According to the separable game definition [39, ch. 11], the cost function in (15) is separable with respect to the parameters beamforming and power control, leading to a separable game with two separate subgames: beamforming subgame and power control subgame.

### 3.1 Noncooperative beamforming subgame

In the proposed noncooperative game, the SUs' power are fixed and individual SUs are capable of adjusting their beamforming in their corresponding strategy spaces to minimize their corresponding cost function. The noncooperative beamforming subgame (NPBS) can be modeled as

$$\mathrm{NPBS} = \left[\Omega,\left\{\widehat{\mathbf{u}}_k\right\},\{J_k(\cdot)\}\right], \quad (16)$$

where the components of NBPS are given as follows.

1. Players $\mathrm{SU_k}$: $k\in\Omega = \{1,2,\cdots,K\}$.
2. Action spaces: $\left\{\widehat{\mathbf{u}}_k\right\}$ is the set of beamforming strategies for $\mathrm{SU_k}$.
3. Cost functions: $\{J_k(\cdot)\}$ is the cost function that maps the $\mathrm{SU_k}$ beamforming spaces for fixed power control.

The cost function of NPBS is defined as

$$\min J_k\left(p_k,\hat{\mathbf{u}}_k\right)|_{\mathbf{P}=\text{fixed}},\quad s.t.\ \hat{\mathbf{u}}_k^T\hat{\mathbf{u}}_k = 1. \quad (17)$$

Here, our aim is to investigate the existence of a Nash equilibrium for NPBS as well as identify the best response strategies for SUs. Before establishing the uniqueness of the Nash equilibrium for NPBS, we state the following formal definitions [40, 41]:

**Definition 1 (Nash equilibrium of NPBS)** : The beamforming matrix $\widehat{\mathbf{U}} = [\widehat{\mathbf{u}}_1, \widehat{\mathbf{u}}_2, \cdots \widehat{\mathbf{u}}_K]$ is a Nash equilibrium of NPBS if, for every $\mathrm{SU_k}$, we have

$$J_k\left(\widehat{\mathbf{u}}_1, \cdots, \widehat{\mathbf{u}}_{k-1}, \widehat{\mathbf{u}}_k, \widehat{\mathbf{u}}_{k+1}, \cdots \widehat{\mathbf{u}}_K\right) \leq J_k\left(\widehat{\mathbf{u}}_1, \cdots, \widehat{\mathbf{u}}_{k-1}, \widehat{\mathbf{u}}_k', \widehat{\mathbf{u}}_{k+1}, \cdots \widehat{\mathbf{u}}_K\right), \quad \forall \widehat{\mathbf{u}}_k' \in \left\{\widehat{\mathbf{u}}_k\right\}. \tag{18}$$

**Definition 2 (Best response for NPBS)** : The best response strategy of $\mathrm{SU_k}$ beamforming to the other SUs is the set

$$B_k^{\hat{\mathbf{u}}_k} = \left\{\widehat{\mathbf{u}}_k \in \left\{\widehat{\mathbf{u}}_k\right\} \middle| J_k\left(\widehat{\mathbf{u}}_1, \cdots, \widehat{\mathbf{u}}_{k-1}, \widehat{\mathbf{u}}_k, \widehat{\mathbf{u}}_{k+1}, \cdots \widehat{\mathbf{u}}_K\right) \leq J_k\left(\widehat{\mathbf{u}}_1, \cdots, \widehat{\mathbf{u}}_{k-1}, \widehat{\mathbf{u}}_k', \widehat{\mathbf{u}}_{k+1}, \cdots \widehat{\mathbf{u}}_K\right), \quad \forall \widehat{\mathbf{u}}_k' \in \left\{\widehat{\mathbf{u}}_k\right\}\right\}. \tag{19}$$

**Definition 3 (Convex game)** : If the best response function of $\mathrm{SU_k}$ is a standard function, then NE in this game will be unique. The corresponding game is convex for non-empty, closed, and bounded convex set $\{\widehat{\mathbf{u}}_k\}$, if the cost function of each $\mathrm{SU_k}$ is in $\widehat{\mathbf{u}}_k$ for every fixed $\widehat{\mathbf{u}}_l$, where $l \neq k$.

For a fixed SU's power, the cost function in (17) is a quadratic form in the beamforming vector $\widehat{\mathbf{u}}_k$. Taking the second-order derivative of the cost function with respect to $\widehat{\mathbf{u}}_k$, we get

$$\frac{\partial^2 J_k}{\partial \hat{\mathbf{u}}_k^2} = 2p_k \mathbf{S}_k, \tag{20}$$

where $\mathbf{S}_k$ is a positive definite matrix. Evidently, the cost function of NPBS is also convex. According to the results proved in Theorem 1 in [42], for concave games, we can extend in a straightforward way to prove the existence of a Nash equilibrium point for convex games [38]. As a consequence, we establish the existence of the equilibrium point for NPBS. In order to solve the best response of the NPBS Nash equilibrium, we defined the $\mathrm{SU_k}$ Lagrangian function as

$$\begin{aligned} L_k^{\hat{\mathbf{u}}_k}\left(\hat{\mathbf{u}}_k, \alpha_k\right) &= J_k + \alpha_k\left(\hat{\mathbf{u}}_k^T \hat{\mathbf{u}}_k - 1\right) \\ &= p_k \hat{\mathbf{u}}_k^T \mathbf{S}_k \hat{\mathbf{u}}_k + \alpha_k\left(\hat{\mathbf{u}}_k^T \hat{\mathbf{u}}_k - 1\right), \end{aligned} \tag{21}$$

where $\alpha_k$ is the Lagrange multipliers. Taking the first-order derivative with respect to the beamformer $\widehat{\mathbf{u}}_k$, it leads to the eigenvector equation

$$\frac{\partial L_k^{\hat{\mathbf{u}}_k}}{\partial \hat{\mathbf{u}}_k} = 2p_k \mathbf{S}_k \hat{\mathbf{u}}_k + 2\alpha_k \hat{\mathbf{u}}_k. \tag{22}$$

Let (22) equal to zero, the best response function of the NPBS Nash equilibrium is given by

$$\mathbf{S}_k \hat{\mathbf{u}}_k = -\frac{\alpha_k}{p_k} \hat{\mathbf{u}}_k, \tag{23}$$

where the best response strategy of NPBS is the eigenvector $\widehat{\nu}_k$, corresponding to the minimum eigenvalue of $\mathbf{S}_k$. Thus, at the Nash equilibrium, all SUs beamforming will be the minimum eigenvectors of $\mathbf{S}_k$. In order to investigate whether the minimum eigenvector strategy is optimal for the $\mathrm{SU_k}$ cost function or not, we use the Taylor series to expand the Lagrangian function around the point which satisfies the necessary Karush-Kuhn-Tucker (KKT) conditions [38, 43]. In this expansion, the term containing the first derivative is equal to zero, provided that the higher order terms are neglected. Therefore, we only prove that the second-order term of (21) in the Taylor expansion is positive and satisfies the KKT conditions. The second derivative of (21) is given by

$$\begin{aligned} D_k^{\hat{\mathbf{u}}_k} &= (-1)\begin{vmatrix} \frac{\partial^2 L_k^{\hat{\mathbf{u}}_k}(\hat{\mathbf{u}}_k, \alpha_k)}{\partial \hat{\mathbf{u}}_k^2} & \frac{\partial^2 L_k^{\hat{\mathbf{u}}_k}(\hat{\mathbf{u}}_k, \alpha_k)}{\partial \hat{\mathbf{u}}_k \partial \alpha_k} \\ \left(\frac{\partial^2 L_k^{\hat{\mathbf{u}}_k}(\hat{\mathbf{u}}_k, \alpha_k)}{\partial \hat{\mathbf{u}}_k \partial \alpha_k}\right)^T & \frac{\partial^2 L_k^{\hat{\mathbf{u}}_k}(\hat{\mathbf{u}}_k, \alpha_k)}{\partial \alpha_k^2} \end{vmatrix} \\ &= (-1)\begin{vmatrix} 2p_k \mathbf{S}_k + 2\alpha_k & 2\hat{\mathbf{u}}_k \\ 2\hat{\mathbf{u}}_k^T & 0 \end{vmatrix} = 4 > 0, \end{aligned} \tag{24}$$

which is positive and satisfies the KKT conditions [32, 34]. Hence, the Nash equilibrium point with respect to the constrained minimization of the $\mathrm{SU_k}$ cost function is optimum.

### 3.2 Noncooperative power control subgame

In the proposed noncooperative game, the SUs' beamforming vectors are assumed to be fixed, and individual SUs adjust their power in their corresponding strategy spaces for the purpose of minimizing their corresponding cost function. The noncooperative beamforming subgame (NPCS) can be modelled as

$$\mathrm{NPCS} = \left[\Omega, \{p_k\}, \{J_k(\cdot)\}\right], \tag{25}$$

where the components of the NPCS are given in the list as follows.

1. Players $\mathrm{SU_k}$: $l \in \Omega = \{1, 2, \cdots, K\}$.
2. Action spaces: $\{p_k\}$ is the set of power control strategies for $\mathrm{SU_k}$.
3. Cost functions: $\{J_k(\cdot)\}$ is the cost function that maps the $\mathrm{SU_k}$ power control spaces for fixed beamforming vectors.

The cost functions of NPCS is defined as

$$\min J_k\left(p_k, \hat{\mathbf{u}}_k\right)\big|_{\hat{\mathbf{u}}_k = \text{fixed}} \quad s.\,t.\; p_k = \gamma_k^* \cdot i_k, \tag{26}$$

where $\gamma_k^*$ is the $\mathrm{SU_k}$ target SINR. In order to investigate the existence of a Nash equilibrium for NPCS and identify the best response strategies for SUs, we can define the $SU_k$ Lagrangian function as

$$L_k^{p_k}(p_k,\beta_k) = J_k + \beta_k\left(p_k - \gamma_k^* \cdot i_k\right) \\ = p_k \hat{\mathbf{u}}_k^T \mathbf{S}_k \hat{\mathbf{u}}_k + \beta_k\left(p_k - \gamma_k^* \cdot i_k\right), \quad (27)$$

where $\beta_k$ is a Lagrange multiplier. Taking the first-order derivative with respect to the multiplier $\beta_k$, we obtain

$$\frac{\partial L_k^{p_k}}{\partial \beta_k} = p_k - \gamma_k^* \cdot i_k = 0, \quad (28)$$

which is a necessary condition for the constrained optimization problem (26), indicating that the $SU_k$-transmitted power should match its target SINR by giving the interference function $i_k$, i.e.,

$$p_k = \gamma_k^* \cdot i_k, \quad (29)$$

where the best response function of NPCS is obtained from (29), and then $\gamma_k^* \cdot i_k$ is the Nash equilibrium point. Similar to (24), we can use the Taylor series to expand the Lagrangian function around the Nash equilibrium point as [34]

$$D_k^{p_k} = (-1)\begin{vmatrix} \frac{\partial^2 L_k^{p_k}(p_k,\beta_k)}{\partial p_k^2} & \frac{\partial^2 L_k^{p_k}(p_k,\beta_k)}{\partial p_k \partial \beta_k} \\ \left(\frac{\partial^2 L_k^{p_k}(p_k,\beta_k)}{\partial p_k \partial \beta_k}\right)^T & \frac{\partial^2 L_k^{p_k}(p_k,\beta_k)}{\partial \beta_k^2} \end{vmatrix} \\ = (-1)\begin{vmatrix} 0 & 1 \\ 1 & 0 \end{vmatrix} = 1 > 0. \quad (30)$$

In this case, the cost function of NPCS is a convex function, which is linear in $p_k$. Thus, following [44], there exists a Nash equilibrium. This is also optimal corresponding to the best response strategy, which is capable of updating power to match the target SINR.

### 3.3 Iterative algorithm

In this section, our proposed algorithm employs incremental updates for beamforming vectors and transmitted powers, which are designed to reduce interference at downlink receivers subject to specified SINR and beamforming constraints. We firstly model the joint beamforming and power control subgame (JBPS) as

$$\text{JBPS} = \left[\Omega, \left\{\widehat{\mathbf{u}}_k \times p_k\right\}, \{J_k(\cdot)\}\right]. \quad (31)$$

To find a Nash equilibrium for JBPS, we must consider the two subgames (NPBS and NPCS) to assure their corresponding best response strategies. However, direct application of the best response strategies is not guaranteed to converge to the optimal Nash equilibrium, although multiple Nash equilibrium points for the JBPS are possible to be achieved. Moreover, these best response strategies may cause large-angle deviation during updating of beamforming vectors and/or abrupt power control changes to meet the target SINRs. It is thus not desirable for a practical system, because it may increase error probability at the receiver or even lead to connection loss from the transmitter to the receiver. From a practical perspective, a more desirable approach can be employed to vary the SUs' beamforming and power control in small increments.

In order to avoid the minimum eigenvector being far away in signal space from the current beamforming vector, we use an incremental update that adapts the user beamforming vector in the direction of the minimum eigenvector. Hence, the beamforming vectors update of $SU_k$ at step $n$ of the algorithm can be given by

$$\hat{\mathbf{u}}_k(n+1) = \frac{\hat{\mathbf{u}}_k(n) + m\beta \mathbf{v}_k(n)}{\left\|\hat{\mathbf{u}}_k(n) + m\beta \mathbf{v}_k(n)\right\|}, \quad (32)$$

where $\beta$ (called beamforming pricing factor) is a constant and $m = \text{sgn}[\,\hat{\mathbf{u}}_k^T(n)\,\mathbf{v}_k(n)]$.

From (29), the transmitted power corresponding to user $k$ should match its target SINR. After the beamforming update in (32), the power value matching the desired target SINR is given by

$$p_k(n+1) = \gamma_k^* \hat{\mathbf{u}}_k(n+1)^T \mathbf{S}_k^{-2} \hat{\mathbf{u}}_k(n+1), \quad (33)$$

In order to avoid abrupt variations, the power control updates of $SU_k$ at step $n$ of the algorithm using a gradient-based approach and the corresponding power control algorithm in small increments are given by

$$p_k(n+1) - p_k(n) = \mu\left(\gamma_k^* i_k(n) - p_k(n)\right), \quad (34)$$

where $\mu$ ($0 < \mu < 1$) (called power pricing factor) is a constant. The joint beamforming and power control update algorithm is given in Table 1.

## 4 Numerical results and simulations

In this section, numerical simulation results are provided to validate the proposed theoretical analysis and examine the performance of our proposed algorithms for various scenarios simultaneously. In CR MIMO-BCs, the channels are subject to Rayleigh fading with zero mean and unity variance. Furthermore, 10,000 channel matrices are generated with Monte Carlo simulations. We consider the CBS with six antennas ($N_t = 6$) and the five SUs each equipped with four antennas ($N_r = 4$), and the noise covariance matrix is defined as $\mathbf{W}_k = 0.1\mathbf{I}_4$. In addition, the power control matrix and the target SINRs are initialized as $\mathbf{P} = 0.1\mathbf{I}_5$ and $\gamma^* = \{2, 1.9, 1.8, 1.7, 1.5\}$, respectively, where the target SINRs for SUs are set to satisfy the admissibility condition in (13). Note that the following parameter values are used in all numerical simulations unless stated otherwise: $\mu = 0.5$, $\beta = 0.5$, $\lambda = 0.1$, and $\varepsilon = 10^{-6}$.

**Table 1** The joint beamforming and power control update algorithm

| |
|---|
| 1. Initial data: beamforming vectors $\hat{\mathbf{u}}_k$, power control $p_k$, channel matrices $\mathbf{H}_k$, and target SINRs $\gamma_k^*$ for $SU_k$, noise covariance matrix $\mathbf{W}$, constants $\beta$, $\mu$ and tolerance $\varepsilon$. |
| 2. Whitening the received signal at the $SU_k$, using the rank of $SU_k$-transformed MIMO channel matrix $\rho_k = \text{rank}\left(\tilde{\mathbf{H}}_k\right)$ to calculate $\hat{\Lambda}_k^{-2}$, calculating $\mathbf{S}_k(n)$ and the eigenvector $\widehat{\mathbf{u}}_k$ corresponding to the minimum eigenvalue of $\mathbf{S}_k$. At the same time, using (32) to update $\hat{\mathbf{u}}_k$. |
| 3. If $\rho_k < N_t$, adding zeros to obtain $N$-dimension matrix $\bar{\mathbf{u}}_k$, and then using $\mathbf{u}_k = \mathbf{V}_k\bar{\mathbf{u}}_k$ to obtain original beamforming vectors. |
| 4. Using (33) to update power. |
| 5. If satisfy $\lvert J_k(n+1) - J_k(n)\rvert \leq \varepsilon$, where $\varepsilon$ ($\varepsilon > 0$) is also a small-valued positive number in practice to represent an infinitely small quantity, iteration continues. Otherwise, go to step 2. |
| 6. Iteration termination if the optimal condition (24) is true, then stop, an optimal Nash equilibrium has been reached. Otherwise, go to step 2. Where checking the optimal condition (24) ensures the optimal Nash equilibrium is reached and the algorithm does not stop in a suboptimal fixed point. |

In Fig. 3, we examine the convergence speed of the proposed algorithm. It is well known the optimal beamforming signal can be obtained through the zero-forcing algorithm[45]. As observed from Fig. 3, after several iterations, the proposed algorithm converges to the optimal SINRs, and it can achieve better performance than zero-forcing algorithm. Evidently, the proposed algorithm has a fast converge speed.

Figures 4, 5, and 6 show the variations of SUs' power, SINRs, and cost functions for variable target SINRs. These features are useful in CR MIMO-BCs, where different QoS requirements may lead to the target SINR variation. We assume that the system starts with $\mathbf{P} = 0.1\mathbf{I}_5$ and $\gamma^* = \{2, 1.9, 1.8, 1.7, 1.5\}$, making a Nash equilibrium (NE) configuration be achieved at iteration 10. We then vary the first SU's target SINR $\gamma_1^* = 2$ to new value $\gamma_1^* = 1$, leading the algorithm to reach a new NE configuration at iteration 35. Finally, we change SINR $\gamma_1^* = 1$ back to its old value $\gamma_1^* = 2$, and the system transitions back to the original set of target values between 30 iterations and 50 iterations. As a consequence, we can observe the system transitions from one optimal configuration to another one for variable target SINRs. The algorithm can also track variable target SINRs or a variable number of active users in the system and is therefore useful for dynamic wireless systems with varying QoS requirements. The beamforming matrices of the proposed algorithm are respectively obtained as follows:

$$\mathbf{S}_1 = \begin{bmatrix} -0.7585 & 0.0837 & -0.3258 & 0.5741 & -0.2577 \\ -0.3080 & 0.2390 & -0.6237 & -0.5845 & 0.0622 \\ 0.2925 & -0.0904 & -0.4672 & 0.1468 & -0.5583 \\ 0.3001 & 0.3993 & 0.1157 & 0.3617 & -0.3332 \\ 0.3601 & -0.8310 & -0.4280 & 0.3427 & 0.5017 \\ -0.1564 & 0.2789 & -0.3000 & 0.2428 & 0.5052 \end{bmatrix},$$
$$\mathbf{S}_2 = \begin{bmatrix} -0.7557 & 0.0850 & -0.3626 & 0.5656 & -0.1272 \\ -0.3072 & 0.2521 & -0.6666 & -0.5906 & -0.0184 \\ 0.3024 & -0.1541 & -0.4034 & 0.1056 & -0.6227 \\ 0.2922 & 0.3103 & 0.1750 & 0.3596 & -0.3516 \\ 0.3706 & -0.8515 & -0.4002 & 0.3415 & 0.5217 \\ -0.1428 & 0.2892 & -0.2659 & 0.2724 & 0.4470 \end{bmatrix}. \quad (35)$$

Figures 7, 8, and 9 show the variations of SUs' power, SINRs, and cost functions for a variable number of SUs. We start with five SUs, with different target SINRs $\gamma^* =$

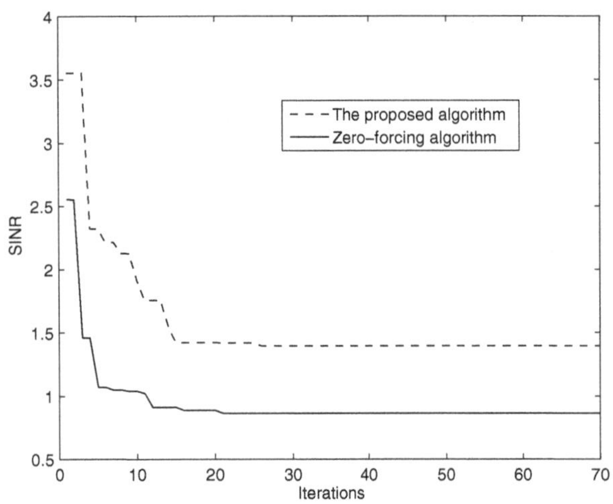


**Fig. 3** The comparison of the proposed algorithm and zero-forcing algorithm for the target SINRs

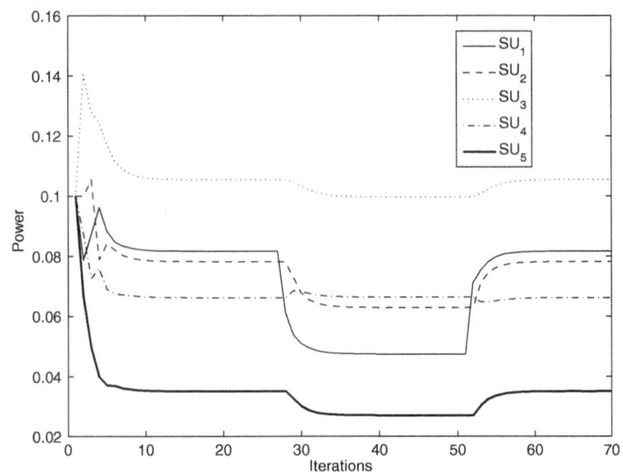


**Fig. 4** The variations of SUs' power versus the number of iterations for variable target SINRs

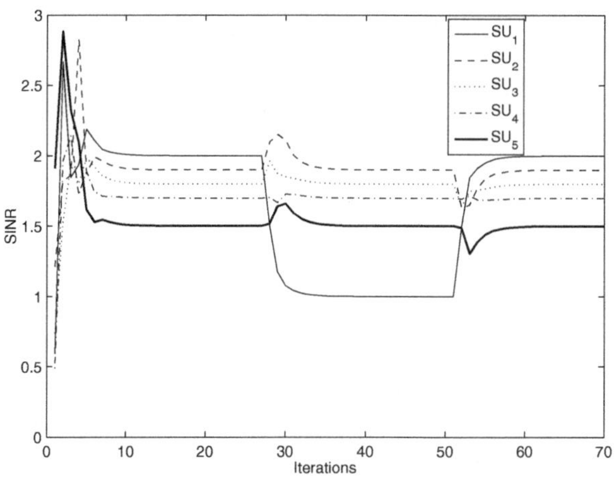


**Fig. 5** The variations of SINRs versus the number of iterations for variable target SINRs

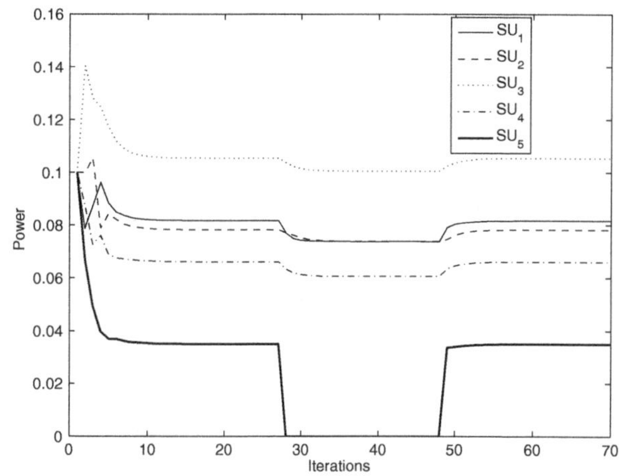


**Fig. 7** The variations of SUs' power versus the number of iterations for a variable number of SUs

$\{2, 1.9, 1.8, 1.7, 1.5\}$ being considered. The SUs' beamforming matrix is initialized randomly, with the SUs' power matrix being taken as $\mathbf{P} = 0.1\mathbf{I}_5$. We then assume $SU_5$ to be inactive. The remaining active SUs keep the same target SINR as $\gamma^* = \{2, 1.9, 1.8, 1.7\}$, which satisfies the admissibility condition in (13). After that, we assume $SU_5$ to be active in the system, so that the number of SUs becomes five again, and the same target SINR as before $\gamma^* = \{2, 1.9, 1.8, 1.7, 1.5\}$ can be obtained. We note that the system remains in a optimal configuration until iteration 28 when user 5 is dropped from the system. This implies an increase of user SINRs above their corresponding targets due to decreasing multiuser interference. Moreover, we can observe that the proposed algorithm finally converges towards the unique NE configuration for a variable number of SUs. The proposed algorithm is illustrated with numerical examples obtained from simulations which illustrate convergence and tracking properties of the algorithm for different scenarios This change in system configuration triggers the adaptation stage of the algorithm, which yields beamforming matrix $\mathbf{S}_3$.

$$\mathbf{S}_3 = \begin{bmatrix} -0.7229 & 0.0788 & -0.3522 & 0.5833 \\ -0.3171 & 0.1763 & -0.6329 & -0.5777 \\ 0.2716 & -0.0232 & -0.4429 & 0.2020 \\ 0.2871 & 0.5249 & 0.1137 & 0.3563 \\ 0.4435 & -0.7910 & -0.4277 & 0.3389 \\ -0.1548 & 0.2468 & -0.2889 & 0.2083 \end{bmatrix}. \quad (36)$$

As is the case with incremental algorithms, the convergence speed of the algorithm depends on the values of the corresponding increments specified by the algorithm constants $\beta$ and $\mu$. We performed extensive simulations

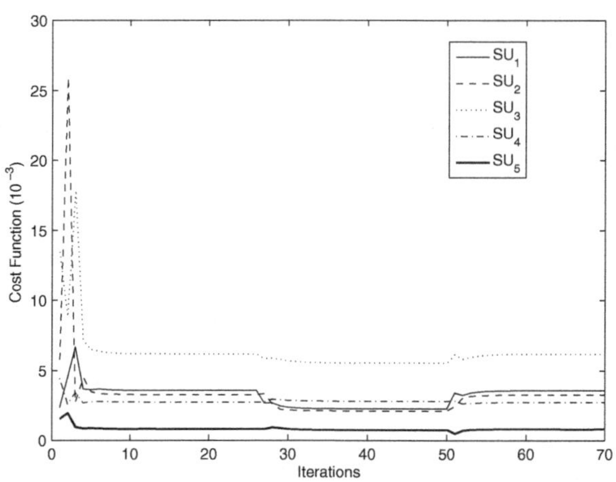


**Fig. 6** The variations of cost functions versus the number of iterations for variable target SINRs

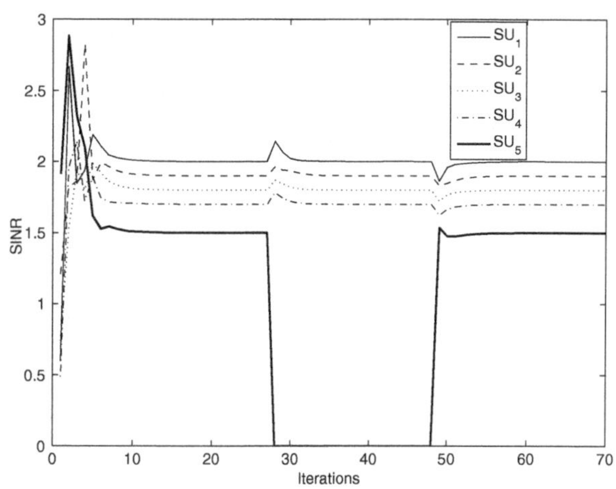


**Fig. 8** The variations of SINRs versus the number of iterations for a variable number of SUs

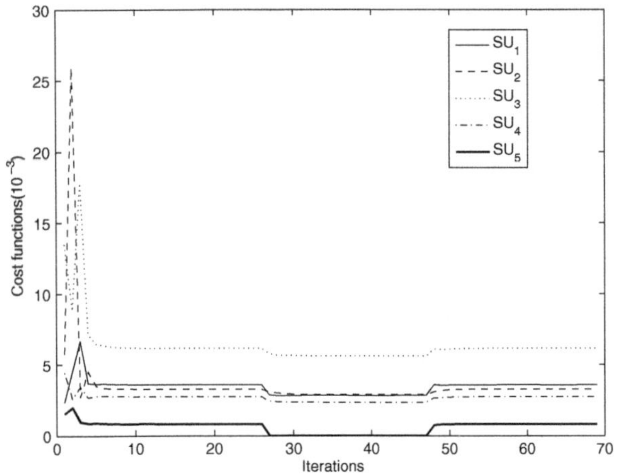


**Fig. 9** The variations of cost functions versus the number of iterations for a variable number of SUs

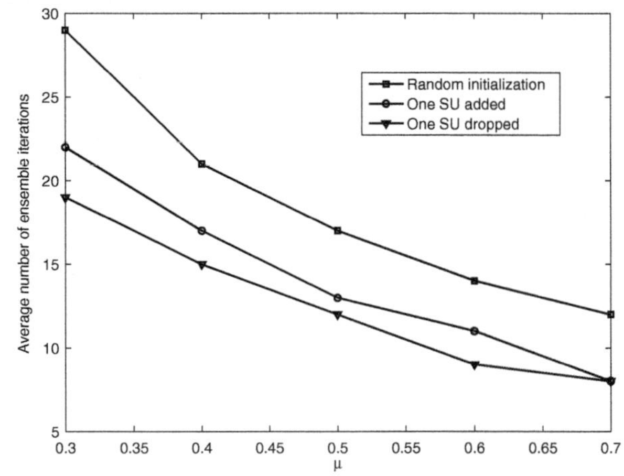


**Fig. 11** The average number of ensemble iterations for convergence to the optimal NE for varying $\mu$ and fixed $\beta = 0.5$

of the proposed algorithm to study convergence to the optimal Nash equilibrium. Figures 10 and 11 illustrate the average number of ensemble iterations for the purpose of converging to the optimal NE with varying $\beta$ and $\mu$. We ran 1000 trials of the algorithm and recorded the number of ensemble iterations for convergence. It is observed that convergence to the optimal NE is mostly determined by the $\mu$ rather than $\beta$. Moreover, it can be seen that the optimal NE is reached in less than 20 ensemble iterations under constants of $\beta = 0.5$ and $\mu = 0.5$ in all considered scenarios.

Figure 12 shows the variations of the total interference to PU versus the pricing factor $\lambda$. As expected, the total interference to PU is a monotonically decreasing function of $\lambda$. This is because SU will pay off a higher price to improve its QoS due to the interference from the other SUs and PU. Furthermore, we can also observe that the total interference to PU is minimized at $\lambda = 0.06$ and the minimum interference of PU will remain unchanged while $\lambda$ increases.

## 5 Conclusions

In this paper, the problem of joint beamforming and power control in CR MIMO-BCs has been formulated as a noncooperative separable game, which can be further divided into a beamforming subgame and power control subgame. We have investigated the best response strategies of NE solutions in terms of its corresponding subgames: in NPBS, players' power control strategies are fixed and they are subject to KKT conditions to update their beamforming strategies. In NPCS, on the other hand, players' beamforming strategies are fixed and they

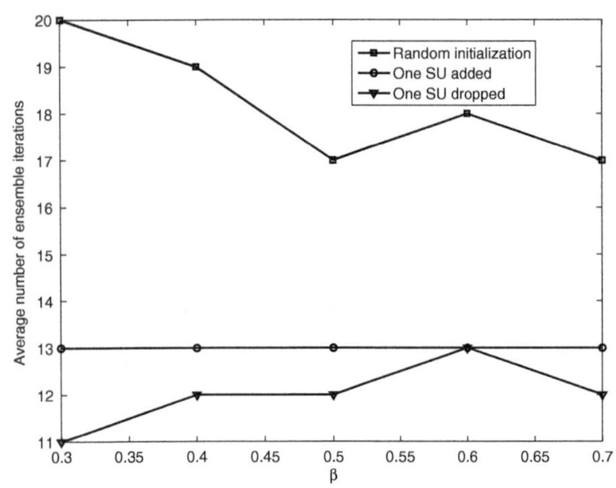


**Fig. 10** The average number of ensemble iterations for convergence to the optimal NE for varying $\beta$ and fixed $\mu = 0.5$

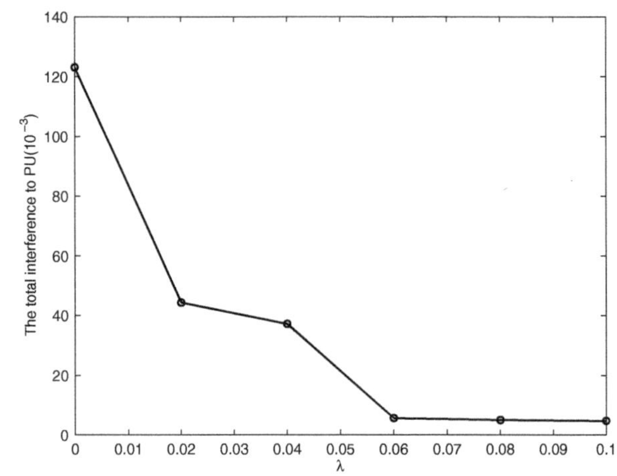


**Fig. 12** The variations of the total interference to PU versus the power pricing factor $\lambda$

update only their power control strategies subject to target SINRs. It is shown that the best response function of NPBS can be formulated, with $SU_k$ beamforming vector corresponding to the minimum eigenvector of the matrix $\mathbf{S}_k$, whereas the best response function of NPCS is derived to enable the transmission with a power control that corresponds to the desired target SINR. Furthermore, we have proposed an iterative algorithm to achieve the optimal Nash equilibrium of JBPS. The proposed algorithm is capable of tracking the variation of target SINRs and/or number of SUs in the system and is therefore useful for dynamic wireless systems with varying QoS requirement. Finally, the proposed algorithm may achieve a faster convergence speed by setting different parameter $\beta$ and $\mu$ values.

## Endnotes

[1] Separable games are a specific class of noncooperative games where the player's cost is a separable function of their strategic choices [34, 39].

[2] The game is noncooperative in the sense that a given player is interested only in minimization of its individual cost function, without paying attention to how its actions affect the other players.

**Acknowledgements**
Thanks are due to Prof. Hailin Xiao for valuable discussion. The authors acknowledge the support from the opening project of Guangxi Key Laboratory of Wireless Wideband Communication and Signal Processing and the opening project of Key Laboratory of Cognitive Radio and Information Processing, Ministry of Education (CRKL150107). The authors also acknowledge the support from the open research fund of the National Mobile Communications Research Laboratory, Southeast University (No. 2015D05).

**Competing interests**
The authors declare that they have no competing interests.

**Author details**
[1]Guangxi Key Laboratory of Wireless Wideband Communication and Signal Processing, Jinji Road, Guilin 541004, China. [2]Library of Guilin University of Electronic Technology, Jinji Road, Guilin 541004, China.

**References**

1. SJ Kim, G Li, GB Giannakis, Multi-band cognitive radio spectrum sensing for quality-of-service traffic. IEEE Trans. Wireless Commun. **10**(10), 3506–3515 (2011)
2. M Hasegawa, H Hirai, K Nagano, H Harada, K Aihara, Optimization for centralized and decentralized cognitive radio networks. Proc. IEEE. **102**(4), 574–584 (2014)
3. S Akin, MC Gursoy, Performance analysis of cognitive radio systems with imperfect channel sensing and estimation. IEEE Trans. Commun. **63**(5), 1554–1566 (2015)
4. Z Zhang, K Long, J Wang, Self-organization paradigms and optimization approaches for cognitive radio technologies: a survey. IEEE Wireless Commun. **20**(2), 36–42 (2013)
5. S Stotas, A Nallanathan, On the throughput and spectrum sensing enhancement of opportunistic spectrum access cognitive radio networks. IEEE Trans. Wireless Commun. **1**(1), 97–107 (2012)
6. L Sboui, Z Rezki, MS Alouini, A unified framework for the ergodic capacity of spectrum sharing cognitive radio systems. IEEE Trans. Wireless Commun. **12**(2), 877–887 (2013)
7. A Liu, Y Liu, H Xiang, W Luo, Polite water-filling for weighted sum-rate maximization in MIMO B-MAC networks under multiple linear constraints. IEEE Trans.Signal Process. **60**(2), 834–847 (2013)
8. F Gao, R Zhang, YC Liang, X Wang, Design of learning based MIMO cognitive radio systems. IEEE Trans. Veh. Technol. **59**(4), 1707–1720 (2010)
9. DN Nguyen, M Krunz, Power minimization in MIMO cognitive networks using beamforming games. IEEE J. Sel. Area in Commun. **31**(5), 916–925 (2013)
10. Y Xu, S Mao, Stackelberg game for cognitive radio networks with MIMO and distributed interference alignment. IEEE Trans. Veh. Technol. **63**(2), 879–892 (2014)
11. CS Vaze, MK Varanasi, The degree-of-freedom regions of MIMO broadcast, interference, and cognitive radio channels with no CSIT. IEEE Trans. Inf. Theory. **58**(8), 5354–5363 (2012)
12. DN Nguyen, M Krunz, Price-based joint beamforming and spectrum management in multi-antenna cognitive radio networks. IEEE J. Sel. Area in Commun. **30**(11), 2295–2305 (2012)
13. S Singh, PD Teal, PA Dmochowski, AJ Coulson, Robust cognitive radio cooperative beamforming. IEEE Trans. Wireless Commun. **13**(11), 6370–6381 (2014)
14. R Zhang, F Gao, YC Liang, Cognitive beamforming made practical: effective interference channel and learning-throughput tradeoff. IEEE Trans. Commun. **58**(2), 706–718 (2010)
15. J Mao, G Xie, J Gao, Y Liu, Energy efficiency optimization for cognitive radio MIMO broadcast Channels. IEEE Commun. Letters. **17**(2), 337–340 (2013)
16. C Xing, S Ma, Y Zhou, Matrix-monotonic optimization for MIMO systems. IEEE Trans. Signal Process. **63**(2), 334–348 (2015)
17. M Beko, Efficient beamforming in cognitive radio multicast transmission. IEEE Trans. Wireless Commun. **11**(11), 4108–4117 (2013)
18. S Burer, RD Monteiro, A nonlinear programming algorithm for solving semidefinite programs via low-rank factorization. Math. Program. **95**(2), 329–357 (2003)
19. YJ Zhang, AMC So, Optimal spectrum sharing in MIMO cognitive radio networks via semidefinite programming. IEEE J. Sel. Area in Commun. **29**(2), 362–373 (2011)
20. Y Huang, DP Palomar, S Zhang, Lorentz-positive maps and quadratic matrix inequalities with applications to robust MISO transmit beamforming. IEEE Trans. Signal Process. **61**(5), 1121–1130 (2013)
21. E Song, Q Shi, M Sanjabi, R Sun, ZQ Luo, in *Proc.IEEE Int. Conf. Acoust., Speech, Signal Process. (ICASSP)*. Robust SINR constrained MISO downlink beamforming: when is semidefinite programming relaxation tight? (2011). 2012(1), 1–11
22. H Du, T Ratnarajah, M Pesavento, CB Papadias, Joint transceiver beamforming in MIMO cognitive radio network via second-order cone programming. IEEE Trans. Signal Processing. **60**(2), 781–792 (2012)
23. JH Noh, SJ Oh, Beamforming in a multi-user cognitive radio system with partial channel state information. IEEE Trans. Wireless Commun. **12**(2), 616–625 (2013)
24. B Gopalakrishnan, ND Sidiropoulos, Cognitive transmit beamforming from binary CSIT. IEEE Trans. Signal Processing. **14**(2), 895–906 (2015)
25. MM Zhao, Y Cai, Q Shi, B Champagne, M Zhao, Robust transceiver design for MISO interference channel with energy harvesting. IEEE Trans. Signal Process. **64**(17), 4618–4633 (2016)
26. MM Zhao, Y Cai, Q Shi, M Hong, B Champagne, Joint transceiver designs for full-duplex K-Pair MIMO interference channel with SWIPT. IEEE Trans. Commun. **65**(2), 890–905 (2017)
27. YJ Kim, HJ Lim, MG Song, GH Im, Power efficient transceiver designs for multi-cell coordination in MIMO cognitive radio networks. IEEE Trans. Commun. **61**(10), 4127–4138 (2013)
28. L Zhang, YC Liang, Y Xin, Joint beamforming and power allocation for multiple access channels in cognitive radio networks. IEEE J. Sel. Areas in Commun. **26**(1), 38–51 (2008)
29. Y He, S Dey, Sum rate maximization for cognitive MISO broadcast channels: beamforming design and large systems analysis. IEEE Trans. Wireless Commun. **13**(5), 2383–2401 (2014)

30. K Huang, R Zhang, Cooperative feedback for multiantenna cognitive radio networks. IEEE Trans. Signal Process. **59**(2), 747–758 (2011)
31. L Zhang, R Zhang, YC Liang, Y Xin, HV Poor, On Gaussian MIMO BC-MAC duality with multiple transmit covariance constraints. IEEE Trans. Infor. Theory. **58**(4), 2064–2078 (2012)
32. G Dartmann, X Gong, W Afzal, G Ascheid, On the duality of the max-min beamforming problem with per-antenna and per-antenna-array power constraints. IEEE Trans. Veh. Technol. **58**(4), 606–619 (2012)
33. S Abraham, DC Popescu, OA Dobre, Joint beamforming and power control in downlink multiuser MIMO system. IEEE Radio Wirel. Symp. (RWS). **15**(3), 444–449 (2010)
34. C Lacatus, DC Popescu, Adaptive interference avoidance for dynamic wireless systems: a game-theoretic approach. IEEE J. Sel. Topics Signal Processing. **1**(1), 189–202 (2007)
35. DC Popescu, C Rose, Interference aviodance and multiaccess vector channels. IEEE Trans. Commun. **55**(8), 1466–1471 (2007)
36. DC Popescu, C Rose, Codeword optimization for uplink CDMA dispersive channel. IEEE Trans. Wireless Commun. **4**(4), 1563–1574 (2005)
37. P Viswanath, V Anantharam, D Tse, Optimal sequences, power control and capacity of spread spectrum systems with multiuser linear receivers. IEEE Trans. Inf. Theory. **45**(6), 1968–1983 (1999)
38. DC Popescu, DB Rawat, O Popescu, M Saquib, Game-theoretic approach to joint transmitter adaptation and power control in wireless systems. IEEE Trans. System, Man, and Cybernetics. **40**(3), 675–682 (2010)
39. JCC McKinsey, *Introduction to the theory of games*. (McGraw-Hill, New York, 1952)
40. HL Xiao, S Ouyang, Power control game in multi-source multi-relay cooperative communication systems with quality-of-service constraint. IEEE Trans. Intelligent Transportation Systems. **16**(1), 41–50 (2015)
41. J Ni, H Xiao, Game Theoretic approach for joint transmit beamforming and power control in cognitive radio MIMO broadcast channels. EURASP J Wireless Commun. Netw. **2016**(4), 1–6 (2016)
42. CW Sung, KW Shum, KK Leung, Stability of distributed power and signature sequence control for CDMA systems—a game-theoretic framework. IEEE Trans. Inf. Theory. **52**(4), 1775–1780 (2006)
43. M Avriel, *Nonlinear programming, analysis and methods, Englewood Cliffs*. (Prentice-Hall, NJ, 1976)
44. J Rosen, Existence and uniqueness of equilibrium point for concave N-person games Econometrica. **33**(3), 550–534 (1965)
45. CK Wen, JC Chen, KK Wong, P Ting, Message passing algorithm for distributed downlink regularized zero-forcing beamforming with cooperative base stations. IEEE Trans. Wirel. Commun. **13**(5), 2920–2930 (2014)

# 9

# A comprehensive wireless sensor network reliability metric for critical Internet of Things applications

Dina Deif and Yasser Gadallah*

**Abstract**

Evaluating the reliability of a wireless sensor network (WSN) deployment is a highly important task especially when the WSN is used for a critical Internet of Things (IoT) application. In this paper, we introduce a novel comprehensive reliability metric to evaluate the reliability of WSN deployments over their intended mission time. Unlike the existing studies on the topic, the proposed metric takes into account that sensor nodes (SNs) are multi-component systems that are subject to different component failures, namely, sensor, transceiver, processor, and battery failures. Consequently, SNs are modeled as three-mode (*on*, *relay*, and *off*) systems instead of the simplistic two-mode (*on* and *off*) model adopted in the existing studies. To calculate the proposed reliability metric in a computationally efficient manner, we develop a search algorithm which generates the complete path set of the given WSN deployment. Extensive experimental results demonstrate the use of the proposed metric in evaluating the reliability of several WSN deployments under different operating conditions. Results also demonstrate the computational efficiency of the developed search algorithm used for calculating the proposed metric and the significant effect of using the proposed three-mode SN model on the accuracy of the evaluated reliability.

**Keywords:** Wireless sensor networks, Reliability, Fault tolerance, Internet of Things, Probability of failure, Combinatorics

## 1 Introduction

In recent years, wireless sensor networks (WSNs) have become a versatile technology for serving a multitude of applications that include residential, industrial, commercial, healthcare, and military applications. As such, WSNs are considered one of the enabling technologies for realizing the Internet of Things (IoT) concept where they play the pivotal role of detecting events and measuring physical and environmental phenomena of interest [1]. Many of the important IoT applications served by WSNs are characterized by being mission-critical, meaning that the failure of the WSN to detect the occurrence of an event or a phenomenon in the targeted region of interest (RoI) will have serious implications [2]. Hence, it is imperative that the WSN functions properly throughout its intended mission time. This poses stringent reliability requirements on the WSN that must be addressed in the design and deployment phase of the network.

The first step in designing a reliable WSN is to be able to evaluate the reliability of a given WSN deployment. The reliability of any multi-component system is formally defined as the "probability that a system will perform satisfactorily during its mission time when used under the stated conditions" [3]. The method by which the reliability of a specific system is evaluated varies according to the type(s) of components the system is composed of, the configuration of the system in terms of how these components are connected to each other, and the state(s) at which the system is defined to have failed. Ultimately, the reliability of the system is a function of the reliability measures of its components and evaluating the reliability of the system as a whole is a probability modeling problem.

In that context, a WSN can be viewed as a multi-component system in which the components are the sensor nodes (SNs) and the sink node(s). The mission time for a WSN can either be its intended lifetime or

* Correspondence: ygadallah@ieee.org
Electronics and Communications Department, School of Sciences and Engineering, The American University in Cairo, AUC Avenue, New Cairo 11835, Egypt

the maximum time interval between scheduled maintenance operations. Hence, the WSN mission time is application-dependent and can vary greatly, ranging from a few days to a few years. If the WSN is composed of different types of SNs with different coverage profiles and capabilities, the WSN is said to be *heterogeneous*. The configuration of the WSN is determined by the way the SNs are deployed in the targeted RoI and the resulting wireless connectivity among them.

In order to identify the states at which a given WSN deployment fails, the functionality of a WSN must first be defined. The functionality of a WSN can be divided into two major elements. The first element is the sensing functionality, which is the ability of a WSN to detect all the targets or phenomena that occur inside the boundaries of the RoI during its mission time. Hence, for a WSN to be functional in terms of sensing, it must provide full coverage for the RoI area (in case of area coverage) or all the targeted locations in the RoI (in case of point coverage) during its mission time. The second element of the WSN functionality is the connectivity functionality, which is the ability of the WSN to deliver sensed data from its sources (i.e., SNs) to the designated destination (i.e., sink node(s)) during its mission time. Hence, for a WSN to be functional in terms of connectivity, any target or a phenomenon detected by one or more SNs has to be recognized at the sink node(s) through multi-hop wireless communication throughout the WSN mission time. Based on this definition of WSN functionality, a WSN is said to have failed if either of its sensing or connectivity functionality elements fail [4].

There are several issues that can affect the reliability of a WSN by compromising its functionality in terms of coverage and/or connectivity. These issues can generally be classified into SN-related and non-SN-related issues. SN-related issues are factors pertinent to the functionality of the deployed SNs, mainly, SN power failure, hardware failures, and software failures. The effect of these issues on the functionality of the network during its mission time (i.e., on the reliability of the network) can be predicted [5] as will be discussed later. These issues can be summarized as follows:

- SN power failure: the majority of the industrial and commercial SNs currently available in the market are battery-powered. Current advances in the fabrication of batteries have recently introduced highly durable batteries for SNs that can last for years (e.g., lithium thionyl chloride batteries (http://www.tadiran batteries.de/eng/products/lithium-thionyl-chloride-batteries/overview.asp)) under certain conditions. Although these batteries can sustain the operation of the SNs for long periods of time, premature battery failures can still occur in practice. This can be attributed to a myriad of reasons such as the deployment of the SNs in harsh environmental conditions (e.g., extreme temperatures or rain), incorrect handling or random failure caused by defective hardware (https://www.omnisense.com/oms_cds/media/008-002-002%20OmniSense%20FMS%20Sensor%20Battery%20Life.pdf).
- SN hardware failures: SNs are subject to random hardware failures. This is attributed to two main reasons. The first one is that most commercial SNs are cost-sensitive, meaning that they are not always built of the highest quality components. The second reason is that SNs are often subjected to harsh environmental conditions which can affect the normal operation of its components [6].
- SN software failures: SNs are prone to random permanent software failures which can render them inactive, i.e., unable to sense or communicate.

On the other hand, non-SN-related issues are factors that are external to the deployed SNs such as wireless link failures (due to fading and external interference) and excessive packet collisions (i.e., internal interference in WSNs adopting contention-based medium access control). The effect of these issues on the overall network reliability is in general difficult to predict. The authors in [7] present a thorough study on the effects of non-SNrelated factors on the quality of wireless links in a WSN and show the complex and highly transient nature of these effects. Furthermore, the effects of the non-SN related issues on the WSN reliability are usually mitigated using measures such acknowledgements and retransmissions [8].

In this paper, we derive a comprehensive WSN reliability metric, which considers the different *SN-related* reliability issues, using a combinatorial approach. We adopt the general assumptions that the WSN is heterogeneous and has an arbitrary deployment configuration (e.g., clustered or flat configuration). The functionality of the WSN is defined in terms of both network coverage and connectivity of the SNs to the sink node(s). We assume that the SNs are subject to four types of failures during the WSN mission time, namely, sensor failure, transceiver failure, processor failure, and power failure. Consequently, SNs are modeled as systems which have three modes of operation, namely, *on*, *relay*, and *off*. To calculate the proposed reliability metric in a computationally efficient manner, we develop a search algorithm which generates the complete paths set of the given WSN deployment.

The rest of this paper is organized as follows. In Section 2, we summarize the existing work on WSN reliability and highlight the contribution of this paper. In Section 3, we briefly discuss some of the fundamental reliability concepts which we will be using throughout this paper. The assumptions and derivation of the proposed reliability metric are presented in Section 4. In Section 5, we present the developed search algorithm which is used to evaluate the reliability of WSN deployments based on the proposed metric. Section 6 presents the experimental results obtained from applying the proposed metric to case study surveillance WSN deployments. Finally, the paper is concluded in Section 7.

## 2 Related work on wireless sensor network reliability

Several studies have addressed the issue of evaluating or estimating the reliability of WSNs. In this section, we review the most significant of these studies and discuss their scope and limitations. Based on this discussion, we highlight the scope and contribution of the reliability metric proposed in this paper. We classify the existing studies on WSN reliability into two major tracks. The first track focuses on evaluating the reliability of a specific aspect of WSN functionality (such as packet transmission reliability) and/or to evaluate the reliability for one or more *parts* of the WSN (such as a single cluster in a cluster-based deployment). The studies which belong to the first track may or may not assume that SNs are subject to random failures and battery energy depletion. On the other hand, the second track focuses on evaluating the reliability of a WSN as a whole, either as a function of time or as a probability over a given network mission time, assuming that SNs are subject to random failures. Studies which belong to the second track define WSN functionality in terms of coverage and/or connectivity and assume that SNs are modeled as a two-mode device (either *on* or *off*) and have a given probability of failure during the mission time of the network.

We begin by reviewing studies which belong to the first track. The studies in [9] and [10] address the problem of evaluating the reliability of SN clusters in WSNs characterized by cluster-based deployments subject to random SN failures. In both studies, the authors assume that the SN clusters are non-overlapping and that each cluster has a designated cluster head which acts as a relay between the SNs in the cluster and the sink node. In [9], the authors define the reliability of a cluster as the probability of successful message delivery between the sink node and the cluster head. The authors in [10] define the reliability of the WSN as the probability that the geographical area of each cluster in the WSN is fully covered by its SNs and that the cluster head has at least one functional direct or multi-hop wireless path to the sink node. Based on this definition, they derive an expression for the reliability of each individual cluster and use a Monte Carlo (MC) simulation approach to estimate it. The main limitation of the studies in [9] and [10] is that the reliability of the WSN as a whole in terms of the reliability of its constituent clusters is not evaluated. In addition, the proposed definitions of reliability cannot be extended to WSNs with different deployment configurations such as flat deployments which are non-hierarchical.

In [11], the authors propose a model for evaluating the reliability of disjoint areas in a WSN subject to two types of failure events, namely, SN failures due to battery depletion and link failures. Their proposed approach depends on dividing the targeted RoI into disjoint areas or target regions. For each region, a reliability model is constructed using a reliability block diagram (RBD) [3], which depends on the number of SNs monitoring the target region, their relative location from the sink node, and the routing protocol used in the network. There are two drawbacks of the proposed reliability modeling proposed in [11]. The first drawback is that the model does not provide a method by which the reliability of the entire WSN deployment can be evaluated in terms of the reliability of its regions. The second drawback is that the reliability modeling is carried out under the assumption that the probabilities of link failures are known and are constant throughout the lifetime of the WSN. This assumption is unrealistic since link quality is affected by numerous factors such as multi-path effects, shadowing (due to static and mobile obstacles), and interference. The effect of these factors on link quality varies significantly and rapidly in time and space [7] and hence, unlike SN-related factors, cannot be reduced to a constant probability of failure throughout WSN mission time. In [12], the authors consider the problem of evaluating the transmission reliability of cluster-based and mesh-based WSN deployments. They define transmission reliability as the ratio of the packets received by a destination node to the whole packets generated by the transmission for a given period of time. They present transmission reliability evaluation models for the uplink and downlink traffic based on the assumptions that SNs are not subject to any hardware failures and that SNs only fail when their initial battery energy is exhausted. Although the time-dependent models presented in [12] can help assess the transmission reliability over time for a given routing strategy, they are limited by the assumption that SNs cannot fail due to random hardware failures unrelated to battery exhaustion. Also, it is not possible to use the study in [12] to calculate or estimate the reliability of the WSN over a given mission time since coverage functionality is not considered.

The study in [13] considers the problem of evaluating the reliability of the sink node decisions in WSNs targeted for intrusion detection applications. The authors model the detection mechanism of the sink node of an intrusion, based on the aggregated data from several SNs in the network, as a weighted voting system (WVS). They assume that both the SNs and the wireless links between the SNs and the sink node have known misdetection probabilities. Based on these assumptions, they derive the reliability of the sink's WVS using the universal generating function (UGF) method. Similar to the study in [12], the scope of the study in [13] does not include evaluating the reliability of the WSN as a whole over a given mission time, since it is restricted to evaluating the reliability of detecting a single target/phenomena based on a fraction of the SNs in the WSN.

On the other hand, the studies in [14, 15] belong to the second track since they address the reliability of SN systems or WSNs as a whole of non-hierarchical deployment configurations subject to random SN failures. In [14], the authors address the problem of evaluating the reliability of WSNs designed for industrial inventory management. They assume that for the purposes of this specific application, the data collected by each SN are stored redundantly on several other SNs to account for random SN failures. Accordingly, the WSN is deemed functional as long as there is a sufficient number of functional SNs that are both connected to each other and to the sink node. Based on this definition of network functionality and the assumption that the WSN deployment is homogeneous, the reliability evaluation problem is reduced to the famous *K*-out-of-*N* reliability problem (http://www.reliabilityanalytics.com/blog/2011/09/02/reliability-modeling-k-out-of-n-configutation/). The authors also present a Monte Carlo (MC) simulation approach similar to that proposed in [10] to estimate the reliability of the WSN at hand. However, the reliability evaluation and estimation approaches proposed in [14] are based on a very restrictive definition of network functionality. Consequently, they cannot be applied to other WSN applications (e.g., surveillance and monitoring applications) where the functionality of the network is dependent not only on the number of SNs connected to the sink node but also on the network coverage. Also, the proposed approaches do not support network heterogeneity which is a major limitation since real-world deployments are often heterogeneous.

The authors in [16] propose a reliability metric for SN systems designed for surveillance purposes subject to random SN failures. They assume an arbitrary deployment configuration where SNs can monitor multiple target locations in the RoI and that each target location can be monitored by multiple SNs. They also assume that the surveillance SN system can be heterogeneous. The reliability of the system is defined as the probability that all target locations are monitored by at least one SN. The authors use a combinatorial approach to formulate the proposed reliability metric and present a search algorithm to calculate the proposed reliability metric in a time-efficient manner. The main limitation of the proposed metric is that system functionality is assumed to be in terms of the degree of target locations coverage only. Connectivity between SNs to form a wireless network is not considered.

The study in [17] propose a method for evaluating the reliability of WSNs designed for industrial IoT applications based on the automatic generation of fault trees (FTs). The proposed method requires the network failure conditions as inputs to enable the generation of the corresponding network FT and compute the network reliability. A network failure condition is defined as a combination of SNs which if fail will lead to the failure of the WSN in terms of network coverage only and not connectivity to the sink. To address the connectivity part of the network functionality, the authors propose a depth-first search algorithm that finds all the paths between SNs belonging to the network failure conditions and the sink node. The study in [17] is extended in [15] by assuming that the WSN is also subject to permanent wireless link failures in addition to SN failures under the same assumptions adopted in [17]. In both studies, the authors in [17] and [15] did not address the computational efficiency of their approach.

In this study, we focus on the second track, i.e., on the problem of evaluating the reliability of the WSN a whole, defined as the probability that the WSN is functional during a given mission time, assuming that SNs are subject to random permanent failures. Based on the above discussion, existing studies in that track all assume that SNs have only two modes of operation, either *on* or *off*. If an SN is *on*, it is assumed to be functional in terms of both sensing its surrounding environment and communicating wirelessly with its neighbors. If it is *off*, then the SN has failed permanently due to one or more of the SN-related reliability issues outlined in Section 1. This representation is not accurate since most commercial SNs are composed of multiple independent chips that carry out different functions, with each having its own probability of failure during the network's mission time.

A more accurate model considers the SN as a multicomponent system [18]. Based on this model, an SN has three modes of operation. These modes of operation are *on*, *relay*, and *off*. The definitions of the *on* and *off* modes are the same as discussed above, while the *relay* mode occurs when the SN is unable to perform its sensory function but it is still able to communicate wirelessly with its neighbors. This mode of operation occurs when the SN's sensor(s) hardware fails while its transceiver, processor, and battery are in working condition. Adopting this SN model has two main advantages. The first one is that it provides a more accurate evaluation of WSN reliability, assuming that the network functionality is adequately defined in terms of both network coverage and connectivity. The addition of the *relay* mode to the SN model provides a more accurate evaluation of WSN reliability because it avoids the *under-evaluation* of the WSN reliability when the conventional two-mode SN model is used. Under-evaluation of the reliability of the network becomes an issue when the reliability is used as a requirement/constraint for WSN deployment. In that case, the network designer aims to deploy sufficient SNs to achieve a minimum level of network reliability while minimizing the number of deployed SNs, i.e., minimizing the deployment cost. In that case, under-evaluating the reliability of a given deployment can lead to an unnecessary increase in the deployment cost. The second advantage is that it enables the network designer to isolate the effect of the quality (i.e., reliability) of the individual components of the deployed SNs (i.e., sensor, transceiver, processor and power unit) on the overall reliability of the WSN. This is discussed in more details in Section 4.1.

In this paper, we derive a comprehensive WSN reliability metric which takes into account the different SN-related reliability issues using a combinatorial approach. Compared to the existing reliability evaluation and estimation approaches, the strengths of our proposed metric can be summarized in the following points:

- Network functionality is defined in terms of both network coverage of a predefined set of target locations in the RoI and connectivity to the designated sink node.
- No specific network deployment configuration is assumed in the proposed model. We assume an arbitrary deployment configuration where each deployed SN may monitor multiple target locations in the RoI and each target location may be monitored by multiple SNs. All SNs can communicate wirelessly with their neighbors, i.e., no imposed communication hierarchy.
- The WSN can be heterogeneous; it can consist of more than one type of SNs, each characterized by a different coverage profile and set of capabilities.
- A more realistic SN model is adopted in the derivation of the proposed metric where an SN has three modes of operation instead of the two-mode model used in the existing studies.
- Each SN type is characterized by four different probabilities of failure during the mission time of the network (namely, sensor, transceiver, processor, and battery probabilities of failure) instead of a single SN probability of failure, as it is the case in the existing studies.
- A search algorithm is developed to calculate the propose reliability metric in a computationally efficient manner.

In this study, we assume that wireless links between SNs are not subject failure. This assumption is justified as follows. Wireless link quality is affected by numerous factors such as multi-path effects, shadowing (due to static and mobile obstacles), and interference. The effect of these factors on link quality varies significantly and rapidly in time and space [7] and hence, unlike SN-related factors, cannot be reduced to a constant probability of failure throughout WSN mission time. On the other hand, permanent wireless link failures are mainly due to a complete failure (i.e., a failure in the transceiver, processor, or battery) in one or both SNs at the ends of the link [19]. In the proposed metric in this paper, this type of failure is taken into consideration since we assume that each of the main SN components are subject to failure with a given probability of failure during the WSN mission time.

## 3 Fundamental reliability concepts

In this section, we discuss some of the fundamental definitions and concepts related to the evaluation of multi-component systems' reliability which we will be using throughout this paper.

### 3.1 Component reliability function and component reliability

The main objective of reliability modeling is to express the reliability of a given system in terms of the reliability measures of its constituent components. There are two main reliability measures for any device or component. The first measure is the reliability function $R_c(t)$, which is used to estimate the probability that the device or component will continue to function beyond a time duration of length $t$ [3]. The second reliability measure is based on the fact that for most practical purposes, a device or component is only required to function during the specified mission time $T_m$ of the system it belongs to. In this case, the reliability function $R_c(t)$ can be substituted by the reliability of the device. The reliability of a device, $R_c$, is simply defined as the probability that the device will continue to function throughout the mission time of the system. Accordingly, the probability of failure of the device during $T_m$ is equal to $1 - R_c(T_m) = 1 - R_c$ [3].

For example, Fig. 1 shows an exponential reliability function, which is one of the simplest functions used in modeling the reliability of electronic components. The exponential reliability function which is $R_{c_e}(t)$ is given by the following equation:

$$R_{c_e}(t) = e^{-\alpha t}, \tag{1}$$

where $\alpha$ is the estimated failure rate of the component per unit of measurement (e.g., hour, year, cycle) and is

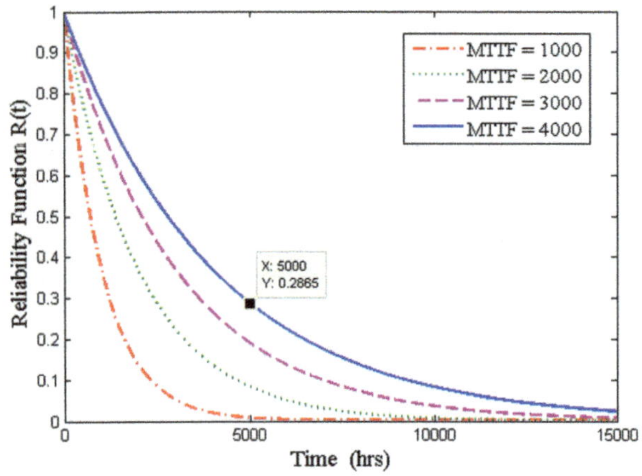


**Fig. 1** Exponential reliability function plot for different values of MTTF (1000, 2000, 3000, and 4000) in hours

equal to the reciprocal of its mean time to fail (MTTF). From the reliability curves in Fig. 1, we can estimate the reliability at $t = 5000$ h for $\alpha = 1/4000$ (i.e., for MTTF = 4000) to be 0.287. This in turn means that there is a $1 - 0.287 = 0.713$ chance that the component will fail during this time interval, i.e., the probability of failure during this time interval is 0.713.

Although the exponential reliability function is commonly used in reliability engineering due to its simplicity, it usually leads to inaccurate estimations of the probabilities of failures. This is because this type of function is based on the assumption that the component has a constant failure rate, which means that its performance does not degrade with time. To obtain a more accurate model for the reliability function of a given electronic device, reliability engineers carry out rigorous reliability testing techniques and/or gather empirical data on the device in service [3]. For example, qualitative and quantitative accelerated reliability testing is used to identify probable hardware failures of SNs and estimate the probability of their occurrence [5].

### 3.2 Combinatorial approach to system reliability evaluation

Combinatorics is a proven useful tool in evaluating and estimating the reliability of complex systems and networks [20, 21]. The fundamental premise of the combinatorial approach to reliability evaluation is that the reliability of any system can be computed by means of evaluating the system's structure function for every possible state of the system. To explain this concept, consider a system $\boldsymbol{S}$ which consists of $n$ components, i.e., $S = \{1, 2, ...., n\}$. Each component can only have two distinct states: it can either be functional or *on* or it can fail or be *off*. Let the binary variable $\pi_i$ be the state indicator of component $i$ as follows:

$$\pi_i = \begin{cases} 1, & \text{if component i is on} \\ 0, & \text{if component i is off} \end{cases} \tag{2}$$

A state $\boldsymbol{\pi}$ of the system $\boldsymbol{S}$ is a description of the states of all its components, hence $\boldsymbol{\pi} = \{\pi_i\}$ for $i = 1, \dots, n$. Let $\boldsymbol{\Pi}$ be the set of all possible states of $\boldsymbol{S}$. The structure function of $\boldsymbol{S}$, denoted $f(\boldsymbol{\pi})$, is a binary function that indicates whether the system is working under a given state according to the following equation:

$$f(\boldsymbol{\pi}) = \begin{cases} 1, & \boldsymbol{S} \text{ is functional} \\ 0, & \boldsymbol{S} \text{ has failed} \end{cases} \tag{3}$$

Based on the above definitions, the reliability of $\boldsymbol{S}$, denoted by $R(\boldsymbol{S})$, can be calculated using the following equation:

$$R(\boldsymbol{S}) = \text{Prob}(f(\boldsymbol{\pi}) = 1) = \sum_{\pi \in \Pi} f(\boldsymbol{\pi}).\text{Prob}(\boldsymbol{\pi}) \tag{4}$$

To calculate $R(\boldsymbol{S})$ using (4), the conditions necessary for $\boldsymbol{S}$ to be functional must be defined and the probability of any system state must be evaluated in terms of the reliabilities (or probabilities of failure) of the system's components, assuming that the system has a specified mission time $T_m$. Theoretically, $f(\boldsymbol{\pi})$ must be evaluated for all the possible system states $\boldsymbol{\pi} \in \boldsymbol{\Pi}$ to calculate $R(\boldsymbol{S})$ using this approach. However, following this extensive method in reliability calculation poses a computational problem for systems of a practical scale. For example, a system composed of 30 components which fail independently has $2^{30}$ states. Therefore, a tremendous amount of time is required to calculate $R(\boldsymbol{S})$ which grows exponentially with the number of components in the system. This computational problem is mitigated by the use of more efficient methods (e.g., reliability block diagram (RBD), fault tree (FT), and search algorithms) that attempt to find all the system's path sets or cut sets [20].

To define a system's path and cut, let $\boldsymbol{S}_1(\boldsymbol{\pi})$ be the set of functioning components, i.e., components in the *on* state, in $\boldsymbol{S}$ for a given system state $\boldsymbol{\pi}$ and $\boldsymbol{S}_0(\boldsymbol{\pi})$ be the set of failed components, i.e., components in the *off* state. $\boldsymbol{S}_1(\boldsymbol{\pi})$ and $\boldsymbol{S}_0(\boldsymbol{\pi})$ can be expressed by the following equations:

$$\boldsymbol{S}_1(\boldsymbol{\pi}) \equiv \{i \mid \pi_i = 1, \ i \in S\}, \tag{5}$$

$$\boldsymbol{S}_0(\boldsymbol{\pi}) \equiv \{i \mid \pi_i = 0, \ i \in S\}, \tag{6}$$

where $\boldsymbol{S}_1(\boldsymbol{\pi}) \cup \boldsymbol{S}_0(\boldsymbol{\pi}) = \boldsymbol{S}$. A state $\boldsymbol{\pi}$ of the system $\boldsymbol{S}$ is called a path if $f(\boldsymbol{\pi}) = 1$. In that case, the corresponding path set is the set $\boldsymbol{S}_1(\boldsymbol{\pi})$, i.e., a path set is the set of components whose simultaneous functional state guarantees that the overall system is functional. On the other hand, a state $\boldsymbol{\pi}$ of the system $\boldsymbol{S}$ is called a cut if $f(\boldsymbol{\pi}) = 0$. In this case, the corresponding cut set is the set $\boldsymbol{S}_0(\boldsymbol{\pi})$. That is, a cut set is the set of components whose simultaneous failure results in the failure of the overall system. If all the path sets or alternatively all the cut sets of a system $\boldsymbol{S}$ are known, we can rewrite (4) as follows:

$$R(\boldsymbol{S}) = \sum\nolimits_{\pi\in\Pi_1} \text{Prob}(\boldsymbol{\pi}) = 1-\sum\nolimits_{\pi\in\Pi_0} \text{Prob}(\boldsymbol{\pi}), \quad (7)$$

where $\boldsymbol{\Pi}_1$ is the set of all the paths of $\boldsymbol{S}$ (i.e., the complete paths set of $\boldsymbol{S}$ ) and $\boldsymbol{\Pi}_0$ is the corresponding set containing all the cuts of $\boldsymbol{S}$ (i.e., the complete cuts set of $\boldsymbol{S}$) such that $\boldsymbol{\Pi}_1 \cup \boldsymbol{\Pi}_0 = \boldsymbol{\Pi}$ . For example, a simple system of $n$ components connected in series has only one path set which is equal to the system set $\boldsymbol{S} = \{1, 2, ...., n\}$ and has $\sum_{k=1}^{n} C_k^n$ cut sets. Therefore, it is simpler to express its reliability as $R(\boldsymbol{S}_{\text{series}}) = \text{Prob}(\boldsymbol{\pi} = \{\pi_i = 1, \forall i = 1, \dots, n\}) = \prod_{i=1}^{n} R_i$, where $R_i$ is the reliability of the $i^{\text{th}}$ component during the system's mission time. On the other hand, a system of $n$ components connected in parallel has only one cut set which is equal to $\boldsymbol{S}$ and has $\sum_{k=1}^{n} C_k^n$ path sets. Hence, the system's reliability can be expressed as $R(\boldsymbol{S}_{\text{parallel}}) = 1-\text{Prob}(\boldsymbol{\pi} = \{\pi_i = 0, \forall i = 1, ..., n\}) = 1-\prod_{i=1}^{n}(1-R_i)$.

## 4 Reliability of wireless sensor networks

In this section, we use the combinatorial approach outlined in Section 3 to derive the reliability of a WSN with an arbitrary deployment configuration. We start by modeling the SN as a multi-component system and identifying its different states and modes of operation. Then, we present the WSN model and define the conditions required for the WSN to be deemed in working condition. Finally, we derive the reliability of the WSN in terms of its structure function and the probabilities of failure of its constituent SNs' hardware components.

### 4.1 Sensor node model

Although SNs vary greatly in terms of their capabilities (e.g., processing power, battery capacity), there are four fundamental chips or components that are common in all SNs [22]: a sensing unit(s) or simply sensor(s), a radio unit or transceiver, a processing and memory unit or processor, and a power unit or battery. The sensor is responsible for the translation of physical phenomena detected/measured in the RoI to electrical signals. The transceiver enables the SN to communicate wirelessly with its neighboring SNs and with the sink node. The processor is responsible for performing all required computations and controlling both the sensor and transceiver. The battery supplies all three components with power. The type and capacity of the SN battery is carefully chosen according to the application and the required mission time of the WSN (http://www.sensorsmag.com/components/a-practical-guide-to-battery-technologies-for-wireless-sensor-networking).

Each of these components is subject to random failure [6], [23] due to several reasons such as faulty hardware, faulty software, harsh environmental conditions, and degradation with time. Accordingly, each of the SN's four main components has a given reliability or, alternatively, a probability of failure during the WSN mission time $T_m$ as defined in Section 3.1. As mentioned earlier in Section 3, the reliabilities of the different components of an SN can be estimated through a standard reliability prediction test provided by the SN vendor or through reliability testing techniques [5].

Since each of the four components can either function or fail, i.e., be in an *on* or *off* state, an SN can theoretically have $2^4$ possible states. To describe these states, let the binary variables $x_s$, $x_t$, $x_p$, and $x_b$ be the state indicators of the sensor, transceiver, processor, and battery, respectively, of an SN as defined in (2). Hence, an SN state $x$ is described using a tuple of these four variables $\{x_s, x_t, x_p, x_b\}$. These variables are not statistically independent; the sensor and transceiver cannot possibly function if either the processor or the battery fails. Therefore, some of the SN states are practically impossible, and hence, their probability of occurrence is zero.

To calculate the probability of occurrence of the other possible states, let $\lambda_s$, $\lambda_t$, $\lambda_p$, and $\lambda_b$ be the probabilities of failure of the sensor, transceiver, processor, and battery, respectively. It should be noted that the estimated probability of failure for any given device or hardware component is obtained regardless of the failure of any other device or component. Hence, $\lambda_s$ and $\lambda_t$ are actually the probability of failure of the sensor and transceiver conditioned on the event that the component is properly controlled (i.e., processor is functional) and powered (i.e., battery is functional). Similarly, $\lambda_p$ is the probability of failure of the processor conditioned on the event that the battery is functional, where as $\lambda_b$ is the unconditional probability that the SN power unit or battery fails during $T_m$. According to the above definitions, the probability of an SN state can be given by the following equations:

$$\begin{aligned} Prob(x) &= Prob(x_s, x_t, x_p, x_b) \\ &= Prob(x_s, x_t|x_p, x_b).Prob(x_p, x_p) \\ &= Prob(x_s|x_p, x_b).Prob(x_t|x_p, x_b).Prob(x_p|x_b).Prob(x_b) \end{aligned} \quad (8)$$

Equation (8) makes use of the fact that the states of the sensor and the transceiver are independent when conditioned on the states of the processor and battery. Figure 2 illustrates the SN's states which have a non-zero

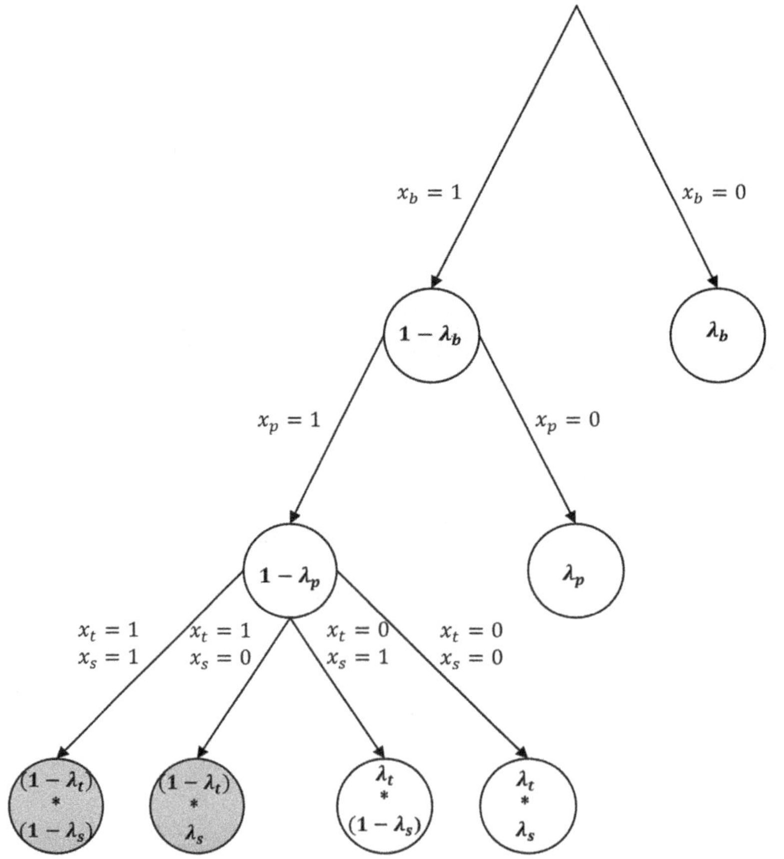


**Fig. 2** SN states: the paths from the top node to a bottom node correspond to the SN states with non-zero probability. The probability of a path, i.e., the probability of a state, is the product of the probabilities in the associated transitions. The SN states corresponding to the *relay* mode and the *on* mode are both marked by a dark shade of gray

probability. It is straightforward to verify that the sum of the probabilities of these states is equal to unity. There are two SN states at which the SN is of use to the WSN. The first state is described by the tuple {1,1 , 1,1}, at which all four components are functional and the SN can both sense its surroundings and communicate wirelessly. This state corresponds to the *on* mode of operation in which the SN is fully functional as defined in Section 1. The second state is described by the tuple {0,1 , 1,1} at which only the sensor(s) failed and the SN can only communicate wirelessly, i.e., acts as a relay node. This state corresponds to the *relay* mode of operation in which the SN is partially functional as defined in Section 1. In all the practically possible remaining states, the SN does not serve the network, and hence, a SN in these states is considered to be in the *off* mode of operation.

### 4.2 Wireless sensor network model

We assume that the targeted RoI of the WSN is a two-dimensional area in which there is a finite set of locations that require some form of monitoring (e.g., motion, image, video) using static SNs. These locations are called *target points* and are denoted by the set $\boldsymbol{T} = \{t_j\}$ for $j = 1 , \ldots , m$. To maintain generality, we do not assume that the target points conform to any regular pattern. Target points are monitored by the SNs in the WSN. We assume that the SNs used in the deployment of the WSN can be of different types (e.g., sound, image) and can have different coverage profiles (e.g., binary disk model, field of view (FoV) model), i.e., the WSN can be heterogeneous in nature. Let the set of deployed SNs be denoted by $\boldsymbol{S} = \{s_i\} , i = 1 , \ldots , n$. We assume an arbitrary deployment configuration in which an SN can monitor multiple target points. We also assume that a target point can be monitored by more than one SN. Therefore, in terms of coverage, the WSN can be modeled as a bipartite graph. Figure 3 shows an example of a WSN consisting of 5 SNs ($n = 5$) monitoring three target points ($m = 3$) and the resulting bipartite graph representation of the network coverage, assuming SNs are characterized by a binary disk coverage model. All

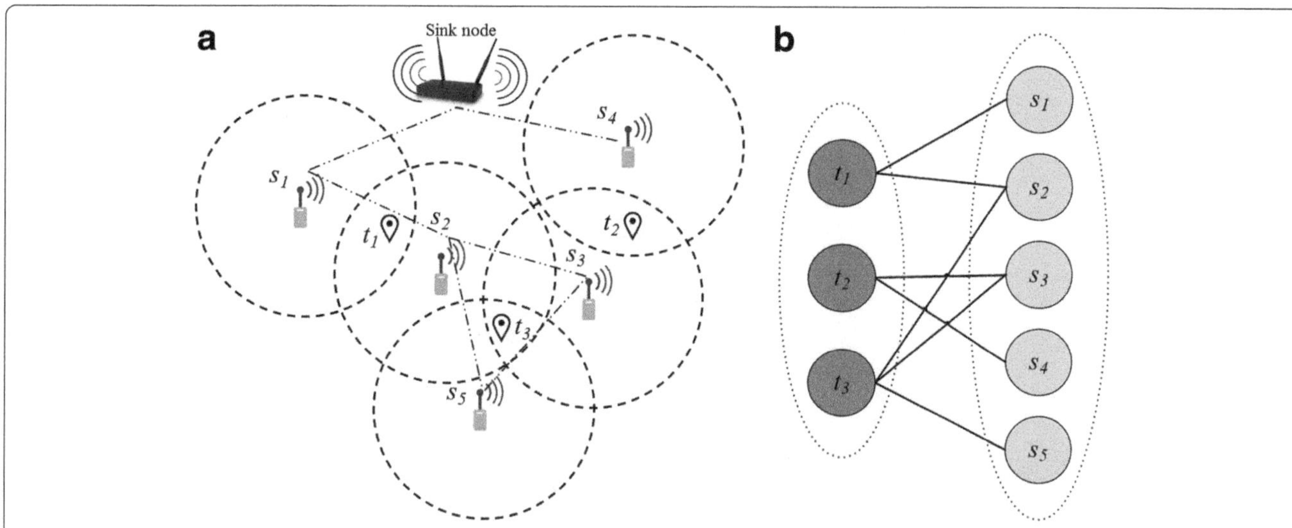


**Fig. 3** **a** A simple WSN consisting of a sink node, 5 SNs ($n = 5$) and three target points ($m = 3$). **b** The coverage of the WSN is modeled as a bipartite graph in which the target point set $\{t_1, t_2, t_3\}$ and the SN set $\boldsymbol{S} = \{s_1, s_2, s_3, s_4, s_5\}$ are the two disjoint sets

sensory data acquired by the SNs should be relayed to a sink node with an arbitrary fixed position in the RoI through wireless multi-hop communications. We assume that all deployed SNs have a fixed communication range, $r_c$. Hence, any two SNs deployed have a wireless communication link if the distance between them is less than or equal to $r_c$. Naturally, it is required that the WSN remains functional in terms of coverage and connectivity throughout its intended mission time $T_m$. To express this mathematically, we use the following definition:

**Definition:** *A WSN is said to be functional in terms of both coverage and connectivity if both of the following two conditions are met:*

1. Each target point $t_j$ for $j = 1, \ldots, m$ is covered by at least one SN with an uncompromised sensing capability, i.e., an SN in the *on* state. Let the set $\boldsymbol{Y}_j$ be the set of SNs in the *on* state that monitor $t_j$. Then this condition can be expressed as, $|\boldsymbol{Y}_j| \neq 0$, $\forall j = 1, \ldots, m$ where |.| denotes the size of a set.
2. Within each $\boldsymbol{Y}_j$, there is at least one SN that has at least one functional path to the sink node. This implies that SNs along that path, including the source SN, have uncompromised communication capabilities, i.e., in either the *on* or the *relay* state. Hence, the events detected at any $t_j$ can be relayed back to the sink node. Let the set $\boldsymbol{Z}_j$ be the set of SNs which belong to $\boldsymbol{Y}_j$ that are connected to the sink node. Hence, $\boldsymbol{Z}_j \subseteq \boldsymbol{Y}_j$. The condition can be expressed as $|\boldsymbol{Z}_j| \neq 0$, $\forall j = 1, \ldots, m$.

In the next subsection, we will use the above definition of WSN functionality conditions in defining the structure function of the WSN, which we defined in Section 3.2.

### 4.3 Wireless sensor network reliability metric derivation

The reliability of the WSN, $\boldsymbol{S}$, denoted by $R(\boldsymbol{S})$, is defined as the probability that the WSN remains functional, in terms of coverage and connectivity, subject to four types of SN component failures during its intended mission time, $T_m$. In order to use the combinatorial approach outlined in Section 3.2 to derive $R(\boldsymbol{S})$, we must define the states of $\boldsymbol{S}$ and its structure function. To define the states of $\boldsymbol{S}$, let $\boldsymbol{X_s}, \boldsymbol{X_t}, \boldsymbol{X_p}$, and $\boldsymbol{X_b}$ be the subsets of SNs in $\boldsymbol{S}$ that have failed sensors, transceivers, processors, and batteries, respectively. Hence, a state of the WSN $\boldsymbol{S}$ is described by the tuple $\boldsymbol{\pi} \equiv \{\boldsymbol{X_s}, \boldsymbol{X_t}, \boldsymbol{X_p}, \boldsymbol{X_b}\}$, where $\boldsymbol{X_s}, \boldsymbol{X_t}, \boldsymbol{X_p}, \boldsymbol{X_b}$ $\boldsymbol{S}$. Therefore, each state $\boldsymbol{\pi}$ is associated with a unique combination of SN components' failures. To calculate the probability of occurrence of a given state, $\boldsymbol{\pi}$, the corresponding state $x_i(\boldsymbol{\pi})$ of each individual SN $s_i \in \boldsymbol{S}$ must be identified. Assuming the components belonging to different SNs fail independently, $Prob(\boldsymbol{\pi})$ can be expressed by:

$$Prob(\boldsymbol{\pi}) = Prob\left(\boldsymbol{X_s}, \boldsymbol{X_t}, \boldsymbol{X_p}, \boldsymbol{X_b}\right) = \prod_{i=1}^{N} Prob(x_i(\boldsymbol{\pi})) \quad (9)$$

Table 1 lists the different values of the probability of an individual SN state $x_i(\boldsymbol{\pi})$ for a given network state $\boldsymbol{\pi}$ based on the SN states illustrated in Fig. 2. The first state in Table 1 corresponds to a SN in the *on* mode, the ninth state corresponds to the *relay* mode, while the rest of the non-zero probability states correspond to the *off* mode as illustrated in Fig. 2.

As defined in Section 3.2, the structure function of a system is a binary indicator of whether the system is functional at a given state. For a WSN, $\boldsymbol{S}$, the structure function indicates whether the network can maintain

**Table 1** Evaluation of the probability of the corresponding individual SN states for a given WSN state=$X_s, X_t, X_p, X_b$}, where "true" and "false" are denoted by 1 and 0, respectively, and $\lambda_s^i$, $\lambda_t^i$, $\lambda_p^i$, $\lambda_b^i$ are the probabilities of failure of the four main components of SN $s_i$

| $s_i \in X_s$ | $s_i \in X_t$ | $s_i \in X_p$ | $s_i \in X_b$ | Prob( $x_i(\pi)$) |
|---|---|---|---|---|
| 0 | 0 | 0 | 0 | $(1-\lambda_s^i)(1-\lambda_t^i)(1-\lambda_p^i)(1-\lambda_b^i)$ |
| 0 | 0 | 0 | 1 | 0 |
| 0 | 0 | 1 | 0 | 0 |
| 0 | 0 | 1 | 1 | 0 |
| 0 | 1 | 0 | 0 | $(1-\lambda_s^i)\lambda_t^i(1-\lambda_p^i)(1-\lambda_b^i)$ |
| 0 | 1 | 0 | 1 | 0 |
| 0 | 1 | 1 | 0 | 0 |
| 0 | 1 | 1 | 1 | 0 |
| 1 | 0 | 0 | 0 | $\lambda_s^i(1-\lambda_t^i)(1-\lambda_p^i)(1-\lambda_b^i)$ |
| 1 | 0 | 0 | 1 | 0 |
| 1 | 0 | 1 | 0 | 0 |
| 1 | 0 | 1 | 1 | 0 |
| 1 | 1 | 0 | 0 | $\lambda_s^i\lambda_t^i(1-\lambda_p^i)(1-\lambda_b^i)$ |
| 1 | 1 | 0 | 1 | 0 |
| 1 | 1 | 1 | 0 | $\lambda_p^i(1-\lambda_b^i)$ |
| 1 | 1 | 1 | 1 | $\lambda_b^i$ |

its functionality conditions in terms of both coverage and connectivity under a specific combination of SNs components failures. Let $\boldsymbol{\Pi}$ be the set of all possible states of $\boldsymbol{S}$. Based on the definition provided in the previous subsection, the structure function of $\boldsymbol{S}$ can be expressed as follows:

$$f(\boldsymbol{\pi}) = \begin{cases} 1, & \text{if } \boldsymbol{Z}_j \subseteq \boldsymbol{Y}_j \neq \phi \quad \forall j = 1, \cdots, m \\ 0, & \text{otherwise} \end{cases} \tag{10}$$

Similar to (3), we can now express the reliability of the WSN $\boldsymbol{S}$ as follows:

$$\begin{aligned} R(\boldsymbol{S}) &= \text{Prob}(\boldsymbol{S} \text{ is functional during } T_m) = \sum_{\pi \in \Pi} [f(\boldsymbol{\pi}).\text{Prob}(\boldsymbol{\pi})] \\ &= \sum_{X_s \subseteq S} \sum_{X_t \subseteq S} \sum_{X_p \subseteq S} \sum_{X_b \subseteq S} \left[ f(X_s, X_t, X_p, X_b).\prod_{i=1}^{N} \text{Prob}(x_i(\boldsymbol{\pi})) \right] \end{aligned} \tag{11}$$

Equation (11) states that the reliability of the WSN is the summation of all the probabilities of the WSN states that have a structure function value of unity (i.e., the probabilities of all the paths of the WSN $\boldsymbol{S}$). Depending on the set of failed components in the state π, the individual probabilities $\text{Prob}(x_i(\pi))$ , $i = 1, \ldots N$ can be calculated using Table 1.

## 5 Reliability metric calculation

From the derived expression of $R(\boldsymbol{S})$ in (11), it is clear that the reliability calculation involves in turn evaluating the structure function of the network, $f(\boldsymbol{\pi})$, for all possible states of $\boldsymbol{S}$ , i.e. for all $\boldsymbol{\pi} \in \boldsymbol{\Pi}$. As explained in Section 3.2, this can pose a computational challenge since WSNs designed for practical purposes are often composed of tens of SNs, resulting in a huge number of possible network states. This problem is further complicated by modeling the SNs as four-component systems which in turn have multiple possible states. For example, a WSN composed of just 30 SNs would have $2^{4 * 30} = 2^{120}$ states. This means that the calculation of the reliability metric in this case would require a prohibitive amount of time. To solve this computational problem, we make use of the following two properties of the WSN, $\boldsymbol{S}$, using the model presented in Section 4.2:

- The majority of the network states have null probabilities, and hence, they do not contribute to the value of $R(\boldsymbol{S})$. This stems from the fact the majority of the individual SN states also have a null probability (i.e., are not practically possible) as shown in Table 1.
- The WSN $\boldsymbol{S}$ has the property of being a monotone/coherent system [21]. This property implies the following. If the failure of a set of SNs' components causes $\boldsymbol{S}$ to fail, then the failure of any set which contains this set will also cause $\boldsymbol{S}$ to fail. For example, if we assume that the SNs $s_1$ and $s_2$ in the WSN depicted in Fig. 3a are both in the *off* mode while the remaining SNs are in the *on* mode, then it

can be readily observed that this would cause $\boldsymbol{S}$ to fail since any phenomenon at target point $t_1$ cannot be detected or communicated to the sink node. This means that network states corresponding to this situation have a structure function value of zero as expressed in (10). Using the monotone property, we can say that the network states that include the SNs $s_1$ and $s_2$ being in the *off* mode and $s_4$ being in the *relay* mode would also have a structure function value of zero without actually evaluating the function.

These two useful properties are used to develop a Breadth-First Search (BFS) algorithm that generates the complete path set of $\boldsymbol{S}$, denoted by $\boldsymbol{\Pi}_1$, and use it to calculate $R(\boldsymbol{S})$ using (11). The general structure of the developed search algorithm is illustrated in Fig. 4. The pseudo-code of the algorithm, which provides execution details, is given in Table 2. The structure of the algorithm can be summarized in the following steps. In step 1, all the required parameters for the calculation of $R(\boldsymbol{S})$ are specified as inputs. This includes the two-dimensional RoI layout, the positions of the target locations within the RoI provided by the set of target points $\boldsymbol{T} = \{t_j\}$, the positions of the deployed SNs provided by $\boldsymbol{S} = \{s_i\}$, the types of the deployed SNs including their coverage profiles and wireless communication ranges, and the probabilities of failure of the SN components associated with each SN type. We assume here that the sink node can be at any fixed arbitrary position in the targeted RoI. We initialize the value of $R(\boldsymbol{S})$ with the probability of the network state $\boldsymbol{\pi}$ which corresponds to all the deployed SNs being in the on mode. Since this network state is an obvious path of $\boldsymbol{S}$, we also initialize the network path set $\boldsymbol{\Pi}_1$ with this state as expressed in 1.c. in Table 2.

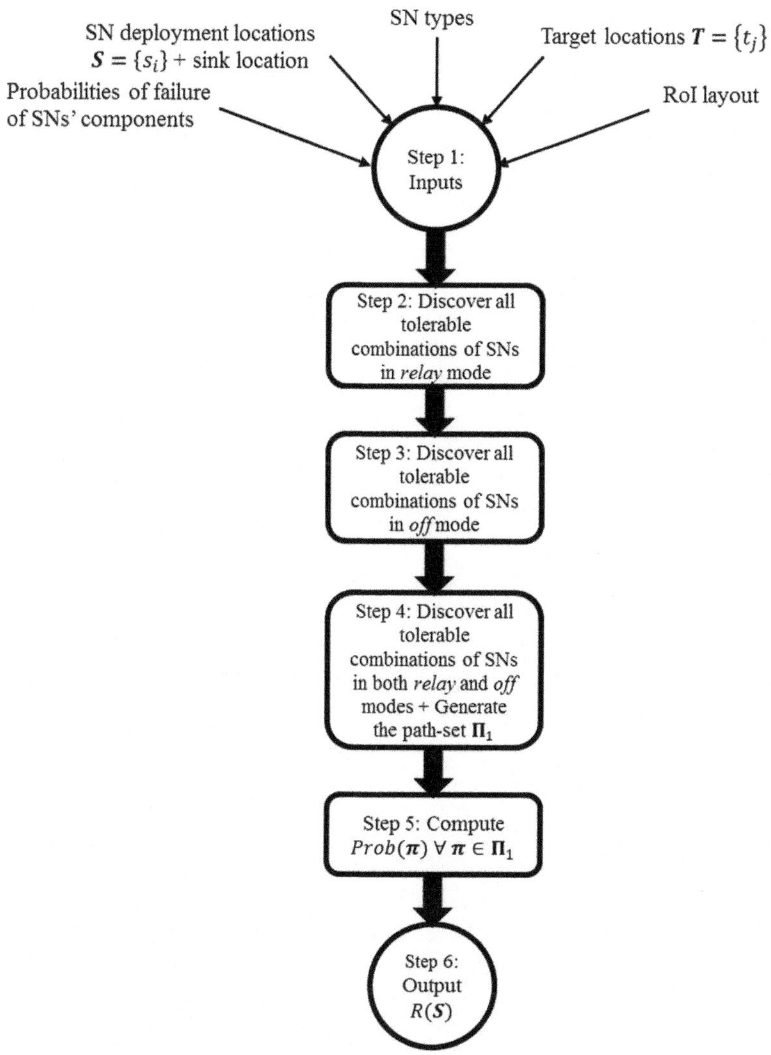


**Fig. 4** The structure of the proposed algorithm for evaluating the WSN reliability $R(\boldsymbol{S})$

**Table 2** Pseudo-code for the proposed algorithm for calculating the reliability of a WSN $S$

| Step | Algorithm for computing WSN reliability $R(S)$ |
|---|---|
| 1.a. | Set all parameters ($\boldsymbol{S} = \{s_i\}$, $\boldsymbol{T} = \{t_j\}$, types of SNs, sink location, $\lambda_s^i$, $\lambda_t^i$, $\lambda_p^i$ and $\lambda_b^i$ for $i = 1, \ldots, n$ and $j = 1, \ldots, m$) |
| 1.b. | Initialize $R = Prob(\boldsymbol{\pi} \mid s_i \in \boldsymbol{S}$ is in *on* mode $\forall i = 1, \ldots, n)$ |
| 1.c. | Initialize $\boldsymbol{\Pi}_1 = \{(\boldsymbol{\pi} \mid s_i \in \boldsymbol{S}$ is in *on* mode $\forall i = 1, \ldots, n)\}$ |
| 2.a. | Let $k$ be the number of SNs in *relay* mode. Initialize $k = 1$. |
| 2.b. | Let $\mathcal{F}_r^k$ be a $k$–combination of SNs in *relay* mode. Let $\boldsymbol{F}_r^k$ be the set of $k$–combinations of SNs in *relay* mode that $\boldsymbol{S}$ can tolerate.<br>Initialize $\mathcal{F}_r^k = \boldsymbol{F}_r^k = \{\varphi\}$. |
| 2.c. | For $i = 1, \ldots, n$ |
| | - Let $s_i$ be in *relay* mode, i.e. $\mathcal{F}_r^k = \{s_i\}$ |
| | - Evaluate $f(\boldsymbol{\pi} \mid \mathcal{F}_r^k)$ using (10)<br>- If $f(\boldsymbol{\pi} \mid \mathcal{F}_r^k) = 1 \rightarrow \boldsymbol{F}_r^k = \boldsymbol{F}_r^k \cup \mathcal{F}_r^k$ |
| | End For loop |
| 2.d. | While $\boldsymbol{F}_r^k \neq \{\varphi\} \rightarrow k = k + 1$, Let $F_{rl^{k-1}} \in \boldsymbol{F}_{r^{k-1}}$,<br>$\mathcal{F}_r^k = \boldsymbol{F}_r^k = \{\varphi\}$ |
| 2.e. | For $l = 1, \ldots, \lvert \boldsymbol{F}_{r^{k-1}} \rvert$ and $i = 1, \ldots, n$ |
| | - Let $\mathcal{F}_r^k = \{ F_{rl^{k-1}}, s_i \}$<br>- Evaluate $f(\pi \mid \mathcal{F}_r^k)$ using (10) |
| | - If $f(\pi \mid \mathcal{F}_r^k) = 1 \rightarrow \boldsymbol{F}_r^k = \boldsymbol{F}_r^k \cup \mathcal{F}_r^k$ |
| 2.f. | End For loops, End While loop |
| 3.a. | Let $k$ be the number of SNs in *off* mode. Initialize $k = 1$. |
| 3.b. | Let $\mathcal{F}_o^k$ be a $k$–combination of SNs in *off* mode. Let $\boldsymbol{F}_o^k$ be the set of $k$–combinations of SNs in *off* mode that $\boldsymbol{S}$ can tolerate.<br>Initialize $\mathcal{F}_o^k = \boldsymbol{F}_o^k = \{\varphi\}$. |
| 3.c. | Repeat step 2.c. for *off* mode, i.e. $\mathcal{F}_o^k = \{s_i\}$ |
| 3.d. | While $\boldsymbol{F}_o^k \neq \{\varphi\} \rightarrow k = k + 1$, Let $F_{ol^{k-1}} \in \boldsymbol{F}_{o^{k-1}}$,<br>$\mathcal{F}_o^k = \boldsymbol{F}_o^k = \{\varphi\}$ |
| 3.e. | Repeat 2.e. using $F_{ol^{k-1}}$ and $\mathcal{F}_o^k$ to get $\boldsymbol{F}_o^k$ |
| 3.f. | End For loops, End While loop |
| 4.a. | Let $\mathcal{F}_r$ and $\mathcal{F}_o$ be a combination of SNs in *relay* and *off* modes respectively.<br>Let $\boldsymbol{F}_r$ and $\boldsymbol{F}_o$ be the sets of all combinations of SNs of in *relay* and *off* mode that that $\boldsymbol{S}$ can tolerate respectively.<br>Let $F_{rl_r} \in \boldsymbol{F}_r$ and $F_{ol_o} \in \boldsymbol{F}_o$ |
| 4.b. | For $l_r = 1, \ldots, \lvert \boldsymbol{F}_r \rvert$ and $l_o = 1, \ldots, \lvert \boldsymbol{F}_o \rvert$ |
| | - Let $\mathcal{F}_r = F_{rl_r}$ *and* $\mathcal{F}_o = F_{ol_o}$<br>- Evaluate $f(\pi \mid \mathcal{F}_r, \mathcal{F}_o)$ using (10) |
| | - If $f(\pi \mid \mathcal{F}_r, \mathcal{F}_o) = 1 \rightarrow \boldsymbol{\Pi}_1 = \boldsymbol{\Pi}_1 \cup \pi$ |
| | End For loops |
| 5.a. | Let $\boldsymbol{\pi}_l \in \boldsymbol{\Pi}_1$ |
| 5.b. | For $l = 1, \ldots, \lvert \boldsymbol{\Pi}_1 \rvert$ |
| | - $R(\boldsymbol{S}) = R(\boldsymbol{S}) + Prob(\boldsymbol{\pi}_l)$ |
| | End For loop |
| 6. | Output: $R(\boldsymbol{S})$ |

Whether any other given state $\boldsymbol{\pi}$ is a path of the network or not (i.e., whether $f(\boldsymbol{\pi}) = 1$ or 0) depends on the WSN configuration/topology. To evaluate $f(\boldsymbol{\pi})$, the two conditions of WSN functionality, namely, coverage and connectivity, are checked according to the same definitions and order presented in Section 4.2. Checking the WSN coverage is straightforward and has the computational complexity of $O(n^*m)$. If one or more of the target points in the RoI is uncovered, then $f(\boldsymbol{\pi}) = 0$ and the connectivity condition does not need to be checked. Checking the WSN connectivity condition is more complex computationally, and it depends on the connectivity matrix between the SNs and the sink. Constructing that matrix has the complexity $O(n^2)$. For every WSN state $\boldsymbol{\pi}$, the connectivity matrix is updated according to SNs' modes in the state $\boldsymbol{\pi}$ and the updated connectivity matrix is used to check the connectivity condition. We carry out this check using the Floyd-

Warshall algorithm (https://en.wikipedia.org/wiki/Floyd%E2%80%93Warshall_algorithm), which can compute the shortest paths (if one exists) between all SNs (in the *on* or *relay* mode) and the sink node with the computational complexity $O(n^3)$. If all the SNs covering any given target point do not have a path to the sink node, $f(\boldsymbol{\pi}) = 0$, otherwise connectivity is intact and $f(\boldsymbol{\pi}) = 1$.

In step 2, the algorithm searches for all the combinations of SNs that can be in the *relay* mode without compromising the functionality of $\boldsymbol{S}$, assuming the remainder of the deployed SNs are in the *on* mode. These SN combinations are referred to as the "tolerable combinations of SNs in the *relay* mode." This means that for the network states corresponding to these SN combinations, the structure function $f(\boldsymbol{\pi})$ expressed in (10) is equal to unity. To perform the required search in step 2, we define $\boldsymbol{F}_r^k$ as the set that holds the tolerable combinations of SNs in *relay* mode of length $k$ starting with $k = 1$ as expressed in 2.a–2.c. in Table 2. For example, consider the WSN depicted in Fig. 3a. The set of single tolerable SNs in the *relay* mode will be given by $\boldsymbol{F}_r^1 = \{\{s_1\}, \{s_2\}, \{s_3\}, \{s_4\}, \{s_5\}\}$. The algorithm then proceeds with the search for an increasing value of $k$ as expressed in 2.d–2.f. in Table 2. For example, the combination $\{s_1, s_3\}$ belongs to $\boldsymbol{F}_r^2$ while $\{s_1, s_2\}$ does not. This search continues until the algorithm reaches a value of $k$ which results in an empty $\boldsymbol{F}_r^k$, i.e., $\boldsymbol{F}_r^k = \{\varnothing\}$. The set of all tolerable combinations of different lengths of SNs in *relay* mode is denoted $\boldsymbol{F}_r$.

In step 3, the algorithm searches for all the combinations of SNs that can be in the *off* mode without compromising the functionality of $\boldsymbol{S}$, assuming the remainder of the SNs is in the *on* mode, i.e. tolerable combinations of SNs in the *off* mode. The search follows the same procedure in step 2. We define $\boldsymbol{F}_o^k$ as the set that holds the tolerable combinations of SNs in the *off* mode of length $k$. Using the same example WSN in Fig. 3a, $\boldsymbol{F}_o^1 = \{\{s_2\}, \{s_3\}, \{s_4\}, \{s_5\}\}$. The combination $\{s_4, s_5\}$ belongs to $\boldsymbol{F}_o^2$ while $\{s_2, s_5\}$ does not. The set of all tolerable combinations of different lengths of SNs in the *off* mode is denoted $\boldsymbol{F}_o$.

In step 4, the algorithm uses the sets $\boldsymbol{F}_r$ and $\boldsymbol{F}_o$ to discover all the pairs of combinations of SNs that can be in the *relay* and *off* modes simultaneously without compromising the functionality of $\boldsymbol{S}$, assuming the remainder of the SNs is in the *on* mode. For example, the combination $\{s_1, s_3\}$ can be in the *relay* mode while $\{s_5\}$ can be in the *off* mode simultaneously without causing the WSN depicted in Fig. 3a to fail. Each of the discovered pairs of combinations corresponds to one or more distinct network path and hence the complete path set $\boldsymbol{\Pi}_1$ is updated accordingly as expressed in 4.b in Table 2. In step 5, the probabilities of the network paths in $\boldsymbol{\Pi}_1$ are calculated using (11) and Table 1. Finally, the reliability of the given WSN $R(\boldsymbol{S})$ is calculated using (7) and given as an output in step 6.

## 6 Case study

### 6.1 Experimental setup

In this section, we apply the reliability metric that we proposed in Section 4 and the search algorithm proposed in Section 5 to evaluate the reliability of a surveillance WSN designed to cover part of an international airport terminal. Figure 5 shows the layout of the airport terminal (http://www.aeroflot.ru/cms/en/before_and_after_fly/terminal_info.), which comprises the RoI of the WSN. Target points, marked on the figure in red, represent the vital locations that need to be placed under image/video surveillance such as arrival checkpoints, entrances, and staircases. The sink node to which all SNs in the WSN should be connected is marked in black.

To obtain our test deployments of the WSN, we use the Variable Length Genetic Algorithm (VLGA) presented in [24]. This optimization algorithm is designed to obtain cost-optimized deployments for heterogeneous WSNs that fulfill specific design objectives using a variable-length chromosome integer-encoded GA. In [24], the only considered design objective is providing coverage for all the target points in the RoI, i.e., providing full-coverage of the set $T = \{t_j\}$ for $j = 1, \ldots, m$. However, since a well-designed surveillance WSN should be functional in terms of coverage and connectivity, as defined in Section 3.2, we modified the VLGA in [24] to add network connectivity to the design objectives. To achieve this, we modify the fitness function that is used to evaluate the fitness of the candidate deployments or chromosomes in [24], as follows:

$$f(c(l)) = -\left(\sum_{i=1}^{l} p_i + w_1 * (m - \text{cov}) + w_2 * con\_test\right) \tag{12}$$

where $\sum_{i=1}^{l} p_i$ is the total cost of the deployment $c(l)$, cov is the number of target points that are covered by $c(l)$, *con_test* is a binary variable that is equal to unity when $c(l)$ is a disconnected deployment (i.e., has isolated SNs from the sink node) and zero otherwise, $w_1$ is the penalty imposed on the fitness for failing to cover a single target point and $w_2$ is a penalty for violating the connectivity constraint. The negative sign is added so that the maximum fitness would correspond to deployments achieving the coverage and connectivity constraints at minimum cost. For further details on the VLGA and the settings of its parameters, we refer the reader to the study in [24]. Both the VLGA and the search algorithm presented in Section 5 for the reliability metric calculation were implemented using MATLAB version 7.8.0

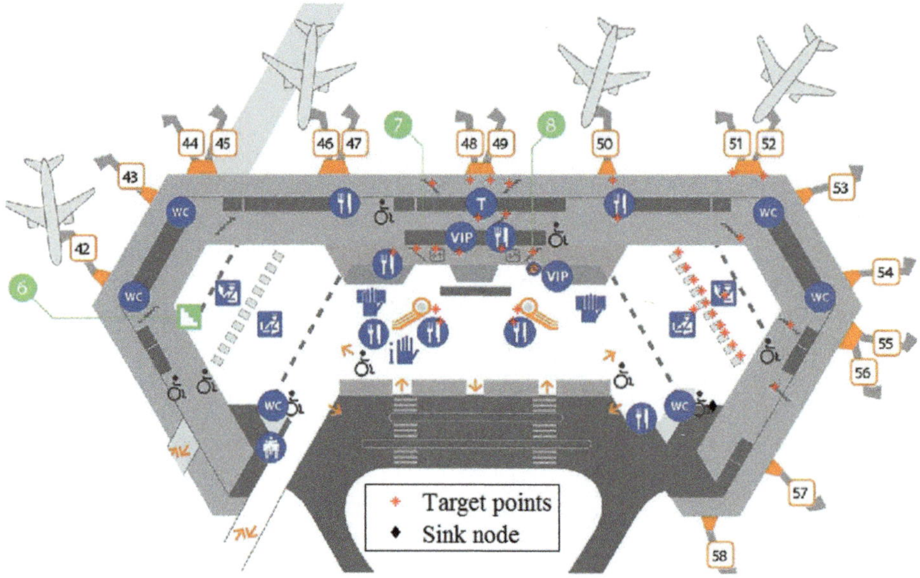


**Fig. 5** Schematic of an international airport terminal with the marked positions of the target points and sink node for case 5 in Table 3

which runs on an Intel processor Core i7-3621QM CPU, 2.1 GHz and 6 GB of RAM.

To demonstrate the proposed metric ability to evaluate the reliability of heterogeneous deployments, we assume that there are two types of SNs available for the deployment of the WSN. The coverage profile parameters and probabilities of failure of the SNs' components for each SN type are listed in Table 3. We assume that both SN types have a coverage range and a communication range of 30 and 40 m, respectively. Although the exact reliability figures for commercial SNs such as Tmote2 and Iris nodes are not publicly available, we estimated the given probabilities of failure using the reliability figures available for Texas Instrument CC2420 IEEE 802.15.4 transceiver (http://www.ti.com/product/CC2420/quality) as a reference point, assuming a WSN mission time of 5 years. We also used the fact that sensor hardware is the SN component most prone to failure [6] and that the premature battery failure rate for the highly durable lithium thionyl chloride batteries recently used for SNs is very low (https://www.omnisense.com/oms_cds/media/008-002-002%20OmniSense%20FMS%20Sensor%20Battery%20Life.pdf).

To thoroughly evaluate the proposed metric in terms of the computational efficiency, it is crucial to assess the effect of the deployment size and complete path set size $|\mathbf{\Pi}_1|$ for a given deployment on the computation time of the proposed algorithm outlined in Table 2. To achieve this objective, we apply the VLGA to obtain cost-optimized deployments for five target point sets of sizes $m = 15, 20, 25, 30,$ and 35. For each deployment scenario, i.e., for each value of $m$, we obtain five deployments of different sizes (i.e., different values of $n$), with the deployment of the smallest size being the most cost-optimal and the deployment with the largest size being the least cost-optimal. Each deployment fulfills the coverage and connectivity network functionality conditions in the case of no SN failures and has a different level of SN redundancy, where the higher the $n$, the higher the redundancy level and the larger the complete path set $\mathbf{\Pi}_1$and vice versa. Data of the resulting 25 deployments, including the size of the deployment ($n$), number of SNs of each type ($n_1$ and $n_2$), and the total deployment cost ($C$), are presented in Table 4.

### 6.2 Results and discussion

To evaluate the computational efficiency of the proposed algorithm outlined in Table 2, we use the algorithm to evaluate the reliability of the WSN deployments in Table 4. For each deployment, Table 4 shows the value of the reliability $R(\boldsymbol{S})$ , the total number of possible network states $|\mathbf{\Pi}|$ (which

**Table 3** Parameters of the SN types used in the deployment of the case-study surveillance WSN

| | FoV | $r_s$ | $r_c$ | $\lambda_s$ | $\lambda_t$ | $\lambda_p$ | $\lambda_b$ | Price($) |
|---|---|---|---|---|---|---|---|---|
| Type 1 | 90° | 30 m | 40 m | $1.0\times10^{-2}$ | $5.0\times10^{-3}$ | $2.0\times10^{-3}$ | $1.0\times10^{-3}$ | 150 |
| Type 2 | 60° | 30 m | 40 m | $1.5\times10^{-2}$ | $5.5\times10^{-3}$ | $2.5\times10^{-3}$ | $1.5\times10^{-3}$ | 100 |

**Table 4** Data of the obtained deployments for the case-study surveillance WSN for the RoI shown in Fig. 5

| Deployment no. | | $n$ | $n_1$ | $n_2$ | $C$ ($) | $R(S)$ | $\|\Pi\|$ | $\|\Pi_1\|$ | $FE$ | $FE/\|\Pi\|$ (%) |
|---|---|---|---|---|---|---|---|---|---|---|
| Scenario 1 $m = 15$ | S1-D1 | 9 | 0 | 9 | 900 | 0.829 | $2^{36}$ | 4 | 45 | $6.55 \times 10^{-8}$ |
| | S1-D2 | 10 | 1 | 9 | 1050 | 0.849 | $2^{40}$ | 28 | 164 | $1.49\times10^{-8}$ |
| | S1-D3 | 11 | 2 | 9 | 1200 | 0.870 | $2^{44}$ | 196 | 723 | $4.11\times10^{-9}$ |
| | S1-D4 | 12 | 3 | 9 | 1350 | 0.891 | $2^{48}$ | $1.37\times10^{3}$ | $4.16\times10^{3}$ | $1.48 \times10^{-9}$ |
| | S1-D5 | 13 | 4 | 9 | 1500 | 0.912 | $2^{52}$ | $9.60\times10^{3}$ | $3.01\times10^{3}$ | $6.68\times10^{-10}$ |
| Scenario 2 $m = 20$ | S2-D1 | 16 | 3 | 13 | 1750 | 0.731 | $2^{64}$ | 16 | 256 | $1.39\times10^{-15}$ |
| | S2-D2 | 17 | 4 | 13 | 1900 | 0.748 | $2^{68}$ | 112 | 881 | $2.98\times10^{-16}$ |
| | S2-D3 | 18 | 4 | 14 | 2000 | 0.756 | $2^{72}$ | 560 | $3.08\times10^{3}$ | $6.52\times10^{-17}$ |
| | S2-D4 | 19 | 5 | 14 | 2150 | 0.774 | $2^{76}$ | $3.92\times10^{3}$ | $1.46\times10^{4}$ | $1.93\times10^{-17}$ |
| | S2-D5 | 20 | 6 | 14 | 2300 | 0.793 | $2^{80}$ | $2.74\times10^{4}$ | $9.08\times10^{4}$ | $7.51\times10^{-18}$ |
| Scenario 3 $m = 25$ | S3-D1 | 21 | 1 | 20 | 2150 | 0.657 | $2^{84}$ | 64 | $1.24\times10^{3}$ | $6.40\times10^{-21}$ |
| | S3-D2 | 22 | 2 | 20 | 2300 | 0.673 | $2^{88}$ | 448 | $4.16\times10^{3}$ | $1.34\times10^{-21}$ |
| | S3-D3 | 23 | 3 | 20 | 2450 | 0.696 | $2^{92}$ | $5.38\times10^{3}$ | $1.97\times10^{4}$ | $3.98\times10^{-22}$ |
| | S3-D4 | 24 | 4 | 20 | 2600 | 0.703 | $2^{96}$ | $3.23\times10^{4}$ | $7.49\times10^{4}$ | $9.45 \times10^{-23}$ |
| | S3-D5 | 25 | 5 | 20 | 2750 | 0.720 | $2^{100}$ | $2.26\times10^{5}$ | $5.37\times10^{5}$ | $4.24\times10^{-23}$ |
| Scenario 4 $m = 30$ | S4-D1 | 25 | 8 | 17 | 2900 | 0.630 | $2^{100}$ | 128 | $2.91\times10^{3}$ | $2.29 \times10^{-25}$ |
| | S4-D2 | 26 | 6 | 20 | 2900 | 0.612 | $2^{104}$ | 192 | $4.51\times10^{3}$ | $2.22\times10^{-26}$ |
| | S4-D3 | 27 | 6 | 21 | 3000 | 0.633 | $2^{108}$ | $2.30\times10^{3}$ | $1.61\times10^{4}$ | $4.95\times10^{-27}$ |
| | S4-D4 | 28 | 7 | 21 | 3150 | 0.649 | $2^{112}$ | $1.61\times10^{4}$ | $6.61\times10^{4}$ | $1.27\times10^{-27}$ |
| | S4-D5 | 29 | 8 | 21 | 3300 | 0.665 | $2^{116}$ | $1.13\times10^{5}$ | $3.55\times10^{5}$ | $4.28\times10^{-28}$ |
| Scenario 5 $m = 35$ | S5-D1 | 28 | 4 | 24 | 3000 | 0.553 | $2^{112}$ | 32 | 876 | $1.69\times10^{-29}$ |
| | S5-D2 | 29 | 6 | 23 | 3200 | 0.555 | $2^{116}$ | 48 | $1.34\times10^{3}$ | $1.61\times10^{-30}$ |
| | S5-D3 | 30 | 7 | 23 | 3350 | 0.568 | $2^{120}$ | 336 | $4.36\times10^{3}$ | $3.28\times10^{-30}$ |
| | S5-D4 | 31 | 8 | 23 | 3500 | 0.589 | $2^{124}$ | $6.38\times10^{3}$ | $3.04\times10^{4}$ | $1.43\times10^{-31}$ |
| | S5-D5 | 32 | 9 | 23 | 3650 | 0.597 | $2^{128}$ | $4.47\times10^{4}$ | $1.57\times10^{5}$ | $4.61\times10^{-32}$ |

is equal to $2^{4*n}$), the size of the deployment complete path set $|\mathbf{\Pi}_1|$, the number of network structure function evaluations $FE$ performed by the algorithm, and the value of the ratio $FE/|\mathbf{\Pi}_1|$ in percentage points. The latter ratio is used as a measure of the computational efficiency of the proposed algorithm. This is because the most computationally expensive sub-routine in the algorithm is the evaluation of the network structure function expressed in (10). For each structure function evaluation, checking the two network functionality conditions, i.e., checking the network coverage of the set of target points and the connectivity to the sink, has a computational complexity of $O(n^*m)$ and $O(n^3)$, respectively. Therefore, the computation time of the algorithm is mainly determined by the number of structure function evaluations denoted by $FE$.

It can be readily observed that the values of $R(\mathbf{S})$, $|\mathbf{\Pi}_1|$, and $FE$ increase steadily with the increase of $n$ in each deployment scenario. This behavior is expected and is attributed to the increase in the level of SN redundancy in the deployment as $n$ increases. An increase in the level of SN redundancy translates to an exponential increase in the number of the paths of the deployment and hence the increase of the reliability $R(\mathbf{S})$. It can also be observed that the number of performed structure function evaluations $FE$ increases significantly with the increase in the level of SN redundancy as a direct result of the exponential increase in $|\mathbf{\Pi}_1|$. However, the value of the ratio $FE/|\mathbf{\Pi}|$ decreases rapidly with the increase of $n$ in each scenario. It can also be observed that the ratio $|\mathbf{\Pi}_1|/FE$ generally increases with the increase of the level of SN redundancy in each of the five tested scenarios. For example, the value of $|\mathbf{\Pi}_1|/FE$ is 27% for the deployment S3-D3 and 43% for S3-D4. These two observations mean that the computational efficiency of the proposed algorithm becomes more prominent with the increase of the SN redundancy level due to the efficiency of its search method for the deployment's paths performed by the algorithm.

It is instructive to examine the two deployments S4-D1 and S4-D2 which are the only exception in Table 4 to the trend discussed above. Although S4-D2 has more SNs than S4-D1 and a larger number of paths $|\mathbf{\Pi}_1|$, it is approximately 2% less reliable than S4-D1. This can be attributed to the higher ratio of more reliable SNs of type 1 to the less reliable SNs of type 2 in the S4-D1 compared to S4-D2. It can also be observed that pends mainly on the SN redundancy level (i.e., the value of $|\mathbf{\Pi}_1|$) relative to the total number of deployed SNs $n$ (which controls the value of the probability of occurrence of the paths in $\mathbf{\Pi}_1$). Since the increase in $n$ in each deployment scenario is very similar, the value of $|\mathbf{\Pi}_1|$ for the deployments of the same order in the different scenarios (e.g., S4-D3 and S5-D3) is comparable. This means that the SN redundancy level relative to $n$ actually decreases with the increase of deployment scenario order, i.e., with the increase of $m$, resulting in a steady decrease in $R(\mathbf{S})$.

To demonstrate the significance of modeling the SNs as three-mode (*on*, *relay*, and *off*) devices, we evaluate the reliability of the deployments presented in Table 4 using the reliability metric proposed in [16], which adopts the conventional two-mode (*on* and *off*) SN model. For a fair comparison, we use our proposed network structure function expressed in (10) (which defines the WSN functionality in terms of both network coverage and connectivity as opposed to network coverage only in [16]). Since the two-mode SN model assumes that a given SN is either in a fully functional (*on* state) or failed (*off* state) state, SNs cannot contribute to the network functionality as relays. Hence, the corresponding probability of the *off* state for a given SN $s_i$ is equal to the probability that any of the four SN components fail, i.e., is equal to unity minus the probability that all of the four SN components are functioning simultaneously (i.e., $1 - (1 - \lambda_s^i)(1 - \lambda_t^i)(1 - \lambda_p^i)(1 - \lambda_t^i)$).

As can be observed from Fig. 6, the value of $R(\mathbf{S})$ evaluated using the two-mode SN model is significantly smaller than that using the proposed three-mode model for all the deployments in Table 4, exceeding 6% for some deployments. This behavior is expected and can be attributed to the fact that the two-mode SN model is an unrealistic model that does not take into account the ability of an SN with a failed sensor to contribute to the functionality of the WSN in practice as a *relay*. Consequently, the size of the resulting paths set is drastically reduced which in turn reduces the value of $R(\mathbf{S})$. It should be explained that the difference between both models in $R(\mathbf{S})$ value for a given deployment is primarily dependent on the number of the tolerable combinations of SNs in the *relay* mode. This is because the higher the number of these combinations, the higher the number of SNs with redundant coverage. Since this coverage redundancy is not accounted for in calculating $R(\mathbf{S})$ using the two-mode SN model, the difference in $R(\mathbf{S})$ between

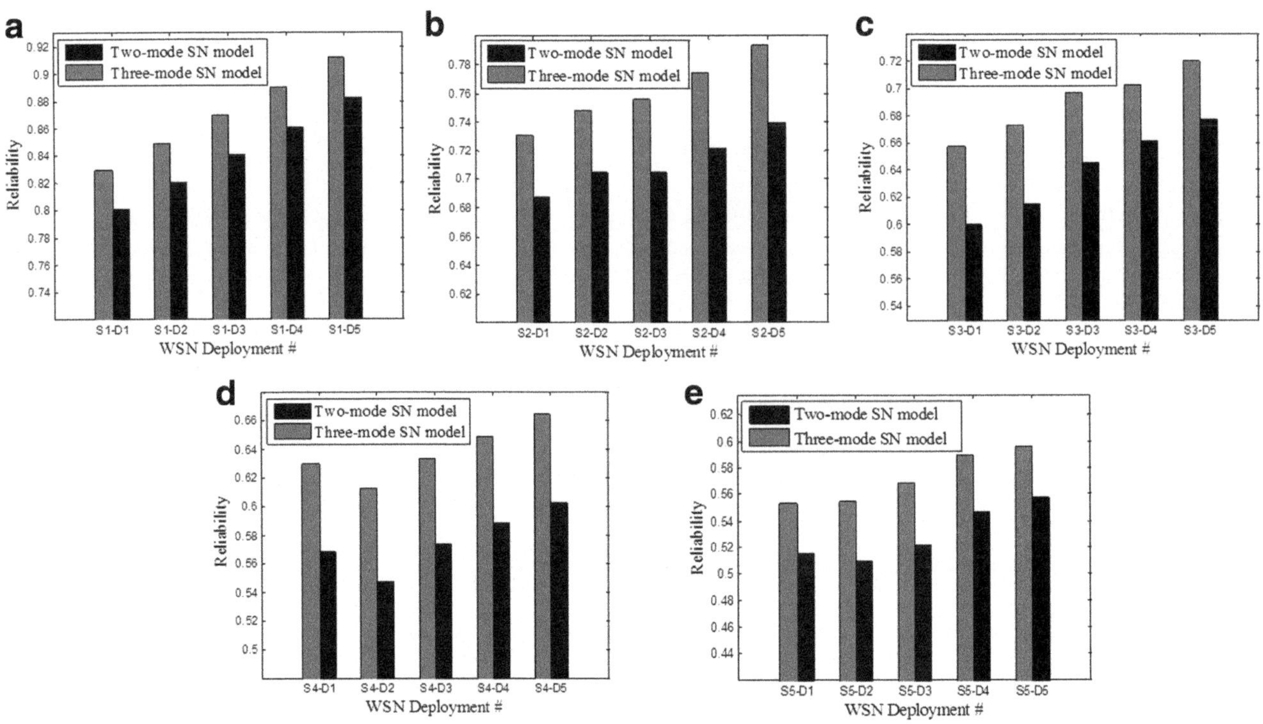


**Fig. 6** Comparison between the reliability of WSN deployments in Table 4 evaluated using the proposed three-mode SN model and the two-model SN model adopted in existing studies in [8], [11], [14], and [16] for the deployment scenarios 1 through 5 (**a**–**e**)

the two models increases with the increase of the level of coverage redundancy. For example, the deployment S2-D5 has a higher level of coverage redundancy than S5-D5. This is reflected in their difference in $R(\boldsymbol{S})$ value between the two models, which is 5.4% for the former and 3.9% for the latter.

In order to examine the sensitivity of $R(\boldsymbol{S})$ of a given deployment to changes in the probabilities of failure of its constituent SNs, we arbitrarily choose one of the deployments in Table 4, deployment S3-D1, and assume all of its 21 SNs are of type 1 only. For this new deployment, we evaluate the reliability at different probabilities of failure ranging from 0.001 to 0.01 for each of the four SN components, assuming the remaining components have the default probabilities of failure given in Table 3. The results obtained are shown in Fig. 7. As expected, the highest value of $R(\boldsymbol{S})$ is obtained when the probabilities of failure of the four SN components are at their minimum value. Figure 7 also shows that $R(\boldsymbol{S})$ is less sensitive to changes in the sensor probability of failure than to changes in the other three components probabilities of failure. This can be attributed to the adopted three-mode SN model, for which the SN can contribute to the network functionality in both the *on* and *relay* modes. In the *relay* mode, the SN sensor is not functional. However, for both modes, the SN battery, processor, and transceiver must be functioning. Hence, the reliability of a given deployment is less affected by the change in the sensor probability of failure compared to that of the other components.

## 7 Conclusions

In this paper, we derived a novel comprehensive reliability metric for heterogeneous WSN deployments of an arbitrary deployment configuration using a combinatorial approach. In deriving the proposed metric, SNs are modeled as three-mode systems that are characterized by four different probabilities of component failure for the sensor, transceiver, processor, and battery. We addressed the computational problem associated with calculating the reliability of deployments at practical scales using the proposed reliability metric by developing a search algorithm that generates the complete set of paths for a given deployment in a time efficient manner. We applied the proposed metric and search algorithm to several deployments of a case-study surveillance WSN under different operational parameters. Results show that the reliability of a given deployment is mainly a function of its level of SN redundancy and probabilities of failure of its constituent SNs' components. Results also demonstrated the computational efficiency of the developed search algorithm. Moreover, the significance of adopting the proposed three-mode SN model on the evaluated value of WSN reliability as opposed to the conventional simplistic two-mode SN model adopted in existing studies can be observed in the results.

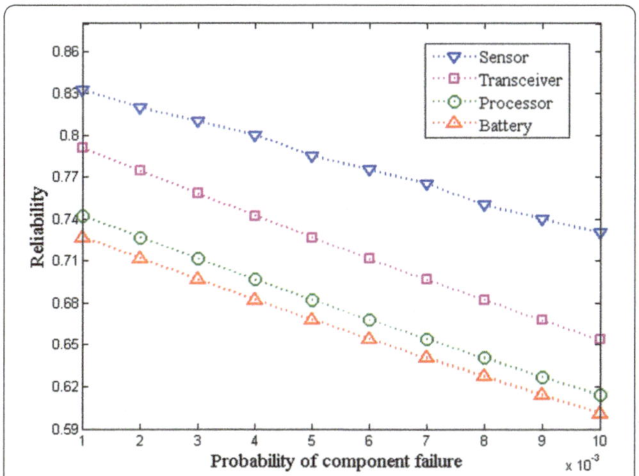


**Fig. 7** Reliability $R(\boldsymbol{S})$ for the deployment S3-D1 in Table 4 at different probabilities of failure of the sensor, transceiver, processor, and battery, assuming all deployed SNs are of type 1

**Acknowledgements**
The authors extend their appreciation to the anonymous reviewers for their helpful and supportive comments towards improving this paper.

**Funding**
This open-access publication of this paper is supported by a research grant provided by the American University in Cairo, under grant number 10300000-04131100002-8500-68,665,000.

**Authors' contributions**
DD and YG conceived the idea and wrote the paper. DD performed the experiments and analyzed the data. YG gave valuable suggestions on the structuring of the paper and assisted in the revising and proofreading. Both authors read and approved the final manuscript.

**Competing interests**
The authors declare that they have no competing interests.

**References**

1. A Flammini, E Sisinni, Wireless sensor networking in the Internet of things and cloud computing era, in Proceedings of European Conf. on Solid-State Transducers (EUROSENSOR 2014), Italy, 2014
2. L. Lei, Y. Kuang, X. Shen, K. Yang, J. Qiao, Z. Zhong, Optimal reliability in energy harvesting industrial wireless sensor networks. IEEE Trans. on Wireless Comm. **15**, 5399–5413 (2016). doi:10.1109/TWC.2016.2558146
3. W Kuo, MJ Zuo, *Optimal reliability modeling: principles and applications*, 1st edn. (John Wiley, Hoboken, 2003), pp. 85–102
4. C. Zhu, C. Zheng, L. Shu, G. Han, A survey on coverage and connectivity issues in wireless sensor networks. Network and Computer Appl. **35**, 619–632 (2012)
5. J Virkki, Y Zhu, Y Meng and L Chen, Reliability of WSN hardware, Int. J. of Embedded Sys. 1 (2012). doi:10.5121/ijesa.2011.1201
6. F Koushanfar, M Potkonjak, A Sangiovanni-Vecentelli, Fault-tolerance in wireless sensor networks, in handbook of sensor networks, 1st edn., ed. by M Ilyas, E Mahgoub (CRC Press, Boca Raton, 2005)

7. N Baccour, A Koubaa, L Mottola, MA Zuniga, H Youssef, CA Boano and A Alves, Radio link quality estimation in wireless sensor networks: a survey, ACM Trans. on Sensor Networks 8 (2012). doi:10.1145/2240116.2240123
8. M.A. Mahmood, K.G. Winston, I. Welch, Reliability in wireless sensor networks: a survey and challenges ahead. Comput. Netw. **79**, 166–187 (2015)
9. H. AboElFotoh, S. Iyengar, K. Chakrabarty, Computing reliability and message delay for cooperative wireless distributed sensor networks subject to random failures. IEEE Trans. Reliab. **54**(145–155) (2005)
10. Y Jin, H Lin, Z Zhang and X Zhang, Estimating the reliability and lifetime of wireless sensor network, in Proceedings of Int. Conf. on Wireless Communications, Networking and Mobile Computing (WiCOM), China, 2008
11. A. Damaso, N. Rosa, P. Maciel, Reliability of wireless sensor networks. MPDI Sensors **14**, 760–785 (2014)
12. X Zhu, Y Lu, J Han and L. Shi, Transmission reliability evaluation for wireless sensor networks, Hindawi J. Of Distributed Sensor Networks 10 (2016)
13. Q. Liu, H. Zhang, Weighted voting system with unreliable links. IEEE Trans. Reliab. **66**(2), 339–350 (2017)
14. C Jaggle, J Neidig, T Grosch and F Dressler, Introduction to model-based reliability evaluation of wireless sensor networks, in Proceedings of Int. Federation of Automatic Control (IFAC) Workshop on Dependable Control of Discrete Systems, Italy, 2009
15. I Silva, R Leandro, D Macedo and LA Guedes, A dependability evaluation tool for the Internet of things', Computers & Electrical Eng. 39, 806–838 (2013)
16. EI Gokce, AK Shrivastava and Y ding, fault tolerance analysis of surveillance sensor systems, IEEE Trans. on Reliab. 62, 478-489 (2013)
17. I. Silva, L.A. Guedes, P. Portugal, F. Vasques, Reliability and availability evaluation of wireless sensor networks for industrial applications. MPDI Sensors **12**, 806–838 (2012)
18. H Van-Trinh, N Julien and P Berruet, On-line self-diagnosis based on power measurement for a wireless sensor node, in Proceedings of the IEEE Workshop on Highly-Reliable Power-Efficient Embedded Designs, China, 2013
19. A.E. Zonouz, L. Xing, V.M. Vokkarane, Y. Sun, in *Proceedings of Annual Reliability and Maintainability Symposium (RAMS)*. A time-dependent link failure model for wireless sensor networks (2014)
20. Y. Shpungin, Combinatorial approach to reliability evaluation of network with unreliable nodes and unreliable edges. Int. J. Comput. Sci. **1**(177–183) (2006)
21. IB Gertsbakh and Y Shpungin, Y, Models of network reliability: analysis, combinatorics, and Monte Carlo, 1st edn. (CRC Press, Boca Raton, 2010), pp. 17-23
22. M Healy, T Newen and E Lewis, Wireless sensor node hardware: a review, IEEE Sensors, 621–624 (2008) doi:10.1109/ICSENS.2008.4716517
23. H Liu, A Nayak and I Stojmenovic, Fault tolerant algorithms/protocols in wireless sensor networks, in Guide to wireless sensor networks, ed. by SC Misra, I Woungang, S Misra (Springer-Verlag, Berlin, Germany, 2009), pp. 261–292
24. D. Deif, Y. Gadallah, *Wireless sensor network deployment using a variable-length genetic algorithm* (Proceedings of the Wireless Communications and Networking Conference (WCNC), Turkey, 2014)

# 10

# Performance analysis of cognitive AF relay networks with multiuser switched diversity in Rayleigh fading channels

Anas M. Salhab* and Salam A. Zummo

**Abstract**

The multiuser switched diversity (MUSwiD) selection schemes are useful in reducing the required channel estimation load in wireless networks. In this paper, we propose and evaluate the performance of cognitive amplify-and-forward (AF) MUSwiD relay networks where a cognitive user is selected among a set of users for data reception. The selection process is performed such that the end-to-end (e2e) signal-to-noise ratio (SNR) of the selected user satisfies a predetermined switching threshold. Such a user that satisfies this threshold is scheduled instead of the best user to receive its message from the secondary source. In the proposed system, we consider a cognitive source, a cognitive relay, a set of cognitive users, and a primary user. In this paper, an upper bound on the e2e SNR of a user is used in deriving of closed-form approximations for the outage probability and average symbol error probability (ASEP) of the studied system in addition to deriving the ergodic channel capacity. To get more about system insights, the performance is studied at the high SNR regime where approximate expressions for the outage probability, SEP, diversity order, and coding gain are derived. The derived analytical and asymptotic expressions are verified by Monte-Carlo simulations, and some numerical examples are provided to illustrate the effect of some parameters such as number of users and switching threshold on the system performance. Findings illustrate that the diversity order of the studied cognitive AF multiuser switched diversity relaying network is the same as its non-cognitive counterpart. Also, results show that the asymptotic results tightly converge to the exact ones, and the analytical bounds are indeed very tight, validating the accuracy of our approach of analysis. Furthermore, findings illustrate that the proposed MUSwiD user selection schemes are efficient in the range of low SNR values, which makes them attractive options for practical implementation in emerging mobile broadband communication systems. In contrast, these selection schemes are shown to be inefficient in the range of high SNR values where the multiuser diversity gain is noticeably degraded when they are implemented.

**Keywords:** Amplify-and-forward, Cognitive relay network, Multiuser switched diversity, Switching threshold

## 1 Introduction

Allowing secondary (unlicensed) users to simultaneously share the spectrum of primary (licensed) users is known as cognitive radio. Cognitive radio has been proposed to improve the spectrum resource utilization efficiency in wireless networks [1]. Several cognitive radio paradigms have been proposed in [2], among, which is the underlay scenario. This paradigm allows users in a secondary cell (secondary users) to utilize the frequency bands of users in a primary cell (primary users) only if the interference is below a certain threshold. Another important technique, which is used to provide diversity and to enhance the quality of wireless channels is the cooperative or relay network [3]. In such network, a relay or a set of relays are employed to help the source node transmitting its message to destination. Currently, a lot of research is being conducted on what is called cognitive relay networks (CRNs), which are also known as spectrum-sharing relay networks. In designing spectrum-sharing underlay systems, a major challenge is to fulfill the two conflicting objectives; protecting the primary user from interference, and satisfying the quality of service (QoS) requirement of secondary

*Correspondence: salhab@kfupm.edu.sa
Electrical Engineering Department, King Fahd University of Petroleum & Minerals, Dhahran 31261, Saudi Arabia

users. Between these objectives, the former is of higher priority, making strict regulation of secondary transmit powers necessary [4].

The outage performance of decode-and-forward (DF) CRNs was evaluated in [5–10]. In these studies, the secondary cell was assumed to include a source, a set of relays, and a destination in addition to a primary user in the neighboring primary cell. Closed-form expressions for the outage probability were derived assuming Rayleigh fading and best-relay or opportunistic relaying scheme. Recently, the outage performance of DF CRNs with the $N^{\text{th}}$-best relay selection scheme was presented in [11] and [12], respectively. Such scheme is efficient in conditions where the best relay is unavailable for relaying and is busy in some scheduling and load balancing duties in other parts of the network. Particularly, in [12], Zhang et al. evaluated a closed-form expression for the outage probability where the relay with the $N^{\text{th}}$ best second hop power metric is selected to forward the secondary source message to destination. This metric is conditioned on the maximum transmit power at the relay and the maximum tolerated interference power level at the primary user. In [13], three relay selection scenarios were proposed, the relay with the best second hop, the relay with worst second hop, and the relay satisfying the minimum level of interference with the primary user. Closed-form expressions for the outage and error probabilities were derived assuming Rayleigh fading channels.

In [14], the outage performance of amplify-and-forward (AF) CRNs was studied with two relay selection criteria, full channel estimation where channel-state-information (CSI) of all links are assumed to be available at the source and partial channel estimation where CSI of relays first hops are required at the source for relay selection. The opportunistic and partial-relay selection schemes were also recently studied in [15] where the outage and symbol error probabilities in addition to ergodic capacity were evaluated for AF CRNs. The outage and error rate performances of an underlay fixed-gain AF CRN with reactive relay selection were evaluated in [16]. Among the relays, which satisfy the interference constraint, the relay with the best end-to-end (e2e) channel is selected to forward the source message to destination. In [17], the error rate performance of an AF CRN was studied using the partial-relay selection scheme in which the relay with the strongest first hop channel was selected as the best relay. As an extension to the previous works on AF CRNs, the outage performance of an AF CRN with multiple primary users was recently presented in [18]. A study that proposes three relaying strategies was presented in [19]. These strategies are: selective AF, selective DF, and partial relay selection AF relaying. Asymptotic expressions for the outage probability and coding gain were provided in addition to studying the effect of imperfect channel estimation on the system performance.

Recently, in addition to deriving the ergodic channel capacity, Bao et al. evaluated in [20] the outage probability and error probability performances of AF CRNs assuming Rayleigh fading channels. An upper bound on the e2e SNR was used in the analysis where the dependence between channels from the source and relays in the secondary cell and primary user in the primary cell were taken into account in all derivations. A study that combines between the overlay and underlay CRNs was presented in [21]. While the primary system is idle, secondary users utilize licensed spectrum with overlay mode, and secondary transmitters are with transmit power limits. While primary system is active, secondary users utilize licensed spectrum with underlay mode, and secondary transmitters are with transmit power limits and peak interference power constraints. Upper bound for outage probability was derived assuming Rayleigh fading channels with the existence of direct link in the secondary cell. The effect of interference caused by a primary user on all receiving nodes in a secondary cell was studied in [22]. Closed-form and asymptotic expressions for the outage probability were derived over Rayleigh and Nakagami-$m$ fading channels and with the DF best-relay selection scheme being used.

Currently, the performance of CRNs with multiple secondary users is attracting a lot of researchers to work on such important topic. In [23], da Costa et al. evaluated the outage performance of DF spectrum-sharing relay networks with multiple primary and secondary users over Nakagami-$m$ fading channels. Opportunistic scheduling was used where the secondary destination of the best channel is allowed to receive the secondary source message. In [24], the secondary user was selected in multiuser CRNs to achieve the largest secondary rate while satisfying primary rate target. The authors showed that the diversity order of the primary system equals number of secondary users and for the secondary system equals number of secondary users plus 1. Closed-form expressions for the outage probability were derived in [25] for multiuser multirelay CRNs with DF and AF relaying schemes over Rayleigh fading channels. The secondary user with best direct link with the secondary source was scheduled to receive from secondary source. The same system studied in [25] was also studied in [26] but with one primary user. Most recently, the outage performance of CRNs with one secondary source, one secondary relay, multiple secondary users, and one primary user was evaluated in [27]. Unlike previous studies, the interference between primary user and the secondary relay and users is considered in this work. Opportunistic scheduling was used to select the secondary user with the best second hop SNR. Most recently, the opportunistic scheduling was used in [28] and [29] to select among secondary users in DF and AF multiple-input multiple-output cognitive relay networks, respectively.

As can be seen, the mostly used user selection scheme in multiuser CRNs is the opportunistic scheduling. A drawback of this scheme is the heavy load of channel estimations it requires to select the best user among the available users where the channels of all users need to be estimated each transmission time. Usually, in systems using this scheme, the SU relay is assumed to possess perfect CSI of the relay-destinations links each transmission time, which necessitates that each SU destination or user to estimate its channel and feed it back to the relay each scheduling period. As the number of users in the system increases, the opportunistic scheduling becomes much complicated, and hence, it is often a practical interest to come up with a low complexity scheduling algorithm with a reasonable system performance. An efficient candidate, which can reduce the number of feedback signals between the SU relay and users and hence reducing the system complexity is the low-complexity MUSwiD selection scheme. In this scheme, each user can trigger a feedback only when the channel quality is greater than a pre-determined threshold. The idea behind the threshold-based feedback scheme is that only the users with good enough channel quality are worth being considered to be scheduled [30, 31]. This idea was first used in space diversity systems where the switch-and-examine diversity combining (SEC) selection scheme was used to select among antennas [32]. In contrast to the opportunistic scheduling, which requires a heavy load of channel estimations or a heavy feedback each transmission time, in the MUSwiD scheme, once a checked user satisfies a predetermined switching threshold, it is scheduled for data reception. In this scheme, the e2e SNR of a secondary user is compared with a certain switching threshold. If it is larger, this user is selected to receive the secondary source message, if not, other user is examined. This process continues till a suitable user is found or the last user is reached. Later, if the last user is found unacceptable, the scheme sticks to it. The MUSwiD user selection schemes proved their effectiveness in reducing the required number of channel estimations, and hence, in reducing the system complexity when compared with the opportunistic user selection scheme [33]. More details on how the MUSwiD selection scheme works are provided in Section 2. Another motivation to this research work is the conduction of comprehensive performance analysis to the considered system model with the use of the MUSwiD low-complexity user selection scheme. The derivations and performance measures are being presented for the first time in this paper.

To the best of our knowledge, the performance of cognitive AF relay networks with multiuser switched diversity over Rayleigh fading channels has not been presented yet. The contributions of our paper over the existing studies are as follows: we propose the MUSwiD user selection scheme for cognitive AF multiuser relay networks in addition to analyzing its performance. In contrast to the opportunistic scheduling, in MUSwiD scheme, the first checked secondary user whose e2e SNR exceeds a predetermined switching threshold is selected to receive the secondary source message. Once a user is selected, no need for communicating further feedback signals between the users and the scheduling unit, which is the SU relay in our study. This results in a noticeable reduction in the required number of channel estimations, saves the power of users, and hence, reduces the system complexity. The proposed user selection scheme becomes mainly efficient as the number of users in the system increases. In such condition, the opportunistic scheduling and other scheduling schemes, which require all users channels to be estimated each transmission time become much complicated. In this paper, we derive closed-form approximations for the outage probability and ASEP for the independent non-identically distributed (i.n.i.d.) and independent identically distributed (i.i.d.) cases of users channels. We also derive the ergodic channel capacity for the case of i.i.d. users channels. First, an upper bound on the e2e SNR of a user is introduced. Then, this bound is used to derive a conditional cumulative distribution function (CDF) of the SNR at the output of the MUSwiD selection scheme combiner, which is then used to evaluate the e2e outage probability, ASEP, and ergodic channel capacity of the system. In order to get more insights about the system performance, we study the performance at the high SNR regime where approximate expressions for the outage probability, ASEP, diversity order, and coding gain are derived and analyzed. Furthermore, the MUSwiD with post-examine selection (MUSwiDps) which is an improved version of the conventional MUSwiD user, selection scheme is introduced and analyzed. The main difference between the MUSwiDps and the MUSwiD schemes is in the case where the last user is reached and found unacceptable. In such condition, the MUSwiDps scheme chooses the best user among all checked users to receive the source message.

This paper is organized as follows: Section 2 presents the system and channel models. The performance evaluation is conducted in Section 3. Section 4 provides the asymptotic performance analysis. Some simulation and numerical results are discussed in Section 5. Finally, Section 6 concludes the paper.

## 2 System and channel models

Consider a dual-hop spectrum-sharing relay network consisting of one secondary user (SU) source S, one AF SU relay R, $K$ SU destinations $\mathsf{D}_k$ ($k = 1, \ldots, K$), and one primary user (PU) receiver P, as shown in Fig. 1. All nodes are assumed to be equipped with single antenna, and the communication is assumed to operate in a half-duplex mode. The direct link between the SU source and SU destinations is assumed to be in a deep fade, and hence, it is neglected in analysis. The communications take place in two phases. In first phase, the SU source sends its message $x$ to relay under a transmit power constraint, which guarantees that the interference with the PU receiver P does not exceed a threshold $\mathcal{I}_{\mathsf{p}}$. As a result, the SU source S must transmit at a power given by $P_{\mathsf{s}} = \mathcal{I}_{\mathsf{p}}/|h_{\mathsf{s,p}}|^2$, where $h_{\mathsf{s,p}}$ is the channel coefficient of the S $\rightarrow$ P link. In the second phase, R amplifies the received message from S with a variable gain $G$ and forwards the amplified message to the $K$ SU destinations. The transmit power at R must also satisfy PU constraint and is defined as $P_{\mathsf{R}} = \mathcal{I}_{\mathsf{p}}/|h_{\mathsf{r,p}}|^2$, where $h_{\mathsf{r,p}}$ is the channel coefficient of the R $\rightarrow$ P link. Hence, the received message at the $k^{\text{th}}$ destination $\mathsf{D}_k$ from the relay R is given by $y_{\mathsf{r},k} = \sqrt{P_{\mathsf{S}}} G h_{\mathsf{r},k} h_{\mathsf{s,r}} x + G h_{\mathsf{r},k} n_{\mathsf{s,r}} + n_{\mathsf{r},k}$, where $h_{\mathsf{s,r}}$ and $h_{\mathsf{r},k}$ are the channel coefficients of the S $\rightarrow$ R and R $\rightarrow$ $\mathsf{D}_k$ links, respectively, $n_{\mathsf{s,r}}$ and $n_{\mathsf{r},k}$ represent the additive white Gaussian noise (AWGN) terms at R and $\mathsf{D}_k$, respectively, with a power of $N_0$. It is assumed that perfect channel information including the interference channel is available at the secondary users[1]. Also, it is assumed in this study that the interference from the primary user is neglected[2]. In a similar manner, the secondary user should possess the interference threshold value of the primary user, which is usually announced by the primary user itself at the beginning of each communication or data transmission mode. This can be done also by using certain energy sensing techniques [34]. We are considering slowly varying block fading channel model. This assumption is suitable for relay networks as in such networks, the main application is data communication in slow movements, which allows for channel estimation and makes it worth it given the slow nature of the fading. Also, the assumption of block fading model where the channel coefficients stay constant over an entire block of communication allows for completing the selection process (scheduling) between users before their channels change again. As we are using a channel-state-information (CSI)-assisted AF relaying, the gain $G$ can be expressed as $G^2 = 1 \Big/ |h_{\mathsf{r,p}}|^2 \left( \frac{|h_{\mathsf{s,r}}|^2}{|h_{\mathsf{s,p}}|^2} + \frac{N_0}{\mathcal{I}_{\mathsf{p}}} \right)$. Thus, the instantaneous e2e SNR of the S $\rightarrow$ R $\rightarrow$ $\mathsf{D}_k$ link can be written as [20]

$$\gamma_{\mathsf{S-R-D}_k} = \frac{\frac{\frac{\mathcal{I}_{\mathsf{p}}}{N_0}|h_{\mathsf{s,r}}|^2}{|h_{\mathsf{s,p}}|^2} \frac{\frac{\mathcal{I}_{\mathsf{p}}}{N_0}|h_{\mathsf{r},k}|^2}{|h_{\mathsf{r,p}}|^2}}{\frac{\frac{\mathcal{I}_{\mathsf{p}}}{N_0}|h_{\mathsf{s,r}}|^2}{|h_{\mathsf{s,p}}|^2} + \frac{\frac{\mathcal{I}_{\mathsf{p}}}{N_0}|h_{\mathsf{r},k}|^2}{|h_{\mathsf{r,p}}|^2} + 1} \leq \gamma_k^{\mathsf{up}} = \min\Bigg( \underbrace{X/Y}_{\gamma_1}, \underbrace{X_k}_{\gamma_{2k}} \Bigg), \quad (1)$$

where $X = \frac{\mathcal{I}_{\mathsf{p}}}{N_0}|h_{\mathsf{s,r}}|^2$, $Y = |h_{\mathsf{s,p}}|^2$, $X_k = \frac{\frac{\mathcal{I}_{\mathsf{p}}}{N_0}|h_{\mathsf{r},k}|^2}{|h_{\mathsf{r,p}}|^2}$. The MUSwiD user scheduling is achieved by selecting the user with the e2e SNR $\gamma_k^{\mathsf{up}}$ satisfies a predetermined switching threshold.

Herein, we assume that all channel coefficients undergo i.n.i.d. Rayleigh fading, and hence, the channel gains $|h_{\mathsf{s,p}}|^2$, $|h_{\mathsf{s,r}}|^2$, $|h_{\mathsf{r},k}|^2$, and $|h_{\mathsf{r,p}}|^2$ follow exponential distribution with mean powers $\Omega_{\mathsf{s,p}}$, $\Omega_{\mathsf{s,r}}$, $\Omega_{\mathsf{r},k}$, and $\Omega_{\mathsf{r,p}}$, respectively. For easy following of the paper, Table 1 summarizes the main symbols of this paper and their description.

Referring to Fig. 2, the MUSwiD selection scheme works as follows: in each transmission or scheduling period, the SU source sends a ready-to-send (RTS) packet to the SU relay through which the relay estimates the first hop channel. Then, the SU relay probes the users in a sequential way so only a single user has an opportunity to send a

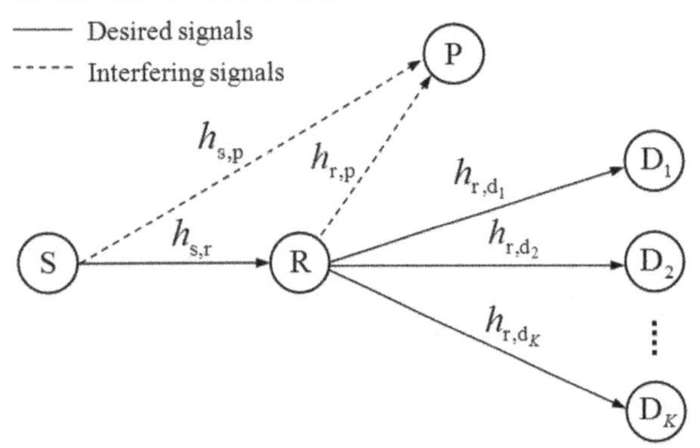


**Fig. 1** Cognitive dual-hop AF network with MUSwiD user selection scheme

**Table 1** The used symbols and their description

| Symbol | Description | Symbol | Description |
|---|---|---|---|
| S | Secondary source | $G$ | Gain of R |
| $\mathsf{D}_k$ | $k^{\text{th}}$ secondary destination | $h_{\mathsf{s,p}}$ | Channel coefficient between S and P |
| R | Secondary relay | $h_{\mathsf{s,r}}$ | Channel coefficient between S and R |
| P | Primary user | $h_{\mathsf{r,p}}$ | Channel coefficient between R and P |
| $\mathcal{I}_{\mathsf{p}}$ | Interference threshold | $h_{\mathsf{r},k}$ | Channel coefficient between R and $\mathsf{D}_k$ |
| $P_{\mathsf{S}}$ | Transmit power at S | $n_{\mathsf{s,r}}$ | Noise at R |
| $x$ | Source symbol | $n_{\mathsf{r},k}$ | Noise at $\mathsf{D}_k$ |
| $P_{\mathsf{R}}$ | Transmit power at R | $N_0$ | Noise power |

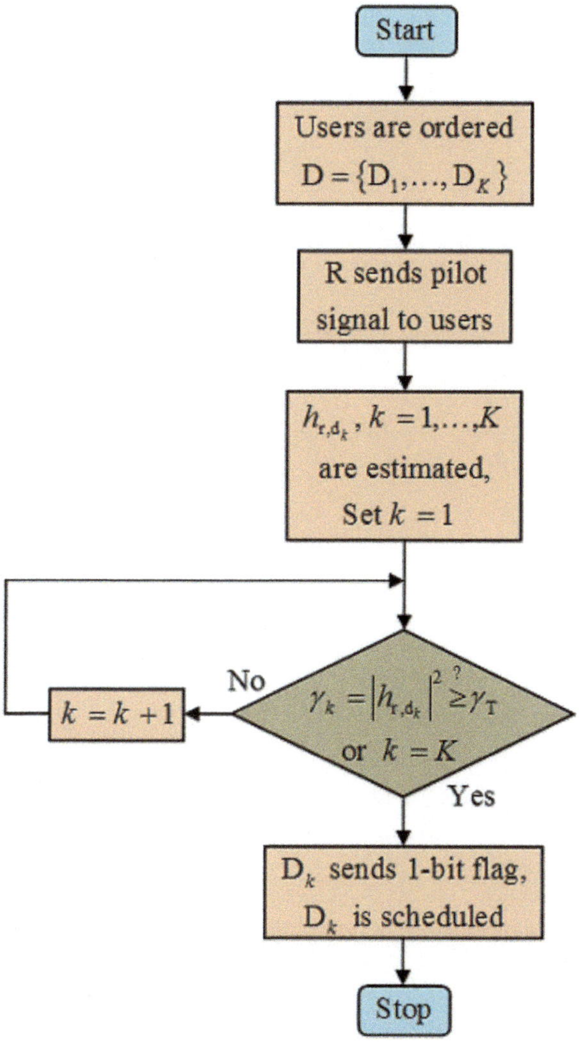


**Fig. 2** Flowchart of the proposed MUSwiD user scheduling

feedback at one time. In order for each user to decide whether to send a feedback or not, a single feedback threshold is used for all the users. This threshold could be assumed to be constant or it could be calculated to optimize a certain performance measure[3]. The order of the users is set by the relay and sent to all users each scheduling period. The second or even the $k^{\text{th}}$ user will not send any feedback signal to the relay unless it does not receive a flag from the previous user in the sequence within a certain time duration[4]. Suppose the users are arranged in a certain order, the first user compares its channel quality with the threshold. If the channel quality is higher than the threshold, the first user sends a feedback to the relay and sends a flag to other users signaling its presence. Otherwise, the first user keeps silent and the second (next) user compares its channel quality against the threshold. Again, if the channel quality of the second user exceeds the threshold, the second user sends a feedback to the relay and sends a flag to other users signaling its presence, otherwise the third user will get a chance. As soon as the relay detects a feedback from any user, it immediately selects the user who sent the feedback for the subsequent data reception and the whole user selection process ends. This process continues till a suitable user is found or all users are examined and found failing to satisfy the threshold. In this case, the MUSwiD scheme selects the last examined user for simplicity. On the other hand, if the last user is reached and found unacceptable in the MUSwiDps scheme, this protocol selects the best user among all checked users. It is worthwhile to mention here that the first hop channel is also considered in the selection process where the e2e channel of each user is represented by the minimum of its first hop and second hop channels through the used SNR bound. Therefore, for a user to be selected for data reception, the minimum of its two hop channels should satisfy the threshold. Despite the effectiveness of the MUSwiD and MUSwiDps user selection schemes in reducing the amount of feedback signaling from users to the relay, there exist some challenges that should be addressed before the multiuser switched diversity selection schemes can lend themselves for practical implementation. In our opinion, these challenges can be summarized in the following points:

- *Fairness.* Achieving fairness among users or achieving maximum sum capacity is a tradeoff issue. Also, maximizing the sum capacity is not always an appropriate optimization criterion for realistic network scenarios since users usually have asymmetric channel statistics. To guarantee fairness among users in the selection process, the time division feedback access can re-arrange the user sequence every scheduling opportunity so that every user can have the same chance of taking the first place in user sequence over an extended period of time. Also, it is worthwhile to mention here that the sum capacity of such systems can be enhanced by assigning different switching thresholds for the different users as discussed in [33]. This issue is beyond the scope of our paper.
- *Centralized scheduling.* In order to avoid any collusion, which may happen when multiple users send feedbacks, a centralized feedback collection method can be used. This method organizes the users to be orthogonal when they send feedbacks, such as time division multiplexing (TDM) where users are separated over time.

The multiuser switched diversity selection schemes proved themselves as a less complicated user selection schemes compared to the opportunistic scheduling from the number of channel estimations-wise. This happens on the expense of the system sum rate, which was shown in [35] to be compensated by significant savings in the CSI feedback load. The authors showed that the MUSwiD user selection schemes provide significant reduction of CSI feedback load at the cost of slight reduction in the achievable multiuser diversity gains, especially, at the low SNR values. This makes the MUSwiD selection schemes an attractive option for practical implementation in emerging mobile broadband communication systems.

## 3 Performance analysis

In this section, we evaluate closed-form approximations for the outage probability and average symbol error probability of the studied system. The channel capacity if also derived and numerically calculated in this section. The outage probability is a measure for the system performance and the possibility of the system to fall in outage. This is clear from the definition of the outage probability, which says that the system could get in outage once the SNR at the destination goes below a predetermined outage threshold $\gamma_{\mathrm{out}}$ or, equivalently the system is unable to achieve adequate reception. The possibility of the destination SNR to fall under a certain threshold exists in any communication system. In regard to deriving the ergodic capacity, as we are considering slowly varying fading channel model, this means that the coherence time of the channel is much larger than the data rate, and hence, the fading status remains constant over a large number of transmitted bits. This means that the average (i.e., ergodic) capacity of the channel could be a suitable measure for the best achievable capacity of the considered system links. In summary, the ergodic capacity we derived in this paper is valid under the assumption that the information symbol is long enough to ensure the long-term ergodic properties of the links.

### 3.1 Outage probability

In this section, we evaluate closed-form approximations for the outage probability for the i.n.i.d. and i.i.d. cases of users channels. The outage probability is defined as the probability that the SNR at the scheduled destination $\gamma_{\mathrm{up}}$ goes below a predetermined outage threshold $\gamma_{\mathrm{out}}$, i.e., $P_{\mathrm{out}} = \Pr\left[\gamma_{\mathrm{up}} \leq \gamma_{\mathrm{out}}\right]$, where Pr[.] denotes the probability operation. Our results on the outage probability of the studied system are summarized in the following lemma and corollary.

**Lemma 1** *The outage probability for cognitive CSI-assisted AF MUSwiD relay system over Rayleigh fading channels is given for the case of i.n.i.d. users channels as*

$$\begin{aligned}
P_{\mathrm{out}} \simeq & \sum_{i=0}^{K-1} \pi_i \lambda_{\mathrm{s,p}} \Bigg\{ \frac{\left(1-\exp\left(-\lambda_{\mathrm{s,p}}\gamma_{\mathrm{T}}\right)\right)}{\lambda_{\mathrm{s,p}}} + \sum_{k=0}^{K-1}(-1)^{k+1} \sum_{\substack{n_0<\ldots<n_k\\ n(.)\neq i}}^{K-1} \\
& \times \prod_{t=0}^{k} \frac{\left(1+\lambda_{2_{n_t}}\gamma_{\mathrm{T}}\right)^{-1}}{\Delta_1} - \left(1+\lambda_{2_i}\gamma_{\mathrm{out}}\right)^{-1} \\
& \times \Bigg[ \frac{\left(1-\exp\left(-\left(\lambda_1\gamma_{\mathrm{out}}+\lambda_{\mathrm{s,p}}\right)\gamma_{\mathrm{T}}\right)\right)}{\lambda_1\gamma_{\mathrm{out}}+\lambda_{\mathrm{s,p}}} + \sum_{k=0}^{K-1}(-1)^{k+1} \\
& \times \sum_{\substack{n_0<\ldots<n_k\\ n(.)\neq i}}^{K-1} \prod_{t=0}^{k} \frac{\left(1-\exp\left(-\Delta_1\gamma_{\mathrm{T}}\right)\right)}{\left(1+\lambda_{2_{n_t}}\gamma_{\mathrm{T}}\right)\Delta_1} \Bigg]\Bigg\} \\
& + \sum_{l=0}^{K-1} \pi_l \lambda_{\mathrm{s,p}} \Bigg( \sum_{q=0}^{K} \frac{(-1)^q}{q!} \sum_{m_1,\ldots,m_q}^{K} \prod_{z=1}^{q} \frac{\left(1-\exp\left(-\Delta_1\gamma_{\mathrm{T}}\right)\right)}{\left(1+\lambda_{2_{m_z}}\gamma_{\mathrm{T}}\right)\Delta_1} \\
& + \sum_{w=0}^{K-1} \pi_{((l-w))_K} \Bigg[ \left(1+\lambda_{2_l}\gamma_{\mathrm{T}}\right)^{-1} \\
& \times \Bigg\{ \frac{\exp\left(-\left(\lambda_1\gamma_{\mathrm{T}}+\lambda_{\mathrm{s,p}}\right)\gamma_{\mathrm{T}}\right)}{\lambda_1\gamma_{\mathrm{T}}+\lambda_{\mathrm{s,p}}} + \sum_{p=0}^{w-1}(-1)^{p+1} \sum_{v_0<\ldots<v_p}^{w-1} \\
& \times \prod_{g=0}^{p} \frac{\exp\left(-\Delta_2\gamma_{\mathrm{T}}\right)}{\Delta_3\Delta_2} \Bigg\} - \left(1+\lambda_{2_l}\gamma_{\mathrm{out}}\right)^{-1} \\
& \times \Bigg\{ \frac{\exp\left(-\left(\lambda_1\gamma_{\mathrm{out}}+\lambda_{\mathrm{s,p}}\right)\gamma_{\mathrm{T}}\right)}{\lambda_1\gamma_{\mathrm{out}}+\lambda_{\mathrm{s,p}}} + \sum_{p=0}^{w-1}(-1)^{p+1} \sum_{v_0<\ldots<v_p}^{w-1} \\
& \times \prod_{g=0}^{p} \frac{\exp\left(-\Delta_4\gamma_{\mathrm{T}}\right)}{\Delta_3\Delta_4} \Bigg\}\Bigg]\Bigg),
\end{aligned} \tag{2}$$

*where* $\Delta_1 = \lambda_1\gamma_{\mathrm{T}} + \lambda_{\mathrm{s,p}}$, $\Delta_2 = \sum_{u=0}^{p} \lambda_{2_{((l-w+v_u))_K}} + \lambda_1\gamma_{\mathrm{T}} + \lambda_{\mathrm{s,p}}$, $\Delta_3 = 1 + \lambda_{2_{((l-w+v_g))_K}}\gamma_{\mathrm{T}}$, *and* $\Delta_4 = \sum_{u=0}^{p} \lambda_{2_{((l-w+v_u))_K}} + \lambda_1\gamma_{\mathrm{out}} + \lambda_{\mathrm{s,p}}$.

*Proof* Please, see Appendix. □

**Corollary 1** *The outage probability for cognitive CSI-assisted AF MUSwiD relay system over Rayleigh fading channels is given for the case of i.i.d. users channels* ($\Omega_{\mathrm{r},1} = \Omega_{\mathrm{r},2} = \ldots = \Omega_{\mathrm{r},K} = \Omega_{\mathrm{r,d}}$) *as*

$$\begin{aligned}
P_{\mathrm{out}} \simeq \lambda_{\mathrm{s,p}} & \Bigg\{ \sum_{g=0}^{K-1} \frac{\binom{K-1}{g}(-1)^g}{\left(1+\lambda_2\gamma_{\mathrm{T}}\right)^g} \Bigg[ \frac{\left(1-\exp\left(-\left(g\lambda_1\gamma_{\mathrm{T}}+\lambda_{\mathrm{s,p}}\right)\gamma_{\mathrm{T}}\right)\right)}{\left(g\lambda_1\gamma_{\mathrm{T}}+\lambda_{\mathrm{s,p}}\right)} \\
& - \frac{\left(1-\exp\left(-\left(\left(\gamma_{out}+g\gamma_T\right)\lambda_1+\lambda_{s,p}\right)\gamma_T\right)\right)}{\left(1+\lambda_2\gamma_{\mathrm{out}}\right)\left(\left(\gamma_{\mathrm{out}}+g\gamma_{\mathrm{T}}\right)\lambda_1+\lambda_{\mathrm{s,p}}\right)} \Bigg] \\
& + \sum_{j=0}^{K-1}\sum_{w=0}^{j} \frac{\binom{j}{w}(-1)^w}{\left(1+\lambda_2\gamma_{\mathrm{T}}\right)^w} \Bigg[ \frac{\exp\left(-\left((j+1)\lambda_1\gamma_{\mathrm{T}}+\lambda_{\mathrm{s,p}}\right)\gamma_T\right)}{\left(1+\lambda_2\gamma_{\mathrm{T}}\right)\left((j+1)\lambda_1\gamma_{\mathrm{T}}+\lambda_{\mathrm{s,p}}\right)} \\
& - \frac{\exp\left(-\left((\gamma_{\mathrm{out}}+j\gamma_{\mathrm{T}})\lambda_1+\lambda_{\mathrm{s,p}}\right)\gamma_T\right)}{\left(1+\lambda_2\gamma_{\mathrm{out}}\right)\left(\left(\gamma_{\mathrm{out}}+j\gamma_{\mathrm{T}}\right)\lambda_1+\lambda_{\mathrm{s,p}}\right)} \Bigg] \\
& + \sum_{q=0}^{K} \frac{\binom{K}{q}(-1)^q}{\left(1+\lambda_2\gamma_{\mathrm{T}}\right)^q} \frac{\exp\left(-\left(q\lambda_1\gamma_{\mathrm{T}}+\lambda_{\mathrm{s,p}}\right)\gamma_{\mathrm{T}}\right)}{\left(q\lambda_1\gamma_{\mathrm{T}}+\lambda_{\mathrm{s,p}}\right)} \Bigg\}.
\end{aligned} \tag{3}$$

*Proof* In evaluating the outage probability in (3), the conditional CDF of $\gamma_{\mathrm{up}}$ is required to be obtained first. Based on the way the MUSwiD selection scheme works, this CDF can be written for the case of i.i.d. users channels as [36]

$$F_{\gamma_{\mathrm{up}}}^{\mathrm{MUSwiD}}\left(\gamma \mid h_{\mathrm{s,p}}\right) = \begin{cases} \left[F_{\gamma^{\mathrm{up}}}\left(\gamma_{\mathrm{T}} \mid h_{\mathrm{s,p}}\right)\right]^{K-1} F_{\gamma^{\mathrm{up}}}\left(\gamma \mid h_{\mathrm{s,p}}\right), & \gamma < \gamma_{\mathrm{T}}; \\ \sum_{j=0}^{K-1}\left[F_{\gamma^{\mathrm{up}}}\left(\gamma \mid h_{\mathrm{s,p}}\right) - F_{\gamma^{\mathrm{up}}}\left(\gamma_{\mathrm{T}} \mid h_{\mathrm{s,p}}\right)\right] & \\ \times \left[F_{\gamma^{\mathrm{up}}}\left(\gamma_{\mathrm{T}} \mid h_{\mathrm{s,p}}\right)\right]^{j} + \left[F_{\gamma^{\mathrm{up}}}\left(\gamma_{\mathrm{T}} \mid h_{\mathrm{s,p}}\right)\right]^{K}, & \gamma \geq \gamma_{\mathrm{T}}, \end{cases} \tag{4}$$

where $F_{\gamma^{\mathrm{up}}}(\gamma \mid h_{\mathrm{s,p}})$ is the CDF of $\gamma_k^{\mathrm{up}}$ conditioned on $h_{\mathrm{s,p}}$, and it can be expressed for the case of i.i.d. first and second hop channels as

$$F_{\gamma_k^{\mathrm{up}}}\left(\gamma \mid h_{\mathrm{s,p}}\right) = 1 - \frac{\exp\left(-\lambda_1 \gamma \left|h_{\mathrm{s,p}}\right|^2\right)}{(1+\lambda_2\gamma)}, \tag{5}$$

where $\lambda_1$ is as defined in the Appendix and $\lambda_2 = \Omega_{\mathrm{r,p}} \Big/ \left(\Omega_{\mathrm{r,d}} \frac{\mathcal{I}_{\mathrm{p}}}{N_0}\right)$. Upon substituting (5) in (4) and following the same procedure as in the Appendix, we get the result in (3). □

Now, we derive the outage probability for the MUSwiDps selection scheme. Based on the way the MUSwiDps selection scheme works, the conditional CDF of $\gamma_{\mathrm{up}}$ can be written for the case of i.i.d. users channels as [36]

$$F_{\gamma_{\mathrm{up}}}^{\mathrm{MUSwiDps}}\left(\gamma \mid h_{\mathrm{s,p}}\right) = \begin{cases} \left[F_{\gamma^{\mathrm{up}}}(\gamma \mid h_{\mathrm{s,p}})\right]^{K-1}, & \gamma < \gamma_{\mathrm{T}}; \\ 1 - \sum_{j=0}^{K-1}\left[F_{\gamma^{\mathrm{up}}}(\gamma_{\mathrm{T}} \mid h_{\mathrm{s,p}})\right]^{j}\left[1 - F_{\gamma^{\mathrm{up}}}(\gamma \mid h_{\mathrm{s,p}})\right], & \gamma \geq \gamma_{\mathrm{T}}. \end{cases} \tag{6}$$

Upon substituting (26) in (6) and following the same procedure as in the Appendix, the outage probability for the studied system with the MUSwiDps selection scheme can be evaluated in a closed-form approximation as

$$\begin{aligned} P_{\mathrm{out}} \simeq \lambda_{\mathrm{s,p}} \Bigg\{ & \frac{\exp\left(-\lambda_{\mathrm{s,p}}\gamma_{\mathrm{T}}\right)}{\lambda_{\mathrm{s,p}}} - \sum_{j=0}^{K-1}\sum_{k=0}^{j} \frac{\binom{j}{k}(-1)^k}{(1+\lambda_2\gamma_{\mathrm{T}})^k} \\ & \times \frac{\exp\left(-\left((k\gamma_{\mathrm{T}}+\gamma_{\mathrm{out}})\lambda_1+\lambda_{\mathrm{s,p}}\right)\gamma_{\mathrm{T}}\right)}{(1+\lambda_2\gamma_{\mathrm{out}})\left((k\gamma_{\mathrm{T}}+\gamma_{\mathrm{out}})\lambda_1+\lambda_{\mathrm{s,p}}\right)} \\ & + \sum_{i=0}^{K}\binom{K}{i}(-1)^i \frac{\left(1-\exp\left(-\left(i\lambda_1\gamma_{\mathrm{out}}+\lambda_{\mathrm{s,p}}\right)\gamma_{\mathrm{T}}\right)\right)}{(1+\lambda_2\gamma_{\mathrm{out}})^i\left(i\lambda_1\gamma_{\mathrm{out}}+\lambda_{\mathrm{s,p}}\right)} \Bigg\}. \end{aligned} \tag{7}$$

### 3.2 Average symbol error probability

In this section, we evaluate closed-form approximations for the SEP for the i.n.i.d. and i.i.d. cases of users channels. The SEP can be expressed in terms of the CDF of $\gamma_{\mathrm{up}}$, $F_{\gamma_{\mathrm{up}}}(\gamma) = P_{\mathrm{out}}(\gamma_{\mathrm{out}} = \gamma)$ as

$$\mathrm{SEP} \simeq \frac{a\sqrt{b}}{2\sqrt{\pi}} \int_0^{\infty} \frac{\exp(-b\gamma)}{\sqrt{\gamma}} F_{\gamma_{\mathrm{up}}}(\gamma)\, d\gamma, \tag{8}$$

where $a$ and $b$ are modulation-specific constants. Our results on the SEP of the studied system are summarized in the following lemma and corollary.

**Lemma 2** *The SEP for cognitive CSI-assisted AF MUSwiD relay system is given for the case of i.n.i.d. users channels as*

$$\begin{aligned} \mathrm{SEP} \simeq \frac{a\sqrt{b}}{2\sqrt{\pi}}\lambda_{\mathrm{s,p}} \Bigg\{ & \sum_{i=0}^{K-1}\pi_i \Bigg( \frac{\sqrt{\pi}}{\sqrt{b}\lambda_{\mathrm{s,p}}} + \sum_{k=0}^{K-1}(-1)^{k+1} \sum_{\substack{n_0<\ldots<n_k \\ n(.)\neq i}}^{K-1} \prod_{t=0}^{k} \\ & \times \frac{\left(1+\lambda_{2_{n_t}}\gamma_{\mathrm{T}}\right)^{-1}}{\vartheta_1} - \Bigg[ \left(-\frac{\lambda_1}{\lambda_{2_i}} + \lambda_{\mathrm{s,p}}\right)^{-1} \\ & \times \Gamma(1/2)(\lambda_{2_i})^{-1/2} \Bigg\{ \exp\left(\frac{b}{\lambda_{2_i}}\right) \Bigg( \Gamma\left(1/2, b/\lambda_{2_i}\right) \\ & - \exp\left(-\left(\lambda_{\mathrm{s,p}} - \frac{\lambda_1}{\lambda_{2_i}}\right)\gamma_{\mathrm{T}}\right) \Gamma\left(1/2, (\lambda_1\gamma_{\mathrm{T}}+b)/\lambda_{2_i}\right) \Bigg) \\ & + \left(\frac{\lambda_{2_i}\lambda_{\mathrm{s,p}}}{\lambda_1}\right)^{-1/2} \exp\left(\frac{b\lambda_{\mathrm{s,p}}}{\lambda_1}\right) \Bigg( \Gamma\left(1/2, (\lambda_1\gamma_{\mathrm{T}}+b)\lambda_{\mathrm{s,p}}/\lambda_1\right) \\ & - \Gamma\left(1/2, b\lambda_{\mathrm{s,p}}/\lambda_1\right) \Bigg) \Bigg\} + \sum_{k=0}^{K-1}(-1)^{k+1} \\ & \times \sum_{\substack{n_0<\ldots<n_k \\ n(.)\neq i}}^{K-1} \prod_{t=0}^{k} \frac{\left(1-\exp\left(-\vartheta_1\gamma_{\mathrm{T}}\right)\right)}{\left(1+\lambda_{2_{n_t}}\gamma_{\mathrm{T}}\right)\vartheta_1} \Bigg] \Bigg) + \sum_{l=0}^{K-1}\pi_l \Bigg( \sum_{q=0}^{K} \frac{(-1)^q}{q!} \\ & \times \sum_{m_1,\ldots,m_q}^{K} \prod_{z=1}^{q} \frac{\exp\left(-\vartheta_1\gamma_{\mathrm{T}}\right)}{\left(1+\lambda_{2_{m_z}}\gamma_{\mathrm{T}}\right)\vartheta_1} + \sum_{w=0}^{K-1}\pi_{((l-w))_K} \\ & \times \Bigg[ \left(1+\lambda_{2_l}\gamma_{\mathrm{T}}\right)^{-1} \Bigg\{ \frac{\exp\left(-\vartheta_1\right)}{\vartheta_1} + \sum_{p=0}^{w-1}(-1)^{p+1} \sum_{v_0,\ldots,v_p}^{w-1} \prod_{g=0}^{p} \\ & \times \frac{\exp\left(-\vartheta_3\gamma_{\mathrm{T}}\right)}{\left(1+\lambda_{2_{((l-w+v_g))_K}}\gamma_{\mathrm{T}}\right)\vartheta_3} \Bigg\} - \Bigg\{ \left(-\frac{\lambda_1}{\lambda_{2_l}} + \lambda_{\mathrm{s,p}}\right)^{-1} \\ & \times \Gamma(1/2)(\lambda_{l2})^{-1/2} \exp\left(-\lambda_{\mathrm{s,p}}\gamma_{\mathrm{T}}\right) \Bigg( \exp\left(\frac{\lambda_1\gamma_{\mathrm{T}}+b}{\lambda_{2_l}}\right) \\ & \times \Gamma\left(1/2, (\lambda_1\gamma_{\mathrm{T}}+b)/\lambda_{2_l}\right) - \left(\frac{\lambda_{2_l}\lambda_{\mathrm{s,p}}}{\lambda_1}\right)^{-1/2} \\ & \times \exp\left(\frac{(\lambda_1\gamma_{\mathrm{T}}+b)\lambda_{\mathrm{s,p}}}{\lambda_1}\right) \Gamma\left(1/2, (\lambda_1\gamma_{\mathrm{T}}+b)\lambda_{\mathrm{s,p}}/\lambda_1\right) \Bigg) \Bigg\} \\ & + \sum_{p=0}^{w-1}(-1)^{p+1} \prod_{g=0}^{p} \frac{\exp\left(-\vartheta_1\gamma_{\mathrm{T}}\right)}{1+\lambda_{2_{((l-w+v_g))_K}}\gamma_{\mathrm{T}}} \\ & \times \Bigg\{ \left(-\frac{\lambda_1}{\lambda_{2_l}} + \lambda_{\mathrm{s,p}}\right)^{-1} \Gamma(1/2)(\lambda_{2_l})^{-1/2} \Bigg( \exp\left(\frac{\lambda_1\gamma_{\mathrm{T}}+b}{\lambda_{2_l}}\right) \\ & \times \Gamma\left(1/2, (\lambda_1\gamma_{\mathrm{T}}+b)/\lambda_{2_l}\right) - \left(\frac{\lambda_{2_l}\lambda_{\mathrm{s,p}}}{\lambda_1}\right)^{-1/2} \\ & \times \exp\left(\frac{(\lambda_1\gamma_{\mathrm{T}}+b)\lambda_{\mathrm{s,p}}}{\lambda_1}\right) \Gamma\left(1/2, (\lambda_1\gamma_{\mathrm{T}}+b)\lambda_{\mathrm{s,p}}/\lambda_1\right) \Bigg) \Bigg\} \Bigg] \Bigg) \Bigg\}, \end{aligned} \tag{9}$$

*where* $\Gamma(.,.)$ *is the incomplete gamma function defined in* ([37], Eq. (8.350.2)), $\vartheta_1 = \lambda_1\gamma_T + \lambda_{s,p}$ *and* $\vartheta_3 = 2\lambda_1\gamma_T + \lambda_{s,p}$.

By replacing $\gamma_{out}$ with $\gamma$ in (2) and using the partial fraction expansion and the integration in (8) and with the help of ([37], Eq. (3.361.2)) and ([37], Eq. (3.383.10)), we get the result in (9).

**Corollary 2** *The SEP for cognitive CSI-assisted AF MUSwiD relay system is given for the case of i.i.d. users channels as*

$$
\begin{aligned}
\mathrm{SEP} \simeq\ & \frac{a\sqrt{b}}{2\sqrt{\pi}}\lambda_{s,p} \Bigg\{ \sum_{g=0}^{K-1} \frac{\binom{K-1}{g}(-1)^g}{(1+\lambda_2\gamma_T)^g} \Bigg[ \frac{\sqrt{\pi}\left(1-\exp\left(-\left(g\lambda_1\gamma_T+\lambda_{s,p}\right)\gamma_T\right)\right)}{\sqrt{b}\left(g\lambda_1\gamma_T+\lambda_{s,p}\right)} \\
& - \frac{\Gamma(1/2)(\lambda_2)^{-1/2}}{\left(\left(-\frac{1}{\lambda_2}+g\gamma_T\right)\lambda_1+\lambda_{s,p}\right)} \times \Bigg\{ \exp\left(b/\lambda_2\right) \Bigg( \Gamma\left(1/2, b/\lambda_2\right) \\
& - \exp\left(-\left(\left(g-\frac{1}{\lambda_2}\right)\lambda_1\gamma_T+\lambda_{s,p}\right)\gamma_T\right)\Gamma\left(1/2,(\lambda_1\gamma_T+b)/\lambda_2\right)\Bigg) \\
& - \frac{\exp\left(\frac{b(g\lambda_1\gamma_T+\lambda_{s,p})}{\lambda_1}\right)}{\left(\lambda_2\left(g\lambda_1+\frac{\lambda_{s,p}}{\lambda_1}\right)\right)^{1/2}} \Bigg( \Gamma\left(1/2, b\left(g\lambda_1\gamma_T+\lambda_{s,p}\right)/\lambda_1\right) \\
& - \Gamma\left(1/2,(\lambda_1\gamma_T+b)\left(g\lambda_1\gamma_T+\lambda_{s,p}\right)/\lambda_1\right)\Bigg)\Bigg\}\Bigg] \\
& + \sum_{j=0}^{K-1}\sum_{w=0}^{j} \frac{\binom{j}{w}(-1)^w}{(1+\lambda_2\gamma_T)^w} \Bigg[ \frac{\sqrt{\pi}\exp\left(-\left((j+1)\lambda_1\gamma_T+\lambda_{s,p}\right)\gamma_T\right)}{\sqrt{b}(1+\lambda_2\gamma_T)\left((j+1)\lambda_1\gamma_T+\lambda_{s,p}\right)} \\
& - \frac{\Gamma(1/2)(\lambda_2)^{-1/2}\exp\left(-\left(j\lambda_1\gamma_T+\lambda_{s,p}\right)\gamma_T\right)}{\left(\left(-\frac{1}{\lambda_2}+j\gamma_T\right)\lambda_1+\lambda_{s,p}\right)} \\
& \times \Bigg\{ \exp\left(\frac{\lambda_1\gamma_T+b}{\lambda_2}\right)\Gamma\left(1/2,(\lambda_1\gamma_T+b)/\lambda_2\right) \\
& - \frac{\exp\left(\frac{\left(j\gamma_T+\frac{\lambda_{s,p}}{\lambda_1}\right)}{(\lambda_1\gamma_T+b)^{-1}}\right)}{\left(\lambda_2\left(j\lambda_1+\frac{\lambda_{s,p}}{\lambda_1}\right)\right)^{1/2}} \Gamma\left(1/2, \frac{\left(j\gamma_T+\frac{\lambda_{s,p}}{\lambda_1}\right)}{(\lambda_1\gamma_T+b)^{-1}}\right)\Bigg\}\Bigg] \\
& + \sum_{q=0}^{K} \frac{\binom{K}{q}(-1)^q}{(1+\lambda_2\gamma_T)^q} \frac{\sqrt{\pi}\exp\left(-\left(q\lambda_1\gamma_T+\lambda_{s,p}\right)\gamma_T\right)}{\sqrt{b}\left(q\lambda_1\gamma_T+\lambda_{s,p}\right)} \Bigg\}.
\end{aligned}
\quad (10)
$$

By replacing $\gamma_{out}$ with $\gamma$ in (3) and using the partial fraction expansion and the integration in (8) and with the help of ([37], Eq. (3.361.2)) and ([37], Eq. (3.383.10)), we get the result in (10).

### 3.3 Ergodic channel capacity

In this section, we evaluate an expression for the ergodic channel capacity for the i.i.d. case of users channels. The ergodic capacity can be expressed in terms of the PDF of $\gamma_{up}$ as

$$
C = \frac{1}{2\ln(2)}\int_0^{\infty} \ln(1+\gamma) f_{\gamma_{up}}(\gamma)d\gamma. \quad (11)
$$

Our result on the ergodic capacity of the studied system is provided in the following lemma.

**Lemma 3** *The ergodic capacity for cognitive CSI-assisted AF MUSwiD relay system over Rayleigh fading channels is given for the case of i.i.d. users channels as*

$$
\begin{aligned}
C \simeq\ & \frac{1}{2\ln(2)} \sum_{g=0}^{K-1} \binom{K-1}{g} (-1)^g (1+\lambda_2\gamma_T)^{-g} \Bigg[ \alpha\ln(\lambda_2)(\lambda_2-1)^{-1} \\
& + \alpha\exp\left(-\left(g\lambda_1\gamma_T+\lambda_{s,p}\right)\gamma_T\right) \\
& \times \Bigg\{ -(\lambda_1\gamma_T+\lambda_2)\int_0^{\infty} \frac{\ln(1+\gamma)\exp(-\lambda_1\gamma_T\gamma)}{(1+\lambda_2\gamma)^2}d\gamma \\
& - \lambda_1\lambda_2\gamma_T\int_0^{\infty}\frac{\ln(1+\gamma)\gamma\exp(-\lambda_1\gamma_T\gamma)}{(1+\lambda_2\gamma)^2}d\gamma\Bigg\} + \lambda_1\beta \\
& \times \int_0^{\infty}\frac{\ln(1+\gamma)}{(\lambda_1\gamma+g\lambda_1\gamma_T+\lambda_{s,p})}d\gamma + \beta\exp\left(-\left(g\lambda_1\gamma_T+\lambda_{s,p}\right)\gamma_T\right) \\
& \times \Bigg\{ -(\lambda_1)^2\gamma_T\int_0^{\infty}\frac{\ln(1+\gamma)\gamma\exp(-\lambda_1\gamma_T\gamma)}{(\lambda_1\gamma+g\lambda_1\gamma_T+\lambda_{s,p})^2}d\gamma \\
& - \left(g(\lambda_1)^2(\gamma_T)^2+\lambda_1\lambda_{s,p}\gamma_T+\lambda_1\right)\Bigg\}\Bigg] + \sum_{j=0}^{K-1}\sum_{w=0}^{j}\frac{\binom{j}{w}(-1)^w}{(1+\lambda_2\gamma_T)^w} \\
& \times \Bigg[ -\exp\left(-\left(j\lambda_1\gamma_T+\lambda_{s,p}\right)\gamma_T\right)\Bigg(\alpha'\Bigg\{-(\lambda_1\gamma_T+\lambda_2) \\
& \times \int_0^{\infty}\frac{\ln(1+\gamma)\exp(-\lambda_1\gamma_T\gamma)}{(1+\lambda_2\gamma)^2}d\gamma - \lambda_1\lambda_2\gamma_T\int_0^{\infty} \\
& \times \frac{\ln(1+\gamma)\gamma\exp(-\lambda_1\gamma_T\gamma)}{(1+\lambda_2\gamma)^2}d\gamma\Bigg\} + \beta'\Bigg\{-(\lambda_1)^2\gamma_T \\
& \times \int_0^{\infty}\frac{\ln(1+\gamma)\gamma\exp(-\lambda_1\gamma_T\gamma)}{(\lambda_1\gamma+j\lambda_1\gamma_T+\lambda_{s,p})^2}d\gamma \\
& - \left(j(\lambda_1)^2(\gamma_T)^2+\lambda_1\lambda_{s,p}\gamma_T+\lambda_1\right)\int_0^{\infty} \\
& \times \frac{\ln(1+\gamma)\exp(-\lambda_1\gamma_T\gamma)}{(\lambda_1\gamma+j\lambda_1\gamma_T+\lambda_{s,p})^2}d\gamma\Bigg\}\Bigg)\Bigg],
\end{aligned}
\quad (12)
$$

*where* $\alpha$, $\beta$, $\alpha'$, *and* $\beta'$ *are as given in the proof below. According to authors knowledge, the integrations in* (12) *have no closed-form solution. Hence, the ergodic channel capacity of the system is numerically evaluated.*

*Proof* Evaluating (12) requires the determination of the PDF of $\gamma_{\text{up}}$. By replacing $\gamma_{\text{out}}$ with $\gamma$ in (3) and using the partial fraction, the CDF $F_{\gamma_{\text{up}}}(\gamma)$ can be written as

$$\begin{aligned}
F_{\gamma_{\text{up}}}(\gamma) \simeq \lambda_{\text{s,p}} \Bigg\{ & \sum_{g=0}^{K-1} \frac{\binom{K-1}{g}(-1)^g}{(1+\lambda_2\gamma_{\text{T}})^g} \Bigg[ \left(1-\exp\left(-\left(g\lambda_1\gamma_{\text{T}}+\lambda_{\text{s,p}}\right)\gamma_{\text{T}}\right)\right) \\
& \times \left(g\lambda_1\gamma_{\text{T}}+\lambda_{\text{s,p}}\right)^{-1} - \alpha \frac{\left(1-\exp\left(-\left((\gamma+g\gamma_{\text{T}})\lambda_1+\lambda_{\text{s,p}}\right)\gamma_{\text{T}}\right)\right)}{(1+\lambda_2\gamma)} \\
& -\beta \frac{\left(1-\exp\left(-\left((\gamma+g\gamma_{\text{T}})\lambda_1+\lambda_{\text{s,p}}\right)\gamma_{\text{T}}\right)\right)}{\left(\lambda_1\gamma+g\lambda_1\gamma_{\text{T}}+\lambda_{\text{s,p}}\right)} \Bigg] \\
& + \sum_{j=0}^{K-1}\sum_{w=0}^{j} \frac{\binom{j}{w}(-1)^w}{(1+\lambda_2\gamma_{\text{T}})^w} \Bigg[ \frac{\exp\left(-\left((j+1)\lambda_1\gamma_{\text{T}}+\lambda_{\text{s,p}}\right)\gamma_{\text{T}}\right)}{(1+\lambda_2\gamma_{\text{T}})\left((j+1)\lambda_1\gamma_{\text{T}}+\lambda_{\text{s,p}}\right)} \\
& - \exp\left(-\left((j\lambda_1\gamma_{\text{T}})+\lambda_{\text{s,p}}\right)\gamma_{\text{T}}\right) \\
& \times \left(\exp\left(-\lambda_1\gamma_{\text{T}}\gamma\right)\left\{\frac{\alpha'}{(1+\lambda_2\gamma)} + \frac{\beta'}{\left((\gamma+j\gamma_{\text{T}})\lambda_1+\lambda_{\text{s,p}}\right)}\right\}\right)\Bigg] \\
& + \sum_{q=0}^{K} \frac{\binom{K}{q}(-1)^q}{(1+\lambda_2\gamma_{\text{T}})^q} \frac{\exp\left(-\left(q\lambda_1\gamma_{\text{T}}+\lambda_{\text{s,p}}\right)\gamma_{\text{T}}\right)}{\left(q\lambda_1\gamma_{\text{T}}+\lambda_{\text{s,p}}\right)} \Bigg\}.
\end{aligned} \tag{13}$$

Solving for $\alpha$, $\alpha'$, $\beta$, and $\beta'$, we get $\alpha = \left(-\frac{\lambda_1}{\lambda_2}+g\lambda_1\gamma_{\text{T}}+\lambda_{\text{s,p}}\right)^{-1}$, $\beta = \left(1-\frac{(g\lambda_1\gamma_{\text{T}}+\lambda_{\text{s,p}})\lambda_2}{\lambda_1}\right)^{-1}$, and $\alpha' = \alpha$, $\beta' = \beta$ with replacing $g$ by $j$. By taking the derivative of (13) with respect to $\gamma$, the PDF $f_{\gamma_{\text{up}}}(\gamma)$ can be obtained as

$$\begin{aligned}
f_{\gamma_{\text{up}}}(\gamma) \simeq \lambda_{\text{s,p}} \Bigg\{ & \sum_{g=0}^{K-1} \frac{\binom{K-1}{g}(-1)^g}{(1+\lambda_2\gamma_{\text{T}})^g} \Bigg[ \alpha \Bigg( \frac{\lambda_2}{(1+\lambda_2\gamma)^2} \\
& - \exp\left(-\left(g\lambda_1\gamma_{\text{T}}+\lambda_{\text{s,p}}\right)\gamma_{\text{T}}\right) \Bigg\{ \frac{\lambda_1\gamma_{\text{T}}\exp\left(-\lambda_1\gamma_{\text{T}}\gamma\right)}{(1+\lambda_2\gamma)} \\
& + \frac{\lambda_2\exp\left(-\lambda_1\gamma_{\text{T}}\gamma\right)}{(1+\lambda_2\gamma)^2} \Bigg\} \Bigg) + \beta \Bigg( \frac{\lambda_1}{\left(\lambda_1\gamma+g\lambda_1\gamma_{\text{T}}+\lambda_{\text{s,p}}\right)^2} \\
& - \exp\left(-\left(g\lambda_1\gamma_{\text{T}}+\lambda_{\text{s,p}}\right)\gamma_{\text{T}}\right) \Bigg\{ \frac{\lambda_1\gamma_{\text{T}}\exp\left(-\lambda_1\gamma_{\text{T}}\gamma\right)}{\left(\lambda_1\gamma+g\lambda_1\gamma_{\text{T}}+\lambda_{\text{s,p}}\right)} \\
& + \frac{\lambda_1\exp\left(-\lambda_1\gamma_{\text{T}}\gamma\right)}{\left(\lambda_1\gamma+g\lambda_1\gamma_{\text{T}}+\lambda_{\text{s,p}}\right)^2} \Bigg\} \Bigg) \Bigg] + \sum_{j=0}^{K-1}\sum_{w=0}^{j} \frac{\binom{j}{w}(-1)^w}{(1+\lambda_2\gamma_{\text{T}})^w} \\
& \times \Bigg[ - \exp\left(-\left((j\lambda_1\gamma_{\text{T}})+\lambda_{\text{s,p}}\right)\gamma_{\text{T}}\right) \\
& \times \Bigg( \alpha' \Bigg\{ -\frac{\lambda_1\gamma_{\text{T}}\exp\left(-\lambda_1\gamma_{\text{T}}\gamma\right)}{(1+\lambda_2\gamma)} - \frac{\lambda_2\exp\left(-\lambda_1\gamma_{\text{T}}\gamma\right)}{(1+\lambda_2\gamma)^2} \Bigg\} \\
& + \beta' \Bigg\{ -\frac{\lambda_1\gamma_{\text{T}}\exp\left(-\lambda_1\gamma_{\text{T}}\gamma\right)}{\left(\lambda_1\gamma+j\lambda_1\gamma_{\text{T}}+\lambda_{\text{s,p}}\right)} - \frac{\lambda_1\exp\left(-\lambda_1\gamma_{\text{T}}\gamma\right)}{\left(\lambda_1\gamma+j\lambda_1\gamma_{\text{T}}+\lambda_{\text{s,p}}\right)^2} \Bigg\} \Bigg) \Bigg] \Bigg\}.
\end{aligned} \tag{14}$$

Upon substituting (14) in (11) and after some mathematical arrangements and with the help of ([37], Eq. (4.291.15)), we get the result in (12). □

## 4 Asymptotic performance analysis

Due to complexity of the achieved expressions in previous sections, it is hard to get more insights about system performance. Therefore, we see that it is important to derive simple expressions for the outage probability and ASEP at the high SNR values where more insights about the system behavior can be achieved. In this section, we evaluate the asymptotic outage performance of the studied system with the MUSwiD and MUSwiDps user selection schemes. At high SNR values, the outage probability can be expressed as $P_{\text{out}} \approx (G_{\text{c}}\text{SNR})^{-G_{\text{d}}}$, where $G_{\text{c}}$ denotes the coding gain of the system and $G_{\text{d}}$ is the diversity order of the system [36]. The parameters on which the diversity order depends will affect the slope of the outage probability curves, and the parameters on which the coding gain depends will affect the position of the curves. In the upcoming analysis, the users are assumed to have identical channels.

As $\frac{\mathcal{I}_{\text{p}}}{N_0} \to \infty$, the CDF in (26) simplifies to $F_{\gamma_k^{\text{up}}}\left(\gamma|h_{\text{s,p}}\right) \approx \lambda_1|h_{\text{s,p}}|^2\gamma$. Upon substituting this CDF in (4) and following the same procedure as in the Appendix, the outage probability at high SNR values can be evaluated with the help of ([37], Eq. (3.351.1)) and ([37], Eq. (3.351.2)) as

$$\begin{aligned}
P_{\text{out}}^{\infty} = {} & (\gamma_{\text{T}})^{K-1}\left(\frac{\lambda_1}{\lambda_{\text{s,p}}}\right)^{K} \gamma_{\text{out}}\,\gamma\left(K+1,\lambda_{\text{s,p}}\gamma_{\text{T}}\right) \\
& + \sum_{j=0}^{K-1} (\gamma_{\text{T}})^{j}\left(\frac{\lambda_1}{\lambda_{\text{s,p}}}\right)^{j+1} \Gamma\left(j+2,\lambda_{\text{s,p}}\gamma_{\text{T}}\right)\left(\gamma_{\text{out}}-\gamma_{\text{T}}\right) \\
& + (\gamma_{\text{T}})^{K}\left(\frac{\lambda_1}{\lambda_{\text{s,p}}}\right)^{K} \Gamma\left(K+1,\lambda_{\text{s,p}}\gamma_{\text{T}}\right),
\end{aligned} \tag{15}$$

where $\gamma(.,.)$ is the incomplete Gamma function defined in ([37], Eq. (8.350.1)).

One can notice that the result in (15) is dominated by the second term, which is still dominant when $j = 0$. Hence, the outage probability for the studied system with the MUSwiD selection scheme can be evaluated at high SNR values as

$$P_{\text{out}}^{\infty} = \left(\frac{\lambda_1}{\lambda_{\text{s,p}}}\right)\Gamma\left(2,\lambda_{\text{s,p}}\gamma_{\text{T}}\right)\left(\gamma_{\text{out}}-\gamma_{\text{T}}\right). \tag{16}$$

Recalling that $\lambda_1 = 1\big/\left(\Omega_{\text{s,r}}\frac{\mathcal{I}_{\text{p}}}{N_0}\right)$, (16) can be further simplified as

$$P_{\text{out}}^{\infty} = \left\{\frac{\Omega_{\text{s,r}}\lambda_{\text{s,p}}}{\Gamma\left(2,\lambda_{\text{s,p}}\gamma_{\text{T}}\right)\left(\gamma_{\text{out}}-\gamma_{\text{T}}\right)}\frac{\mathcal{I}_{\text{p}}}{N_0}\right\}^{-1}. \tag{17}$$

It is clear from (17) that the system with the MUSwiD selection scheme has a diversity order of 1 and a coding gain that is function of $\Omega_{\text{s,r}}$, $\lambda_{\text{s,p}}$, $\gamma_{\text{out}}$, and $\gamma_{\text{T}}$.

The asymptotic SEP for the MUSwiD user selection scheme can be obtained by replacing $\gamma_{\text{out}}$ by $\gamma$ in (16) and then substituting the result in (8). Upon doing this and with the help of ([37], Eq. (3.351.3)), we get

$$\mathrm{SEP}^{\infty} = \frac{a\sqrt{b}\lambda_1 \Gamma\left(2, \lambda_{s,p}\gamma_T\right)}{2\sqrt{\pi}\lambda_{s,p}} \left[\frac{\Gamma\left(\frac{3}{2}\right)}{b^{3/2}} - \frac{\Gamma\left(\frac{1}{2}\right)}{b^{1/2}}\gamma_T\right]. \tag{18}$$

Recalling that $\lambda_1 = 1/\left(\Omega_{s,r}\frac{\mathcal{I}_p}{N_0}\right)$, (18) can be further simplified as

$$\mathrm{SEP}^{\infty} = \left\{\left(\frac{a\sqrt{b}\Gamma\left(2, \lambda_{s,p}\gamma_T\right)}{2\sqrt{\pi}\lambda_{s,p}\Omega_{s,r}} \left[\frac{\Gamma\left(\frac{3}{2}\right)}{b^{3/2}} - \frac{\Gamma\left(\frac{1}{2}\right)}{b^{1/2}}\gamma_T\right]\right)^{-1} \frac{\mathcal{I}_p}{N_0}\right\}^{-1}. \tag{19}$$

Now, we derive the asymptotic outage probability for the studied system with the MUSwiDps user selection scheme. Upon substituting the CDF $F_{\gamma_k^{\mathrm{up}}}\left(\gamma|h_{s,p}\right) \approx \lambda_1|h_{s,p}|^2\gamma$ in (6) and following the same procedure as in the Appendix, the outage probability at high SNR values can be evaluated with the help of ([37], Eq. (3.351.1)) and ([37], Eq. (3.351.2)) as

$$\begin{aligned} P_{\mathrm{out}}^{\infty} &= \left(\frac{\lambda_1}{\lambda_{s,p}}\right)\Gamma\left(2, \lambda_{s,p}\gamma_T\right)\left(\gamma_{\mathrm{out}} - \gamma_T\right) \\ &+ \gamma_T\left(\frac{\lambda_1}{\lambda_{s,p}}\right)^2 \gamma_{\mathrm{out}}\Gamma\left(3, \lambda_{s,p}\gamma_T\right) \\ &+ \left(\frac{\lambda_1}{\lambda_{s,p}}\right)^K \left(\gamma_{\mathrm{out}}\right)^K \gamma\left(K+1, \lambda_{s,p}\gamma_T\right). \end{aligned} \tag{20}$$

Again, one can notice that the result in (20) is dominated by the first term, and hence, the outage probability for the studied system with the MUSwiDps selection scheme can be evaluated at high SNR values as

$$P_{\mathrm{out}}^{\infty} = \left(\frac{\lambda_1}{\lambda_{s,p}}\right)\Gamma\left(2, \lambda_{s,p}\gamma_T\right)\left(\gamma_{\mathrm{out}} - \gamma_T\right). \tag{21}$$

Recalling that $\lambda_1 = 1/\left(\Omega_{s,r}\frac{\mathcal{I}_p}{N_0}\right)$, (21) can be further simplified as

$$P_{\mathrm{out}}^{\infty} = \left\{\frac{\Omega_{s,r}\lambda_{s,p}}{\Gamma\left(2, \lambda_{s,p}\gamma_T\right)\left(\gamma_{\mathrm{out}} - \gamma_T\right)} \frac{\mathcal{I}_p}{N_0}\right\}^{-1}. \tag{22}$$

It is clear that the asymptotic outage probability of the MUSwiDps user selection scheme is exactly the same as that of the MUSwiD scheme. Accordingly, the asymptotic SEP of the MUSwiDps scheme will be equal to that of the MUSwiD scheme as obtained in (19).

As can be seen from the results in (17) and (22), the coding gain of the system with the MUSwiD and MUSwiDps user selection schemes is affected by several parameters as $\Omega_{s,r}$, $\lambda_{s,p}$, $\gamma_{\mathrm{out}}$, and $\gamma_T$; while the diversity order is constant at 1. This is clear in the numerical examples where all curves of different $K$ asymptotically converge to the same behavior and result in a diversity order of 1. It is expected from the way the MUSwiD and MUSwiDps selection schemes operate that the gain achieved in system performance due to having more SU destinations to happen at the SNR values that are comparable to $\gamma_T$. As in this case, the switching rate will increase and the probability of having better users increases also. At the same time, as the asymptotic analysis is done at high SNR values, it is expected to have most of the users being acceptable the whole time and thus, the first checked user is being selected in the two selection schemes. This means all curves of different $K$ asymptotically converge to same behavior, and this explains why the system with the two selection schemes has the same diversity order and the same coding gain as will be shown in numerical examples. In addition, it can be noticed from (17) and (22) that the diversity order of cognitive CSI-assisted AF MUSwiD relay network is the same as that of its non-cognitive counterpart. Specifically, it is equal to 1 and is independent of the primary network. It is worthwhile to mention here again that the importance of the proposed MUSwiD and MUSwiDps user selection schemes is in the reduced number of channel estimations and feedback load they require each transmission time compared to the opportunistic scheduling. Furthermore, from the asymptotic results, we conclude that the MUSwiD user selection schemes are inefficient at the range of high SNR values where the diversity order of the system becomes fixed and equal 1 when they are implemented. Accordingly, it is preferable to apply such low-complexity user scheduling schemes in systems that operate at the low range of SNR values as in the emerging mobile broadband communication systems. This fact about the applicability of the MUSwiD selection schemes was also proved and mentioned in [35].

The optimum switching threshold can be numerically calculated to minimize the e2e outage probability or symbol error probability. Due to complexity of the achieved expressions, any further manipulation with them to find the optimum threshold will increase the system complexity. Alternatively, we present here a simple method that can be used to get approximate but accurate values of the optimum switching threshold. In general, a good choice of the switching threshold is to have it near the average value of the e2e SNR, which can be upper bounded by the minimum of its two hops. Therefore, the optimum switching threshold can be calculated using $\min\left(\frac{\frac{\mathcal{I}_p}{N_0}|h_{s,r}|^2}{|h_{s,p}|^2}, \frac{\frac{\mathcal{I}_p}{N_0}|h_{r,k}|^2}{|h_{r,p}|^2}\right)$. This method give excellent results as will be seen in the coming section.

## 5 Simulation and numerical results

In this section, we illustrate the validity of the achieved analytical and asymptotic expressions via a comparison with Monte-Carlo simulations. We also provide some numerical examples to show the effect of some system parameters such as number of users, switching threshold, and outage threshold on the system performance.

Figure 3 portrays the outage probability versus SNR for the studied system with the MUSwiD selection scheme for different number of users $K$. It is clear from this figure that the asymptotic results perfectly converge to the analytical results as well as the exact ones. Also, it is obvious that the used bound on the e2e SNR is indeed very tight; especially, at the high SNR region. Furthermore, we can see from this figure that the MUSwiD selection scheme has nearly the same performance as the best-user selection scheme for very low SNR region; whereas, as we go further in increasing SNR, the best-user selection scheme is clearly outperforming the MUSwiD scheme, as expected. In addition, we can see from this figure that for the MUSwiD scheme as $K$ increases, the system performance becomes more enhanced; especially, at the range of SNR values that are comparable to the switching threshold $\gamma_T$. More importantly, for $K = 2$, 3, and 4, it is obvious that at both low and high SNR values, all curves asymptotically converge to the same behavior and no gain is achieved in system performance with having more users. This is expected since when $\gamma_T$ takes values much smaller or much larger than the average SNR, the system asymptotically converges to the case of two users, and hence, having more users will have no effect on the system performance. Also, it is clear that the curves in this figure greatly match the asymptotic or high SNR results where the diversity order was shown to be constant and equal 1. This shows that the MUSwiD and MUSwiDps selection schemes are inefficient at the range of high SNR values where the diversity order is noticeably degraded when they are implemented. Finally, the effectiveness of the MUSwiD and MUSwiDps schemes is in the reduction of CSI feedback load they offer compared to the opportunistic scheduling. In order to achieve this effectiveness with a slight reduction in the multiuser diversity gain, these user selection schemes should be implemented in the range of low SNR values such as in emerging mobile broadband communications systems.

Figure 4 studies the outage performance versus SNR for the studied system with the MUSwiD and MUSwiDps user selection schemes for different values of first hop power $\Omega_{s,r}$. As can be seen from this figure, the MUSwiD and MUSwiDps selection schemes behave exactly the same in such kind of systems. Also, it is clear from the asymptotic results of this figure that the MUSwiD and MUSwiDps schemes have the same diversity order and the same coding gain. Finally, as $\Omega_{s,r}$ increases, the coding gain increases and better the achieved performance.

Figure 5 shows the outage performance versus outage threshold for the studied system with the MUSwiD selection scheme for different values of SNR $\mathcal{I}_p/N_0$. As expected, as $\gamma_{out}$ increases, the worse the achieved performance. Also, it is obvious from this figure that the best performance is achieved at the largest value of SNR.

Figure 6 illustrates the accuracy of the proposed method of finding the optimum switching threshold compared to

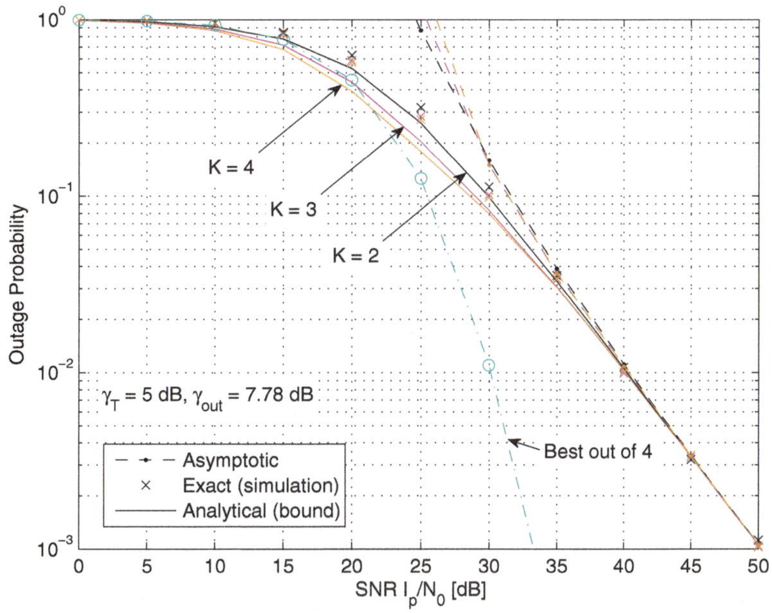


**Fig. 3** Outage probability vs SNR of cognitive AF relay network with MUSwiD user selection scheme for different values of $K$ and $\Omega_{s,p} = 30$, and $\Omega_{s,r} = 0.8$, $\Omega_{r,p} = 0.1$, $\Omega_{r,k} = 0.7$ for $k = 1, \ldots, 4$

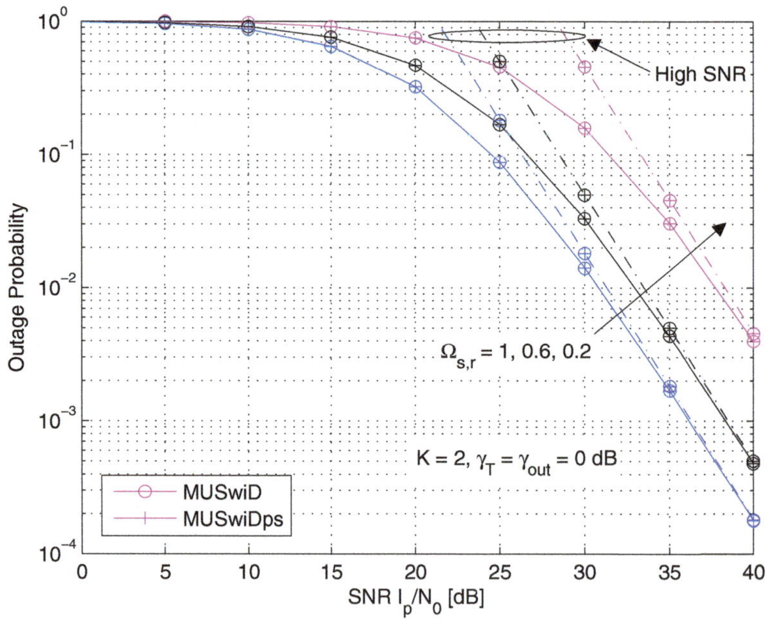


**Fig. 4** Outage probability vs SNR of cognitive AF relay network with MUSwiD and MUSwiDps user selection schemes for different values of $\Omega_{s,r}$ and $\Omega_{s,p} = 95$, $\Omega_{r,p} = 0.97$, $\Omega_{r,k} = 1$ for $k = 1, 2$

the numerical method where the e2e outage probability is minimized. It is clear from this figure that the proposed method of finding approximate values for the optimum switching threshold $\gamma_T$ gives a very close performance to that of the numerical method. This happens over a wide range of SNR values, which is 0–20 dB.

Figure 7 studies the effects of the switching threshold $\gamma_T$ and number of users $K$ on the error probability performance of the studied system with the MUSwiD selection scheme. As can be seen from this figure, increasing $K$ leads to a significant gain in system performance; especially, in the range of $\gamma_T$ values that are comparable to the

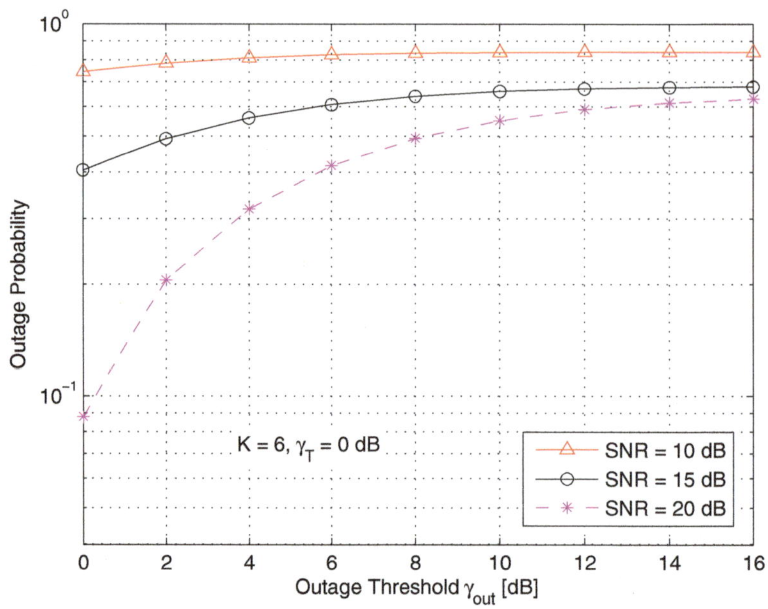


**Fig. 5** Outage probability vs outage threshold of cognitive AF relay network with MUSwiD user selection scheme for different values of $\mathcal{I}_p/N_0$ and $\Omega_{s,p} = 60$, and $\Omega_{s,r} = 0.8$, $\Omega_{r,p} = 0.9$, $\Omega_{r,k} = 0.7$ for $k = 1, \ldots, 6$

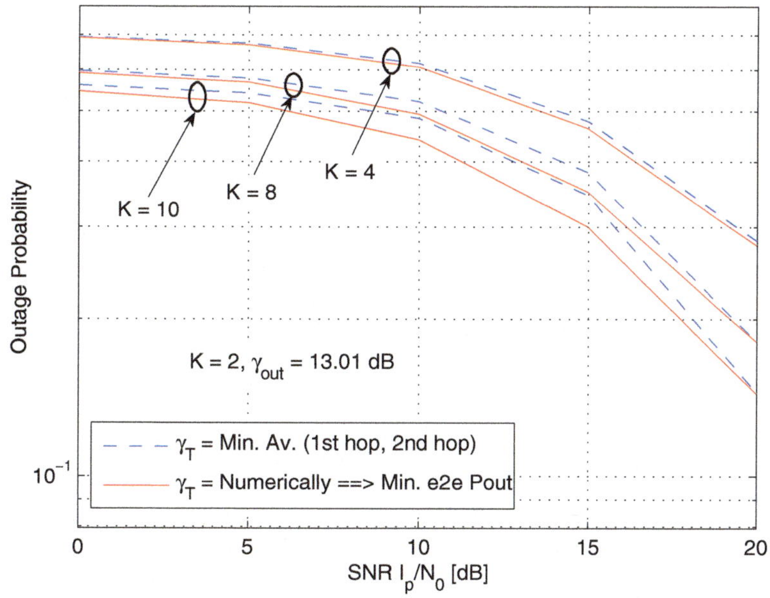


**Fig. 6** Outage probability vs SNR of cognitive AF relay network with MUSwiD user selection scheme for different values of $\gamma_T$ and $\Omega_{s,p} = 7$, and $\Omega_{s,r} = 0.91$, $\Omega_{r,p} = 0.73$, $\Omega_{r,k} = 0.82$ for $k = 1, 2$

average value of $\gamma_k^{up}$. On the other hand, as $\gamma_T$ becomes much smaller or much larger than the average value of $\gamma_k^{up}$, the improvement in performance decreases, as all curves asymptotically converge to the case of two users. This is due to the fact that, if the average value of $\gamma_k^{up}$ is very small compared to $\gamma_T$, all users will be unacceptable most of the time. Whereas, if it is very high compared to $\gamma_T$, all users will be acceptable and one user will be scheduled most of the time. Thus, in both cases, having more users will not lead to any gain in the behavior.

Figure 8 shows the error probability performance versus number of users $K$ for the studied system with the MUSwiD user selection scheme for different values of SNR $\mathcal{I}_p/N_0$. It can be seen from this figure that the

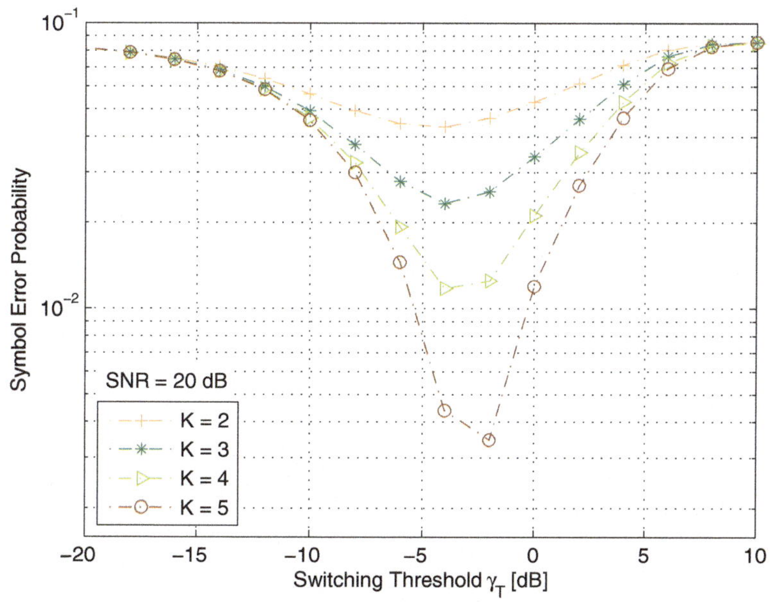


**Fig. 7** Symbol error probability vs switching threshold of cognitive AF relay network with MUSwiD user selection scheme for different values of $K$ and $\Omega_{s,p} = 20$, and $\Omega_{s,r} = 0.1$, $\Omega_{r,p} = 0.3$, $\Omega_{r,k} = 0.2$ for $k = 1, \ldots, 5$

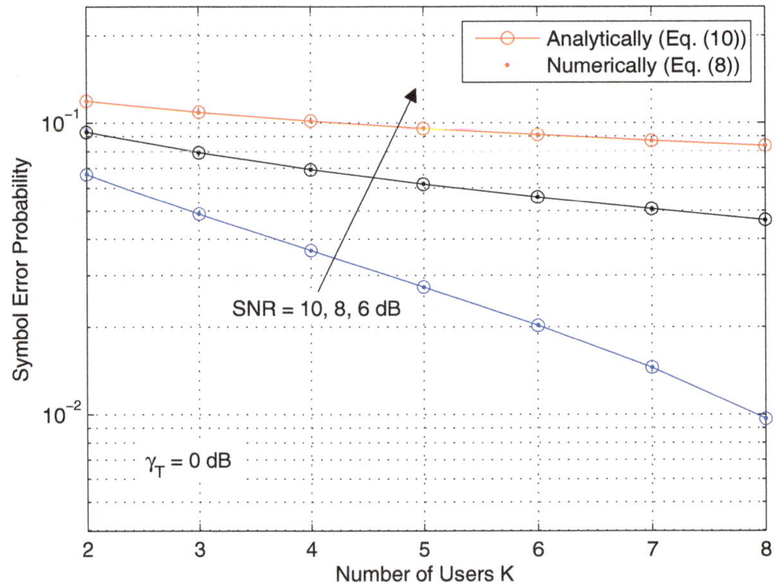


**Fig. 8** Symbol error probability vs number of users of cognitive AF relay network with MUSwiD user selection scheme for different values of $\mathcal{I}_p/N_0$ and $\Omega_{s,p} = 20$, and $\Omega_{s,r} = 0.8$, $\Omega_{r,p} = 0.6$, $\Omega_{r,k} = 0.8$ for $k = 1, \ldots, 8$

considered relay network still achieves performance gain and the error probability decreases when $K$ increases, but the slope of the curves depends on the SNR values. Also, this figure validates the derived analytical expression of the average symbol error probability where a perfect fitting can be seen between the analytical result in (10) and its numerical solution that is achieved using (8).

Figure 9 portrays the ergodic channel capacity versus SNR for the studied system with the MUSwiD user selection scheme for different values of $\Omega_{s,r}$. Due to complexity of the derived expression of channel capacity, it is numerically calculated and plotted in this figure. As expected, the ergodic channel capacity enhances as the quality of first hop channel enhances.

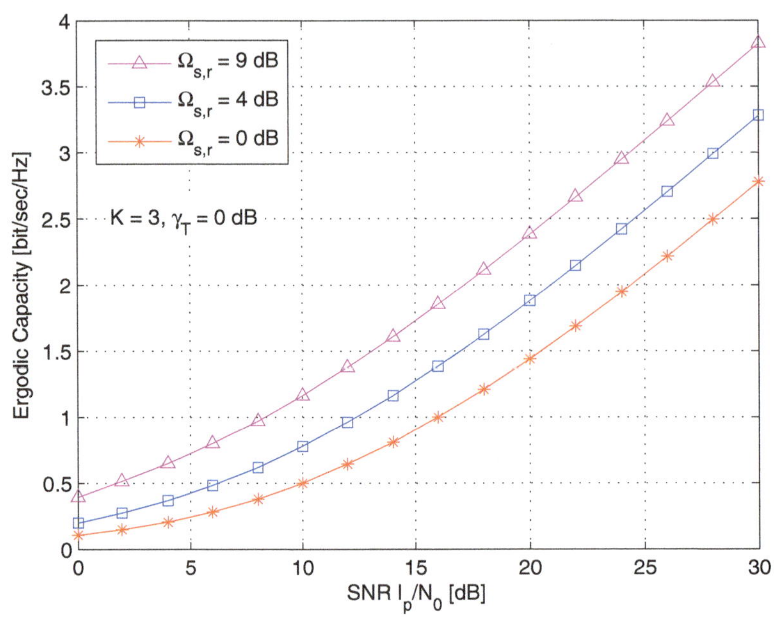


**Fig. 9** Channel capacity vs SNR of cognitive AF relay network with MUSwiD user selection scheme for different values of $\Omega_{s,r}$ and $\Omega_{s,p} = 10$, and $\Omega_{s,r} = 1$, $\Omega_{r,p} = 2$, $\Omega_{r,k} = 3$ for $k = 1, \ldots, 3$

Figure 10 illustrates the validity of the PDF $f_{\gamma_{up}}(\gamma)$ used in calculating the channel capacity. It is clear from this figure that the derived PDF of $\gamma_{up}$ is a valid and a non-negative function. Also, we can see that the area under the curves of all considered cases in this figure is unity.

Figure 11 illustrates the average number of channel estimations versus switching threshold $\gamma_T$ for the MUSwiD and MUSwiDps user selection schemes in comparison with the opportunistic or best-user selection scheme for the case of four users. The average number of channel estimations for the MUSwiD and MUSwiDps schemes are provided in [38]. In this comparison, we assume that all channels are available for the relay or the central unit in order to conduct the selection among users. We can see from this figure that as the channels of all users are required for its operation, the opportunistic or best-user selection scheme is always of need for four channel estimations. On the other hand, we can see from this figure that the conventional MUSwiD selection scheme needs to estimate at most four channels because when the first three users are found unacceptable, the last checked user will be scheduled by the central unit regardless of its quality. Therefore, the MUSwiD scheme requires less channel estimations than the MUSwiDps scheme where in order to find the best user among all checked users, the channels of all users need to be estimated in the MUSwiDps scheme. Also, we can notice from this figure that as $\gamma_T$ increases, the average number of channel estimations of users increases since it is more difficult to find a user with an acceptable quality.

## 6 Conclusions

In this paper, we evaluated the performance of a cognitive CSI-assisted AF relay network with the MUSwiD user selection scheme. Closed-form approximations for the outage probability and average symbol error probability were derived for the i.n.i.d. and i.i.d. cases of users channels. The ergodic channel capacity was also numerically evaluated. Furthermore, the system outage and error probability performances were evaluated at high SNR values where simple expressions for the outage probability, symbol error probability, diversity order, and coding gain were derived and analyzed for the MUSwiD and MUSwiDps user selection schemes. Monte-Carlo simulations proved the accuracy of the achieved analytical and asymptotic results. Findings illustrated that the proposed MUSwiD user selection schemes are efficient in the range of low SNR values, which makes them attractive options for practical implementation in emerging mobile broadband communication systems. In contrast, these selection schemes are inefficient in the range of high SNR values where the multiuser diversity gain is noticeably degraded when they are implemented. Furthermore, results illustrated that the gain achieved in performance due to having more users happens in the range of SNR values that are comparable to the switching threshold.

## Endnotes

[1] Secondary users can know the channel information of the primary user by either a direct reception of pilot

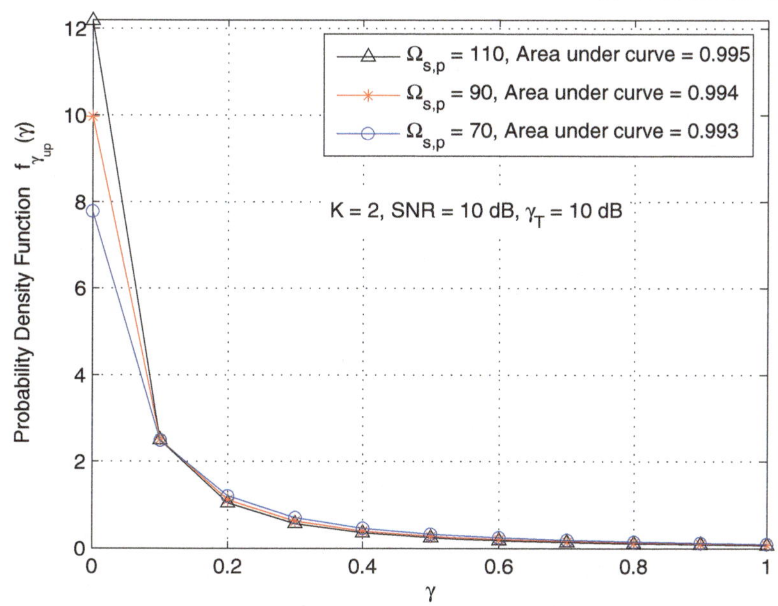


**Fig. 10** The PDF $f_{\gamma_{up}}(\gamma)$ vs $\gamma$ of cognitive AF relay network with MUSwiD user selection scheme for different values of $\Omega_{s,p}$ and $\Omega_{s,r} = 0.91$, $\Omega_{r,p} = 0.73$, $\Omega_{r,k} = 0.82$ for $k = 1, 2$

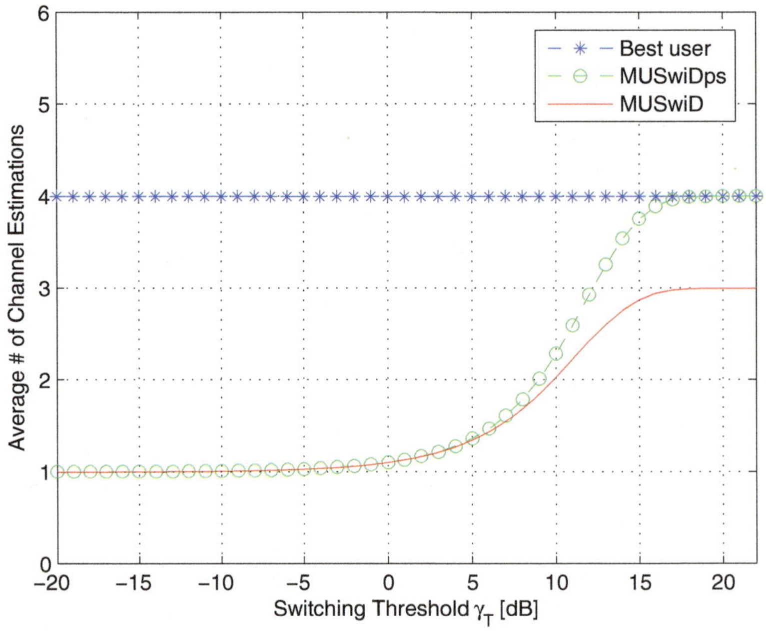


**Fig. 11** Average number of channel estimations of the MUSwiD and MUSwiDps user selection schemes in comparison with the best-user or opportunistic user selection scheme with $K = 4$ and an average power/user path = 10 dB

signals from a primary user [39], or by exchange of channel information between primary and secondary users through a band manager [40].

[2] This assumption is possible if the primary transmitter is located far from the secondary user [41], or the interference is represented in terms of noise when the primary transmitter's signal is generated by random Gaussian codebooks [42].

[3] In this paper, the switching threshold is numerically calculated to optimize the e2e outage probability. Also, a simple method is proposed in Section 4 to obtain approximate but accurate values of the optimum switching threshold.

[4] The time duration of the feedback channel is not long, and hence, the MUSwiD scheduling scheme does not cause additional delay to the scheduling process [35].

## Appendix

### Proof of Lemma 1

In this Appendix, we derive the outage probability for the studied system when the secondary users have i.n.i.d. channels. In order to evaluate the outage probability, the CDF of $\gamma_{\text{up}}$ is required to be obtained first. Herein, we first apply the conditional statistics on the fading channel from S to P. The CDF of $\gamma_k^{\text{up}}$ conditioned on $h_{\text{s,p}}$ can be written as

$$F_{\gamma_k^{\text{up}}}\left(\gamma|h_{\text{s,p}}\right) = 1-\left(1-F_{\gamma_1}\left(\gamma|h_{\text{s,p}}\right)\right)\left(1-F_{\gamma_{2_k}}\left(\gamma|h_{\text{s,p}}\right)\right). \tag{23}$$

It is easy to see that

$$F_{\gamma_1}\left(\gamma|h_{\text{s,p}}\right) = 1-\exp\left(-\lambda_1\gamma|h_{\text{s,p}}|^2\right), \tag{24}$$

$$\begin{aligned} F_{\gamma_{2_k}}\left(\gamma|h_{\text{s,p}}\right) &= \int_0^\infty F_{|h_{\text{r},k}|^2}\left(\frac{N_0\gamma}{\mathcal{I}_{\text{p}}}x\right)f_{|h_{\text{r,p}}|^2}(x)dx \\ &= 1-\left(1+\lambda_{2_k}\gamma\right)^{-1}, \end{aligned} \tag{25}$$

where $\lambda_1 = 1\Big/\left(\Omega_{\text{s,r}}\frac{\mathcal{I}_{\text{p}}}{N_0}\right)$ and $\lambda_{2_k} = \Omega_{\text{r,p}}\Big/\left(\Omega_{\text{r},k}\frac{\mathcal{I}_{\text{p}}}{N_0}\right)$.

Upon substituting (24) and (25) in (23), we get

$$F_{\gamma_k^{\text{up}}}\left(\gamma|h_{\text{s,p}}\right) = 1-\frac{\exp\left(-\lambda_1\gamma|h_{\text{s,p}}|^2\right)}{\left(1+\lambda_{2_k}\gamma\right)}. \tag{26}$$

Base on the way the MUSwiD selection scheme works, the conditional CDF at the output of this scheme can be written as [36]

$$F^{\text{MUSwiD}}_{\gamma_{\text{up}}}\left(\gamma|h_{\text{s,p}}\right) = \begin{cases} \sum_{i=0}^{K-1}\pi_i F_{\gamma_i^{\text{up}}}(\gamma|h_{\text{s,p}})\prod_{\substack{k=0\\k\neq i}}^{K-1}F_{\gamma_k^{\text{up}}}(\gamma_{\text{T}}|h_{\text{s,p}}), & \gamma<\gamma_{\text{T}}; \\ \sum_{l=0}^{K-1}\pi_l\Big(\prod_{q=1}^{K}F_{\gamma_q^{\text{up}}}(\gamma_{\text{T}}|h_{\text{s,p}}) & \\ +\sum_{w=0}^{K-1}\pi_{((l-w))_K}\left[F_{\gamma_l^{\text{up}}}(\gamma|h_{\text{s,p}})-F_{\gamma_l^{\text{up}}}(\gamma_{\text{T}}|h_{\text{s,p}})\right] & \\ \times\prod_{p=0}^{w-1}F_{\gamma^{\text{up}}_{((l-w+p))_K}}(\gamma_{\text{T}}|h_{\text{s,p}})\Big), & \gamma\geq\gamma_{\text{T}}, \end{cases} \tag{27}$$

where $K$ is the number of SU destinations and $\gamma_T$ is a predetermined switching threshold, $\pi_i, i = 0, \ldots, K-1$ are the stationary distribution of a $K$-state Markov chain and it is the probability that the $i^{\text{th}}$ destination or user is chosen, and $((l-w))_K$ denotes $l-w$ modulo $K$. It is given by

$$\pi_i = \left[ \sum_{j=0}^{K-1} \left( \frac{F_{\gamma^{\text{up}}_{K-1}}(\gamma_T|h_{\text{s,p}}) \left(1 - F_{\gamma_j^{\text{up}}}(\gamma_T|h_{\text{s,p}})\right)}{F_{\gamma_j^{\text{up}}}(\gamma_T|h_{\text{s,p}}) \left(1 - F_{\gamma^{\text{up}}_{K-1}}(\gamma_T|h_{\text{s,p}})\right)} \right) \right]^{-1} \times \frac{F_{\gamma^{\text{up}}_{K-1}}(\gamma_T|h_{\text{s,p}}) \left(1 - F_{\gamma_i^{\text{up}}}(\gamma_T|h_{\text{s,p}})\right)}{F_{\gamma_i^{\text{up}}}(\gamma_T|h_{\text{s,p}}) \left(1 - F_{\gamma^{\text{up}}_{K-1}}(\gamma_T|h_{\text{s,p}})\right)}. \tag{28}$$

Upon substituting (26) in (27), we get

$$F_{\gamma_{\text{up}}}(\gamma|h_{\text{s,p}}) = \begin{cases} \sum_{i=0}^{K-1} \pi_i \left(1 - \frac{\exp(-\lambda_1 \gamma |h_{\text{s,p}}|^2)}{(1+\lambda_{2_i}\gamma)}\right) \underbrace{\prod_{\substack{k=0 \\ k\neq i}}^{K-1} \left(1 - \frac{\exp(-\lambda_1 \gamma_T |h_{\text{s,p}}|^2)}{(1+\lambda_{2_k}\gamma_T)}\right)}_{\mathcal{P}_1}, & \gamma < \gamma_T; \\ \sum_{l=0}^{K-1} \pi_l \left\{ \underbrace{\prod_{q=1}^{K} \left(1 - \frac{\exp(-\lambda_1 \gamma_T |h_{\text{s,p}}|^2)}{(1+\lambda_{2_q}\gamma_T)}\right)}_{\mathcal{P}_2} + \sum_{w=0}^{K-1} \pi_{((l-w))_K} \left[ \left(1 - \frac{\exp(-\lambda_1 \gamma |h_{\text{s,p}}|^2)}{(1+\lambda_{2_l}\gamma)}\right) - \left(1 - \frac{\exp(-\lambda_1 \gamma_T |h_{\text{s,p}}|^2)}{(1+\lambda_{2_l}\gamma_T)}\right) \right] \times \underbrace{\prod_{p=0}^{w-1} \left(1 - \frac{\exp(-\lambda_1 \gamma_T |h_{\text{s,p}}|^2)}{\left(1+\lambda_{2_{((l-w+p))_K}}\gamma_T\right)}\right)}_{\mathcal{P}_3} \right\}, & \gamma \geq \gamma_T. \end{cases} \tag{29}$$

With the help of the product identities in [43] and [20], the terms $\mathcal{P}_1$, $\mathcal{P}_2$, and $\mathcal{P}_3$ in (29) can be simplified as follows:

$$\mathcal{P}_1 = 1 + \sum_{k=0}^{K-1} (-1)^{k+1} \sum_{\substack{n_0<\ldots<n_k \\ n_{(.)}\neq i}}^{K-1} \prod_{t=0}^{k} \left(1 + \lambda_{2_{n_t}} \gamma_T\right)^{-1} \times \exp\left(-\lambda_1 \gamma_T |h_{\text{s,p}}|^2\right), \tag{30}$$

where $\sum_{\substack{n_0<\ldots<n_k \\ n_{(.)}\neq i}}^{K-1}$ is a short-hand notation for $\sum_{\substack{n_0=0 \\ n_0\neq i}}^{K-k-1} \sum_{\substack{n_1=n_0+1 \\ n_1\neq i}}^{K-k} \cdots \sum_{\substack{n_k=n_{k-1}+1 \\ n_k\neq i}}^{K-1}$.

$$\mathcal{P}_2 = \sum_{q=0}^{K} \frac{(-1)^q}{q!} \sum_{m_1,\ldots,m_q}^{K} \prod_{z=1}^{q} \left(1 + \lambda_{2_{m_z}} \gamma_T\right)^{-1} \times \exp\left(-\lambda_1 \gamma_T |h_{\text{s,p}}|^2\right), \tag{31}$$

where $\sum_{m_1,\ldots,m_q}^{K}$ is a short-hand notation for $\sum_{\substack{m_1=\ldots=m_q=1 \\ m_1\neq\ldots\neq m_q}}^{K-k-1}$.

$$\mathcal{P}_3 = 1 + \sum_{p=0}^{w-1} (-1)^{p+1} \sum_{v_0<\ldots<v_p}^{w-1} \prod_{g=0}^{p} \left(1 + \lambda_{2_{((l-w+v_g))_K}} \gamma_T\right)^{-1} \times \exp\left(-\lambda_1 \gamma_T |h_{\text{s,p}}|^2\right), \tag{32}$$

where $\sum_{v_0<\ldots<v_p}^{w-1}$ is a short-hand notation for $\sum_{v_0=0}^{w-p-1} \sum_{v_1=v_0+1}^{w-p} \cdots \sum_{v_p=v_{p-1}+1}^{w-1}$.

Up to now, the outage probability can be expressed as

$$P_{\text{out}} \simeq \int_0^{\infty} F_{\gamma_{\text{up}}}(\gamma|h_{\text{s,p}}) f_{|h_{\text{s,p}}|^2}(y)dy. \tag{33}$$

Upon substituting (30), (31), and (32) in (29) and using (33), the outage probability can be evaluated in a closed-form approximation as in (2).


**Acknowledgements**

This work is supported by Deanship of Scientific Research (DSR) at King Fahd University of Petroleum & Minerals (KFUPM) through project #: RG1408-1,2.


**Competing interests**

The authors' declare that they have no competing interests.

**References**


1. S Haykin, Cognitive radio: Brain-empowered wireless communications. IEEE J. Sel. Areas Commun. **23**(2), 201–220 (2005)
2. A Goldsmith, S Jafar, I Maric, S Srinivasa, Breaking spectrum gridlock with cognitive radios: an information theoretic perspective. Proc. IEEE. **97**(5), 894–914 (2009)
3. JN Laneman, DNC Tse, GW Wornell, Cooperative diversity in wireless networks: efficient protocals and outage behavior. IEEE Trans. Inf. Theory. **50**(12), 3062–3080 (2004)
4. VNQ Bao, TQ Duong, C Tellambura, On the performance of cognitive underlay multihop networks with imperfect channel state information. IEEE Trans. Commun. **61**(12), 4864–4873 (2013)
5. Y Guo, G Kang, N Zhang, W Zhou, P Zhang, Outage performance of relay-assisted cognitive-radio system under spectrum-sharing constraints. Elect. Lett. **46**(2), 182–184 (2010)
6. J Lee, H Wang, JG Andrews, D Hong, Outage probability of cognitive relay networks with interference constraints. IEEE Trans. Wirel. Commun. **10**(2), 390–395 (2011)
7. L Luo, P Zhang, G Zhang, J Qin, Outage performance for cognitive relay networks with underlay spectrum sharing. IEEE Commun. Lett. **15**(7), 710–712 (2011)
8. Z Yan, X Zhang, W Wang, Exact outage performance of cognitive relay networks with maximum transmit power limits. IEEE Commun. Lett. **15**(12), 1317–1319 (2011)
9. Z El-Moutaouakkil, K Tourki, KA Qaraqe, S Saoudi, in *Proc. IEEE Veh. Technol. Conf. (VTC'12-Fall)*. Exact outage probability analysis for relay-aided underlay cognitive communications (IEEE, Canada, 2012), pp. 1–5
10. JP Hong, B Hong, TW Ban, W Choi, On the cooperative diversity gain in underlay cognitive radio systems. IEEE Trans. Commun. **60**(1), 209–219 (2012)

11. VNQ Bao, DT Tran, in *Proc. The 2012 IEEE Int'l Conf. on Adv. Tech. for Commun*. Performance analysis of cognitive underlay DF relay protocol with $K^{th}$ best partial relay selection (IEEE, Vietnam, 2012), pp. 130–135
12. X Zhang, Z Yan, Y Gao, W Wang, On the study of outage performance for cognitive relay networks (CRN) with the Nth best-relay selection in Rayleigh fading channels. IEEE Wireless Commun. Lett. **2**(1), 110–113 (2013)
13. H Chamkhia, MO Hasna, R Hamila, SI Hussain, in *Proc. IEEE Int'l Symp. on Personal, Indoor & Mobile Radio Commun. (PIMRC'12)*. Performance analysis of relay selection schemes in underlay cognitive networks with decode and forward relaying (IEEE, Australia, 2012), pp. 1552–1558
14. D Li, Outage probability of cognitive radio networks with relay selection. IET Commun. **5**(18), 2730–2735 (2011)
15. M Xia, S Aïssa, Cooperative AF relaying in spectrum-sharing systems: performance analysis under average interference power constraints and Nakagami-$m$ fading. IEEE Trans. Commun. **60**(6), 170–176 (2012)
16. SI Hussain, M-S Alouini, K Qaraqe, M Hasna, in *Proc. IEEE Int'l Conf. on Commun. (ICC'12)*. Reactive relay selection in underlay cognitive networks with fixed gain relays (IEEE, Canada, 2012), pp. 1784–1788
17. KB Fredj, S Aïssa, Performance of amplify-and-forward systems with partial relay selection under spectrum-sharing constraints. IEEE Trans. Wirel. Commun. **11**(2), 500–504 (2012)
18. J Chen, Z Li, J Si, H Huang, R Gao, Outage probability analysis of partial AF relay selection in spectrum-sharing scenario with multiple primary users. Elect. Lett. **48**(19), 1211–1212 (2012)
19. H Ding, J Ge, DB da Costa, Z Jiang, Asymptotic analysis of cooperative diversity systems with relay selection in a spectrum-sharing scenario. IEEE Trans. Veh. Technol. **60**(2), 457–472 (2011)
20. VNQ Bao, TQ Duong, DB da Costa, GC Alexandropoulos, A Nallanathan, Cognitive amplify-and-forward relaying with best relay selection in non-identical Rayleigh fading. IEEE Commun. Lett. **17**(3), 475–478 (2013)
21. Z Yan, X Zhang, W Wang, in *Proc. IEEE Wire. Commun. and Net. Conf. (WCNC'11)*. Outage performance of relay assisted hybrid overlay/underlay cognitive radio systems (IEEE, Mexico, 2011), pp. 1920–1925
22. JB Kim, D Kim, Outage probability and achievable diversity order of opportunistic relaying in cognitive secondary radio networks. IEEE Trans. Commun. **60**(9), 2456–2466 (2012)
23. DB da Costa, M Elkashlan, PL Yeoh, N Yang, MD Yacoub, in *Proc. 23rd IEEE Int'l Symp. on Personal Indoor & Mobile Radio Commun. (PIMRC'12)*. Dual-hop cooperative spectrum sharing systems with multi-primary users and multi-secondary destinations over Nakagami-$m$ fading (IEEE, Australia, 2012), pp. 1577–1581
24. C Zhai, W Zhang, Adaptive spectrum leasing with secondary user scheduling in cognitive radio networks. IEEE Trans. Wirel. Commun. **12**(7), 3388–3398 (2013)
25. FRV Guimarães, DB da Costa, TA Tsiftsis, CC Cavalcante, GK Karagiannidis, Multi-user and multi-relay cognitive radio networks under spectrum sharing constraints. IEEE Trans. Veh. Technol. **63**(1), 433–439 (2014)
26. EEB Olivo, DPM Osorio, DB da Costa, JCSS Filho, in *Proc. 24th IEEE Int'l Symp. on Personal Indoor & Mobile Radio Commun. (PIMRC'13)*. Outage analysis of a spectrally efficient scheme for multiuser cognitive relaying networks with spectrum sharing constraints (IEEE, UK, 2013), pp. 997–1001
27. Y Huang, F Al-Qahtani, C Zhong, Q Wu, in *Proc. 2013 IEEE Int'l Conf. on Wireless Communications and Signal Processing*. Outage analysis of spectrum sharing relay systems with multi-secondary destinations in the presence of primary user's interference (IEEE, China, 2013), pp. 1–6
28. AH Abd El-Malek, F Al-Qahtani, SA Zummo, H Alnuweiri, in *Proc. IEEE Wire. Commun. and Net. Conf. (WCNC'15)*. MIMO multiuser cognitive relay network in spectrum sharing environment with antenna correlation over Rayleigh fading channels (IEEE, New Orleans, 2015), pp. 504–509
29. J Yang, TQ Duong, M Elkashlan, X Lei, X Gao, in *Proc. IEEE Int'l Conf. on Commun. (ICC'15)*. Multiuser scheduling for cognitive MIMO with channel estimation errors and feedback delay (IEEE, London, 2015), pp. 7737–7742
30. B Holter, MS Alouini, GE Oien, HC Yang, in *Proc. IEEE Veh. Technol. Conf.* Multiuser switched diversity transmission, vol. 3 (IEEE, Los Angeles, 2004), pp. 2038–2043
31. H Nam, MS Alouini, Multiuser switched diversity scheduling systems with per-user threshold. IEEE Trans. Commun. **58**(5), 1321–1326 (2010)
32. H-C Yang, M-S Alouini, Performance analysis of multibranch switched diversity systems. IEEE Trans. Commun. **51**(5), 782–794 (2003)
33. L Yang, M-S Alouini, Performance analysis of multiuser selection diversity. IEEE Trans. Veh. Technol. **55**(6), 1848–1861 (2006)
34. Y Zhang, J Zheng, H-H Chen, *Cognitive radio networks: architectures, protocols, and standards*, 1st edn. (CRC Press, Taylor & Francis Group, 2010)
35. M Shaqfeh, H Alnuweiri, M-S Alouini, Multiuser switched diversity scheduling schemes. IEEE Trans. Commun. **60**(9), 2499–2510 (2012)
36. MK Simon, M-S Alouini, *Digital communication over fading channels*, 2nd edn. (Wiley, 2005)
37. IS Gradshteyn, IM Ryzhik, *Tables of integrals, series and products*, 6th edn. (Acadamic Press, San Diago, 2000)
38. H-C Yang, M-S Alouini, *Order statistics in wireless communications: diversity, adaptation, and scheduling in MIMO and OFDM Systems*, 1st edn. (Cambridge University Press, 2011)
39. T Ban, W Choi, B Jung, D Sung, Multi-user diversity in a spectrum sharing system. IEEE Trans. Wireless Commun. **8**(1), 102–106 (2009)
40. A Ghasemi, E Sousa, Fundamental limits of spectrum-sharing in fading environments. IEEE Trans. Wireless Commun. **6**(2), 649–658 (2007)
41. X Kang, Y Liang, A Nallanathan, in *Proc. IEEE Veh. Technol. Conf.* Optimal power allocation for fading channels in cognitive radio networks: delay-limited capacity and outage capacity (IEEE, Singapore, 2008), pp. 1544–1548
42. R Etkin, A Parekh, D Tse, Spectrum sharing for unlicensed bands. IEEE J. Sel. Areas Commun. **25**(3), 517–528 (2007)
43. PS Bithas, GK Karagiannidis, NC Sagias, PT Mathiopoulos, SA Kotsopoulos, GE Corazza, Performance analysis of a class of GSC receivers over nonidentical Weibull fading channels. IEEE Trans. Veh. Technol. **54**(6), 1963–1970 (2005)

# 11

# Very-low-SNR cognitive receiver based on wavelet preprocessed signal patterns and neural network

Husam Y. Alzaq* and B. Berk Ustundag

**Abstract**

A pattern-based cognitive communication system (PBCCS) that optimizes non-periodic RF waveforms for security applications is proposed. PBCCS is a cross-layer approach that merges the channel encoding and modulation. The transmitter encodes sequences of bits into continuous signal patterns by selecting the proper symbol glossaries. The cognitive receiver preprocesses the received signal by extracting a limited set of wavelet features. The extracted features are fed into an artificial neural network (ANN) to recover the digital data carried by the distorted symbol. The PBCCS system offers a flexible management for robustness against a high noise level and increases the spectral efficiency. In this study, the spectral efficiency and robustness of a PBCCS scheme for an additive white Gaussian noise (AWGN) channel is investigated. The results show that at an SNR of −5 dB, a 3-bit glossary achieves a bit error rate (BER) of $10^{-5}$. Also, the link spectral efficiency (LSE) of the proposed system is 2.61 bps/Hz.

**Keywords:** Cognitive radio (CR), Pattern-based cognitive communication system (PBCCS), Artificial neural network, Wavelet decomposition, Digital signal processing (DSP), Very-low-SNR

## 1 Introduction

The efficiency of bandwidth utilization takes an important role in spectrum management [1, 2]. Due to fixed spectrum assignment policies and its inadequate to meet an unexpected increase in the number of higher-data-rate devices, the spectrum is inefficiently used. Cognitive radio (CR) [3–6] was proposed as a promising solution to alleviate the spectrum scarcity problem through dynamic management of the available spectrum. The pioneer work of Mitola et al. [3] led to an efficient utilization of the spectral bandwidth by allowing the secondary user (SU), who is not serviced, to detect and access the primary network spectrum gaps. CR allows detection of the state of the spectrum to adjust its own system parameters (transmission power, frequency band, throughput and modulation scheme) in real time [7]. The result is that the utilization of the spectral bandwidth is performed with the software flexibility in an adaptive manner with respect to the system parameters.

However, efficient spectral bandwidth usage under the influence of higher noise is not the major consideration of CR. Claude Shannon [8] showed that the SNR is a leading factor that influences the link spectral efficiency (LSE), $\eta = C/B$, (in bps/Hz). SNR also limits the channel capacity. Therefore, the utilization of spectral bandwidth and the robustness to high SNR level are the keys to maximize the channel capacity.

Thus, a pattern-based cognitive communication system (PBCCS) was introduced to optimize the overall spectral efficiency with respect to SNR [9, 10]. It is inspired by the recognition capability of humans to concentrate on a single conversation irrespective of the surrounding loudness. If human ears hear sounds from different sources, the brain chooses to pay attention to a particular voice amongst a whole range of sound streams in an environment. Similar to human cognitive capabilities, the communication system in PBCCS selectively recognizes and recovers the communication signal(s) a into known symbol(s), even within the same frequency range.

*Correspondence: alzaq@itu.edu.tr
Department of Computer Engineering, Faculty of Computer Engineering, Istanbul Technical University, Maslak, Ayazaga, 34469 Istanbul, Turkey

### 1.1 Related work

Conventional cognitive radio is equipped with various techniques for making wireless systems more flexible and robust to channel variation. Mitola, in his dissertation [11], stated that, although many aspects of wireless networks are artificial, they may still be enhanced by machine learning (ML). Recently, machine learning algorithms have become one of the key enabling features of cognitive radio in many applications. In previous literature, many techniques and algorithms have been applied to the cognitive radio engine [12, 13], such as the artificial neural network (ANN), hidden Markov model (HMM), fuzzy logic control, meta-heuristic algorithms (evolutionary/genetic algorithm) and rule-based systems [14–16].

On the other hand, the PBCCS structure is an extension for CR, which is a cross-layer architecture. The control unit of PBCCS is located in the data-link layer and communicates with the external glossary space that manages the transmission process. Table 1 summarizes the similarities and differences between PBCCS and cognitive radio.

Generally, ANN has been adapted in the cognitive radio engine for various modulation classification, known as automatic modulation recognition (AMR) or automatic modulation classification (AMC). For instance, ANN was implemented for channel sensing [17–19] and spectrum prediction [17, 20], etc. To enhance ANN classification accuracy, ANN is usually combined with the extracted features from the received signal, which allows the engine to have the capability to identify the modulation scheme at low SNR levels. Cyclic spectral analysis [17], wavelet cyclic features [21], temporal feature-based modulation [22, 23], carrier frequency and baud rate [24], and continuous wavelet transform (CWT) [25] are some examples of these features. Dahap et al. [26] proposed a digital recognition algorithm that uses six features extracted from instantaneous information and signal spectrum to discriminate between different modulated signals. Table 2 shows a brief summary of these approaches. However, most of the aforementioned approaches have been used to classify low-order modulation technique, such as 2-ASK and 2-PSK. In addition, a prior signal information such as the carrier frequency is required. Additionally, if these approaches classify high-order modulation schemes, they construct a large ANN. The PBCCS is a kind of AMC that has not only the ability to classify high-order modulation, but also can encode the received analog signal at very low SNR.

In this work, we choose to use the ANN model at the PBCCS receiver owing to its powerful capabilities. ANN can predict the correct class of the received signal even if the input signals have not been seen before, which allow the model to learn from training dataset and generalize the model to any received signal. Moreover, ANN is a nonlinear model and hence can predict the nonlinear received signal better than the linear model. Finally, the ANN parallel processing and the appropriate simple structure are two important properties for realizing ANN on hardware.

Furthermore, we have implemented a cognitive radio solution, which offers flexibility between the available spectrum and SNR. This solution has the capability to balance between LSE and the overall channel capacity under a very low SNR. It constructs optimal communication symbols, which compensate for the difference in data rates under various noise levels. In addition, the PBCCS system integrates the modulator and channel encoder through a cross-layer approach. The binary data is encoded into the appropriate waveform according to the selected glossary. Each binary word is assigned to the artificially constructed patterns. The transmitter selects the appropriate set of patterns that maximize LSE.

### 1.2 Contribution

In our previous work [27], we have experimentally investigated the performance of PBCCS in an additive white Gaussian noise (AWGN) channel by employing 2-level Daubechies-2 wavelet (DB2) as a discrete wavelet transformation (DWT) to preprocess the received signal. With

**Table 1** Comparison between pattern based cognitive communication system and cognitive radio

| | PBCCS | Cognitive radio |
|---|---|---|
| Objective | To improve data transmission performance under bandwidth limitations by maximizing the LSE value with respect to SNR level. | To efficiently manage the spectrum by using the unoccupied spectrum band when it is not used by the licensed user. |
| Transmitter side | It selects one signal pattern from the glossary space in adaptive manner. | It uses the frequency bandwidth in an adaptive manner with respect to channel availability. |
| Receiver side | The distorted signal is recovered by a trained ANN. Recovering the information without separate demodulation, decoding and error recovery operations. | Receiver must have a wide-band front-end to detect spectrum holes. It should take into account the adaptive modulation feature, since some receivers employs AMC. Many ML algorithms can be used. |
| Modulation technique | AFPSK[a] | QAM, QPSK, BPSK, . . . etc. |
| Cognition manager | Glossary selector (GS) | Cognitive engine(CE) |

[a] Amplitude frequency phase shift keying

**Table 2** ANN within cognitive radio

| Reference | Brief summary |
|---|---|
| [21] | Wavelet cyclic feature has been proposed to reduce the complexity of calculating classical cyclic spectrum and a FFNN[a] has been used to classify the received signals into BPSK[b], QPSK[c], MSK,[d] and 2-FSK[e]. Cons: limited to low-order modulation schemes. |
| [22] | Based on the instantaneous temporal features (the maximum value of the spectral power density, the standard deviation of the direct and absolute instantaneous phase values and the standard deviation of the normalized instantaneous amplitude), the authors have proposed a FFNN and probabilistic ANN to classify the received signals into 2- and 4-ASK[f], BPSK, QPSK, 2-, and 4-FSK, 8-PSK[g], 16-QAM[h]. Cons: Large ANN architecture and requires prior information on some specific parameters to guarantee the highest accuracy and reliable recognition. |
| [23] | Based on the instantaneous temporal features (the maximum value of the spectral power density, the standard deviation of the direct and absolute instantaneous phase and the standard deviation of the normalized instantaneous amplitude), the authors have proposed a simple FFNN to classify the received signals into five classes, namely; 2- and 4-ASK, BPSK, and QPSK. Cons: low-order modulations were considered. This approach requires prior information on some specific parameters to guarantee the highest accuracy and reliable recognition. |
| [25] | Based on the extracted CWT instantaneous features (the mean, variance and central moments values), the authors have proposed an ANN to classify the modulation scheme into k-ASK, k-PSK, k-FSK, k-QAM, OOK,[i] and MSK. |
| [26] | Based on the extracted instantaneous temporal features, the authors proposed a rule-based approach to discriminate between 15 modulation schemes (AM[j], FM[k], DSB[l], LSB[m], USB[n], VSB[o], combined AM–FM, CW, Noise, 2-, and 4-ASK, 2- and 4-PSK, 2-, and 4-FSK). Cons: due to limited number of features and signal sensitivity, the approach was unable to classify the same modulation schemes of different order. |
| [37] | FFNN, radial basis function ANN and multi-class support vector machine (SVM) have been suggested to classify the modulation technique of the received signal into 2- and 4-FSK, 4-ASK, 8-ASK, 2-PSK, 4-PSK, 8-PSK, V32, 8-, 16-, 32-, and 64-QAM. Cons: it requires prior information on specific parameters to guarantee the highest accuracy and reliable recognition. |
| [38] | An expert discrete wavelet adaptive network based on fuzzy inference system has been proposed for classifying the digital modulated signals into 8-ASK, 8-FSK, 8-PSK, and 8-QAM. Cons: very large ANN structure, with four hidden layers. |
| [39] | A system that is only based on wavelet transform has been developed, where a comparison between signals and templates in wavelet domain has been adapted to classify the received signals into 2-ASK, 2-FSK, and BPSK. Cons: binary digital modulation schemes were considered. It also requires prior signal information, such as, carrier frequency and symbol duration. |
| [40] | A system that is based solely on DWT and signals statistics was used to classify the modulated received signals into 16-QAM, QPSK and BPSK. Cons: degradation of performance at SNR appeared when the ANN was trained on signals with lower SNR. |

[a]FFNN: Feed-forward neural network
[b] BPSK: Binary phase shift keying
[c]QPSK: Quadrature phase shift keying
[d]MSK: Minimum shift keying
[e]*k*-FSK: *k*—bit Frequency shift keying
[f]*k*-ASK: *k*—bit Amplitude shift keying
[g]*k*-PSK: *k*—bit Phase shift keying
[h]*k*-QAM: *k*—bit Quadrature amplitude modulation
[i]OOK: On-off keying
[j]AM: Amplitude modulation
[k]FM: Frequency modulation
[l]DSB: Double sideband modulation
[m]LSB: Lower sideband modulation
[n]USB: Upper sideband modulation
[o]VSB: Vestigial sideband

regards to this, in the current work, learning the appropriate communication patterns for the cognitive receiver and the influence factors on the received symbols are being studied. The main contributions of this paper are in fourfold.

- We analyzed various DWT approaches, which have an influence on the recognition rate of the ANN.
- We studied the effect of using 4- and 5-level DWT, which reduce the size of the ANN.
- We analyzed various back-propagation learning algorithms, which have an influence on system performance as well as the speed of learning.
- Finally, we showed that the space complexity of the receiver exhibits a reduced ANN structure in terms of inputs and the hidden layer. As fewer resources were used, the receiver could be implemented with fewer hardware units.

### 1.3 Paper organization

The rest of this paper is organized in the following way. Section 2 describes the structure of the PBCCS model and its blocks in detail. It also gives a short introduction on wavelet and neural networks. In Section 3, we evaluate the performance of the PBCCS system. In Section 4, we conclude the paper and recommend directions for future work.

## 2 PBCCS structure

The proposed system consists of two main parts, the transmitter and receiver. Basically, the system employs pattern-based encoding at the transmitter and a wavelet-preprocessed artificial neural network based decoder at the receiver. In this section, we describe the details of the individual parts of PBCCS.

### 2.1 The transmitter of PBCCS

The transmitter of PBCCS is responsible for three tasks: 1) selecting the appropriate glossary with respect to the SNR level, 2) encoding the user data, and 3) transmitting the signal through the antenna. In PBCCS, the modulation is performed by using the sinusoidal pattern envelope construction (SPEC) algorithm [10]. The SPEC algorithm is used to prevent unwanted extra spectral usage, and it guarantees that the signal's pattern ends at its initial point to ensure a zero-power density in average and has no high-frequency components.

The SPEC algorithm has two essential parameters—namely, "depth", and "level". Depth determines the length of the pattern in terms of the time—i.e., number of periods. Meanwhile, level identifies a value for each feature of the signal pattern. It represents the maximum and the minimum values of any signal characteristics (amplitude (A), frequency (F) or phase (P) at depth *i*. All possible outcomes in the SPEC algorithm are due to the changes in the A, F, and P features of the signal.

Figure 1a shows the block diagram of the transmitter, which consists of an encoder, a glossary space, and a glossary selector. The glossary is an information encoding method. It is composed of different patterns generated by the SPEC algorithm. The SPEC is responsible for combining different sinusoidal waveforms to form a symbol as shown in Eq. (1). The generated signal of each symbol has different waveforms that change over time in terms of amplitude, frequency, and phase.

$$m_i(t) = \sum_{i=0}^{J} a_i * x_i(t) * cos\left(2 * \pi * f_i * t + \phi_i\right) \tag{1}$$

where $m_i(t)$ is the $i^{th}$ pattern; $a_i$ ,$f_i$ and $\phi_i$ are the amplitude, frequency and phase of the signal, respectively; $J$ is the total number of sinusoidal waves that the pattern contains (determined by the depth parameter); and $x_i(t)$ is defined as

$$x_i(t) = \begin{cases} 1 \text{ if } 2 * \pi * i \leq t < 2\pi * (i+1) + 2\pi, \\ 0 \text{ otherwise} \end{cases} \tag{2}$$

Each block of $k$ input data bits $d_i, i = 1, 2, \ldots, 2^k$ is mapped to pattern $m_i(t)$. All symbols, $m_i(t)$, are combined to form a $k$-bit glossary space. Figure 2 illustrates a 3-bit glossary space, containing eight signal patterns. In this work, the performance of 3-bit, 4-bit, and 5-bit glossary spaces, with 8, 16, and 32 patterns, respectively, is studied.

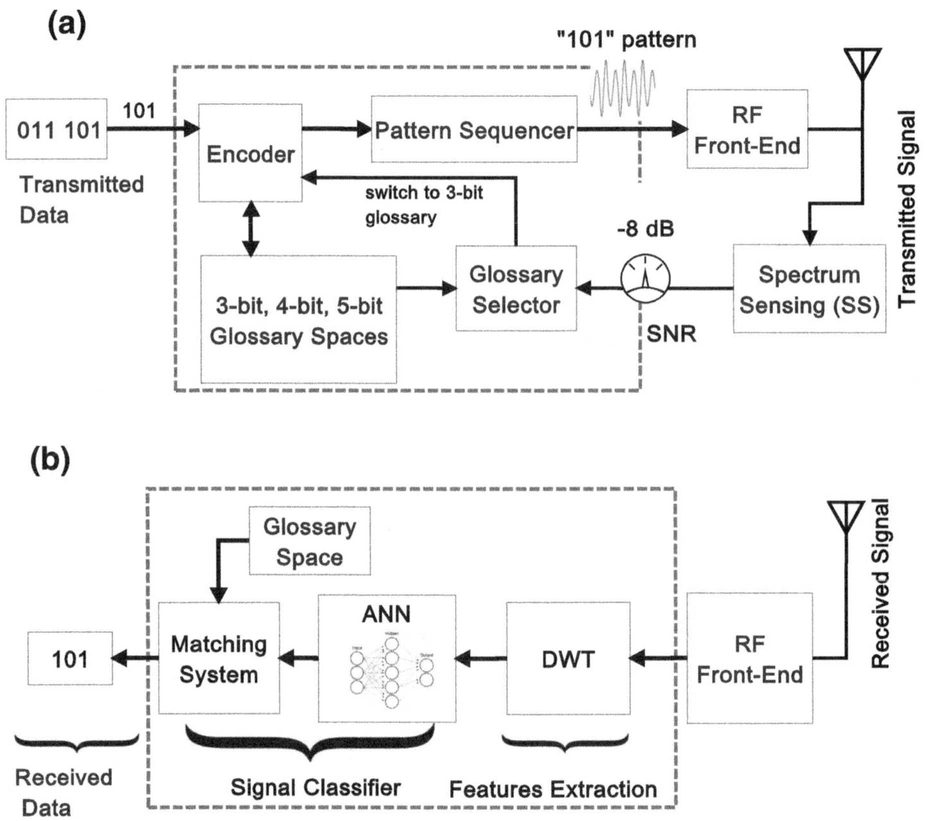


**Fig. 1** The block diagram of the PBCCS. The transmitter (**a**) PBCCS Transmitter Block diagram selects a 3-bit glossary space, so the corresponding "101" waveform is transmitted. The receiver (**b**) PBCCS Receiver Block diagram receives a distorted signal and produces the corresponding bits

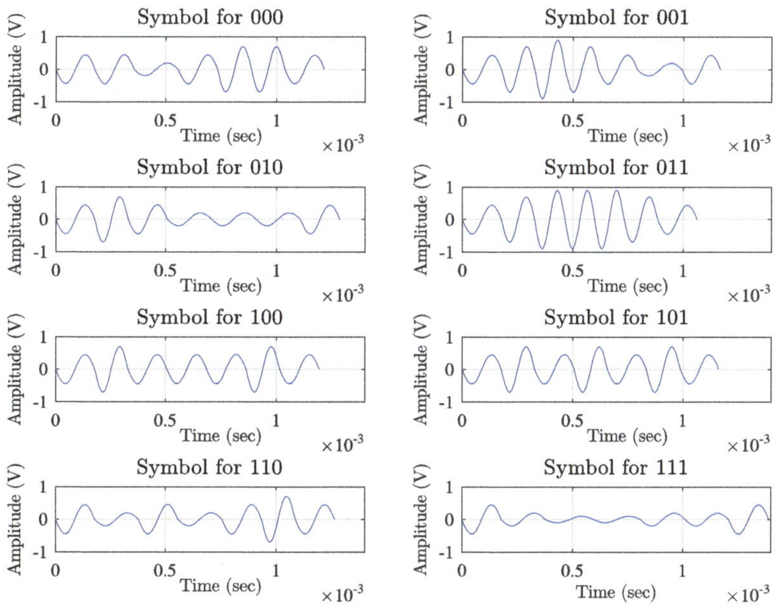


**Fig. 2** The 3-bit glossary space with 5 levels, each of which has 7 patterns (depth is set to 6). The signal's amplitude is [0.1, 0.2, 0.45, 0.7, 0.9] V, frequency is [4.600, 5.000, 5.600, 6.600, 7.400] kHz, and phase shift is $[-\pi/2\ -\pi/4\ 0\ \pi/4\ \pi/2]$ rad

The transmitted symbol, $m(t)$, contains a sequence of known data symbols, $m_1(t), m_2(t), \ldots$. According to the selected glossary, the pattern that matches the data index is selected and applied to the RF front-end.

The glossary selector is the core component in the transmitter's design, because it selects the most appropriate glossary from the glossary space as part of the adaptation process. It takes the glossary space information and channel spectral situation—i.e., the SNR value from the environment, as an input to determine the most proper glossaries set in the glossary space by computing the maximum likelihood value. For example, Fig. 1 shows that the measured SNR is -8 dB. Therefore, the glossary selector switches to a 3-bit glossary and maps '101' to the sixth pattern (shown in Fig. 2).

## 2.2 The receiver of PBCCS

The main modules of the receiver are the discrete wavelet transform unit and ANN, as illustrated in Fig. 1b. The aim of using ANN at the receiver is to predict the original bits of the distorted received signal. The receiver does not construct a similar analog signal or estimate its parameter. Instead, it classifies the input samples to a known pattern, so that the correct bits can be inferred. In the following subsections, we briefly describe the functionality of each part of the receiver.

### 2.2.1 *Features extraction and reduction*

One of the aspects of signal classification is the selection of proper classification features. The goal of feature extraction is to obtain a set of features that can discriminate different received signals. In this work, the discrete wavelet transform (DWT) [28] is used to extract the signal features.

The discrete wavelet transform is a linear signal processing technique that transforms a signal $r(t)$ from the time domain to the "wavelet" domain—i.e., wavelet coefficients. A transformation from the time domain to the "wavelet" domain is analogous to the Fourier transform. The key difference between wavelet transform and Fourier transform is that wavelets are local in both time (via translation) and frequency (via dilation), whereas Fourier analysis is local only in frequency but not in time. Because the generated waveforms contain numerous nonstationary or transitory characteristics, which are often the most important parts of signals, Fourier analysis is unsuitable to describe such characteristics. Moreover, the received pattern signal can be represented by a compact form and hold most features that distinguish it from other patterns. As a result, the wavelet analysis is appropriate to capture the changes in the pattern's frequency over time and achieves better lossy compression, which dramatically reduces the size of ANN.

In general, the received signal, $r(t)$, can be modeled in the AWGN channel as follows:

$$r(t) = m_s(t) + w_n(t) \tag{3}$$

where $m_s(t)$ denotes the original pattern signal, and $w_n(t)$ is white Gaussian noise with normal distribution. $r(t)$ is discretized in time so that $n-$point discrete signal $r[n]$ is constructed. The DWT is defined by Eq. (4) as follows:

$$W(j,k) = \sum_j \sum_k r(k)\psi_{jk}(n), \qquad j,k \in \mathbb{Z} \tag{4}$$

$$\psi_{jk}(n) = 2^{-j/2}\psi\left(2^{-j}n - k\right) \tag{5}$$

where $W(j,k)$ are the wavelet transform coefficients; $\psi_{jk}(n)$ is the mother wavelet, and $j$ and $k$ are the scale parameter and shift parameter, respectively. In practice, it is inconvenient to apply Eq. (4) in calculating the coefficients. The DWT can be implemented as a series of high-pass and low-pass filters, called the recursive wavelet transform, which decomposes the signal $x[n]$ into two parts. The wavelet decomposition depends mainly on orthonormal filter banks. Figure 3 shows the signal decomposition by a two-channel wavelet structure, where $x[n]$ is the input signal, $h[n]$ is the high-pass filter, $g[n]$ is the low-pass filter, and $\downarrow 2$ is the down-sampling by a factor of two. In this way, each filter creates a series of coefficients that represent and compact the original information of the signal.

Mathematically, a signal $x[n]$ is composed of high and low frequencies as shown in Eq. (6). It shows that the obtained signal can be represented by using half the coefficients because they are decimated by 2.

$$x[k] = x_{high}[k-1] + x_{low}[k-1] \tag{6}$$

The filtered and decimated output of low-pass part is recursively passed through identical wavelet filter banks to add the dimension of varying resolution at every stage. Equations (7) and (8) are mathematical expressions of filtering a signal through a digital high-pass filter $h[n]$, and low-pass filter $g[n]$. This operation corresponds to convolution with an impulse response of $k-$tap filters.

$$y_{high}[k] = \sum_n x[n]\,.h[2k-n] \tag{7}$$

$$y_{low}[k] = \sum_n x[n]\,.g[2k-n] \tag{8}$$

where $n$ becomes $2n$ representing the down-sampling process. The output of the low-pass filter, $y_{low}[k]$, provides approximation signal, whereas the output of the high-pass filter, $y_{high}[k]$, provides detailed signal. In addition, Eqs. (7) and (8) show that using DWT can not only greatly reduce the number of input nodes, but also effectively expresses the features of the received signal, thereby enhancing the ability of neural networks to recognize the signal.

A variety of different wavelet families has been proposed in the literature. In this work, the simplest wavelet family—i.e., the Haar wavelet and triangle wavelet are selected [28]. In addition, the 4- and 10-coefficient wavelet family (the second and the fifth orders of Daubechies wavelet—i.e., DB2, DB5), and the 6-coefficient wavelet family coiflets (C6) proposed by Daubechies are used [29]. The decomposition low-pass filter parameters of the Haar, DB2 and DB5 wavelet are shown in Fig. 4.

### 2.2.2 *Recognition layer*

After extracting the proper features from the received signal, classifying these patterns into appropriate classes is the final step to recognize the symbol. In this work, the artificial neural network (ANN) [30, 31] is considered as a recognition layer to recover the transmitted data, and it forms the cognitive part of the PBCCS receiver.

The main reason for choosing the preprocessed wavelet with ANN is its high sensitivity for recognizing the amplitude, frequency and phase changes in the communication signal. The output that is produced by the ANN is

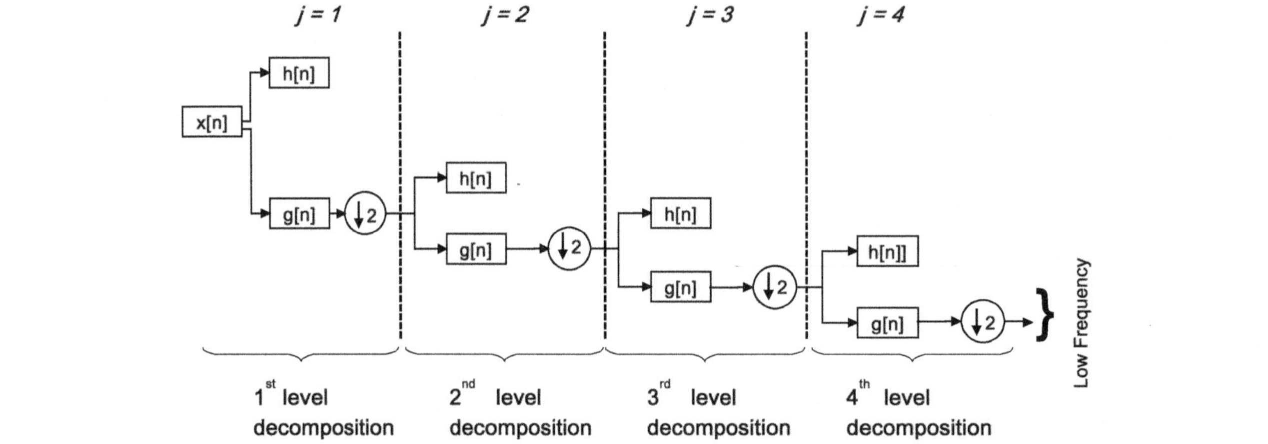


**Fig. 3** The block diagram of a two-channel four-level DWT decomposition ($J = 4$) that decomposes a discrete signal into two parts

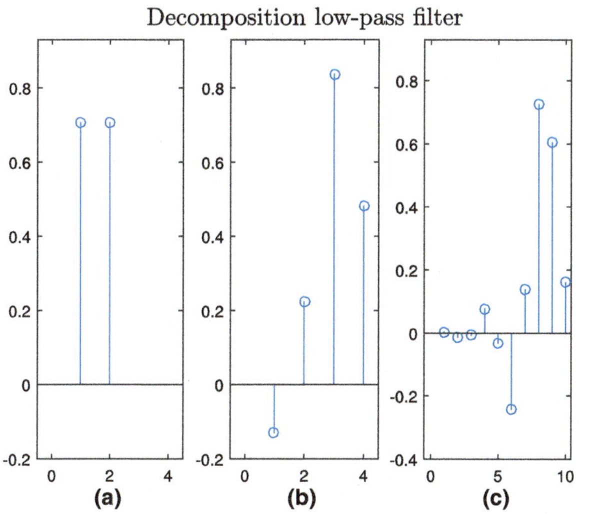


**Fig. 4** Low-pass filter parameters of different wavelet families. **a** Haar, **b** DB2, **c** DB5

decoded in the final stage into the correct bit sequence, as shown in Fig. 5.

In this work, the most common ANN model, namely multilayer perceptron (MLP), is used. MLP is a type of feed-forward neural network (FFNN) model that maps the input data onto a set of appropriate outputs. It consists of at least three layers—i.e., the input layer, one or more hidden layers and an output layer. The network is fully connected from one layer to the next as a directed acyclic graph (Fig. 5). Each neuron is capable of multiplying the inputs by its weight and sum up the results. In other words, the neuron operations are performed by multipliers and adders.

Mathematically, for $n$ arbitrary distinct received samples $(x_i, t_i)$, where $x_i$ is the extracted features' vector from the received signal, $x_i = [x_{i1}, x_{i2}, \ldots, x_{in}]^T \in \mathbb{R}^n$ and $t_i$ is the target vector, $t_i = [t_{i1}, t_{i2}, \ldots, t_{im}]^T \in \mathbb{R}^m$. $n$ and $m$ are the size of the input feature and the target vectors, respectively. The target vector $t_i$ represents the actual sequence of bits that the recognition layer must produce.

A single hidden layer of a FFNN (Fig. 5) with an activation function, $g(.)$ and $k$ hidden neurons is mathematically modeled as

$$g\left(\sum_{i=1}^{n} w_{ji} \cdot x_i + b_j\right) = y_j, \quad j = 1, 2, \ldots, k \tag{9}$$

where $w_{ji} = \left[w_{j1}, w_{j2}, \ldots, w_{jk}\right]^T$ is the weight vector connecting the $j^{th}$ hidden neurons with the inputs and $b_j$ is the bias value of the $j^{th}$ hidden neuron. The bias allows the sigmoid function curve to be shifted horizontally along the input axis while leaving the shape of the function unchanged. $w_{ji} \cdot x_i$ denotes the inner product of $w_{ji}$ and $x_j$.

The result of the $j^{th}$ output neuron can be mathematically computed as shown in Eq. (10):

$$g\left(\sum_{i=1}^{k} \beta_{ji} \cdot y_i + b_j\right) = O_j, \quad j = 1, 2, \ldots, M \tag{10}$$

where $M$ is the total number of output neurons. $\beta_j i = \left[\beta_{j1}, \beta_{j2}, \ldots, \beta_{jm}\right]^T$ is the weight vector connecting the $j^{th}$ hidden neurons and output neurons and $b_j$ is the bias value of the $j^{th}$ output neuron.

Because the final goal is to find the relation between the input $x_i$ and the target $t_i$, the activation function $g(.)$ can approximate these $n$ received samples with zero mean

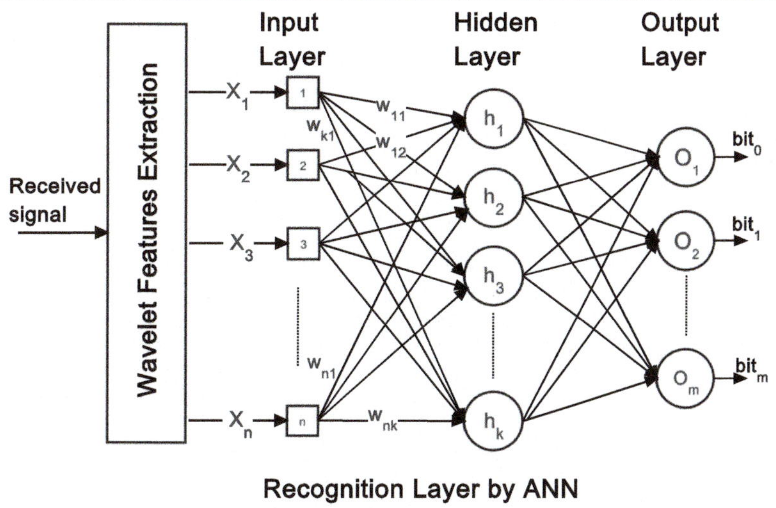


**Fig. 5** Feed-forward ANN (FFNN) with one hidden layer and an output layer. The extracted features are fed into ANN with *n* input nodes, and *k* and *m* are the number of hidden and output neurons, respectively. The output bits are the bits that were recognized by the FFNN

error, such that $\sum_{j=1}^{N} \|t_i - y_i\| = 0$. Thus, there exist $\beta_i, w_i$ and $b_i$ such that

$$\sum_{i=1}^{n} \beta_i g\left(w_i \cdot x_j + b_i\right) = t_j, \quad j = 1, 2, \ldots, m \tag{11}$$

The back-propagation algorithm (BP) [30] is used to compute the weights and biases of the ANN by minimizing the error function in weight space using gradient descent.

The received sequence of bits, $\hat{a} = \hat{a_1}, \hat{a_2}, \ldots, \hat{a_m}$ is approximated from the output neurons as shown in Eq. (12):

$$\hat{a_j} = \begin{cases} 1 & \text{if } O_j \geq 0.5, \\ 0 & \text{otherwise} \end{cases} \tag{12}$$

where $O_j$ is the $j^{th}$ the output neuron.

As a final point, the structure of the FFNN should be modular and simple, so that the hardware architecture can be efficiently realized on floating point digital signal processor (DSP) or a field programmable gate array (FPGA). As there are various combinations of designing a FFNN, an efficient design should include the following aspects

- Input neurons: the number of neurons is equivalent to the sample size of the DWT vector.
- Hidden neurons: due to low complexity and high applicability perspective [10], a single hidden layer with few number of neurons should be used. The number of neurons will be determined by the cross validation method.
- Output neurons: The number of output neurons should be identical to the size of the glossary space—i.e., total number of patterns. However, each glossary differs in the number of symbols that it represents. For instance, the 3−bit glossary has 8 symbols while 4−bit glossary has 16 symbols. Therefore, the size of the glossary can be added to determine the width of the output sequence bits.

## 3 Experimental results

In this section, the performance of the proposed PBCCS based on the extracted features (discussed in Section 2) is verified in an AWGN channel. At the end of this section, the proposed approach complexity is presented.

### 3.1 Simulation settings

Simulations were carried out to transmit $2^k$ different symbols ($k$-bit glossary) at various SNR levels. The SNR levels were in the range of $[-15, 25]$ dB. The ANN parameters are shown in Table 4. All experiments were performed using Matlab software.

### 3.2 Constructing the glossary

Each test pattern is constructed with five sub-signal. According to the SPEC algorithm, patterns are generated with bandwidth-limited switching among different frequency, phase, and amplitude levels. The features of the communication signal (frequency, phase, and amplitude) are chosen from five different levels, listed in Table 3. The depth is set to 6, which indicates that each signal symbol is constructed from 7 sub-patterns.

Synthetic signals that represent the symbols was validated on hardware. For this purpose, we used an FMC150 [32] daughter card attached to a Xilinx ML605 board [33]. The FMC150 is a dual 14-bit channel ADC and dual channel 16-bit DAC FMC daughter card. Figure 6 shows the synthesized signal by the hardware against a simulated signal generated by Matlab. There are 7 sub-patterns in each signal. The frequencies of each pattern are 1.25, 5, 6.25, 5, 1.25, 2.5, and 1.25 MHz. The amplitude and phase are identical to the values presented in Table 3. The frequency of the simulated signal might slightly differ from the real one owing to the register precision of the FMC150—i.e., the measured frequency of each sub-pattern is 1.229, 4.9020, 6.1728, 4.9020, 1.229, 2.457, and 1.229 MHz. The mean voltage of the baseband signal is $2.9710 * 10^{-4}$, which is approximately zero.

### 3.3 Learning process and model evaluation

Before assessing the system, two datasets were generated for each symbol. These sets should be used for learning (training) and testing (evaluating). The learning set is used to derive the model offline, whereas the test set is used to estimate model's accuracy online, as shown in Fig. 7. Symbols that were used during the learning stage would not be involved in the testing stage. Moreover, the learning dataset should be larger than the testing dataset, so that the model can be built from a large space of sampled signals. For example, the dataset of the 3-bit glossary contains 248, 000 symbols and 74, 400 symbols for learning and testing, respectively. In other words, 70% of the total symbols were used for learning.

To assess the accuracy of the wavelet filter banks and ANN model, we use the success rate of the recognition symbol as an indication of the effectiveness of the receiver to correctly recognize the received symbols. It indicates the capability of the trained model to classify the received symbols in the testing set, which were not seen before.

**Table 3** Base signal specifications

| Feature | -L2 | -L1 | L0 | L1 | L2 |
|---|---|---|---|---|---|
| Frequency (kHz) | 4.600 | 5.000 | 5.600 | 6.600 | 7.400 |
| Phase (rad) | $-\pi/2$ | $-\pi/4$ | 0 | $\pi/4$ | $\pi/2$ |
| Amplitude (V) | 0.1 | 0.2 | 0.45 | 0.7 | 0.9 |

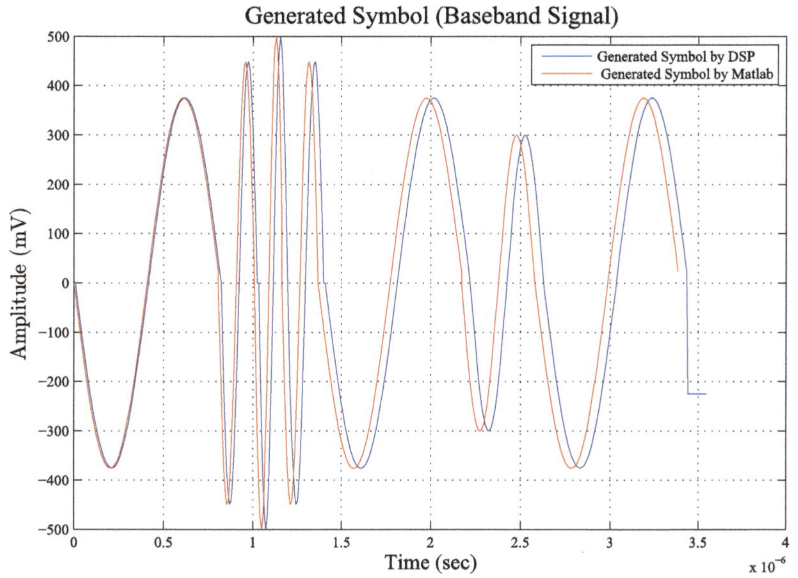


**Fig. 6** Synthetic signal (baseband signal) by using Matlab compared with the real synthetic signal generated by the hardware. $F_s = 122.88$ MHz

It also expresses the probability of correct classification, which is computed as follows:

$$Success\ Rate = \frac{1}{N}\Sigma_i^N x_i * 100 \quad (13)$$

where $N$ is the total number of test symbols and $x_i$ is an indicator whose value is 1 if the $i^{th}$ symbol is correctly received. In other words, *Success Rate* measures the symbol error rate (SER).

In addition to the success rate, the BER performance is used to assess the accuracy of the whole system. It expresses the number of bit errors per second divided by the total number of transferred bits.

### 3.4 Classification performance of different wavelet families

In this section, we study the effect of applying different wavelet families on the performance of the receiver for 3-bit glossary. The received signal has 880 samples. The number of input nodes was subsequently reduced from 220 to 55 and 27 to identify the most relevant input features to ANN by employing 2-, 4-, and 5-level DWT decomposition ($j = 2, 4$ and 5). The ANN model was then trained for different numbers of neurons in the hidden layer. These experiments were repeated 10 times, and the success rate was averaged.

Figure 8a illustrates the effect of using the Haar wavelet for various numbers of neurons in the input and hidden layers. It shows that the average success rate of the model is more than 90% for all scenarios. By using 27 neurons at the input layer, the success rate is greater than 94% with 12 neurons or more at the hidden layer.

In Fig. 8b, the effect of using a triangle wavelet is shown, which is similar to Fig. 8a. However, the success rate of the model with 55 neurons outperforms the model that has 220 inputs. The improvement of the input size reduction to 55 inputs by using DB2 is illustrated in Fig. 8c. The success rate is greater than 97% with more than 12 hidden neurons (with the exception of the case with 27 inputs).

DB5 wavelet has similar performance compared to the DB2, as shown in Fig. 8d. Similar result was also obtained by using a coiflets (C6) wavelet as shown in Fig. 8e. The figure shows that with 55 input nodes and 10 hidden nodes, the performance is improved compared with the experiment of 220 input nodes.

To analyze the previous results, the success rates of PBCCS are elaborated in the $[-20, 15]$ dB range for various network configurations in the hidden layer (Fig. 9). When the SNR is equal to or greater than $-12$ dB, the

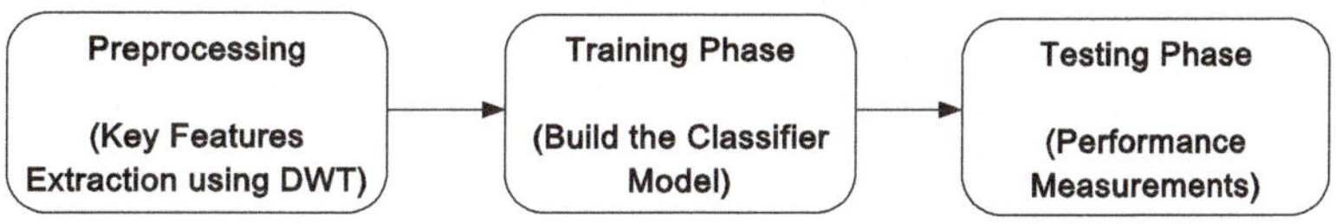


**Fig. 7** The functional blocks of the experiments

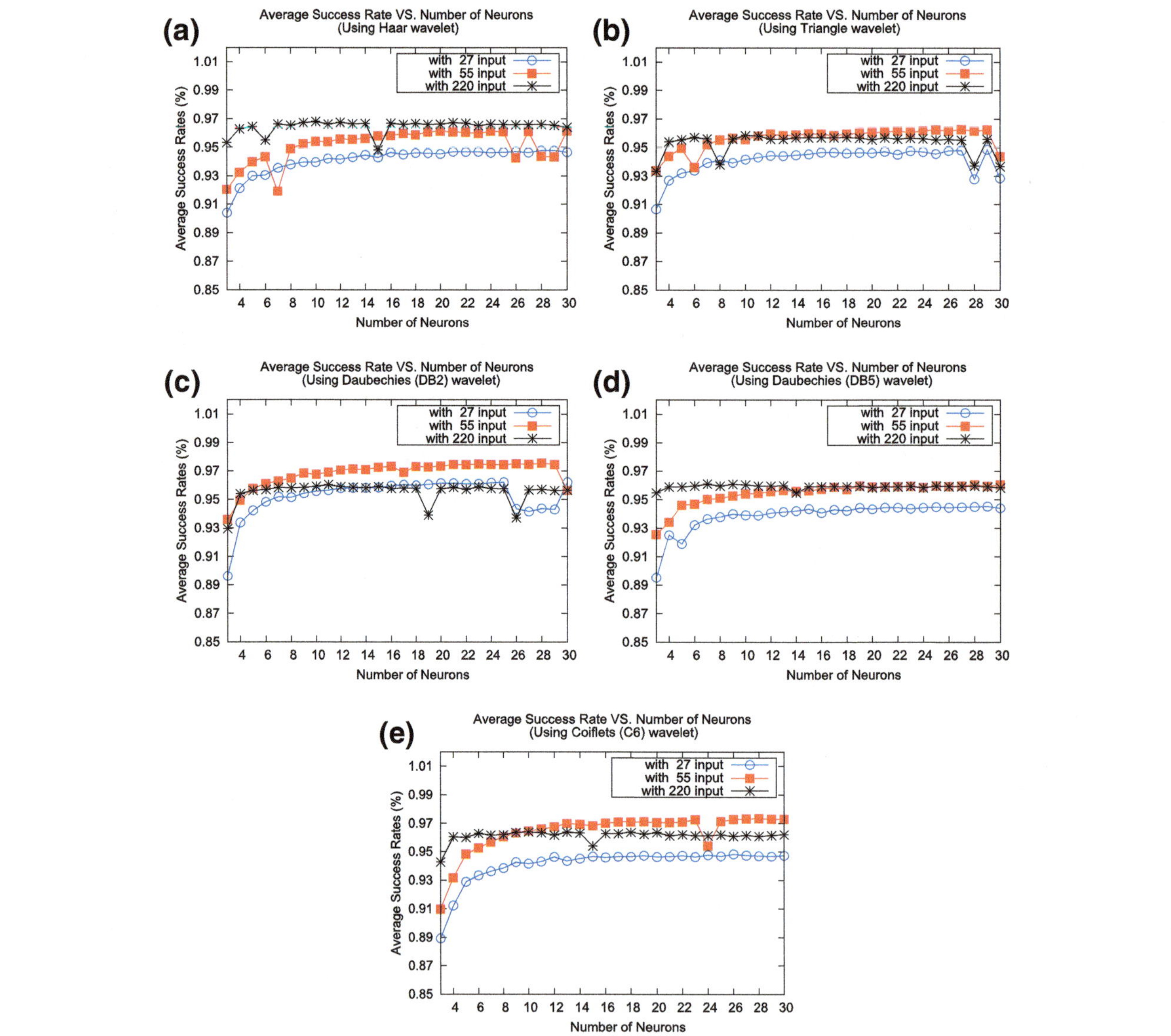


**Fig. 8** The probability of correct classification (average success rate) vs. number of hidden neurons by using different wavelet transforms and various input size in an AWGN channel. **a** Haar wavelet, **b** Triangle wavelet, **c** Daubechies (DB2) wavelet, **d** Daubechies (DB5) wavelet, **e** Coiflets (C6) wavelet

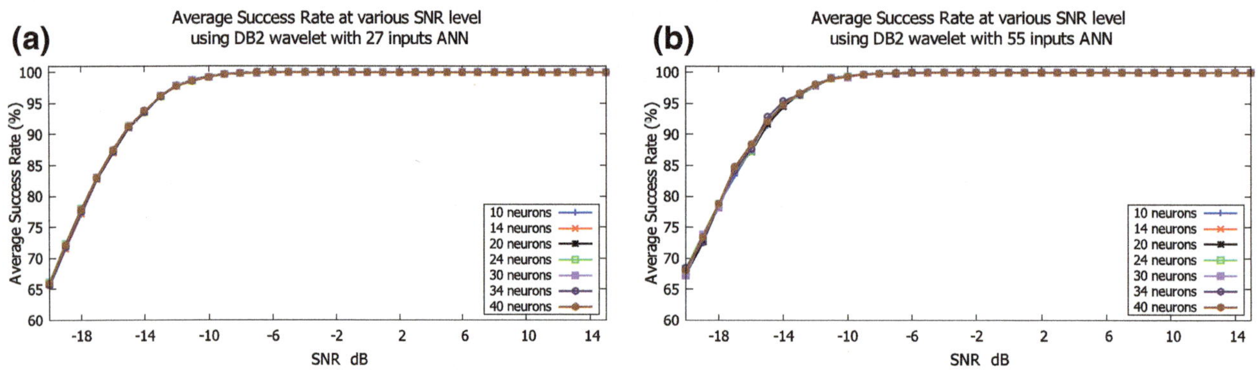


**Fig. 9** The probability of correct classification (success rates) in AWGN channel when applying a DB2 wavelet connected to FFNN with 10, 14, 20, 24, 30, 34, and 40 hidden neurons. **a** Using 5-Level DB2 DWT (27 coefficients), **b** Using 4-Level DB2 DWT (55 coefficients)

success rate is greater than 99.0% for all configurations, which means that there is no error in the received symbols ($SER = 0.1$). Below −12 dB, the success rate linearly decreases to 65%. In all cases, using 10 hidden neurons or more layer does not improve the success rate.

In summary, we found that for 3-bit glossary the DB2 wavelet has better performance compared with other wavelet families studied in our tests. It is also found that with a 27−input ANN, the performance is better than that when using many extracted features. Furthermore, the use of 14 hidden neurons or more has a similar recognition rate to a network that contains 8 hidden neurons. As a result, an ANN that is based on 5-level DWT can be realized with 27 inputs and as minimum as 14 hidden neurons. This reduction will use fewer resources during the hardware design realization.

### 3.5 Classification performance of various learning algorithms

In this section, we examined four popular back-propagation learning algorithms—namely, (Levenberg-Marquardt (LM), scaled conjugate gradient (SCG), gradient descent with momentum and adaptive learning rate back-propagation (GDX) and Bayesian regularization back-propagation (BR) in terms of speed and the number of iterations to achieve the same performance goal (MSE)=0.01. The ANN parameters are set according to Table 4. For LM algorithm, the maximum number of iterations was set to 25 to limit the learning time as shown in Table 5. It is worth mentioning that we choose the best parameters for LM algorithms by experimental study.

The average training accuracies of each algorithms over various ANN structures are shown in Fig. 10a. It indicates that all learning algorithms achieve 96%. The LM algorithm is failed in some experiments because the algorithm reaches the maximum number of iterations (Fig. 10b). Figure 10c shows that as the number of hidden units increases, the learning time of both LM and BR algorithms is significantly increased, whereas the learning time of both SCG and GDX algorithms is constant and less than 20 m. The average learning time of BR and LM with 40 hidden units network are 3 and 2 h, while the number of iterations is less than 25 and 100 iterations, which means that both algorithms have high computation complexity.

The average MSE versus number of hidden neurons are shown in Fig. 10d. Smaller values that are close to zero are better because they indicate that the MLP had fitted the data well. SCG, GDX, and BR learning algorithms have better performance compared with LM algorithm because the number of iterations of LM was limited to 25 iterations. Increasing the number iterations will improve the MSE values but has dramatically effect on learning time.

**Table 4** ANN parameters

| Parameters | Value |
|---|---|
| Number of layers | 2 |
| Number of input nodes | 220, 55, 27 |
| Number of output neurons | 3, 4, 5 |
| Activation function | Sigmoid |
| Training algorithm | Scaled conjugate gradient |
| Gradient error level | $1 \times 10^{-6}$ |
| Performance function | Mean squared error |
| Number of neurons in hidden layer | Varied from 8 to 40 |
| Maximum number of epochs to train | 3000 |

**Table 5** ANN parameters

| Parameter | Value |
|---|---|
| Maximum number of epochs to train | 25 |
| Initial $\mu$ value | 0.1 |
| Decrease factor ($\alpha$) | 0.2 |
| Increase factor ($\beta$) | 6 |
| Minimum performance gradient | $10^{-5}$ |
| Performance function | Mean squared error |

### 3.6 System performance with *k*-bit glossary spaces in AWGN Channel

Based on the previous results, DB2 wavelet decomposition was applied to extract 27 features from the received signal. We simulated the BER by applying $3*10^5$ at each SNR level and measuring the number of uncorrected received bits. In Fig. 11, the SNR curves illustrate the 3-, 4-, and 5-bit glossary performance with various hidden neurons. The total area under each curve indicates the overall system performance under different noise and data bit error rate (BER) levels. It means that the carve with minimum area has a better performance. It is depicted in Fig. 11a that the BER is $10^{-5}$ at −5 dB for most ANN configurations. Similarly, Fig. 11b presents that the curve for 20 neurons has SNR of approximately −2 dB at $10^{-5}$. Also, it shows that the system performance improves with increasing number of neurons due to the fact that the recognition capability could be enhanced as the number of hidden neurons is increased. Figure 11c shows similar performance as shown in Fig. 11b except that SNR is approximately 4 dB at $10^{-5}$.

Figure 11 shows a strange behavior as the BER carve reaches $10^{-5}$, where some errors increases again. We expect that the ANN could not distinguish between the samples either because the learning process is not enough or it overfits the data.

In summary, we found that the best performance of 4- and 5-bit glossary space is achieved when the number of hidden neurons is between 20 and 40 neurons. This means that the BER of an ANN with a hidden layer of 20

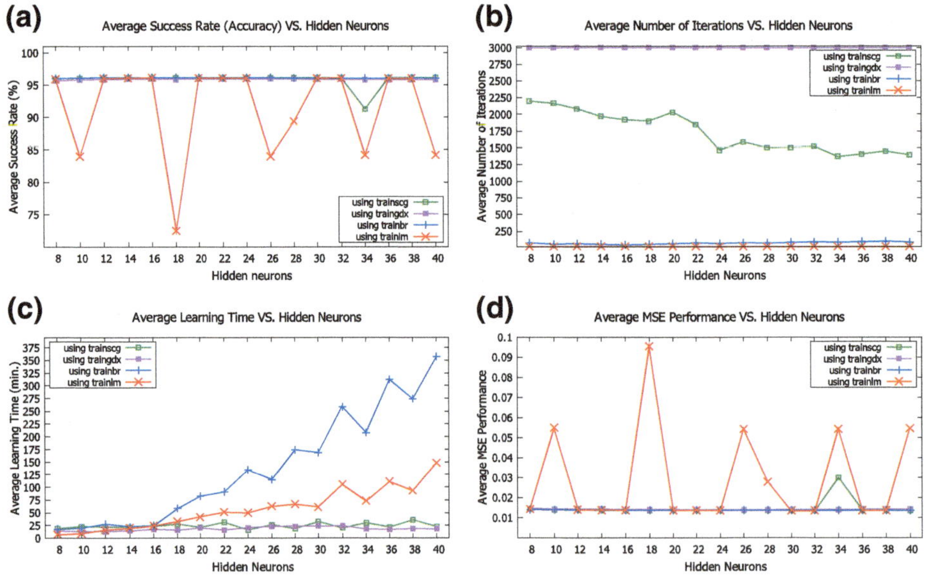


**Fig. 10** Comparison between SCG, LM, GDX and BR learning algorithms with respect to **a** Average success rate (accuracy). **b** Average number of iterations. **c** Average Learning time. **d** Average MSE performance

neurons is approximately equal to an ANN configuration with higher hidden units. This phenomenon indicates that an increasing number of hidden neurons does not always improve the performance. As a result, the best PBCCS performance can be realized with a fixed-structure ANN of 27 inputs nodes and 20 hidden neurons for 3-, 4-, and 5-bit glossaries.

### 3.7 Spectral efficiency

The ability of the system to balance between the spectral efficiency under a very low SNR is an advantage of the proposed scheme. For each glossary, we construct a random signal of 2000 symbols to measure the average data rate and the occupied bandwidth. Table 6 shows the spectral efficiency, $\eta$, of the previous constructed glossaries.

The most significant feature of the proposed method is its competence to operate in a region beyond the Shannon boundary. The Shannon boundary is drawn in Fig. 12 by using the Shannon limit theorem (equation (6.5—49), [2]), which is rewritten as follows:

$$\frac{E_b}{N_0} > \frac{2^{\eta} - 1}{\eta} \tag{14}$$

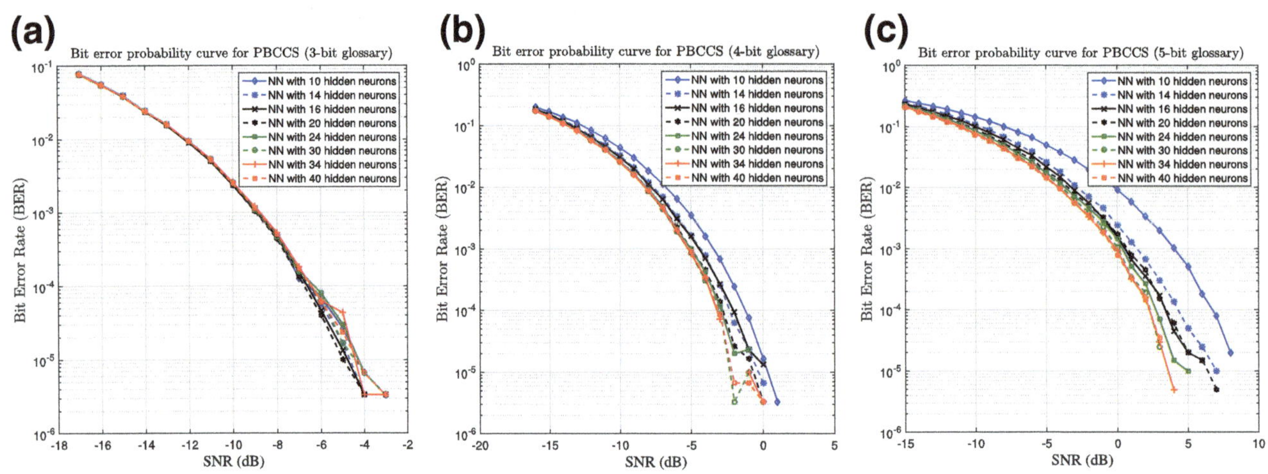


**Fig. 11** Bit error rate for three different glossaries in an AWGN channel. Each curve represents the number of neurons at the hidden layer. **a** with 3-bit glossary, **b** with 4-bit glossary, **c** with 5-bit glossary

**Table 6** Spectral efficiency of PBCCS at $P_e = 10^{-5}$ bit error probability

| Glossary | 3-bit | 4-bit | 5-bit |
|---|---|---|---|
| M (symbols) | 8 | 16 | 32 |
| SNR value (dB) | -5 | -2 | 4 |
| Data rate (kbps) | 2.508 | 3.553 | 4.6 |
| Average BW (kHz) | 0.960 | 0.881 | 0.9135 |
| LSE,$\eta$, (bps/Hz) | 2.61 | 4.0318 | 5.03 |

where,

$$\eta < \log_2\left(1 + \frac{E_b R}{N_0 W}\right) = \log_2\left(1 + \eta\frac{E_b}{N_0}\right) \tag{15}$$

Equation 14 states the condition of reliable communication in terms of bandwidth efficiency, $\eta$, and power efficiency in terms of $E_b/N_0$. Figure 12 shows the minimum value of $E_b/N_0 = -1.59$ for which reliable communication is possible. The marks in the figure show the best working point regarding the glossary at a $10^{-5}$ bit error rate and $M-$QAM. This figure depicts that the 4-bit glossary and 16-QAM have the same spectral efficiency, but the 4-bit glossary is more robust to high levels of AWGN than 16-QAM. Similarly, 5-bit glossary is more robust than 32−QAM at the same spectral efficiency, $\eta = 5$.

However, the most interesting finding is that the 3−bit glossary breaks down the Shannon limits. This phenomenon occurs because the ANN can perfectly discriminate among the eight signals of the 3−bit glossary. In the learning step, the ANN model has constructed a model that allows the PBCCS to classify any unseen signal, in a manner analogous to memory. Another reason for breaking down the Shannon limit is the usage of non-periodic signals. Shannon's law is restricted to periodic signals [34]. The PBCCS constructs a non-periodic and uncorrelated communication waveforms that provide a manageable SNR capability between high noise redundancy and high data bandwidth requirements under observed spectrum conditions.

It is worth mentioning that this result depends on the difference between the bandwidth of recovered digital data based on a priori information in the glossary and the raw physical data bandwidth inside the communication medium. In addition, the synchronization overhead between the transmitted symbols is not considered.

### 3.8 Bit error rate comparison

A comparative analysis between the PBCCS and of the matched filter is shown in Fig. 13. The total area under each curve indicates the overall system performance under different noise and data BER levels. According to Fig. 13, 3−,4−,5-bit glossary constructed by PBCCS modulation have 20 dB, 21 dB and 22 dB better performances at $10^{-5}$ BER than 8-QAM, 16-QAM, and 32-QAM, respectively. It is worth noting that each $k$-bit symbol space has $2^k$ symbols as $m$-QAM because $m = 2^k$. Also, 3−,4−,5-bit glossary outperforms the 4−,8−,16-PSK by 16, 17, and 23 dB.

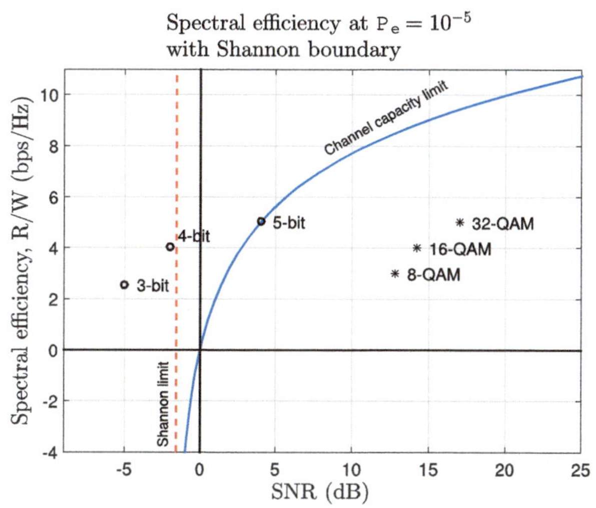


**Fig. 12** Comparison between PBCCS and $m$-QAM in an AWGN channel at $P_e = 10^{-5}$ bit error probability. The neural network contains 20 hidden neurons

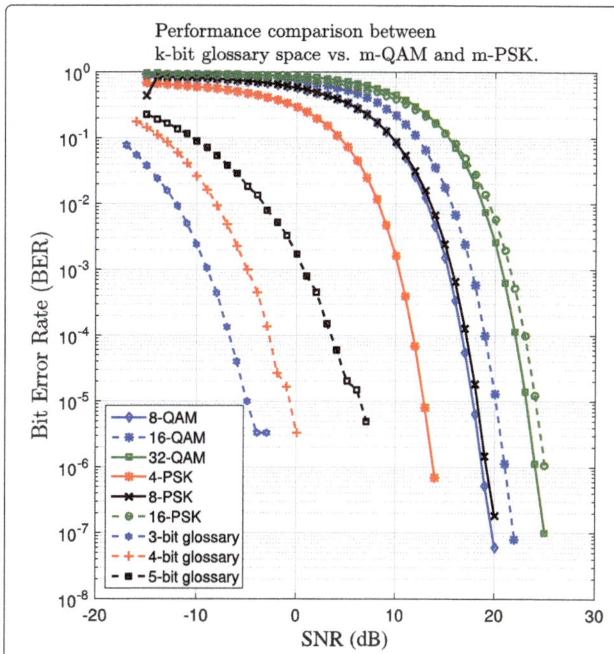


**Fig. 13** The simulated performance of PBCCS and $m$-QAM in an AWGN channel. The neural network contains 20 hidden neurons

The bit error probability for 16-QAM is given by (equation (4.2—85), [2]):

$$P_{b_{16-QAM}} = 3Q\left(\sqrt{\frac{4E_b}{5N_o}}\right) - \frac{9}{4}\left[Q\left(\sqrt{\frac{4E_b}{5N_o}}\right)\right]^2 \quad (16)$$

where $Q(x)$ is the $Q$-function. $Q(.)$ is a monotonically decreasing function of its argument; the probability of error decreases as the ratio $\frac{4E_b}{5N_o}$ increases. This means that the decision boundary of the QAM technique depends on increasing the signal energy, which makes the binary signals dissimilar. However, the PBCSS depends not only on the signal energy but also on the extracted features from the received signal by DWT and the nonlinear model of ANN, which enable the PBCCS model to reduce the probability of error.

### 3.9 System performance comparison with AMC

Because this work and similar works have used different methodologies in signal type recognition, direct comparison with these works is difficult. Table 7 compares the PBCCS with some of AMC techniques. A brief summary of each recognition approach was given in Table 2. It is clear that the proposed PBCCS outperforms the previous approaches (at -11 dB) because the symbols that are stored in the glossary was designed with properties that allow the ANN to discriminate between them.

### 3.10 Receiver space complexity

After simulating the proposed approach, the next step is to verify the design on real hardware such as a field-programmable gate array (FPGA) and application-specific integrated circuit (ASIC). We prefer using FPGA because the parallel structure of an ANN and the similarity of neurons makes its design simple and straightforward.

Each FPGA comes with limited resources, which poses challenges for real implementation. Space complexity gives an indication of the number of used functional units. It approximates the numbers of connections, multipliers and adders that the real design will occupy when it is implemented on an FPGA.

The transmitter could be implemented by means of memory. Each glossary requires one memory (blocked RAM (BRAM)), which is indexed by the transmitted bits. For example, Xilinx Virtex VI xc6vlx240t-1ff1156 has 416 BRAMs, each storing 36 Kb [33]. Figure 14 shows the block diagram of the PBCCS transmitter, where the preamble was used for the synchronization. It shows only 3-bit glossary with 13 × 16 BRAM.

At the receiver, each neuron of the implemented ANN has a set of multipliers that are used to multiply the

**Table 7** Comparison between different works in terms of features, ANN model and the achieved SNR with the recognition accuracy

| Reference | Application | Applied features | Type of ANN | Recognition accuracy (%) |
|---|---|---|---|---|
| [22] | AMC | Instantaneous temporal feature-based | FANN (5,19,8) and PANN (5,1800,8) | Overall success rate at -5 db are 65.63% and 55.5% respectively. |
| [23] | AMC | Instantaneous temporal feature-based modulation | FANN (4,7,5) | the overall success rate at -5 dB is 99.65%. |
| [25] | AMC | Continuous wavelet transform (CWT) | N/A | The overall success rate at 0 dB is 99.6% (using 10 features). |
| [26] | AMC | Instantaneous information and signal spectrum | N/A | The overall success rate at 3 dB is 98.6% (using 10 features). |
| [37] | AMC | Combination of the higher order moments, higher order cumulants and instantaneous characteristics of digital modulations | Radial basis function (RBF) probabilistic neural network (PNN) | The overall success rate at -3 dB is 87.50%. The overall success rate at -3 dB is 86.45%. |
| [38] | AMC | 7-level DWT | Adaptive Network Based Fuzzy Inference Systems of 5 hidden layers | The overall success rate using DB2 at -5 dB is 98%. |
| [39] | AMC | Haar Wavelet Transform | N/A | The overall success rate at -7 dB is 99.71%. |
| [40] | AMC | Haar Wavelet Transform | N/A | The overall success rate at 5 dB is 97.93%. |
| [27] | Modulation classification and signal encoding | 1-level DB2 DWT | FFNN(30,14,3) | The overall success rate a -11 dB for 3-bit glossaries is 96.0%. |
| PBCCS | Modulation classification and signal encoding | 5-level DB2 DWT | FFNN(27,14,3), FFNN(27,14,4), FFNN(27,14,5) | The overall success rate at -11 dB for 3-bit, 4-bit and 5-bit glossaries are 99.0%, 90.3% and 72.79%, respectively. |

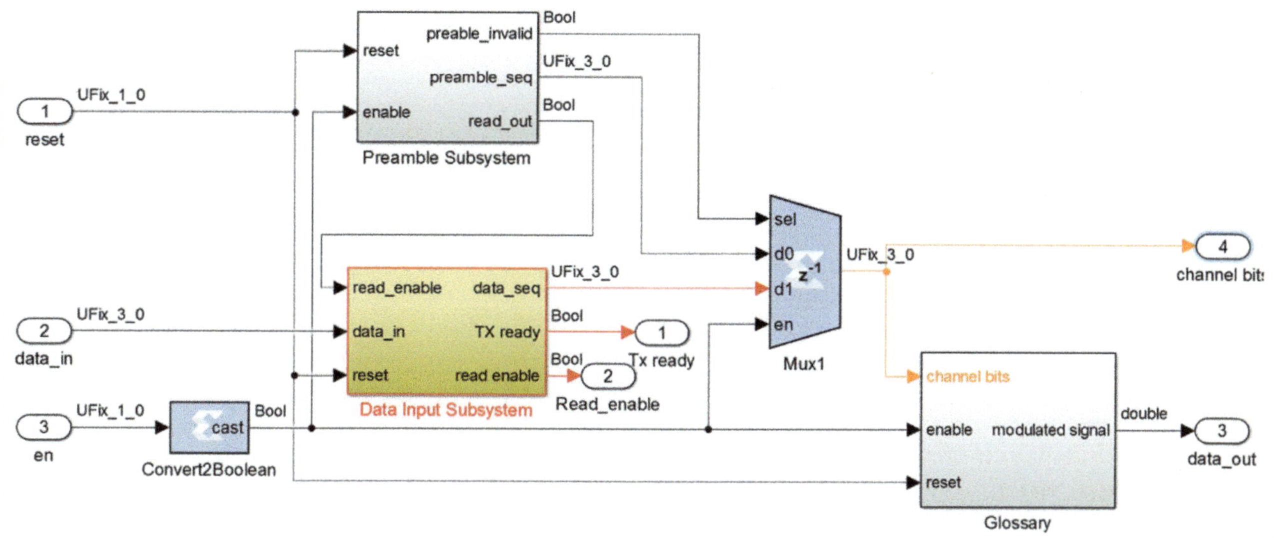


**Fig. 14** The block diagram of the PBCCS transmitter

weight of the connections with the received data values. For example, if there are 26 2-input nodes, then each hidden neuron requires 26 multipliers and 26 2-input adders (including one adder to the bias). Because multipliers are more expensive than the adders and each FPGA comes with a limited number of them, they, in fact, significantly influence the design. For instance, Xilinx Virtex VI xc6vlx240t-1ff1156 has 768 multipliers (named DSP48E1) [33].

Table 8 shows a comparison between the architecture of the proposed approach compared with the one that is based on time delay ANN (TDNN) [10]. It shows that the total number of multipliers in the FFNN is drastically reduced compared with the TDNN. Moreover, the 3-bit glossary space requires $26 * 20 * 3 = 1560$ connections. This indicates the number of internal connections among the input, hidden and output layers of any used neural network, whereas TDNN requires $120 * 20 * 3 = 7200$ connections.

In addition to the impact of the neural network, the wavelet decomposition has effect on the space complexity

**Table 8** Space complexity comparison between the proposed approach and time delay ANN [10]

| Parameters | Wavelet and ANN | | Time delay ANN | |
|---|---|---|---|---|
| | $k = 3$ | $k = 4$ | $k = 3$ | $k = 4$ |
| Number of layers | 2 | 2 | 2 | 2 |
| bit size of glossary space | 3 | 4 | 3 | 4 |
| Input neurons | 27 | 27 | 120 | 120 |
| Hidden neurons | 20 | 20 | 20 | 50 |
| Output neurons | 3 | 4 | 3 | 4 |
| Number of multipliers | 600 | 620 | 2460 | 2480 |

of the receiver. Similar to the finite impulse response (FIR) filter, the wavelet decomposition convolves wavelet coefficients with the received signals. This convolution process requires as many multiplication resources as the number of filter taps. For example, DB2 and DB5 can be implemented as 4-tap and 10-tap FIR filters, respectively. Furthermore, it is an iterative process—i.e., the output of one stage is an input to the next stage (Fig. 3). The direct implementation DWT, known as the multiply-accumulate structure (MAC), requires as many resources as the number of stages times the numbers of filter taps—i.e., 5-level DB2 requires 20 multipliers. An alternative but efficient implementation could be achieved by means of the distributed arithmetic algorithm (DAA) [35, 36]. DAA efficiently realizes the sum-of-products computation by means of memory (LUT), adders and shift registers, without employing any multipliers. That is, the total number of multipliers at the receiver will not be affected by the DWT implementation.

## 4 Conclusions

PBCCS was designed to increase the spectral efficiency by constructing a secure and non-periodic communication signals. In addition, PBCCS minimizes the bit error rate through optimized signal patterns that are decoded solely by DWT preprocessed signals and artificial neural network.

In this article, we analyzed the performance of ANN to recover original transmitted symbols using wavelets as a feature extractor. We applied different wavelet decomposition techniques to study their effects. Several experiments were conducted to find the most appropriate wavelet family for PBCCS. The results obtained are intended to be a guidance tool in selecting the most appropriate operating point of the glossary selector with the discrete wavelet family at the receiver. We found out that the DB2 wavelet decomposition filter shows better performance compared with the other studied wavelet families. Thanks to DWT, a simple ANN structure was constructed with few hidden neurons, which is impossible for a third-party to predict. In addition, we studied the effect of various back-propagation learning algorithm. We could conclude that in terms of learning time and performance, both SCG and GDX are better to handle large dataset that includes thousands of signals. Finally, because of robustness to stationary noise, the proposed approach has a great advantage of less bit error, unlike the standard modulation techniques, which has higher bit error rate.

The simulation results also reveal that by using 5-level DWT and a neural network, SNR values of -5 dB, -2 and 4 dB are achieved at a BER of $10^{-5}$ for 3-bit, 4-bit and 5-bit glossary spaces, respectively. The advantage is obvious, because the transmitter can adapt the bit rate according to SNR values. Therefore, adaptive glossary and its performance can be considered in a future work.

An initial evaluation of hardware implementation was demonstrated, and the applicability of the proposed modulation technique and the recognition layer were discussed. In brief, according to our preliminary works on the FPGA platform, the system can be realized with limited level glossaries in the existing technology. The next future step of this work is validating the simulation and preliminary laboratory testbed results under real application and environmental conditions.

**Acknowledgements**
The authors would like to thank the anonymous reviewers for their valuable comments and suggestions to improve the quality of the article.

**Funding**
There are no sources of funding body reported for this manuscript.

**Authors' contributions**
All authors contributed to the work. All authors read and approved the final manuscript.

**Competing interests**
The authors declare that they have no competing interests.

**References**

1. DM Dobkin, *RF Engineering for Wireless Networks: Hardware, Antennas, and Propagation (Communications Engineering)*. (Newnes, USA, 2004)
2. JG Proakis, M Salehi, *Digital Communication*, 5th edn. (McGraw-Hill Education, New York, 2008)
3. J Mitola, JGQ Maguire, Cognitive radio: making software radios more personal. Pers. Commun. IEEE. **6**(4), 13–18 (1999). doi:10.1109/98.788210
4. S Haykin, Cognitive radio: brain-empowered wireless communications. Selected Areas Commun. IEEE J. **23**(2), 201–220 (2005). doi:10.1109/JSAC.2004.839380
5. S Haykin, *Cognitive Dynamic Systems: Perception-action Cycle, Radar and Radio*. (Cambridge University Press, New York, 2012)
6. T Yucek, H Arslan, A survey of spectrum sensing algorithms for cognitive radio applications. Commun. Surv. Tutorials IEEE. **11**(1), 116–130 (2009). doi:10.1109/SURV.2009.090109
7. IF Akyildiz, W-Y Lee, MC Vuran, S Mohanty, NeXt generation/dynamic spectrum access/cognitive radio wireless networks: a survey. Intl. J. Comput. Telecommun. Netw. **50**(13), 2127–2159 (2006). doi:10.1016/j.comnet.2006.05.001
8. CE Shannon, A mathematical theory of communication. Bell Syst. Technical J. **27**(3), 379–423 (1948). doi:10.1002/j.1538-7305.1948.tb01338.x
9. B Ustundag, O Orcay, in *Cognitive Radio Oriented Wireless Networks and Communications, 2008. CrownCom 2008. 3rd International Conference On*. Pattern Based Encoding for Cognitive Communication, (2008), pp. 1–6. doi:10.1109/CROWNCOM.2008.4562494
10. B Ustundag, O Orcay, A pattern construction scheme for neural network-based cognitive communication. Entropy. **13**(1), 64–81 (2011). doi:10.3390/e13010064
11. J Mitola, *Cognitive Radio: An Integrated Agent Architecture for Software Defined Radio. Doctor of technology*. (Royal Institute Technology (KTH), Stockholm, 2000)
12. N Ahad, J Qadir, N Ahsan, Neural networks in wireless networks: techniques, applications and guidelines. J. Netw. Comput. Appl. **68**, 1–27 (2016). doi:10.1016/j.jnca.2016.04.006
13. N Abbas, Y Nasser, KE Ahmad, Recent advances on artificial intelligence and learning techniques in cognitive radio networks. EURASIP J. Wirel. Commun. Netw. **2015**(1), 174 (2015). doi:10.1186/s13638-015-0381-7
14. A He, KK Bae, TR Newman, J Gaeddert, K Kim, R Menon, L Morales-Tirado, JJ Neel, Y Zhao, JH Reed, WH Tranter, A survey of artificial intelligence for cognitive radios. Veh. Technol. IEEE Trans. **59**(4), 1578–1592 (2010). doi:10.1109/TVT.2010.2043968
15. M Alshawaqfeh, X Wang, AR Ekti, MZ Shakir, K Qaraqe, E Serpedin, in *Cognitive Radio Oriented Wireless Networks: 10th International Conference, CROWNCOM 2015, Doha, Qatar, April 21-23. Revised Selected Papers*, ed. by M Weichold, M Hamdi, ZM Shakir, M Abdallah, KG Karagiannidis, and M Ismail. A Survey of Machine Learning Algorithms and their Applications in Cognitive Radio (Springer, Cham, 2015), pp. 790–801
16. M Bkassiny, Y Li, SK Jayaweera, A survey on machine-learning techniques in cognitive radios. IEEE Commun. Surv. Tutorials. **15**(3), 1136–1159 (2013). doi:10.1109/SURV.2012.100412.00017
17. A Fehske, J Gaeddert, J Reed, in *New Frontiers in Dynamic Spectrum Access Networks, 2005. DySPAN 2005. 2005 First IEEE International Symposium On*. A new Approach to Signal Classification Using Spectral Correlation and Neural Networks, (2005), pp. 144–150. doi:10.1109/DYSPAN.2005.1542629
18. S Baban, D Denkoviski, O Holland, L Gavrilovska, H Aghvami, in *Personal Indoor and Mobile Radio Communications (PIMRC), 2013 IEEE 24th International Symposium On*. Radio Access Technology Classification for Cognitive Radio Networks, (2013), pp. 2718–2722. doi:10.1109/PIMRC.2013.6666608
19. S Baban, O Holland, H Aghvami, in *Wireless Communication Systems (ISWCS 2013), Proceedings of the Tenth International Symposium On*. Wireless Standard Classification in Cognitive Radio Networks Using Self-Organizing Maps, (Ilmenau, 2013), pp. 1–5
20. X Xing, T Jing, W Cheng, Y Huo, X Cheng, Spectrum prediction in cognitive radio networks. Wirel. Commun. IEEE. **20**(2), 90–96 (2013). doi:10.1109/MWC.2013.6507399
21. L Zhou, H Man, in *Vehicular Technology Conference (VTC Fall), 2013 IEEE 78th*. Wavelet Cyclic Feature Based Automatic Modulation Recognition Using Nonuniform Compressive Samples, (2013), pp. 1–6. doi:10.1109/VTCFall.2013.6692456
22. MMT Abdelreheem, MO Helmi, in *Telecommunications (BIHTEL), 2012 IX International Symposium On*. Digital Modulation Classification through Time and Frequency Domain Features using Neural Networks, (2012), pp. 1–5. doi:10.1109/BIHTEL.2012.6412073
23. JJ Popoola, R Van Olst, in *AFRICON, 2011*. Application of Neural Network for Sensing Primary Radio Signals in a Cognitive Radio Environment, (2011), pp. 1–6. doi:10.1109/AFRCON.2011.6072009

24. A Attar, A Sheikhi, A Zamani, in *Telecommunications and Networking - ICT 2004. Lecture Notes in Computer Science*, ed. by J de Souza, P Dini, and P Lorenz. Communication System Recognition by Modulation Recognition, vol. 3124 (Springer, Berlin Heidelberg, 2004), pp. 106–113
25. M Walenczykowska, A Kawalec, Type of modulation identification using wavelet transform and neural network. J. Pol. Acad. Sci. **64**(1), 257–261 (2016). doi:10.1515/bpasts-2016-0028
26. BI Dahap, L HongShu, M Ramadan, in *The Proceedings of the Second International Conference on Communications, Signal Processing, and Systems. Lecture Notes in Electrical Engineering*, ed. by B Zhang, J Mu, W Wang, Q Liang, and Y Pi. Simple and Efficient Algorithm for Automatic Modulation Recognition for Analogue and Digital Signals, vol. 246 (Springer, Switzerland, 2014), pp. 345–357
27. H Alzaq, BB Ustundag, in *European Wireless 2015; 21th European Wireless Conference, Proceedings Of.* Wavelet Preprocessed Neural Network Based Receiver for Low SNR Communication System, (Budapest, 2015), pp. 1–6
28. S Mallat, *A Wavelet, Tour of Signal Processing, The Sparse Way*, 3rd edn. (Academic Press, Philadelphia, 2008)
29. I Daubechies, *Ten Lectures on Wavelets. Society for Industrial and Applied Mathematics*, (Philadelphia, 1992)
30. S Haykin, *Neural Networks and Learning Machines*, 3rd edn. (Prentice-Hall, Inc., Upper Saddle River, 2008)
31. YF Hassan, Rough sets for adapting wavelet neural networks as a new classifier system. Appl. Intell. **35**(2), 260–268 (2011). doi:10.1007/s10489-010-0218-3
32. 4DSP, Design and system integration for digital signal processing. 4DSP - FMC150. Available at http://www.4dsp.com/FMC150.php/
33. Xilinx Inc., Virtex-6 FPGA ML605 Evaluation Kit. Available at http://www.xilinx.com/products/boards-and-kits/ek-v6-ml605-g.html
34. J Prothero, *The Shannon Law for non-periodic channels. Technical Report R-2012-1*. (Astrapi Corporation, Washington, D.C., 2012)
35. A Peled, B Liu, A new hardware realization of digital filters. Acoust. Speech Signal Process. IEEE Trans. **22**(6), 456–462 (1974). doi:10.1109/TASSP.1974.1162619
36. SA White, Applications of distributed arithmetic to digital signal processing: a tutorial review. ASSP Mag. IEEE. **6**(3), 4–19 (1989). doi:10.1109/53.29648
37. A Ebrahimzadeh, R Ghazalian, Blind digital modulation classification in software radio using the optimized classifier and feature subset selection. Eng. Appl. Artif. Intell. **24**(1), 50–59 (2011). doi:10.1016/j.engappai.2010.08.008
38. E Avci, D Hanbay, A Varol, An expert discrete wavelet adaptive network based fuzzy inference system for digital modulation recognition. Expert Syst. Appl. **33**(3), 582–589 (2007). doi:10.1016/j.eswa.2006.06.001
39. S Hassanpour, AM Pezeshk, F Behnia, in *2015 38th International Conference on Telecommunications and Signal Processing (TSP)*. A Robust Algorithm Based on Wavelet Transform for Recognition of Binary Digital Modulations, (2015), pp. 508–512. doi:10.1109/DASC.2012.6382368
40. D Digdarsini, M Kumar, G Khot, T Ram, V Tank, in *2014 International Conference on Signal Processing and Integrated Networks (SPIN)*. FPGA Implementation of Automatic Modulation Recognition System for Advanced SATCOM System, (2014), pp. 464–469. doi:10.1109/SPIN.2014.6776998

# 12

# Artificial noise-aided biobjective transmitter optimization for service integration in multi-user MIMO broadcast channel

Weidong Mei, Zhi Chen*, Jun Fang and Shaoqian Li

**Abstract**

This paper considers an artificial noise (AN)-aided transmit design for multi-user MIMO systems with integrated services. Specifically, two sorts of service messages are combined and served simultaneously: one multicast message intended for all receivers and one confidential message intended for only one receiver and required to be perfectly secure from other unauthorized receivers. Our interest lies in the joint design of input covariances of the multicast message, confidential message, and artificial noise (AN), such that the achievable secrecy rate and multicast rate are simultaneously maximized. This problem is identified as a secrecy rate region maximization (SRRM) problem in the context of physical-layer service integration. Since this biobjective optimization problem is inherently complex to solve, we put forward two different scalarization methods to convert it into a scalar optimization problem. First, we propose to prefix the multicast rate as a constant, and accordingly, the primal biobjective problem is converted into a secrecy rate maximization (SRM) problem with quality of multicast service (QoMS) constraint. By varying the constant, we can obtain different Pareto optimal points. The resulting SRM problem can be iteratively solved via a provably convergent difference-of-concave (DC) algorithm. In the second method, we aim to maximize the weighted sum of the secrecy rate and the multicast rate. Through varying the weighted vector, one can also obtain different Pareto optimal points. We show that this weighted sum rate maximization (WSRM) problem can be recast into a primal decomposable form, which is amenable to alternating optimization (AO). Then, we compare these two scalarization methods in terms of their overall performance and computational complexity via theoretical analysis as well as numerical simulation, based on which new insights can be drawn.

**Keywords:** Physical-layer service integration, Artificial noise, Convex optimization, Secrecy rate region

## 1 Introduction

### 1.1 Background

Recently, *physical-layer service integration* (PHY-SI), a technique of combining multicast service and confidential service into one integrated service for one-time transmission at the physical layer, has received much attention in wireless communications. For one thing, PHY-SI caters to the demand for high transmission rate and secure communication, which has been identified as the key targets that need to be effectively addressed by fifth generation (5G) wireless systems [1]. Besides, compared to the conventional upper-layer-based approach, PHY-SI enables coexisting services to share the same resources by solely exploiting the physical characteristics of wireless channels, thereby significantly increasing the spectral efficiency. This property makes PHY-SI a prominent approach to satisfy the ever-increasing need for radio spectrum. The technique of PHY-SI could also find a wide range of applications in the commercial and military areas. For example, many commercial applications, e.g., advertisement, digital television, and Internet telephony, are supposed to provide personalized service customization. As a consequence, confidential service and public service are collectively provided to satisfy the demand of different user groups. In battlefield scenarios, it is essential to propagate commands with different security levels to the frontline. The public information should be

*Correspondence: chenzhi@uestc.edu.cn
National Key Laboratory of Science and Technology on Communications, University of Electronic Science and Technology of China, No. 2006, Xiyuan Ave, West Hi-Tech Zone, Chengdu 611731, China

distributed to all soldiers, while the confidential information can only be accessed by specific soldiers. Such emerging applications lead to a crucial problem: how to establish the security of confidential service while not compromising the quality of public service?

### 1.2 Related works

Let us first have a very brief review on physical-layer security, a technique that lays foundation for the research on PHY-SI. The broadcast nature of wireless medium makes privacy an inherent concern. Physical layer security technique is playing an increasingly important role in wireless communication recently. It can secure communications information-theoretically at the physical layer without using secret keys whose distribution or management may become difficult in, e.g., ad hoc wireless networks. Different strategies against eavesdropping have been developed with various levels of channel state information (CSI) available to the transmitter (see the comprehensive overview in [2–6]). Liu and Poor first coined the term *confidential broadcasting* in [7, 8] and established the corresponding secrecy capacity region. In confidential broadcasting, a transmitter broadcasts multiple confidential messages to all receivers. Each confidential message is intended for one specified receiver but required to be perfectly secret from the others. Some efforts have been made in, e.g., [9, 10] to maximize the sum secrecy rate in the scenario of confidential broadcasting.

The study of PHY-SI can be traced back to Csiszár and Körner's seminal work in [11]. In the basic model of PHY-SI, a transmitter sends a common message to two receivers and simultaneously sends a confidential message intended only for one receiver and kept perfectly secure from the other one. Under discrete memoryless broadcast channel (DMBC) setup, Csiszár and Körner gave a closed-form expression of the maximum rate region that can be applied reliably under the secrecy constraint (i.e., the secrecy capacity region). In recent years, this kind of approach has gained renewed interest, especially that in various multi-antenna scenarios, such as Gaussian broadcast channels [12–15] and bidirectional relay channels [16, 17]. In [12], the authors extended the results in [11] to a general MIMO Gaussian case by adopting the channel-enhancement argument. Further, the works [13, 14] considered the case with two confidential messages intended for two different receivers. The resulting secrecy capacity region is proved to be attainable by combining the secret dirty-paper coding (S-DPC) with Gaussian superposition coding. Furthermore, in [16] and [17], Wyrembelski and Boche amalgamated broadcast service, multicast service, and confidential service in bidirectional relay networks, in which a relay adds an additional multicast message for all nodes and a confidential message for only one node besides establishing the conventional bidirectional communication. Nonetheless, the main goal of the aforementioned papers is just to obtain capacity results or to characterize coding strategies that lead to certain rate regions [18]. For implementation efficiency, it is also important to treat physical layer service integration from a signal processing point of view. In particular, optimal or complexity-efficient transmit strategies have to be characterized, so that the achieved performance could reach/approach the boundary of the secrecy capacity region. Such strategies are usually given by optimization problems, which generally turn out to be nonconvex. Along with this comes the fact that most works on PHY-SI end once a certain characterization of a rate region is derived.

Recently, to fill in the gap between the previous information-theoretic results and practical implementation, there is growing interest in analyzing PHY-SI from a signal processing point of view. In [12], the authors proposed a *re-parameterizing* method to devise transmit strategies for achieving the secrecy boundary performance. However, this method is only applicable to a very simple two-user MISO scenario. Besides, it involves solving a sequence of convex feasibility problems, which is computationally expensive. To improve it, the work [19] proposed a *quality-of-service (QoS)-based* method to seek the boundary-achieving transmit strategies. Its basic idea is to establish the tradeoff between the secrecy rate and the multicast rate by maximizing the secrecy rate while ensuring the multicast rate above a given threshold. This method is demonstrated as effective in characterizing the secrecy boundary and thus triggered research endeavors on extending the result to a more general and realistic setting. Notable results include the extension to the multi-user [20] and imperfect CSI [21, 22] settings. Even so, relatively less work focussed on the MIMO channel setup, due to the intractability of the associated optimization problems. In [23], the authors circumvented the intractability by proposing a generalized singular value decomposition (GSVD)-based transmission scheme. Using GSVD, multicast message and confidential message can be perfectly decoupled and the resulting problem is easier to handle. However, this result is not applicable to the general multi-user MIMO case. In addition, it is also interesting to incorporate artificial noise (AN) into consideration, as such technique has been shown to be effective in enhancing transmission security [24–28]. Specifically, the authors in [24–27], and [28] respectively showed that AN is of paramount importance to physical-layer security when there exist multiple eavesdroppers in the network, when the CSI of eavesdropper(s) is imperfectly known at the transmitter, and/or when eavesdroppers are randomly located in the network.

### 1.3 Main contributions

In this paper, we delve into the AN-aided transmit precoding design in PHY-SI under a general multi-user MIMO case. Specifically, two sorts of service messages are combined and promulgated at the same time: a multicast message intended for all receivers, and a confidential message intended for merely one authorized receiver. The confidential message must be kept perfectly secure from all other unauthorized receivers. Meanwhile, AN is employed to degrade the potential eavesdropping of the unauthorized receivers. This paper aims to jointly optimize the input covariance matrices of the multicast message, confidential message, and AN, to maximize the achievable secrecy and multicast rates simultaneously, or, equivalently, to maximize the achievable secrecy rate region. This secrecy rate region maximization (SRRM) problem turns out to be a biobjective optimization problem. Since the re-parameterizing method is invalid in a general MIMO case, we develop two effective scalarization methods to convert the biobjective problem into an easier-to-handle scalar version for characterizing its Pareto boundary.

1. In the first method, we propose to fix the multicast rate as a constant. Through varying the value of the constant, this method could yield different secrecy boundary points. Since the Pareto optimal points must reside on the boundary of the achievable rate region, this method is bound to provide a complete set of the Pareto optimal points. Though the resultant secrecy rate maximization (SRM) problem is nonconvex by nature, we show this problem actually falls into the context of difference-of-concave (DC) programming [29]. Hence, it can be handled by classical DC algorithm with convergence guarantee.
2. As for the second method, a weighted sum-based scalarization is introduced. The crux of this scalarization method is to optimize the weighted sum of the two objectives with different weight vectors. By varying the weight vector, this method gives rise to different Pareto optimal solution. To solve this weighted sum rate maximization (WSRM) problem, we reveal its hidden decomposability by recasting it as an equivalent form amenable to alternating optimization (AO). AO algorithm is naturally employed to solve the WSRM problem. It can be proved that this AO algorithm must converge to one stationary point of the WSRM problem.
3. It is particularly worth mentioning that though the DC and AO algorithms have been applied to address the issue of physical-layer security before in, e.g., [24, 30, 31], none of these works considered integrating an additional multicast message. Our paper is an initial attempt to study the application of DC and AO to the emerging PHY-SI system, which turns out to be a harder task than its counterpart in physical-layer security due to the coexisting multicast service.

Then, we compare these two sorts of scalarization methods in terms of their overall performance and computational complexity. The comparison results reveal that the first method is more efficacious in finding all Pareto optimal points than the second one. The advantage of the second method lies in its problem structure, which provides the service provider a solution to maximizing the overall revenue. Besides, we show that the DC algorithm is more time-efficient at low transmit power than the AO algorithm. Interestingly, the numerical results indicate that at high transmit power, the AO algorithm becomes the more time-efficient one.

### 1.4 Organization and notations

This paper is organized as follows. Section 2 provides the system model description and problem formulation. The optimization aspects of our formulated problems are addressed in Sections 3 and 4, corresponding to the first and the second scalarization methods, respectively. The comparison results are given in Section 5. Section 6 presents simulation results to show the efficacy of our proposed methods. Finally, conclusions are drawn in Section 7.

The notation of this paper is as follows. Bold symbols in capital letter and small letter denote matrices and vectors, respectively. $(\cdot)^H$, rank$(\cdot)$, and Tr$(\cdot)$ represent conjugate transpose, rank, and trace of a matrix, respectively. $\mathbb{R}_+$ and $\mathbb{H}_+^n$ denote the set of nonnegative real numbers and of $n$-by-$n$ Hermitian positive semidefinite (PSD) matrices. The $n \times n$ identity matrix is denoted by $\mathbf{I}_n$. $\mathbf{x} \sim \mathcal{CN}(\mu, \mathbf{\Omega})$ denotes that $\mathbf{x}$ is a complex circular Gaussian random vector with mean $\mu$ and covariance $\mathbf{\Omega}$. $\mathbf{A} \succeq \mathbf{0}$ ($\mathbf{A} \succ \mathbf{0}$) implies that $\mathbf{A}$ is a Hermitian positive semidefinite (definite) matrix. $\|\cdot\|$ represents the vector Euclidean norm. $K$ represents a proper cone, and $K^*$ represents a dual cone associated with $K$.

## 2 System model and problem formulation

We consider the downlink of a multiuser system in which a multi-antenna transmitter serves $K$ receivers, and each receiver is equipped with multiple antennas. Assume that all receivers have ordered the multicast service and receiver 1 further ordered the confidential service[1]. To enhance the security performance, the transmitter utilizes a fraction of its transmit power to send artificially generated noise to interfere the unauthorized receivers (eavesdroppers), i.e., receiver 2 to receiver $K$. We assume in this paper that all receivers are static and that all the communication links undergo slow frequency-flat fading.

The received signal at receiver $k$ is modeled as

$$\mathbf{y}_k = \mathbf{H}_k\mathbf{x} + \mathbf{z}_k, k = 1, 2, \cdots, K \quad (1)$$

where $\mathbf{H}_k \in \mathbb{C}^{N_{r,k}\times N_t}$ is the channel response between the transmitter and receiver $k$; $N_t$ and $N_{r,k}$ are the number of transmit antennas employed by the transmitter and $k$th receiver, respectively. $\mathbf{z}_k$ is independent identically distributed (i.i.d.) complex Gaussian noise with zero mean and unit variance. $\mathbf{x} \in \mathbb{C}^{N_t}$ is the coded transmit message, which consists of three independent components, i.e.,

$$\mathbf{x} = \mathbf{x}_0 + \mathbf{x}_c + \mathbf{x}_a, \quad (2)$$

where $\mathbf{x}_0$ is the multicast message intended for all receivers, $\mathbf{x}_c$ is the confidential message intended for receiver 1, and $\mathbf{x}_a$ is the artificial noise. We assume $\mathbf{x}_0 \sim \mathcal{CN}(\mathbf{0}, \mathbf{Q}_0)$, $\mathbf{x}_c \sim \mathcal{CN}(\mathbf{0}, \mathbf{Q}_c)$ [12], where $\mathbf{Q}_0$ and $\mathbf{Q}_c$ are the transmit covariance matrices. The AN $\mathbf{x}_a$ follows a distribution $\mathbf{x}_a \sim \mathcal{CN}(\mathbf{0}, \mathbf{Q}_a)$, where $\mathbf{Q}_a$ is the AN covariance. The CSI on all links is assumed to be perfectly known at the corresponding transmitter and receivers in that all receivers have to register in the network for subscribing the multicast service. In practice, the CSI at the receivers can be obtained from the channel estimation of the downlink pilots. CSI at the transmitter can be acquired via uplink channel estimation in time division duplex (TDD) systems. The design of a high-quality channel estimation scheme is beyond the scope of this paper. Note that the full CSI assumption is commonly adopted in the area of physical layer security/multicasting, especially in MIMO channels [10, 24, 30, 32–36].

For ease of exposition, let us define $\mathcal{K} \triangleq \{1, 2, \ldots, K\}$ and $\mathcal{K}_e \triangleq \mathcal{K}/\{1\}$, which denote the indices of all receivers and of all unauthorized receivers, respectively. Denote $R_0$ and $R_c$ as the achievable rates associated with the multicast and confidential messages, respectively. Then, an achievable secrecy rate region $R_s(\{\mathbf{H}_k\}_{k\in\mathcal{K}}, P)$ is given as the set of nonnegative rate pairs $(R_0, R_c)$ satisfying [12]

$$\begin{aligned} R_0 &\le \min_{k\in\mathcal{K}} C_{m,k}(\mathbf{Q}_0, \mathbf{Q}_c, \mathbf{Q}_a) \\ R_c &\le C_b(\mathbf{Q}_c, \mathbf{Q}_a) - \max_{k\in\mathcal{K}_e} C_{e,k}(\mathbf{Q}_c, \mathbf{Q}_a), \end{aligned} \quad (3)$$

where

$$\begin{aligned} C_{m,k} &= \log\left|\mathbf{I} + \left(\mathbf{I} + \mathbf{H}_k(\mathbf{Q}_c + \mathbf{Q}_a)\mathbf{H}_k^H\right)^{-1}\mathbf{H}_k\mathbf{Q}_0\mathbf{H}_k^H\right|, \\ C_b &= \log\left|\mathbf{I} + \left(\mathbf{I} + \mathbf{H}_1\mathbf{Q}_a\mathbf{H}_1^H\right)^{-1}\mathbf{H}_1\mathbf{Q}_c\mathbf{H}_1^H\right|, \\ C_{e,k} &= \log\left|\mathbf{I} + \left(\mathbf{I} + \mathbf{H}_k\mathbf{Q}_a\mathbf{H}_k^H\right)^{-1}\mathbf{H}_k\mathbf{Q}_c\mathbf{H}_k^H\right|, \end{aligned}$$

and $\mathrm{Tr}(\mathbf{Q}_0 + \mathbf{Q}_c + \mathbf{Q}_a) \le P$ with $P$ being total transmit power budget at the transmitter.

The secrecy rate region (3) implies that all receivers first decode their common multicast message by treating the confidential message as noise, and then receiver 1 acquires a clean link for the transmission of its exclusive confidential message, where there is no interference from the multicast message.

To maximize this achievable secrecy rate region, our goal is to find the boundary-achieving $\mathbf{Q}_0$, $\mathbf{Q}_a$ and $\mathbf{Q}_c$, which are also known as Pareto optimal solutions to this SRRM problem. Specifically, we must first solve the following optimization problem, which is a biobjective maximization problem with cone $K = K^* = \mathbb{R}_+^2$,

$$\begin{aligned} &\max_{\mathbf{Q}_0,\mathbf{Q}_c,\mathbf{Q}_a} \left(\text{w.r.t. } \mathbb{R}_+^2\right) \left(\min_{k\in\mathcal{K}} C_{m,k}, C_b - \max_{k\in\mathcal{K}_e} C_{e,k}\right) \\ &\text{s.t. } \mathrm{Tr}(\mathbf{Q}_0 + \mathbf{Q}_c + \mathbf{Q}_a) \le P, \quad (4a) \\ &\quad\ \mathbf{Q}_0 \succeq \mathbf{0}, \mathbf{Q}_c \succeq \mathbf{0}, \mathbf{Q}_a \succeq \mathbf{0}, \quad (4b) \end{aligned}$$

where, with a slight abuse of notations but for notational simplicity, the explicit dependence of $C_{m,k}$, $C_b$ and $C_{e,k}$ on $(\mathbf{Q}_0, \mathbf{Q}_c, \mathbf{Q}_a)$ is omitted. Since the SRRM problem is a biobjective maximization problem, it is necessary to harness some methods of scalarization to convert it into an easier-to-handle scalar version.

**Remark 1** *It is also viable to consider the scenario where all receivers order the confidential service and all confidential messages are propagated concurrently by the transmitter, i.e., the integration of multicasting and confidential broadcasting. The merit of this scheme lies in its higher spectral efficiency and low latency. However, this comes at the expense of much higher operational complexity at the transmitter, especially when the number of users increases. Thus, our considered PHY-SI scheme is particularly desired in delay-tolerant applications or when the transmitter possesses limited computational capacity for security-related computations.*

## 3 A DC-based approach to the SRRM problem

In this section, we develop our first scalarization method to solve (4). The resultant scalar problem is a secrecy rate maximization (SRM) with imposed quality of multicast service (QoMS) constraints.

### 3.1 Scalarization

In particular, our method is to move the part of multicast rate maximization to the constraint, i.e., we fix at the time being the multicast rate as a constant $\tau_{ms} \ge 0$. As a result, the biobjective SRRM problem (4) will be degraded into a scalar maximization problem, which is shown in (5).

$$\begin{aligned} R(\tau_{ms}) = &\max_{\mathbf{Q}_0,\mathbf{Q}_c,\mathbf{Q}_a} C_b(\mathbf{Q}_c, \mathbf{Q}_a) - \max_{k\in\mathcal{K}_e} C_{e,k}(\mathbf{Q}_c, \mathbf{Q}_a) \\ &\text{s.t. } \min_{k\in\mathcal{K}} C_{m,k}(\mathbf{Q}_0, \mathbf{Q}_c, \mathbf{Q}_a) = \tau_{ms}, \quad (5a) \\ &\quad\ \mathrm{Tr}(\mathbf{Q}_0 + \mathbf{Q}_c + \mathbf{Q}_a) \le P, \quad (5b) \\ &\quad\ \mathbf{Q}_0 \succeq \mathbf{0}, \mathbf{Q}_a \succeq \mathbf{0}, \mathbf{Q}_c \succeq \mathbf{0}. \quad (5c) \end{aligned}$$

In (5), $R(\tau_{ms})$ is the optimal objective value, $\tau_{ms}$ can be interpreted as the preset requirement on the multicast rate, and accordingly, the constraint (5a) can be interpreted as a QoMS constraint. To guarantee the feasibility of problem (5), $\tau_{ms}$ cannot exceed a threshold $\tau_{\max}$ given by

$$\tau_{\max} = \max_{\mathbf{Q}_0 \succeq \mathbf{0}, \mathrm{Tr}(\mathbf{Q}_0) \le P} \min_{k \in \mathcal{K}} \log \left| \mathbf{I} + \mathbf{H}_k \mathbf{Q}_0 \mathbf{H}_k^H \right|. \quad (6)$$

The value of $\tau_{\max}$ can be numerically obtained by solving (6) via the convex optimization solver CVX [37].

This sort of scalarization method, in fact, enables us to find one boundary point $(\tau_{ms}, R(\tau_{ms}))$ of the secrecy rate region $R_s(\{\mathbf{H}_k\}_{k\in\mathcal{K}}, P)$ by solving (5). All boundary points of $R_s(\{\mathbf{H}_k\}_{k\in\mathcal{K}}, P)$ can be found if we traverse all possible $\tau_{ms}$'s lying within $[0, \tau_{\max}]$ and store the corresponding optimal objective values. Since the Pareto optimal solution to (4) must reside on the boundary of $R_s(\{\mathbf{H}_k\}_{k\in\mathcal{K}}, P)$, i.e., the Pareto optimal set of (4) is a *subset* of the boundary set of $R_s(\{\mathbf{H}_k\}_{k\in\mathcal{K}}, P)$, all Pareto optimal solution to (5) can also be found by this means.

However, problem (5) is nonconvex. Especially, the determinant equality constraint (5a) is very difficult to handle. To circumvent this difficulty, we pay our attention to the following relaxed problem of (5), in which the equality constraint (5a) is replaced by the inequality constraint (7a).

$$\tilde{R}(\tau_{ms}) = \max_{\mathbf{Q}_0, \mathbf{Q}_c, \mathbf{Q}_a} C_b(\mathbf{Q}_c, \mathbf{Q}_a) - \max_{k \in \mathcal{K}_e} C_{e,k}(\mathbf{Q}_c, \mathbf{Q}_a)$$

$$\text{s.t.} \min_{k\in\mathcal{K}} C_{m,k}(\mathbf{Q}_0, \mathbf{Q}_c, \mathbf{Q}_a) \ge \tau_{ms}, \quad (7a)$$

$$\mathrm{Tr}(\mathbf{Q}_0 + \mathbf{Q}_c + \mathbf{Q}_a) \le P, \quad (7b)$$

$$\mathbf{Q}_0 \succeq \mathbf{0}, \mathbf{Q}_a \succeq \mathbf{0}, \mathbf{Q}_c \succeq \mathbf{0}. \quad (7c)$$

Apparently, any optimal solution to (5) is feasible to (7) in the sense that replacing (5a) with (7a) yields a larger feasible solution set. Hence, problem (7) has $R(\tau_{ms}) \le \tilde{R}(\tau_{ms})$ in general. Interestingly, we show that $R(\tau_{ms}) = \tilde{R}(\tau_{ms})$ can always be achieved without loss of optimality to (7).

**Lemma 1** *Problem (7) is a tight relaxation to problem (5). In other words, the rate pair $(\tau_{ms}, \tilde{R}(\tau_{ms}))$ must be a boundary point of $R_s(\{\mathbf{H}_k\}_{k\in\mathcal{K}}, P)$.*

*Proof* The proof can be easily accomplished by construction. Suppose that the constraint (7a) is satisfied with strict inequality, we can always multiply $\mathbf{Q}_0$ by a scalar $\nu$ ($\nu < 1$) to make (7a) active, yet without decreasing the objective value of (7) and violating the total power constraint (7b). This fact implies that there always exists an optimal solution to (7) such that the constraint (7a) is satisfied with equality and thus accomplishes the proof. □

Lemma 1 implies that problem (7) admits an optimal $(\mathbf{Q}_0^*, \mathbf{Q}_c^*, \mathbf{Q}_a^*)$ with $\min_{k\in\mathcal{K}} C_{m,k}(\mathbf{Q}_0^*, \mathbf{Q}_c^*, \mathbf{Q}_a^*) = \tau_{ms}$. Hence, $(\mathbf{Q}_0^*, \mathbf{Q}_c^*, \mathbf{Q}_a^*)$ is also optimal to (5). The proof of Lemma 1 reveals that such an optimal $(\mathbf{Q}_0^*, \mathbf{Q}_c^*, \mathbf{Q}_a^*)$ can always be constructed algorithmically based on the following procedures.

**Corollary 1** *Suppose that $(\mathbf{Q}_0^*, \mathbf{Q}_c^*, \mathbf{Q}_a^*)$ is an optimal solution returned by solving problem (7). If $\min_{k\in\mathcal{K}} C_{m,k}(\mathbf{Q}_0^*, \mathbf{Q}_c^*, \mathbf{Q}_a^*) = \tau_{ms}$, then output $(\mathbf{Q}_0^*, \mathbf{Q}_c^*, \mathbf{Q}_a^*)$ as an optimal solution of problem (5). Otherwise, solve the following equation with regard to $\nu$, i.e., $\min_{k\in\mathcal{K}} C_{m,k}(\nu\mathbf{Q}_0^*, \mathbf{Q}_c^*, \mathbf{Q}_a^*) = \tau_{ms}$, via bisection search within the unit interval $[0, 1]$, and output $(\nu\mathbf{Q}_0^*, \mathbf{Q}_c^*, \mathbf{Q}_a^*)$ as an optimal solution of problem (5).*

Next, we will point out two special cases, under which problem (7) is *equivalent* to problem (5) or, equivalently, any optimal solution to (7) is achieved with constraint (7a) active. This is described in the following proposition.

**Proposition 1** *Suppose that the system configurations satisfy either one of the following conditions:*

*Condition 1: The number of antennas at the transmitter is larger than that at the authorized receiver, i.e., $N_t > N_{r,1}$.*

*Condition 2: The number of antennas at the transmitter is larger than the sum of the antenna number at the unauthorized receivers, i.e., $N_t > \sum_{k\in\mathcal{K}_e} N_{r,k}$.*

*Then, the rate pair $(\tau_{ms}, \tilde{R}(\tau_{ms}))$ must be a Pareto optimal point of (4), and all Pareto optimal points of (4) can be obtained by solving (5) with different $\tau_{ms}$'s lying within the interval $[0, \tau_{\max}]$.*

*Proof* The proof can be found in Appendix A. □

**Remark 2** *Proposition 1 bridges the Pareto optimal points of (4) to the boundary points of $R_s(\{\mathbf{H}_k\}_{k\in\mathcal{K}}, P)$. When either Condition 1 or Condition 2 is satisfied, all Pareto optimal points of (4) are also the boundary points of $R_s(\{\mathbf{H}_k\}_{k\in\mathcal{K}}, P)$ and vice versa.*

### 3.2 DC iterative algorithm

We now focus on solving the relaxed problem (7) derived in the last subsection. Problem (7) still remains nonconvex due to its objective function and constraint (7a). To deal with it, we first equivalently transform it into its epigraph form by introducing a slack variable $\eta$, i.e.,

$$R(\tau_{ms}) = \max_{\mathbf{Q}_0,\mathbf{Q}_c,\mathbf{Q}_a,\eta} C_b(\mathbf{Q}_c,\mathbf{Q}_a) - \eta$$
$$\text{s.t. } C_{e,k}(\mathbf{Q}_c,\mathbf{Q}_a) \le \eta, \forall k \in \mathcal{K}_e \quad (8a)$$
$$C_{m,k}(\mathbf{Q}_0,\mathbf{Q}_c,\mathbf{Q}_a) \ge \tau_{ms}, \forall k \in \mathcal{K} \quad (8b)$$
$$\mathrm{Tr}(\mathbf{Q}_0 + \mathbf{Q}_c + \mathbf{Q}_a) \le P, \quad (8c)$$
$$\mathbf{Q}_0 \succeq \mathbf{0}, \mathbf{Q}_a \succeq \mathbf{0}, \mathbf{Q}_c \succeq \mathbf{0}. \quad (8d)$$

Next, we will show that problem (8) constitutes a DC-type programming problem, which can be iteratively solved by the DC algorithm.

To begin with, we reformulate the capacity functions $C_b(\mathbf{Q}_c,\mathbf{Q}_a)$, $C_{e,k}(\mathbf{Q}_c,\mathbf{Q}_a)$, and $C_{m,k}(\mathbf{Q}_0,\mathbf{Q}_c,\mathbf{Q}_a)$ into a DC-type form, given by

$$\begin{aligned} C_b(\mathbf{Q}_c,\mathbf{Q}_a) &= \phi_1(\mathbf{Q}_c,\mathbf{Q}_a) - \varphi_1(\mathbf{Q}_a), \\ C_{e,k}(\mathbf{Q}_c,\mathbf{Q}_a) &= \phi_k(\mathbf{Q}_c,\mathbf{Q}_a) - \varphi_k(\mathbf{Q}_a), \\ C_{m,k}(\mathbf{Q}_0,\mathbf{Q}_c,\mathbf{Q}_a) &= \eta_k(\mathbf{Q}_0,\mathbf{Q}_c,\mathbf{Q}_a) - \phi_k(\mathbf{Q}_c,\mathbf{Q}_a), \end{aligned} \quad (9)$$

in which we define

$$\begin{aligned} \phi_k(\mathbf{Q}_c,\mathbf{Q}_a) &= \log\left|\mathbf{I} + \mathbf{H}_k(\mathbf{Q}_c + \mathbf{Q}_a)\mathbf{H}_k^H\right|, \\ \varphi_k(\mathbf{Q}_a) &= \log\left|\mathbf{I} + \mathbf{H}_k\mathbf{Q}_a\mathbf{H}_k^H\right|, \\ \eta_k(\mathbf{Q}_0,\mathbf{Q}_c,\mathbf{Q}_a) &= \log\left|\mathbf{I} + \mathbf{H}_1(\mathbf{Q}_c + \mathbf{Q}_a + \mathbf{Q}_0)\mathbf{H}_1^H\right|. \end{aligned} \quad (10)$$

Substituting (9) into problem (8), we obtain

$$R(\tau_{ms}) = \max_{\mathbf{Q}_0,\mathbf{Q}_c,\mathbf{Q}_a,\eta} \phi_1(\mathbf{Q}_c,\mathbf{Q}_a) - \varphi_1(\mathbf{Q}_a) - \eta$$
$$\text{s.t. } \varphi_k(\mathbf{Q}_a) - \phi_k(\mathbf{Q}_c,\mathbf{Q}_a) + \eta \ge 0, \forall k \in \mathcal{K}_e \quad (11a)$$
$$\eta_k(\mathbf{Q}_0,\mathbf{Q}_c,\mathbf{Q}_a) - \phi_k(\mathbf{Q}_c,\mathbf{Q}_a) \ge \tau_{ms}, \forall k \in \mathcal{K} \quad (11b)$$
$$\mathrm{Tr}(\mathbf{Q}_0 + \mathbf{Q}_c + \mathbf{Q}_a) \le P, \quad (11c)$$
$$\mathbf{Q}_0 \succeq \mathbf{0}, \mathbf{Q}_a \succeq \mathbf{0}, \mathbf{Q}_c \succeq \mathbf{0}. \quad (11d)$$

Since $\phi_k(\mathbf{Q}_c,\mathbf{Q}_a)$, $\varphi_k(\mathbf{Q}_a)$ and $\eta_k(\mathbf{Q}_0,\mathbf{Q}_c,\mathbf{Q}_a)$ are all concave w.r.t. $(\mathbf{Q}_0,\mathbf{Q}_c,\mathbf{Q}_a)$, one can easily notice that the objective function of (5) and constraints (11a) and (11b) are all in a difference-of-concave form. This property makes problem (4) fall into the context of DC program [29], which can be iteratively solved via DC algorithm.

Our next endeavor is to show the DC approach to (11) mathematically. Its basic idea is to locally linearize the nonconcave parts in (11) at some feasible point via Taylor series expansion (TSE) and then iteratively solve the linearized problem. To this end, we introduce the TSE via the following lemma.

**Lemma 2** (Chu et al. [31]) *An affine Taylor series approximation of a function $f(\mathbf{X}) : \mathbb{R}^{M\times N} \to \mathbb{R}$ can be expressed at $\tilde{\mathbf{X}}$ as below.*

$$f(\mathbf{X}) \approx f(\tilde{\mathbf{X}}) + \mathrm{vec}\left(f'(\mathbf{X})\right)^H \mathrm{vec}(\mathbf{X} - \tilde{\mathbf{X}}). \quad (12)$$

The TSE above enables us to reformulate the primal nonconcave parts of (11) into a linear form. In particular, by applying Lemma 2 and the fact $\partial\left(\log|\mathbf{X}|\right) = \mathrm{Tr}\left(\mathbf{X}^{-1}\partial\mathbf{X}\right)$, $\varphi_1(\mathbf{Q}_a)$ can be approximated as

$$\begin{aligned} \varphi_1(\mathbf{Q}_a) &= \log\left|\mathbf{I} + \mathbf{H}_1\mathbf{Q}_a\mathbf{H}_1^H\right| \\ &\approx \varphi_1(\tilde{\mathbf{Q}}_a) + (\mathrm{vec}\,(\mathbf{S}))^H \mathrm{vec}\left(\mathbf{Q}_a - \tilde{\mathbf{Q}}_a\right) \\ &\overset{(a)}{=} \varphi_1(\tilde{\mathbf{Q}}_a) + \mathrm{Tr}\left[\mathbf{S}(\mathbf{Q}_a - \tilde{\mathbf{Q}}_a)\right] \\ &\triangleq \tilde{\varphi}_1(\mathbf{Q}_a) \end{aligned} \quad (13)$$

in the objective function of (11), where $\tilde{\mathbf{Q}}_a$ is a given transmit covariance matrix,

$$\mathbf{S} \triangleq \mathbf{H}_1^H\left(\mathbf{I} + \mathbf{H}_1\tilde{\mathbf{Q}}_a\mathbf{H}_1^H\right)^{-1}\mathbf{H}_1$$

and the equality $(a)$ is due to the fact that $\mathrm{Tr}(\mathbf{A}^H\mathbf{B}) = (\mathrm{vec}(\mathbf{A}))^H\mathrm{vec}(\mathbf{B})$ for appropriate dimensions of $\mathbf{A}$ and $\mathbf{B}$. Likewise, $\phi_k(\mathbf{Q}_c,\mathbf{Q}_a)$, appearing in the constraints (11a) and (11b), can be approximated as

$$\begin{aligned} \phi_k(\mathbf{Q}_c,\mathbf{Q}_a) &= \log\left|\mathbf{I} + \mathbf{H}_k(\mathbf{Q}_c + \mathbf{Q}_a)\mathbf{H}_k^H\right| \\ &\approx \phi_k(\tilde{\mathbf{Q}}_c,\tilde{\mathbf{Q}}_a) + \mathrm{Tr}\left[\mathbf{U}(\mathbf{Q}_c - \tilde{\mathbf{Q}}_c)\right] \\ &\quad + \mathrm{Tr}\left[\mathbf{U}(\mathbf{Q}_a - \tilde{\mathbf{Q}}_a)\right] \\ &\triangleq \tilde{\phi}_k(\mathbf{Q}_c,\mathbf{Q}_a), \end{aligned} \quad (14)$$

in which $\mathbf{U}$ is determined by

$$\mathbf{U} = \mathbf{H}_k^H\left(\mathbf{I} + \mathbf{H}_k(\tilde{\mathbf{Q}}_c + \tilde{\mathbf{Q}}_a)\mathbf{H}_k^H\right)^{-1}\mathbf{H}_k. \quad (15)$$

Based on the approximations above, the original QoMS-constrained SRM problem (11) can be reformulated as

$$\bar{R}(\tau_{ms}) = \max_{\mathbf{Q}_0,\mathbf{Q}_c,\mathbf{Q}_a,\eta} \phi_1(\mathbf{Q}_c,\mathbf{Q}_a) - \tilde{\varphi}_1(\mathbf{Q}_a) - \eta$$
$$\text{s.t. } \varphi_k(\mathbf{Q}_a) - \tilde{\phi}_k(\mathbf{Q}_c,\mathbf{Q}_a) + \eta \ge 0, \forall k \in \mathcal{K}_e \quad (16a)$$
$$\eta_k(\mathbf{Q}_0,\mathbf{Q}_c,\mathbf{Q}_a) - \tilde{\phi}_k(\mathbf{Q}_c,\mathbf{Q}_a) \ge \tau_{ms}, \forall k \in \mathcal{K} \quad (16b)$$
$$\mathrm{Tr}(\mathbf{Q}_0 + \mathbf{Q}_c + \mathbf{Q}_a) \le P, \quad (16c)$$
$$\mathbf{Q}_0 \succeq \mathbf{0}, \mathbf{Q}_a \succeq \mathbf{0}, \mathbf{Q}_c \succeq \mathbf{0}. \quad (16d)$$

where $\bar{R}(\tau_{ms})$ is the optimal objective value of (8), serving as an approximation to $R(\tau_{ms})$. According to the relationship between a concave function and its Taylor series expansion, it is immediate to get

$$\begin{aligned} &\varphi_1(\mathbf{Q}_a) \le \tilde{\varphi}_1(\mathbf{Q}_a), \forall \mathbf{Q}_a \succeq \mathbf{0}, \\ &\phi_k(\mathbf{Q}_c,\mathbf{Q}_a) \le \tilde{\phi}_k(\mathbf{Q}_c,\mathbf{Q}_a), \forall \mathbf{Q}_a \succeq \mathbf{0}, \mathbf{Q}_c \succeq \mathbf{0}. \end{aligned} \quad (17)$$

As a consequence, any feasible solution to (16) should also be feasible to (11), and $\bar{R}(\tau_{ms}) \le R(\tau_{ms})$ must hold.

This approximated problem (16) is convex with regard to (w.r.t.) $(\mathbf{Q}_0,\mathbf{Q}_c,\mathbf{Q}_a)$ and hence $(\mathbf{Q}_0,\mathbf{Q}_c,\mathbf{Q}_a)$ can be iteratively obtained by solving problem (16) via the off-the-shelf interior-point algorithm, i.e., CVX. We summarize our proposed iterative algorithm for solving (11) in

Algorithm 1. To acquire the secrecy rate region, we need to traverse $\tau_{ms}$ lying within the interval $[0, \tau_{\max}]$ and store the corresponding objective value of (16).

**Algorithm 1** Iterative method for solving (11)

1: Initiate $n = 0$ and choose an arbitrary starting point $(\tilde{\mathbf{Q}}_{c,n}, \tilde{\mathbf{Q}}_{a,n})$ feasible to (16)
2: **Repeat**
3: Solve (16) with $\tilde{\mathbf{Q}}_c = \tilde{\mathbf{Q}}_{c,n}$ and $\tilde{\mathbf{Q}}_a = \tilde{\mathbf{Q}}_{a,n}$, and obtain $(\mathbf{Q}_c^*, \mathbf{Q}_a^*)$, which is the optimal solution of (16);
4: Update $\tilde{\mathbf{Q}}_{c,n+1} = \mathbf{Q}_c^*, \tilde{\mathbf{Q}}_{a,n+1} = \mathbf{Q}_a^*$;
5: Update $n = n + 1$;
6: **Until the convergence conditions are satisfied.**
7: Output $\tilde{\mathbf{Q}}_{c,n}$ and $\tilde{\mathbf{Q}}_{a,n}$.

**Remark** 3 *In Algorithm 1, the initialization of $(\tilde{\mathbf{Q}}_{c,0}, \tilde{\mathbf{Q}}_{a,0})$ plays a crucial role in influencing the total iteration times. Let us define $(\mathbf{Q}_c^i, \mathbf{Q}_a^i)$ as the output solution in ith traversal of $\tau_{ms}$. The following "warmstart operation" could be adopted to initialize $(\tilde{\mathbf{Q}}_{c,0}, \tilde{\mathbf{Q}}_{a,0})$ for achieving a fast convergence rate:*

***Warmstart operation**: We start the traversal of $\tau_{ms}$ from $\tau_{ms} = \tau_{\max}$. In the first traversal, $\tilde{\mathbf{Q}}_{c,0}$ and $\tilde{\mathbf{Q}}_{a,0}$ are both initialized as $\mathbf{0}$. In the ith $(i > 1)$ traversal, $(\tilde{\mathbf{Q}}_{c,0}, \tilde{\mathbf{Q}}_{a,0})$ is initialized as the solution output by Algorithm 1 in the $(i-1)$th traversal.*

### 3.3 Convergence analysis

As one can see, the basic merit of DC lies in its tractability, which caters to the numerical optimization using the parser-solver. As an additional merit, the proposed DC approach has a theoretically provable guarantee on its solution convergence, which will be demonstrated in the following proposition.

**Proposition 2** *Every limit point of $(\mathbf{Q}_0^*, \mathbf{Q}_c^*)$ is a stationary point of problem* (7)

*Proof* The proof is a direct application of ([29], Th 10) and thus omitted here for simplicity. □

## 4 An AO-based approach to the SRRM problem

In this section, we develop our other scalarization method, referred to as weighted-sum method, to solve (4). The resulting problem is essentially a WSRM problem, which can be solved via an AO-based approach. Here, we should point out that the application of AO to SRM problem has been observed in some existing papers, i.e., [24]. Nonetheless, the AO algorithm we used in this section is a nontrivial extension of that in [24]. Specifically, the objective function in [24] only contains a single secrecy rate term. While in our considered scenario, an extra multicast rate term is incorporated, which brings some new issues, say, the convergence proof, that should be tackled.

### 4.1 Scalarization

The basic idea of the weighted-sum method is to introduce a so-called weight vector [38] that is positive in the dual cone $K^* = \mathbb{R}_+^2$ and then to transform the primal vector optimization problem into a scalar optimization problem. By varying the vector, we can obtain different Pareto optimal solutions of (4).

To put into context, the Pareto boundary of (3) can be characterized by the solution of

$$\begin{aligned} \max_{\mathbf{Q}_0,\mathbf{Q}_c,\mathbf{Q}_a,R_0,R_c} \quad & R_0 + \lambda_c R_c \\ \text{s.t.} \quad & R_0 \le \min_{k\in\mathcal{K}} C_{m,k}(\mathbf{Q}_0, \mathbf{Q}_c, \mathbf{Q}_a) \\ & R_c \le C_b(\mathbf{Q}_c, \mathbf{Q}_a) - \max_{k\in\mathcal{K}_e} C_{e,k}(\mathbf{Q}_c, \mathbf{Q}_a) \\ & \text{(4a)–(4b) satisfied,} \end{aligned} \tag{18}$$

in which $\lambda_c \in [0, +\infty)$ and $\boldsymbol{\lambda} = [1, \lambda_c]$ are our introduced weight vector. In general, the optimal $(R_0, R_c)$ to (5) is the point where a straight line with slope $-1/\lambda_c$ is tangent to the Pareto boundary. Before proceeding, let us first point out some special cases of problem (5).

1. When $\boldsymbol{\lambda} = [1, 1]$, the optimal $(R_0, R_c)$ turns out to be the so-called utilitarian point, also referred to as "sum-rate" point in communications.
2. The single-service points are the two points where $R_0 = 0$ and where $R_c = 0$, respectively. When $R_0 = 0$, problem (5) is degraded into a conventional AN-aided SRM problem in MIMO wiretap channel. When $R_c = 0$, the maximum $R_0$ can be derived by solving the same convex optimization problem as (6).

### 4.2 AO iterative algorithm

We are now in a position to determine the tractable approaches to the WSRM problem (18). First, one can notice that by discarding $R_0$ and $R_c$ as slack variables, problem (18) is equivalent to the following optimization problem.

$$R(\lambda_c) = \max_{\mathbf{Q}_0,\mathbf{Q}_c,\mathbf{Q}_a} \lambda_c(C_b - \max_{k\in\mathcal{K}_e} C_{e,k}) + \min_{k\in\mathcal{K}} C_{m,k}$$

$$\text{s.t.} \quad \mathrm{Tr}(\mathbf{Q}_0 + \mathbf{Q}_c + \mathbf{Q}_a) \le P, \tag{19a}$$

$$\mathbf{Q}_0 \succeq \mathbf{0}, \mathbf{Q}_c \succeq \mathbf{0}, \mathbf{Q}_a \succeq \mathbf{0}. \tag{19b}$$

The obstacle of solving (19) mainly lies in the nonsmoothness of its objective function, which negates the use of many derivative-related iterative algorithms. As a result, we next develop a derivative-free AO iterative algorithm to solve (19). To this end, we will first need to transform the WSRM problem (19) into a form amenable to AO.

**Lemma 3** (Li et al. [24]) *Let* $\mathbf{E} \in \mathbb{C}^{N\times N}$ *be any matrix satisfying* $\mathbf{E} \succ \mathbf{0}$*. Define the function* $f(\mathbf{S}) = -\mathrm{Tr}(\mathbf{SE}) + \log|\mathbf{S}| + N$*. Then,*

$$\log\left|\mathbf{E}^{-1}\right| = \max_{\mathbf{S}\in\mathbb{C}^{N\times N}, \mathbf{S}\succeq \mathbf{0}} f(\mathbf{S}), \tag{20}$$

*and the optimal solution to the right-hand side (RHS) of (33) is* $\mathbf{S}^* = \mathbf{E}^{-1}$.

Applying Lemma 3 to $C_b$, $C_{e,k}$, and $C_{m,k}$, one can obtain

$$\begin{aligned} C_b &= \max_{\mathbf{S}_1\succeq\mathbf{0}} \varphi_b(\mathbf{Q}_c, \mathbf{Q}_a, \mathbf{S}_1), \\ C_{e,k} &= \min_{\mathbf{S}_k\succeq\mathbf{0}} \varphi_{e,k}(\mathbf{Q}_c, \mathbf{Q}_a, \mathbf{S}_k), \forall k \in \mathcal{K}_e, \\ C_{m,k} &= \max_{\mathbf{U}_k\succeq\mathbf{0}} \varphi_{m,k}(\mathbf{Q}_0, \mathbf{Q}_c, \mathbf{Q}_a, \mathbf{U}_k), \forall k \in \mathcal{K}, \end{aligned} \tag{21}$$

where we define

$$\begin{aligned} \varphi_b(\mathbf{Q}_c, \mathbf{Q}_a, \mathbf{S}_1) &= -\mathrm{Tr}(\mathbf{S}_1\left(\mathbf{I} + \mathbf{H}_1\mathbf{Q}_a\mathbf{H}_1^H\right) + \log|\mathbf{S}_1| + N_{r,1} \\ &\quad + \log\left|\mathbf{I} + \mathbf{H}_1(\mathbf{Q}_a + \mathbf{Q}_c)\mathbf{H}_1^H\right|, \\ \varphi_{e,k}(\mathbf{Q}_c, \mathbf{Q}_a, \mathbf{S}_k) &= -\log|\mathbf{S}_k| - \log\left|\mathbf{I} + \mathbf{H}_k\mathbf{Q}_a\mathbf{H}_k^H\right| - N_{r,k} \\ &\quad + \mathrm{Tr}\left(\mathbf{S}_k\left(\mathbf{I} + \mathbf{H}_k(\mathbf{Q}_a + \mathbf{Q}_c)\mathbf{H}_k^H\right)\right), \\ \varphi_{m,k}(\mathbf{Q}_0, \mathbf{Q}_c, \mathbf{Q}_a, \mathbf{U}_k) &= -\mathrm{Tr}(\mathbf{U}_k\left(\mathbf{I} + \mathbf{H}_k(\mathbf{Q}_c + \mathbf{Q}_a)\mathbf{H}_k^H\right) + \log|\mathbf{U}_k| \\ &\quad + \log\left|\mathbf{I} + \mathbf{H}_k(\mathbf{Q}_0 + \mathbf{Q}_c + \mathbf{Q}_a)\mathbf{H}_k^H\right| + N_{r,k}, \end{aligned} \tag{22}$$

in which $\{\mathbf{S}_k\}_{k\in\mathcal{K}}$ and $\{\mathbf{U}_k\}_{k\in\mathcal{K}}$ are slack variables satisfying $\mathbf{S}_k \succeq \mathbf{0}$ and $\mathbf{U}_k \succeq \mathbf{0}$ for $\forall k \in \mathcal{K}$.

Following the matrix manipulations in [24], we have

$$\begin{aligned} &\max_{k\in\mathcal{K}_e} \min_{\mathbf{S}_k} \varphi_{e,k}(\mathbf{Q}_c, \mathbf{Q}_a, \mathbf{S}_k) \\ &= \min_{\{\mathbf{S}_k\}_{k\in\mathcal{K}_e}} \max_{k\in\mathcal{K}_e} \varphi_{e,k}(\mathbf{Q}_c, \mathbf{Q}_a, \mathbf{S}_k), \end{aligned} \tag{23}$$

and

$$\begin{aligned} &\min_{k\in\mathcal{K}} \max_{\mathbf{U}_k} \varphi_{m,k}(\mathbf{Q}_0, \mathbf{Q}_c, \mathbf{Q}_a, \mathbf{U}_k) \\ &= \max_{\{\mathbf{U}_k\}_{k\in\mathcal{K}}} \min_{k\in\mathcal{K}} \varphi_{m,k}(\mathbf{Q}_0, \mathbf{Q}_c, \mathbf{Q}_a, \mathbf{U}_k). \end{aligned} \tag{24}$$

Substituting (21) into (19) and making use of (23) and (24), one can check that problem (19) is equivalent to the following optimization problem.

$$\begin{aligned} R(\lambda_c) = &\max_{\substack{\mathbf{Q}_0,\mathbf{Q}_c,\mathbf{Q}_a,\\ \{\mathbf{S}_k,\mathbf{U}_k\}_{k\in\mathcal{K}}}} f(\mathbf{Q}_0, \mathbf{Q}_c, \mathbf{Q}_a, \{\mathbf{S}_k, \mathbf{U}_k\}_{k\in\mathcal{K}}) \\ &\text{s.t.} \quad \mathrm{Tr}(\mathbf{Q}_0 + \mathbf{Q}_c + \mathbf{Q}_a) \le P, \qquad (25a) \\ &\qquad\ \ \mathbf{Q}_0 \succeq \mathbf{0}, \mathbf{Q}_c \succeq \mathbf{0}, \mathbf{Q}_a \succeq \mathbf{0}, \qquad (25b) \end{aligned}$$

in which we define

$$\begin{aligned} &f(\mathbf{Q}_0, \mathbf{Q}_c, \mathbf{Q}_a, \{\mathbf{S}_k, \mathbf{U}_k\}_{k\in\mathcal{K}}) = \\ &\quad \lambda_c[\varphi_b(\mathbf{Q}_c, \mathbf{Q}_a, \mathbf{S}_1) - \max_{k\in\mathcal{K}_e} \varphi_{e,k}(\mathbf{Q}_c, \mathbf{Q}_a, \{\mathbf{S}_k\}_{k\in\mathcal{K}_e})] \\ &\quad + \min_{k\in\mathcal{K}} \varphi_{m,k}(\mathbf{Q}_0, \mathbf{Q}_c, \mathbf{Q}_a, \{\mathbf{U}_k\}_{k\in\mathcal{K}}). \end{aligned} \tag{26}$$

The upshot of this reformation is that problem (19) becomes primal decomposable. Specifically, problem (25) is convex w.r.t. *either* $(\mathbf{Q}_0, \mathbf{Q}_c, \mathbf{Q}_a)$ *or* $(\{\mathbf{S}_k\}_{k\in\mathcal{K}}, \{\mathbf{U}_k\}_{k\in\mathcal{K}})$. Hence, AO is naturally employed to solve (25). With $(\mathbf{Q}_0, \mathbf{Q}_c, \mathbf{Q}_a)$ fixed, the optimal solution of $(\{\mathbf{S}_k\}_{k\in\mathcal{K}}, \{\mathbf{U}_k\}_{k\in\mathcal{K}})$ admits an analytical expression, according to Lemma 3, given by

$$\mathbf{S}_1^* = \left(\mathbf{I} + \mathbf{H}_1\mathbf{Q}_a\mathbf{H}_1^H\right)^{-1}, \tag{27a}$$

$$\mathbf{S}_k^* = \left(\mathbf{I} + \mathbf{H}_k(\mathbf{Q}_a + \mathbf{Q}_c)\mathbf{H}_k^H\right)^{-1}, \forall k \in \mathcal{K}_e, \tag{27b}$$

$$\mathbf{U}_k^* = \left(\mathbf{I} + \mathbf{H}_k(\mathbf{Q}_a + \mathbf{Q}_c)\mathbf{H}_k^H\right)^{-1}, \forall k \in \mathcal{K}, \tag{27c}$$

in which we utilize the fact that $\{\mathbf{S}_k\}_{k\in\mathcal{K}}$ and $\{\mathbf{U}_k\}_{k\in\mathcal{K}}$ are decoupled among $\varphi_b$, $\varphi_{e,k}$, and $\varphi_{m,k}$. Comparatively, with $(\{\mathbf{S}_k\}_{k\in\mathcal{K}}, \{\mathbf{U}_k\}_{k\in\mathcal{K}})$ fixed, the optimal solution of $(\mathbf{Q}_0, \mathbf{Q}_c, \mathbf{Q}_a)$ can be obtained by solving a convex optimization problem as below, i.e.,

$$\begin{aligned} &(\mathbf{Q}_0^*, \mathbf{Q}_c^*, \mathbf{Q}_a^*) = \\ &\quad \arg \max_{(\mathbf{Q}_0,\mathbf{Q}_c,\mathbf{Q}_a)\in\mathcal{F}} f(\mathbf{Q}_0, \mathbf{Q}_c, \mathbf{Q}_a, \{\mathbf{S}_k, \mathbf{U}_k\}_{k\in\mathcal{K}}), \end{aligned} \tag{28}$$

where $\mathcal{F}$ denotes the feasible set of (19), which is convex.

The whole AO process for solving (25) is given in Algorithm 2. In line 6 of Algorithm 2, the convex subproblem can be solved via CVX. Following the similar warmstart operation introduced in Remark 3, the iteration times of Algorithm 2 can be significantly decreased.

---

**Algorithm 2** AO algorithm for solving (25)

1: Initiate $n = 1$, and $(\mathbf{Q}_c^0, \mathbf{Q}_a^0) \in \mathcal{F}$;
2: **Repeat**
3: $\quad \mathbf{S}_1^n = \left(\mathbf{I} + \mathbf{H}_1\mathbf{Q}_a^{n-1}\mathbf{H}_1^H\right)^{-1}$;
4: $\quad \mathbf{S}_k^n = \left(\mathbf{I} + \mathbf{H}_k\left(\mathbf{Q}_a^{n-1} + \mathbf{Q}_c^{n-1}\right)\mathbf{H}_k^H\right)^{-1}, \forall k \in \mathcal{K}_e$;
5: $\quad \mathbf{U}_k^n = \left(\mathbf{I} + \mathbf{H}_k\left(\mathbf{Q}_a^{n-1} + \mathbf{Q}_c^{n-1}\right)\mathbf{H}_k^H\right)^{-1}, \forall k \in \mathcal{K}$;
6: $\quad (\mathbf{Q}_0^n, \mathbf{Q}_c^n, \mathbf{Q}_a^n) = \arg \max_{(\mathbf{Q}_0,\mathbf{Q}_c,\mathbf{Q}_a)\in\mathcal{F}} f\left(\mathbf{Q}_0, \mathbf{Q}_c, \mathbf{Q}_a, \{\mathbf{S}_k^n\}_{k\in\mathcal{K}}, \{\mathbf{U}_k^n\}_{k\in\mathcal{K}}\right)$;
7: $\quad n = n + 1$;
8: **Until the convergence conditions are satisfied.**
9: Output $(\mathbf{Q}_0^n, \mathbf{Q}_c^n, \mathbf{Q}_a^n)$.

---

### 4.3 Convergence analysis

It can be verified that the AO algorithm produces a non-decreasing objective value of (25). Besides, the following convergence result is always guaranteed.

**Proposition 3** *Suppose that* $(\mathbf{Q}_0^n, \mathbf{Q}_c^n, \mathbf{Q}_a^n)$ *is the solution generated by the AO algorithm in nth iteration, then the sequence* $\{(\mathbf{Q}_0^n, \mathbf{Q}_c^n, \mathbf{Q}_a^n)\}_n$ *must converge to one stationary point (i.e., Karush-Kuhn-Tucker (KKT) point) of the primal WSRM problem (19).*

*Proof* The proof can be found in Appendix B. □

## 5 Comparison of the proposed methods

In the previous sections, we present two tractable convex formulations of the SRRM problem (4). This naturally leads to the question about the relative performance of the two formulations. In the following subsections, we address this question by comparing their performance and computational complexity in solving (4).

### 5.1 Performance analysis

As introduced in the preceding sections, the QoMS-based scalarization can yield a complete set of boundary points of $R_s(\{\mathbf{H}_k\}_{k\in\mathcal{K}}, P)$, which contains all Pareto optimal points of (4). The resulting scalar problem (7) aims to maximize the secrecy rate and meanwhile maintain the QoMS above a given threshold. Predictably, the use of AN should be effective merely at low QoMS region, since AN exerts a negative effect on the multicasting performance. To guarantee the high demand for QoMS, AN has to be prohibitive at high QoMS region. This QoMS-constrained SRM is a generalization of traditional SRM in physical-layer security and provides the transmitter with some insights in how to tradeoff the security performance and the multicasting performance.

As for the weighted-sum scalarization method, the necessary condition for it to find all Pareto optimal points is that the secrecy rate region should be convex. Besides, its performance is also dependent on the precision of $\lambda_c$. The traversal of $\lambda_c$ should span from zero to an extremely large number with appropriate step, so that each Pareto optimal points can be detected. Nonetheless, the weighted-sum problem structure has an interesting pricing interpretation from the field of economics. To elaborate a little further, let us define $p_0$ and $p_c$ as the unit price for the secrecy rate and the multicast rate, respectively, charged by the service provider. To maximize its revenue, the service provider should be concerned about how to solve the WSRM problem in (18) with setting $\lambda_c = p_c/p_0$. The use of AN could also be explained in this context. It is evident to see when $p_0 \gg p_c$, the revenue from multicasting transmission would dominate the objective function of (18), and thus, eliminating AN would be helpful in increasing the overall revenue.

In all, these two scalarization methods are suitable for different application scenarios and provide different insights. Nonetheless, the QoMS-based scalarization could yield all Pareto optimal points, while the weighted-sum scalarization might only yield some of them, dependent on the shape of the secrecy rate region.

**Remark 4** *Besides the QoMS-based and weighted-sum scalarization methods, some other scalarization methods have been proposed in literature to find the complete Pareto set for biobjective optimization, e.g., the weighted Tchebycheff method* [39]*. However, to implement this method, one has to first obtain the single-service point of the confidential message (cf. (29)) and then solve a highly nonconvex max-min optimization problem.*

$$R_c^{\max} = \max_{\mathbf{Q}_c \succeq \mathbf{0}, \mathrm{Tr}(\mathbf{Q}_c) \le P} \log\left|\mathbf{I} + \mathbf{H}_1\mathbf{Q}_c\mathbf{H}_1^H\right| - \max_{k\in\mathcal{K}} \log\left|\mathbf{I} + \mathbf{H}_k\mathbf{Q}_c\mathbf{H}_k^H\right|. \quad (29)$$

*Unfortunately, problem* (29) *is nonconvex, and so the optimal solution to* (29) *may not be obtained, which invalidates the use of the weighted Tchebycheff method.*

### 5.2 Complexity analysis

The major computational complexity of the two scalarization methods comes from solving the problems (16) and (28). While both of problems (16) and (28) are convex, they are not in a standard semidefinite programming (SDP) form, owing to the logarithm functions therein. To solve them, a successive approximation method embedded with a primal-dual interior-point method (IPM) is employed, say by CVX. As is known, the arithmetic complexity for the generic primal-dual IPM to solve a standard SDP is $\mathcal{O}(\max\{m,n\}^4 n^{1/2}\log(1/\varepsilon))$ [40], in which $m$, $n$, and $\varepsilon$ represent the number of linear constraints, the dimension of the positive semidefinite cone, and the solution accuracy, respectively. Therefore, the complexity of solving (16) or (28) is $\mathcal{O}(L_{SA}\max\{2K, N_t\}^4 N_t^{1/2}\log(1/\varepsilon))$, where $L_{SA}$ denotes the number of successive approximations used. Since we are not aware of the relation between $L_{SA}$ and $N_t$, this complexity expression is rather rough.

However, by utilizing the following approximation [41]:

$$\log\left|\mathbf{I} + \mathbf{HQH}^H\right| = \mathrm{Tr}\left(\mathbf{HQH}^H\right) + \mathcal{O}\left(\|\mathbf{HQH}^H\|\right), \quad (30)$$

all logarithm terms in problems (16) and (28) can be approximated by a trace function at *low* transmit power. This approximation further converts the convex problems (16) and (28) into SDP ones, which makes it possible to acquire a more accurate big-O expression of the computational complexity for low transmit power.

Specifically, consider (16), which has three linear matrix inequality (LMI) constraints of size $N_t$, and $2K$ LMI constraints of size 1 after introducing the approximation (30). Moreover, for (16), the number of decision variables is on the order $n_1 = 3N_t^2 + 1$. Then, when a generic path-following IPM is used to solve problem (16), the total arithmetic computation cost is on the order of [42]

$$T_1 = \sqrt{2K + 3N_t}\phi(n_1),$$
$$\phi(n_1) = n_1\left(2K + 3N_t^3\right) + n_1^2\left(2K + 3N_t^2\right) + n_1^3 \quad (31)$$

with $n_1 = \mathcal{O}\left(3N_t^2 + 1\right)$.

On the other hand, for solving (28), we need to introduce two additional slack variables to move the maximum and minimum terms in the objective function of (28) to the constraints. Hence, the number of decision variables is on the order of $n_2 = 3N_t^2 + 2$, and (28) also has three LMI constraints of size $N_t$, and $2K$ LMI constraints of size 1. The total arithmetic computation cost for solving (28) is in the order of

$$\begin{aligned} T_2 &= \sqrt{2K + 3N_t}\phi(n_2), \\ \phi(n_2) &= n_2\left(2K + 3N_t^3\right) + n_2^2\left(2K + 3N_t^2\right) + n_2^3 \end{aligned} \tag{32}$$

with $n_2 = \mathcal{O}\left(3N_t^2 + 2\right)$.

Comparing (31) and (32), one can note that the total arithmetic computation cost of solving the two problems is comparable, with $T_2$ slightly greater than $T_1$ due to $n_2 > n_1$. This observation implies that the QoMS-based scalarization is more time-efficient at low transmit power. This is also consistent with our following simulation results, as we shall see in Section 6.

## 6 Numerical results

In this section, we provide numerical results to illustrate the secrecy rate region derived from the two proposed methods, compared with two other existing strategies. The first one is the no-AN transmission, i.e., prefixing $\mathbf{Q}_a$ as $\mathbf{0}$ in problem (4). Thus, its achieved secrecy rate region can also be derived via the DC and AO algorithms. Another one is the traditional service integration using time division multiple address (TDMA), which assigns the confidential message and multicast message to two orthogonal time slots. Its maximum secrecy rate and multicast rate can be obtained by seeking the single-service points of $R_s(\{\mathbf{H}_k\}_{k\in\mathcal{K}}, P)$. For the fairness of comparison, the secrecy rate and multicast rate achieved by this TDMA-based strategy should be **halved**[17].

In the first subsection, the convergence results of both algorithms are presented. The second subsection gives the comparison between these two algorithms in terms of achievable performance and computational complexity.

### 6.1 Convergence results

In this subsection, we assume $N_t = 5$, $N_{r,k} = 3$ for all $k \in \mathcal{K}$, and $K = 4$. The channel matrices are randomly generated from an i.i.d. complex Gaussian distribution with zero mean and unit variance. According to Proposition 1, since $N_t > N_{r,1}$, the optimal solution to (7) is attained when the constraint (7a) holds with equality.

First, we evaluate the convergence of the DC algorithm. Especially, we are concerned about whether the primal constraint (7a) is violated by our approximation. Setting $\tau_{ms}$ as 2 bps/Hz, Fig. 1 shows the convergence of the multicast rate in the iteration with different transmit power. $\tilde{\mathbf{Q}}_{c,0}$ and $\tilde{\mathbf{Q}}_{a,0}$ are both initiated as $\mathbf{0}$. The algorithm stops iterating when the difference between two successive values of $\bar{R}(\tau_{ms})$ returned by the algorithm is less than or equal to $10^{-4}$. One can observe that the multicast rates ultimately converge to our predefined multicast rate with a limited number of iterations in all tested transmit powers. This observation indicates the efficacy of TSE in approximating the multicast rate. Then, we also plot the achieved secrecy rates and the approximated secrecy rates in Fig. 2. The general observation of Fig. 1 is also applicable to Fig. 2.

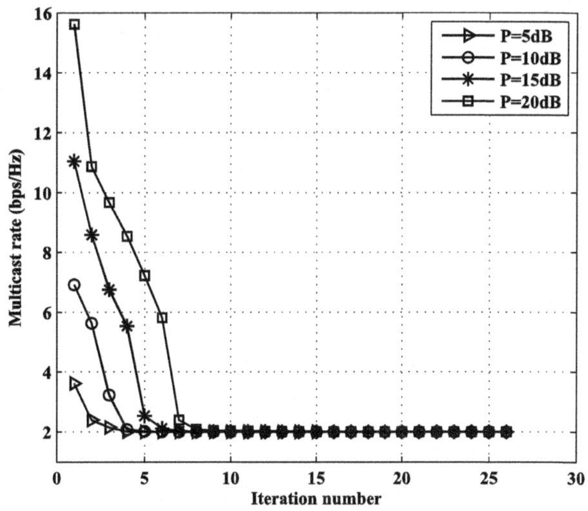


**Fig. 1** DC algorithm: convergence of the multicast rate

The convergence results of the AO algorithm are presented in Fig. 3. In Fig. 3, we set $\lambda_c = 1$ to seek the sum-rate point. $\mathbf{Q}_c^0$ and $\mathbf{Q}_a^0$ are both initialized as $(P/(2N_t))\mathbf{I}_{N_t}$. The algorithm stops iterating when the difference between two successive values of $\bar{R}(\lambda_c)$ is less than

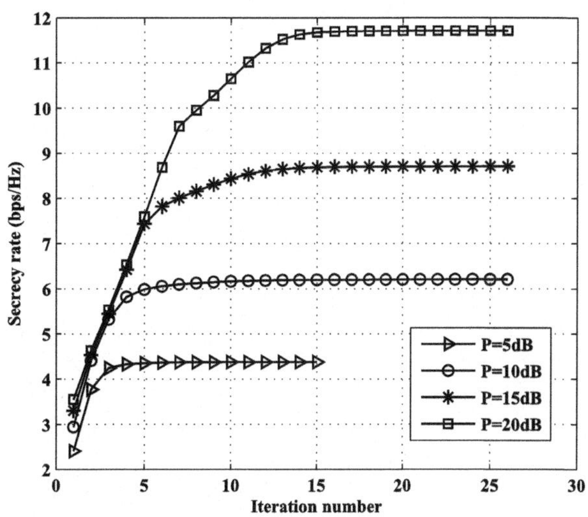


**Fig. 2** DC algorithm: convergence of the secrecy rate

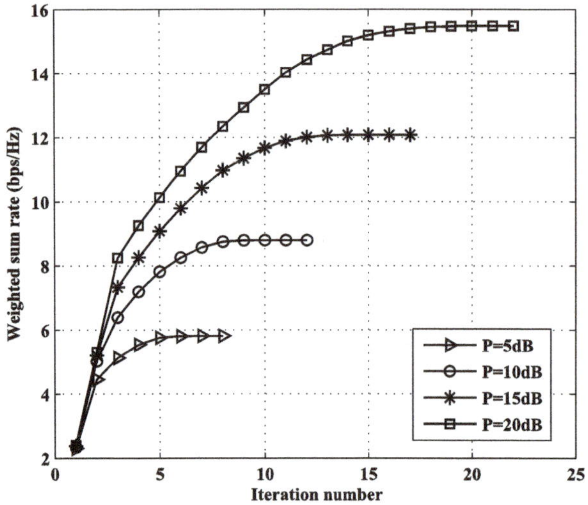


**Fig. 3** AO algorithm: convergence of the weighted sum rate

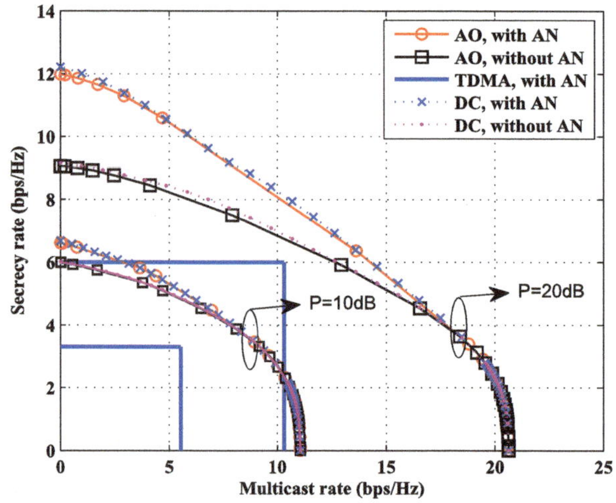


**Fig. 4** Secrecy rate regions with and without AN (Config 1)

or equal to $10^{-4}$. As one can observe from Fig. 3, the achieved weighted sum rate is monotonically increasing and finally converges with a limited number of iterations in all tested transmit powers. In addition, we find out that the AN covariance matrix $\mathbf{Q}_a$ output by AO is no longer diagonal. This implies that the associated AN design is spatially selective rather than isotropic, which blocks the eavesdroppers much more effectively. One can also note that the increase in the weighted sum rate is particularly remarkable when the transmit power is high. After all, higher transmit power means that the transmitter can allocate more power to the confidential message transmission, while not compromising the multicast performance. The extra power allocated to the confidential message can be used to generate more interference at the eavesdropper and/or strengthen the signal reception at the intended receiver, whereby more remarkable improvement is observed.

## 6.2 Performance comparison

In this subsection, we focus on two sorts of system configuration. The first one is the same as that in the last subsection. Besides, we consider another sort of system configuration: $N_t = N_{r,1} = 4$, $N_{r,k} = 5$ for all $k \in \mathcal{K}_e$, and $K = 4$. Under the second system configuration, *neither* Condition 1 *nor* Condition 2 is satisfied.

First, we will show the secrecy rate regions achieved by the first system configuration. Overall results are shown in Fig. 4, with $P$ set as 10 and 20 dB, respectively. Figure 4 reveals two general trends. First, our AN-aided scheme achieves a secrecy rate region larger than the no-AN one. The striking gap indicates the efficacy of AN in expanding the secrecy rate region. However, the gap between these two strategies dramatically reduces when $R_0$ increases. This phenomenon agrees with our conjecture in Section 5.1. The second observation is that our proposed strategies, though only attain a lower bound on $R_s(\{\mathbf{H}_k\}_{k\in\mathcal{K}}, P)$, is sufficient to achieve significantly larger secrecy rate regions than the TDMA-based one. This observation also implies that PHY-SI is an effective approach to improve the spectral efficiency. Then, let us compare the achievable performance of the two proposed scalarization methods. One can notice that the performance gap between these two methods is negligible in the tested system configuration, especially when $P = 10$ dB.

Figure 5 plots the secrecy rate regions achieved by the second system configuration. Still, the secrecy rate region with AN is larger than the one without AN and the one achieved by TDMA. Besides, we can observe two very interesting phenomena. First, when we increase the transmit power from 10 to 20 dB, the secrecy rate regions practically expand in the horizontal direction. That is, under the second system configuration, the increasing transmit power mainly contributes to the multicast message transmission, rather than the confidential message transmission. This can be interpreted from the transmit degree of freedom (d.o.f.). The total d.o.f. of unauthorized receivers is $\sum_{k=2}^{K-1} N_{r,k} = 15$, much higher than the transmit d.o.f. $N_t = 4$. The high d.o.f. at the unauthorized receivers leads to the d.o.f. bottleneck at the transmitter and thus compromises the overall secrecy performance. Second, one can notice that when $P = 20$ dB,

**1)** There exist some boundary points residing on a line, marked by the red dashed lines, that are not Pareto optimal to (4). Apparently, these points cannot be detected by the weighted-sum scalarization but can be easily detected by the QoMS-based scalarization.

**2)** The QoMS-based scalarization detects more Pareto optimal points than the weighted-sum scalarization. This

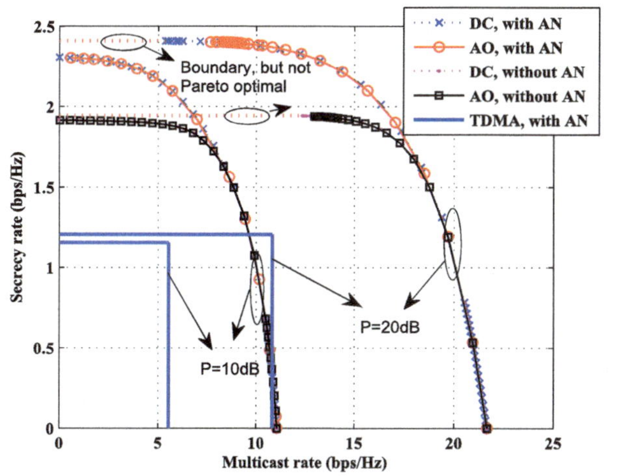


**Fig. 5** Secrecy rate regions with and without AN (Config 2)

is attributed to the insensitivity of the weighted-sum scalarization to the points residing on an approximately horizontal boundary. To detect these boundary points, one has to precisely adjust the value of $\lambda_c$ to get different tangent points.

### 6.3 Complexity comparison

Finally, we tabulated the averaged running times of DC and AO for obtaining a boundary point in Table 1 under the same setting as Fig. 4. As seen, the DC algorithm runs faster than the AO algorithm when the transmit power is low. This phenomenon is consistent with our preceding analysis in Section 5.2. However, at high transmit power, the DC algorithm scales nearly exponentially with $P$ and gradually spends more time converging in each iteration than the AO algorithm. This observation indicates that the two proposed scalarization methods might exhibit a performance-complexity tradeoff at high transmit power.

## 7 Conclusions

In this paper, we considered the AN-aided transmit design for multiuser MIMO broadcast channel with confidential service and multicast service. The transmit covariance matrices of confidential message, multicast message, and AN were designed to maximize the achievable secrecy rate and achievable multicast rate simultaneously. To deal with this biobjective optimization problem, two different sorts of scalarization were introduced to transform this SRRM problem into a scalar optimization problem. In the QoMS-based scalarization, the scalar problem is an SRM problem with QoMS constraints, while in the weighted-sum scalarization, the scalar problem is a WSRM problem. DC and AO algorithms were utilized to solve the QoMS-constrained SRM problem and the WSRM problem, respectively. Both algorithms can converge to a stationary point of the primal problems. Further, we gave a detailed comparison between the two proposed scalarization methods. The comparison results indicated that at low transmit power, the QoMS-based scalarization is superior to the weighted-sum one in terms of achievable performance and computational complexity. On the other hand, at high transmit power, these two methods exhibit a tradeoff between achievable performance and computational complexity. Numerical results also confirmed the effectiveness of AN in expanding the secrecy rate region.

As a future direction, it would be interesting to analyze the robust service integration scheme to combat the possible CSI uncertainties caused by channel aging and to take into account some application-specific requirements in 5G wireless communication system, e.g., the mobility of terminals and the overhead in CSI acquisition.

**Table 1** Averaged running times (in seconds)

| Power (dB) | | | | | | |
|---|---|---|---|---|---|---|
| Method | 0 | 4 | 8 | 12 | 16 | 20 |
| DC | 6.07 | 8.89 | 12.91 | 17.35 | 21.18 | 30.84 |
| AO | 7.57 | 11.58 | 11.04 | 12.61 | 13.61 | 17.11 |

## Endnote

[1] In this paper, we assume that only one receiver orders the confidential service within a single time slot. In practice, this assumption is valid under the case where the confidential service is provided to all receivers in a *round-robin* manner, i.e., the time slots are assigned to each subscriber of the confidential service in equal portions and in circular order.

## Appendix A: proof of Proposition 1

First, we claim that problem (7) has a following interesting property provided that Condition 1 or Condition 2 is satisfied.

**Property 1** *The maximum objective value of problem* (7), $R(\tau_{ms})$, *is obtained only when the equality in* (7a) *holds.*

*Proof* The proof of Property 1 can be accomplished by contradiction. Assume that the maximum value of problem (7) is obtained at the solution $(\hat{\mathbf{Q}}_0, \hat{\mathbf{Q}}_c, \hat{\mathbf{Q}}_a)$ and the equality in (7a) does not hold, i.e.,

$$\min_{k\in\mathcal{K}} \log|\mathbf{I} + \left(\mathbf{I} + \mathbf{H}_k(\hat{\mathbf{Q}}_c + \hat{\mathbf{Q}}_a)\mathbf{H}_k^H\right)^{-1}\mathbf{H}_k\hat{\mathbf{Q}}_0\mathbf{H}_k^H| > \tau_{ms}.$$

Our next step is to construct a new solution $(\bar{\mathbf{Q}}_0, \bar{\mathbf{Q}}_c, \bar{\mathbf{Q}}_a)$ from $(\hat{\mathbf{Q}}_0, \hat{\mathbf{Q}}_c, \hat{\mathbf{Q}}_a)$, which achieves a larger objective value and satisfies the constraint (7a) with equality. Let us first elaborate upon the construction method under Condition 1.

1) *Case for Condition 1*: Specifically, we multiply $\hat{\mathbf{Q}}_0$ by a scaling factor $\xi$ $(0 < \xi < 1)$, add a positive semidefinite (PSD) matrix $\mathbf{E} = \rho\mathbf{I} - \rho\mathbf{H}_1^H(\mathbf{H}_1\mathbf{H}_1^H)^{-1}\mathbf{H}_1$ to $\hat{\mathbf{Q}}_a$, and keep $\hat{\mathbf{Q}}_c$ constant, i.e., $\bar{\mathbf{Q}}_0 = \xi\hat{\mathbf{Q}}_0$, $\bar{\mathbf{Q}}_a = \hat{\mathbf{Q}}_a + \mathbf{E}$ and $\bar{\mathbf{Q}}_c = \hat{\mathbf{Q}}_c$, where the coefficient $\rho$ controls the power of $\mathbf{E}$. Note that $\mathbf{E}$ is the orthogonal complement projector of $\mathbf{H}_1^H$, and its existence is guaranteed by Condition 1. To keep the total transmit power constant, the coefficient $\rho$ should be chosen to satisfy $(1-\xi)\text{Tr}(\hat{\mathbf{Q}}_0) = \text{Tr}(\mathbf{E}) = \rho(N_t - N_{r,1})$, that is, $\rho = \frac{(1-\xi)\text{Tr}(\hat{\mathbf{Q}}_0)}{N_t - N_{r,1}}$. To proceed, we need the following lemma.

**Lemma 4** (Weingarten et al. [43]) *For matrices* $\mathbf{A}, \Delta \succeq \mathbf{0}$ *and* $\mathbf{B} \succ \mathbf{0}$, *the following inequality hold:*

$$\frac{|\mathbf{A}+\mathbf{B}|}{|\mathbf{B}|} \geq \frac{|\mathbf{A}+\mathbf{B}+\Delta|}{|\mathbf{B}+\Delta|}. \tag{33}$$

□

Then, by applying Lemma 1, one can obtain

$$\begin{aligned}
&C_{m,k}(\hat{\mathbf{Q}}_0, \hat{\mathbf{Q}}_c, \hat{\mathbf{Q}}_a)\\
&= \log|\mathbf{I} + \left(\mathbf{I} + \mathbf{H}_k(\hat{\mathbf{Q}}_c + \hat{\mathbf{Q}}_a)\mathbf{H}_k^H\right)^{-1}\mathbf{H}_k\hat{\mathbf{Q}}_0\mathbf{H}_k^H|\\
&> \log|\mathbf{I} + (\mathbf{I} + \mathbf{H}_k(\bar{\mathbf{Q}}_c + \bar{\mathbf{Q}}_a)\mathbf{H}_k^H)^{-1}\mathbf{H}_k\bar{\mathbf{Q}}_0\mathbf{H}_k^H|\\
&= C_{m,k}(\bar{\mathbf{Q}}_0, \bar{\mathbf{Q}}_c, \bar{\mathbf{Q}}_a)
\end{aligned} \tag{34}$$

for any $k \in \mathcal{K}$. Thus, by adjusting the value of $\xi$, the equality in (7) could be achieved.

To proceed, we will show that a larger objective value could always be achieved by $(\bar{\mathbf{Q}}_0, \bar{\mathbf{Q}}_c, \bar{\mathbf{Q}}_a)$. By reapplying Lemma 1, it is easy to get

$$\begin{aligned}
&C_{e,k}(\bar{\mathbf{Q}}_c, \bar{\mathbf{Q}}_a)\\
&= \log|\mathbf{I} + (\mathbf{I} + \mathbf{H}_k\bar{\mathbf{Q}}_a\mathbf{H}_k^H)^{-1}\mathbf{H}_k\bar{\mathbf{Q}}_c\mathbf{H}_k^H|\\
&= \log|\mathbf{I} + \left(\mathbf{I} + \mathbf{H}_k(\hat{\mathbf{Q}}_a + \mathbf{E})\mathbf{H}_k^H\right)^{-1}\mathbf{H}_k\hat{\mathbf{Q}}_c\mathbf{H}_k^H|\\
&< \log|\mathbf{I} + \left(\mathbf{I} + \mathbf{H}_k\hat{\mathbf{Q}}_a\mathbf{H}_k^H\right)^{-1}\mathbf{H}_k\hat{\mathbf{Q}}_c\mathbf{H}_k^H|,\\
&= C_{e,k}(\hat{\mathbf{Q}}_c, \hat{\mathbf{Q}}_a), \forall k \in \mathcal{K}_e.
\end{aligned} \tag{35}$$

Meanwhile, due to $\mathbf{H}_1\mathbf{E}\mathbf{H}_1^H = \mathbf{0}$, it is easy to see

$$C_b(\bar{\mathbf{Q}}_c, \bar{\mathbf{Q}}_a) = C_b(\hat{\mathbf{Q}}_c, \hat{\mathbf{Q}}_a). \tag{36}$$

Combining (35) with (36), we obtain

$$\begin{aligned}
&C_b(\bar{\mathbf{Q}}_c, \bar{\mathbf{Q}}_a) - \max_{k\in\mathcal{K}_e} C_{e,k}(\bar{\mathbf{Q}}_c, \bar{\mathbf{Q}}_a)\\
&\quad > C_b(\hat{\mathbf{Q}}_c, \hat{\mathbf{Q}}_a) - \max_{k\in\mathcal{K}_e} C_{e,k}(\hat{\mathbf{Q}}_c, \hat{\mathbf{Q}}_a),
\end{aligned} \tag{37}$$

i.e., a larger objective value can be found with $(\bar{\mathbf{Q}}_0, \bar{\mathbf{Q}}_c, \bar{\mathbf{Q}}_a)$. This fact is contrary to the primal assumption.

2) *Case for Condition 2*: The only difference between the proof for Condition 1 and Condition 2 lies in the construction method of $(\bar{\mathbf{Q}}_0, \bar{\mathbf{Q}}_c, \bar{\mathbf{Q}}_a)$. To begin with, let us first define a matrix $\mathbf{H}_{ua} \triangleq [\mathbf{H}_2^H, \mathbf{H}_3^H, \cdots, \mathbf{H}_K^H]^H \in \mathbb{C}^{\sum_{k\in\mathcal{K}_e} N_{r,k}\times N_t}$, which stacks all of the unauthorized receivers' channel matrices. Then, we multiply $\hat{\mathbf{Q}}_0$ by a scaling factor $\xi$ $(0 < \xi < 1)$, add a PSD matrix $\mathbf{E} = \rho\mathbf{I} - \rho\mathbf{H}_{ua}^H(\mathbf{H}_{ua}\mathbf{H}_{ua}^H)^{-1}\mathbf{H}_{ua}$ to $\hat{\mathbf{Q}}_c$, and keep $\hat{\mathbf{Q}}_a$ constant, i.e., $\bar{\mathbf{Q}}_0 = \xi\hat{\mathbf{Q}}_0$, $\bar{\mathbf{Q}}_c = \hat{\mathbf{Q}}_c + \mathbf{E}$ and $\bar{\mathbf{Q}}_a = \hat{\mathbf{Q}}_a$, where the coefficient $\rho$ controls the power of $\mathbf{E}$. $\mathbf{E}$ is the orthogonal complement projector of $\mathbf{H}_{ua}^H$, the existence of which is guaranteed by Condition 2. The coefficient $\rho$ should be chosen to satisfy $\rho = \frac{(1-\xi)\text{Tr}(\hat{\mathbf{Q}}_0)}{N_t - \sum_{k\in\mathcal{K}_e} N_{r,k}}$ to keep the total transmit power constant.

Again, by exploiting Lemma 1 and carrying out some matrix manipulations, one can verify that $(\bar{\mathbf{Q}}_0, \bar{\mathbf{Q}}_c, \bar{\mathbf{Q}}_a)$ can achieve a larger objective value than $(\hat{\mathbf{Q}}_0, \hat{\mathbf{Q}}_c, \hat{\mathbf{Q}}_a)$ with the constraint (7a) active. This fact contradicts the primal assumption.

Summarizing the conclusions drawn from the two cases above, we have accomplished the proof of Property 1.

Property 1 makes the proof of ([19], Theorem 1) fully applicable to the proposition here. The remaining parts of the proof can be found in [19] and are omitted here for simplicity.

## Appendix B: proof of Proposition 3

Firstly, we introduce slack variables $\alpha$ and $\beta$ to reexpress (8) as

$$\begin{aligned}
\max_{\mathbf{Q}_0,\mathbf{Q}_c,\mathbf{Q}_a,\alpha,\beta}\quad & \lambda_c(C_b - \beta) + \alpha\\
\text{s.t.}\quad & C_{e,k} \leq \beta, \forall k \in \mathcal{K}_e, & \text{(38a)}\\
& C_{m,k} \geq \alpha, \forall k \in \mathcal{K}, & \text{(38b)}\\
& \text{(19a)–(19b) are satisfied.}
\end{aligned}$$

Equivalently, it suffices to prove that every limit point $(\tilde{\mathbf{Q}}_0, \tilde{\mathbf{Q}}_c, \tilde{\mathbf{Q}}_a)$ of the iterates generated by the AO algorithm, together with $\tilde{\alpha} = \min_{k\in\mathcal{K}} C_{m,k}(\tilde{\mathbf{Q}}_0, \tilde{\mathbf{Q}}_c, \tilde{\mathbf{Q}}_a)$ and $\tilde{\beta} = \max_{k\in\mathcal{K}_e} C_{e,k}(\tilde{\mathbf{Q}}_0, \tilde{\mathbf{Q}}_c, \tilde{\mathbf{Q}}_a)$, is a KKT point of (38).

Due to the compactness of $(\mathbf{Q}_0, \mathbf{Q}_c, \mathbf{Q}_a)$, there must exist a subsequence, denoted by

$$\left\{\left(\mathbf{Q}_0^{n_l}, \mathbf{Q}_c^{n_l}, \mathbf{Q}_a^{n_l}, \{\mathbf{S}_k^{n_l}, \mathbf{U}_k^{n_l}\}_{k=1}^K\right)\right\}_l,$$

such that $\left\{\left(\mathbf{Q}_0^{n_l}, \mathbf{Q}_c^{n_l}, \mathbf{Q}_a^{n_l}, \{\mathbf{S}_k^{n_l}\}_{k=1}^K, \{\mathbf{U}_k^{n_l}\}_{k=1}^K\right)\right\}_l$ converges to a limit point $\left(\tilde{\mathbf{Q}}_0, \tilde{\mathbf{Q}}_c, \tilde{\mathbf{Q}}_a, \{\tilde{\mathbf{S}}_k\}_{k=1}^K, \{\tilde{\mathbf{U}}_k\}_{k=1}^K\right)$ as $l \to \infty$. Next, our proof is composed of two steps. First, we will show that the limit point

$$\left(\tilde{\mathbf{Q}}_0, \tilde{\mathbf{Q}}_c, \tilde{\mathbf{Q}}_a, \{\tilde{\mathbf{S}}_k, \tilde{\mathbf{U}}_k\}_{k=1}^K\right)$$

satisfies the following properties.

$$\tilde{\mathbf{S}}_1 = \arg\max_{\mathbf{S}_1 \succeq \mathbf{0}} \varphi_b(\tilde{\mathbf{Q}}_c, \tilde{\mathbf{Q}}_a, \mathbf{S}_1), \tag{39a}$$

$$\tilde{\mathbf{S}}_k = \arg\min_{\mathbf{S}_k \succeq \mathbf{0}} \varphi_{e,k}(\tilde{\mathbf{Q}}_c, \tilde{\mathbf{Q}}_a, \mathbf{S}_k), \forall k \in \mathcal{K}_e \tag{39b}$$

$$\tilde{\mathbf{U}}_k = \arg\max_{\mathbf{U}_k \succeq \mathbf{0}} \varphi_{m,k}(\tilde{\mathbf{Q}}_0, \tilde{\mathbf{Q}}_c, \tilde{\mathbf{Q}}_a, \mathbf{U}_k), \forall k \in \mathcal{K} \tag{39c}$$

$$(\tilde{\mathbf{Q}}_0, \tilde{\mathbf{Q}}_c, \tilde{\mathbf{Q}}_a) = \arg \max_{(\mathbf{Q}_0, \mathbf{Q}_c, \mathbf{Q}_a) \in \mathcal{F}} f\left(\mathbf{Q}_0, \mathbf{Q}_c, \mathbf{Q}_a, \{\tilde{\mathbf{S}}_k, \tilde{\mathbf{U}}_k\}_{k=1}^K\right). \tag{39d}$$

Second, we will check the KKT conditions of problems (39a)–(39d) to build a bridge between (39) and the KKT conditions of problem (38).

**Step 1.** By noting that

$$S_1^{n_l} = \arg\max_{\mathbf{S}_1 \succeq \mathbf{0}} \varphi_b\left(\mathbf{Q}_c^{n_l-1}, \mathbf{Q}_a^{n_l-1}, \mathbf{S}_1\right), \tag{40a}$$

$$S_k^{n_l} = \arg\min_{\mathbf{S}_k \succeq \mathbf{0}} \varphi_{e,k}\left(\mathbf{Q}_c^{n_l-1}, \mathbf{Q}_a^{n_l-1}, \mathbf{S}_k\right), \forall k \in \mathcal{K}_e \tag{40b}$$

$$\mathbf{U}_k^{n_l} = \arg\max_{\mathbf{U}_k \succeq \mathbf{0}} \varphi_{m,k}\left(\mathbf{Q}_0^{n_l-1}, \mathbf{Q}_c^{n_l-1}, \mathbf{Q}_a^{n_l-1}, \mathbf{U}_k\right), \forall k \in \mathcal{K} \tag{40c}$$

$$(\mathbf{Q}_0^{n_l}, \mathbf{Q}_c^{n_l}, \mathbf{Q}_a^{n_l}) = \arg \max_{(\mathbf{Q}_0, \mathbf{Q}_c, \mathbf{Q}_a) \in \mathcal{F}} f\left(\mathbf{Q}_0, \mathbf{Q}_c, \mathbf{Q}_a, \{\mathbf{S}_k^{n_l}, \mathbf{U}_k^{n_l}\}_{k=1}^K\right), \tag{40d}$$

we have

$$\varphi_b\left(\mathbf{Q}_c^{n_l-1}, \mathbf{Q}_a^{n_l-1}, \mathbf{S}_1^{n_l}\right) \geq \varphi_b\left(\mathbf{Q}_c^{n_l-1}, \mathbf{Q}_a^{n_l-1}, \mathbf{S}_1\right), \forall \mathbf{S}_1 \succeq \mathbf{0} \tag{41a}$$

$$\varphi_{e,k}\left(\mathbf{Q}_c^{n_l-1}, \mathbf{Q}_a^{n_l-1}, \mathbf{S}_k^{n_l}\right) \leq \varphi_{e,k}\left(\mathbf{Q}_c^{n_l-1}, \mathbf{Q}_a^{n_l-1}, \mathbf{S}_k\right), \forall \mathbf{S}_k \succeq \mathbf{0}, \forall k \in \mathcal{K}_e \tag{41b}$$

$$\varphi_{m,k}\left(\mathbf{Q}_0^{n_l-1}, \mathbf{Q}_c^{n_l-1}, \mathbf{Q}_a^{n_l-1}, \mathbf{U}_k^{n_l}\right) \geq \varphi_{m,k}\left(\mathbf{Q}_0^{n_l-1}, \mathbf{Q}_c^{n_l-1}, \mathbf{Q}_a^{n_l-1}, \mathbf{U}_k\right), \forall \mathbf{U}_k \succeq \mathbf{0}, \forall k \in \mathcal{K}, \tag{41c}$$

and for any $(\mathbf{Q}_0, \mathbf{Q}_c, \mathbf{Q}_a) \in \mathcal{F}$, the following inequality holds, i.e.,

$$\begin{aligned} & f\left(\mathbf{Q}_0, \mathbf{Q}_c, \mathbf{Q}_a, \{\mathbf{S}_k^{n_l}\}_{k=1}^K, \{\mathbf{U}_k^{n_l}\}_{k=1}^K\right) \\ & \leq f\left(\mathbf{Q}_0^{n_l}, \mathbf{Q}_c^{n_l}, \mathbf{Q}_a^{n_l}, \{\mathbf{S}_k^{n_l}\}_{k=1}^K, \{\mathbf{U}_k^{n_l}\}_{k=1}^K\right) \\ & \leq f\left(\tilde{\mathbf{Q}}_0, \tilde{\mathbf{Q}}_c, \tilde{\mathbf{Q}}_a, \{\tilde{\mathbf{S}}_k\}_{k=1}^K\right), \{\tilde{\mathbf{U}}_k\}_{k=1}^K\right), \end{aligned} \tag{42}$$

where the second inequality of (42) holds for the reason that AO algorithm yields non-descending objective values. Then, letting $l \to \infty$ in (41) and (42) will lead to (39a)–(39d).

**Step 2.** Then, it follows from (39a) to (39d) and the positive definiteness of $\{\tilde{\mathbf{S}}_k\}_{k=1}^K$ and $\{\tilde{\mathbf{U}}_k\}_{k=1}^K$ that

$$\nabla_{\mathbf{S}_1} \varphi_b(\tilde{\mathbf{Q}}_c, \tilde{\mathbf{Q}}_a, \tilde{\mathbf{S}}_1) = \mathbf{0}, \tilde{\mathbf{S}}_1 \succeq \mathbf{0}, \tag{43a}$$

$$\nabla_{\mathbf{S}_k} \varphi_{e,k}(\tilde{\mathbf{Q}}_c, \tilde{\mathbf{Q}}_a, \tilde{\mathbf{S}}_k) = \mathbf{0}, \tilde{\mathbf{S}}_k \succeq \mathbf{0}, \forall k \in \mathcal{K}_e \tag{43b}$$

$$\nabla_{\mathbf{U}_k} \varphi_{m,k}(\tilde{\mathbf{Q}}_0, \tilde{\mathbf{Q}}_c, \tilde{\mathbf{Q}}_a, \tilde{\mathbf{U}}_k) = \mathbf{0}, \tilde{\mathbf{U}}_k \succeq \mathbf{0}, \forall k \in \mathcal{K}. \tag{43c}$$

By carrying out some matrix manipulations to (43), it is easy to obtain that

$$\tilde{\mathbf{S}}_1 = \left(\mathbf{I} + \mathbf{H}_1 \tilde{\mathbf{Q}}_a \mathbf{H}_1^H\right)^{-1} \succeq \mathbf{0}, \tag{44a}$$

$$\tilde{\mathbf{S}}_k = \left(\mathbf{I} + \mathbf{H}_k (\tilde{\mathbf{Q}}_a + \tilde{\mathbf{Q}}_c) \mathbf{H}_k^H\right)^{-1} \succeq \mathbf{0}, \forall k \in \mathcal{K}_e \tag{44b}$$

$$\tilde{\mathbf{U}}_k = \left(\mathbf{I} + \mathbf{H}_k (\tilde{\mathbf{Q}}_a + \tilde{\mathbf{Q}}_c) \mathbf{H}_k^H\right)^{-1} \succeq \mathbf{0}, \forall k \in \mathcal{K}. \tag{44c}$$

Meanwhile, by introducing slack variables $\alpha$ and $\beta$, (39d) is shown to be equivalent to

$$\begin{aligned} \max_{\mathbf{Q}_0, \mathbf{Q}_c, \mathbf{Q}_a, \alpha, \beta} \quad & \lambda_c(\varphi_b(\mathbf{Q}_c, \mathbf{Q}_a, \tilde{\mathbf{S}}_1) - \beta) + \alpha \\ \text{s.t.} \quad & \varphi_{e,k}(\mathbf{Q}_c, \mathbf{Q}_a, \tilde{\mathbf{S}}_k) \leq \beta, \forall k \in \mathcal{K}_e, && (45a) \\ & \varphi_{m,k}(\mathbf{Q}_0, \mathbf{Q}_c, \mathbf{Q}_a, \tilde{\mathbf{U}}_k) \geq \alpha, \forall k \in \mathcal{K}, && (45b) \\ & (\mathbf{Q}_0, \mathbf{Q}_c, \mathbf{Q}_a) \in \mathcal{F}. && (45c) \end{aligned}$$

It is easy to see that $(\tilde{\mathbf{Q}}_0, \tilde{\mathbf{Q}}_c, \tilde{\mathbf{Q}}_a)$, together with $\tilde{\beta} = \max_{k \in \mathcal{K}_e} \varphi_{e,k}(\tilde{\mathbf{Q}}_c, \tilde{\mathbf{Q}}_a, \tilde{\mathbf{S}}_k)$ and $\tilde{\alpha} = \min_{k \in \mathcal{K}} \varphi_{m,k}(\tilde{\mathbf{Q}}_0, \tilde{\mathbf{Q}}_c, \tilde{\mathbf{Q}}_a, \tilde{\mathbf{U}}_k)$, is an optimal solution of problem (11). Consequently, $(\tilde{\mathbf{Q}}_0, \tilde{\mathbf{Q}}_c, v\mathbf{Q}_a, \tilde{\beta}, \tilde{\alpha})$ satisfy the KKT conditions of (45), shown in (46).

$$\begin{aligned} & \lambda_c \nabla_{\mathbf{Q}_c} \varphi_b(\tilde{\mathbf{Q}}_c, \tilde{\mathbf{Q}}_a, \tilde{\mathbf{S}}_1) - \sum_{k \in \mathcal{K}_e} \rho_k \nabla_{\mathbf{Q}_c} \varphi_{e,k}(\tilde{\mathbf{Q}}_c, \tilde{\mathbf{Q}}_a, \tilde{\mathbf{S}}_k) \\ & + \sum_{k \in \mathcal{K}} \mu_k \nabla_{\mathbf{Q}_c} \varphi_{m,k}(\tilde{\mathbf{Q}}_0, \tilde{\mathbf{Q}}_c, \tilde{\mathbf{Q}}_a, \tilde{\mathbf{U}}_k) - \gamma \mathbf{I} + \mathbf{C} = \mathbf{0}, \\ & \lambda_c \nabla_{\mathbf{Q}_a} \varphi_b(\tilde{\mathbf{Q}}_c, \tilde{\mathbf{Q}}_a, \tilde{\mathbf{S}}_1) - \sum_{k \in \mathcal{K}_e} \rho_k \nabla_{\mathbf{Q}_a} \varphi_{e,k}(\tilde{\mathbf{Q}}_c, \tilde{\mathbf{Q}}_a, \tilde{\mathbf{S}}_k) \\ & + \sum_{k \in \mathcal{K}} \mu_k \nabla_{\mathbf{Q}_a} \varphi_{m,k}(\tilde{\mathbf{Q}}_0, \tilde{\mathbf{Q}}_c, \tilde{\mathbf{Q}}_a, \tilde{\mathbf{U}}_k) - \gamma \mathbf{I} + \mathbf{A} = \mathbf{0}, \\ & \sum_{k \in \mathcal{K}} \mu_k \nabla_{\mathbf{Q}_0} \varphi_{m,k}(\tilde{\mathbf{Q}}_0, \tilde{\mathbf{Q}}_c, \tilde{\mathbf{Q}}_a, \tilde{\mathbf{U}}_k) - \gamma \mathbf{I} + \mathbf{B} = \mathbf{0}, \\ & \varphi_{e,k}(\tilde{\mathbf{Q}}_c, \tilde{\mathbf{Q}}_a, \tilde{\mathbf{S}}_k) \leq \tilde{\beta}, \forall k \in \mathcal{K}_e \\ & \rho_k(\varphi_{e,k}(\tilde{\mathbf{Q}}_c, \tilde{\mathbf{Q}}_a, \tilde{\mathbf{S}}_k) - \tilde{\beta}) = 0, \forall k \in \mathcal{K}_e \\ & \varphi_{m,k}(\tilde{\mathbf{Q}}_0, \tilde{\mathbf{Q}}_c, \tilde{\mathbf{Q}}_a, \tilde{\mathbf{U}}_k) \geq \tilde{\alpha}, \forall k \in \mathcal{K} \\ & \mu_k(\varphi_{m,k}(\tilde{\mathbf{Q}}_0, \tilde{\mathbf{Q}}_c, \tilde{\mathbf{Q}}_a, \tilde{\mathbf{U}}_k) - \tilde{\alpha}) = 0, \forall k \in \mathcal{K} \\ & \textstyle\sum_{k=1}^K \rho_k = 1, \sum_{k=1}^K \mu_k = 1, \\ & \mathbf{A} \succeq \mathbf{0}, \mathbf{B} \succeq \mathbf{0}, \mathbf{C} \succeq \mathbf{0}, \\ & \gamma \geq 0, \rho_k \geq 0, \forall k \in \mathcal{K}_e, \mu_k \geq 0, \forall k \in \mathcal{K}, \\ & \mathrm{Tr}(\tilde{\mathbf{Q}}_0 + \tilde{\mathbf{Q}}_c + \tilde{\mathbf{Q}}_a) \leq P, \tilde{\mathbf{Q}}_0 \succeq \mathbf{0}, \tilde{\mathbf{Q}}_c \succeq \mathbf{0}, \tilde{\mathbf{Q}}_a \succeq \mathbf{0}, \\ & \gamma(\mathrm{Tr}(\tilde{\mathbf{Q}}_0 + \tilde{\mathbf{Q}}_c + \tilde{\mathbf{Q}}_a) - P) = 0, \\ & \mathrm{Tr}(\mathbf{B}\tilde{\mathbf{Q}}_0) = 0, \mathrm{Tr}(\mathbf{C}\tilde{\mathbf{Q}}_c) = 0, \mathrm{Tr}(\mathbf{A}\tilde{\mathbf{Q}}_a) = 0. \end{aligned} \tag{46}$$

In (46), $\left(\{\rho_k\}_{k \in \mathcal{K}_e}, \{\mu_k\}_{k \in \mathcal{K}}, \gamma, \mathbf{A}, \mathbf{B}, \mathbf{C}\right)$ are all dual variables pertaining to the constraints in (45).

To proceed, by applying Danskin's theorem [44], one can verify that the following equalities must hold.

$$\begin{aligned}
\nabla_{\mathbf{Q}_c} C_b(\tilde{\mathbf{Q}}_c, \tilde{\mathbf{Q}}_a) &= \nabla_{\mathbf{Q}_c} \varphi_b(\tilde{\mathbf{Q}}_c, \tilde{\mathbf{Q}}_a, \tilde{\mathbf{S}}_1),\\
\nabla_{\mathbf{Q}_a} C_b(\tilde{\mathbf{Q}}_c, \tilde{\mathbf{Q}}_a) &= \nabla_{\mathbf{Q}_a} \varphi_b(\tilde{\mathbf{Q}}_c, \tilde{\mathbf{Q}}_a, \tilde{\mathbf{S}}_1),\\
\nabla_{\mathbf{Q}_c} C_{e,k}(\tilde{\mathbf{Q}}_c, \tilde{\mathbf{Q}}_a) &= \nabla_{\mathbf{Q}_c} \varphi_{e,k}(\tilde{\mathbf{Q}}_c, \tilde{\mathbf{Q}}_a, \tilde{\mathbf{S}}_k),\\
\nabla_{\mathbf{Q}_a} C_{e,k}(\tilde{\mathbf{Q}}_c, \tilde{\mathbf{Q}}_a) &= \nabla_{\mathbf{Q}_a} \varphi_{e,k}(\tilde{\mathbf{Q}}_c, \tilde{\mathbf{Q}}_a, \tilde{\mathbf{S}}_k),\\
\nabla_{\mathbf{Q}_c} C_{m,k}(\tilde{\mathbf{Q}}_0, \tilde{\mathbf{Q}}_c, \tilde{\mathbf{Q}}_a) &= \nabla_{\mathbf{Q}_c} \varphi_{m,k}(\tilde{\mathbf{Q}}_0, \tilde{\mathbf{Q}}_c, \tilde{\mathbf{Q}}_a, \tilde{\mathbf{U}}_k),\\
\nabla_{\mathbf{Q}_a} C_{m,k}(\tilde{\mathbf{Q}}_0, \tilde{\mathbf{Q}}_c, \tilde{\mathbf{Q}}_a) &= \nabla_{\mathbf{Q}_a} \varphi_{m,k}(\tilde{\mathbf{Q}}_0, \tilde{\mathbf{Q}}_c, \tilde{\mathbf{Q}}_a, \tilde{\mathbf{U}}_k),\\
\nabla_{\mathbf{Q}_0} C_{m,k}(\tilde{\mathbf{Q}}_0, \tilde{\mathbf{Q}}_c, \tilde{\mathbf{Q}}_a) &= \nabla_{\mathbf{Q}_0} \varphi_{m,k}(\tilde{\mathbf{Q}}_0, \tilde{\mathbf{Q}}_c, \tilde{\mathbf{Q}}_a, \tilde{\mathbf{U}}_k).
\end{aligned} \tag{47}$$

Then, substituting (44b) and (44c) into $\varphi_{e,k}(\tilde{\mathbf{Q}}_c, \tilde{\mathbf{Q}}_a, \tilde{\mathbf{S}}_k)$ and $\varphi_{m,k}(\tilde{\mathbf{Q}}_0, \tilde{\mathbf{Q}}_c, \tilde{\mathbf{Q}}_a, \tilde{\mathbf{U}}_k)$, one can obtain

$$\begin{aligned}
C_{e,k}(\tilde{\mathbf{Q}}_c, \tilde{\mathbf{Q}}_a, \tilde{p}_m) &= \varphi_{e,k}(\tilde{\mathbf{Q}}_c, \tilde{\mathbf{Q}}_a, \tilde{p}_m, \tilde{\mathbf{S}}_k), \forall k \in \mathcal{K}_e\\
C_{m,k}(\tilde{\mathbf{Q}}_0, \tilde{\mathbf{Q}}_c, \tilde{\mathbf{Q}}_a) &= \varphi_{e,k}(\tilde{\mathbf{Q}}_0, \tilde{\mathbf{Q}}_c, \tilde{\mathbf{Q}}_a, \tilde{\mathbf{U}}_k), \forall k \in \mathcal{K}.
\end{aligned} \tag{48}$$

Finally, by plugging (47) and (48) into (46), we obtain

$$\begin{aligned}
&\lambda_c \nabla_{\mathbf{Q}_c} C_b(\tilde{\mathbf{Q}}_c, \tilde{\mathbf{Q}}_a) - \sum_{k \in \mathcal{K}_e} \rho_k \nabla_{\mathbf{Q}_c} C_{e,k}(\tilde{\mathbf{Q}}_c, \tilde{\mathbf{Q}}_a) - \gamma \mathbf{I}\\
&+\mathbf{C} + \sum_{k \in \mathcal{K}} \mu_k \nabla_{\mathbf{Q}_c} C_{m,k}(\tilde{\mathbf{Q}}_0, \tilde{\mathbf{Q}}_c, \tilde{\mathbf{Q}}_a) = \mathbf{0},\\
&\lambda_c \nabla_{\mathbf{Q}_a} C_b(\tilde{\mathbf{Q}}_c, \tilde{\mathbf{Q}}_a) - \sum_{k \in \mathcal{K}_e} \rho_k \nabla_{\mathbf{Q}_a} C_{e,k}(\tilde{\mathbf{Q}}_c, \tilde{\mathbf{Q}}_a) - \gamma \mathbf{I}\\
&+\mathbf{A} + \sum_{k \in \mathcal{K}} \mu_k \nabla_{\mathbf{Q}_a} C_{m,k}(\tilde{\mathbf{Q}}_0, \tilde{\mathbf{Q}}_c, \tilde{\mathbf{Q}}_a) = \mathbf{0},\\
&\sum_{k \in \mathcal{K}} \mu_k \nabla_{\mathbf{Q}_0} C_{m,k}(\tilde{\mathbf{Q}}_0, \tilde{\mathbf{Q}}_c, \tilde{\mathbf{Q}}_a) - \gamma \mathbf{I} + \mathbf{B} = \mathbf{0},\\
&\rho_k(C_{e,k}(\tilde{\mathbf{Q}}_c, \tilde{\mathbf{Q}}_a) - \tilde{\beta}) = 0, \forall k \in \mathcal{K}_e\\
&C_{m,k}(\tilde{\mathbf{Q}}_0, \tilde{\mathbf{Q}}_c, \tilde{\mathbf{Q}}_a) \geq \tilde{\alpha}, \forall k \in \mathcal{K}\\
&\mu_k(C_{m,k}(\tilde{\mathbf{Q}}_0, \tilde{\mathbf{Q}}_c, \tilde{\mathbf{Q}}_a) - \tilde{\alpha}) = 0, \forall k \in \mathcal{K}.
\end{aligned} \tag{49}$$

Remarkably, (49), together with the last six lines of (46), represents the KKT conditions of the WSRM problem (38). This fact completes the proof.

**Acknowledgements**
This work was supported in part by the National Natural Science Foundation of China under grant 61571089 and by the High-Tech Research and Development (863) Program of China under grant 2015AA01A707.

**Competing interests**
The authors declare that they have no competing interests.

**References**

1. JG Andrews, S Buzzi, W Choi, SV Hanly, A Lozano, ACK Soong, JC Zhang, What will 5G be? IEEE J. Sel. Areas Commun. **32**(6), 1065–1082 (2014)
2. Y-S Shiu, SY Chang, H-C Wu, SC-H Huang, H-H Chen, Physical layer security in wireless networks: a tutorial. IEEE Wirel. Commun. **18**(2), 66–74 (2011)
3. B He, X Zhou, TD Abhayapala, Wireless physical layer security with imperfect channel state information: a survey (2013). http://arxiv.org/abs/1307.4146. Accessed June 2013
4. Y-WP Hong, P-C Lan, C-CJ Kuo, Enhancing physical-layer secrecy in multiantenna wireless systems: an overview of signal processing approaches. IEEE Signal Process. Mag. **30**(5), 29–40 (2013)
5. A Mukherjee, SA Fakoorian, J Huang, AL Swindlehurst, et al, Principles of physical layer security in multiuser wireless networks: a survey. IEEE Commun. Surv. Tuts. **16**(3), 1550–1573 (2014)
6. Y Liu, H-H Chen, L Wang, Physical layer security for next generation wireless networks: theories, technologies, and challenges. IEEE Commun. Surv. Tuts. **19**(1), 347–376 (2017)
7. R Liu, HV Poor, Secrecy capacity region of a multi-antenna Gaussian broadcast channel with confidential messages. IEEE Trans. Inf. Theory. **55**(3), 1235–1249 (2009)
8. R Liu, T Liu, HV Poor, S Shamai, Multiple-input multiple-output Gaussian broadcast channels with confidential messages. IEEE Trans. Inf. Theory. **56**(9), 4215–4227 (2010)
9. SAA Fakoorian, AL Swindlehurst, On the optimality of linear precoding for secrecy in the MIMO broadcast channel. IEEE J. Sel. Areas Commun. **31**(9), 1701–1713 (2013)
10. D Park, Weighted sum rate maximization of MIMO broadcast and interference channels with confidential messages. IEEE Trans. Wirel. Commun. **15**(3), 1742–1753 (2016)
11. I Csiszár, J Körner, Broadcast channels with confidential messages. IEEE Trans. Inf. Theory. **24**(3), 339–348 (1978)
12. HD Ly, T Liu, Y Liang, Multiple-input multiple-output Gaussian broadcast channels with common and confidential messages. IEEE Trans. Inf. Theory. **56**(11), 5477–5487 (2010)
13. E Ekrem, S Ulukus, Capacity region of gaussian MIMO broadcast channels with common and confidential messages. IEEE Trans. Inf. Theory. **58**(9), 5669–5680 (2012)
14. R Liu, T Liu, HV Poor, S Shamai, in *Proc. IEEE Int. Symp. Inf. Theory (ISIT'2010)*. MIMO Gaussian broadcast channels with confidential and common messages (IEEE, Austin, 2010), pp. 2578–2582
15. R Liu, T Liu, HV Poor, S Shamai (Shitz), New results on multiple-input multiple-output broadcast channels with confidential messages. IEEE Trans. Inf. Theory. **59**(3), 1346–1358 (2013)
16. RF Wyrembelski, H Boche, in *Proc. IEEE Global Communication Conf. Workshops*. Service integration in multiantenna bidirectional relay networks: public and confidential messages (IEEE, Houston, 2011), pp. 884–888
17. RF Wyrembelski, H Boche, Physical layer integration of private, common, and confidential messages in bidirectional relay networks. IEEE Trans. Wirel. Commun. **11**(9), 3170–3179 (2012)
18. R Schaefer, H Boche, Physical layer service integration in wireless networks: signal processing challenges. IEEE Signal Process. Mag. **31**(3), 147–156 (2014)
19. W Mei, Z Chen, J Fang, Secrecy capacity region maximization in Gaussian MISO channels with integrated services. IEEE Signal Process. Lett. **23**(8), 1146–1150 (2016)
20. W Mei, L Li, Z Chen, C Huang, in *Proc. IEEE Global Conf. Signal Info. Process. (GlobalSIP)*. Artificial-noise aided transmit design for multi-user MISO systems with integrated services (IEEE, Orlando, 2015), pp. 1382–1386
21. W Mei, Z Chen, C Huang, in *Proc. IEEE ICASSP*. Robust artificial-noise aided transmit design for multi-user MISO systems with integrated services (IEEE, Shanghai, 2016), pp. 3856–3860
22. W Mei, L Li, Z Chen, C Huang, In *Proc. IEEE Int. Conf. Commun.* Artificial-noise aided transmit design for outage constrained service integration, (Kuala Lumpur, 2016), pp. 1–7
23. W Mei, Z Chen, J Fang, GSVD-based precoding in MIMO systems with integrated services. IEEE Signal Process. Lett. **23**(11), 1528–1532 (2016)
24. Q Li, M Hong, H-T Wai, Y-F Liu, W-K Ma, Z-Q Luo, Transmit solutions for MIMO wiretap channels using alternating optimization. IEEE J. Sel. Areas Commun. **31**(9), 1714–1727 (2013)
25. Q Li, W-K Ma, Spatially selective artificial-noise aided transmit optimization for MISO multi-Eves secrecy rate maximization. IEEE Trans. Signal Process. **61**(10), 2704–2717 (2013)
26. Z Chu, K Cumanan, Z Ding, M Johnston, SY Le Goff, Robust outage secrecy rate optimizations for a MIMO secrecy channel. IEEE Wirel. Commun. Lett. **4**(1), 86–89 (2015)

27. Z Chu, H Xing, M Johnston, SY Le Goff, Secrecy rate optimizations for a MISO secrecy channel with multiple multiantenna eavesdroppers. IEEE Trans. Wirel. Commun. **15**(1), 283–297 (2016)
28. T-X Zheng, H-M Wang, J Yuan, D Towsley, MH Lee, Multi-antenna transmission with artificial noise against randomly distributed eavesdroppers. IEEE Trans. Commun. **63**(11), 4347–4362 (2015)
29. GR Lanckriet, BK Sriperumbudur, in *Proc. Advances Neural Inf. Process. Syst.* On the convergence of the concave-convex procedure (NIPS Foundation, Vancouver, 2009), pp. 1759–1767
30. B Fang, Z Qian, W Shao, W Zhong, Precoding and artificial noise design for cognitive MIMOME wiretap channels. IEEE Trans. Veh. Technol. **65**(8), 6753–6758 (2016)
31. Z Chu, K Cumanan, Z Ding, M Johnston, SY Le Goff, Secrecy rate optimizations for a MIMO secrecy channel with a cooperative jammer. IEEE Trans. Veh. Technol. **64**(5), 1833–1847 (2015)
32. J Yang, I-M Kim, DI Kim, Optimal cooperative jamming for multiuser broadcast channel with multiple eavesdroppers. IEEE Trans. Wirel. Commun. **12**(6), 2840–2852 (2013)
33. SX Wu, W-K Ma, AM-C So, Physical-layer multicasting by stochastic transmit beamforming and Alamouti space-time coding. IEEE Trans. Signal Process. **61**(17), 4230–4245 (2013)
34. H Zhu, N Prasad, S Rangarajan, Precoder design for physical layer multicasting. IEEE Trans. Signal Process. **60**(11), 5932–5947 (2012)
35. W Lee, H Park, HB Kong, JS Kwak, I Lee, A new beamforming design for multicast systems. IEEE Trans. Veh. Technol. **62**(8), 4093–4097 (2013)
36. B Du, Y Jiang, X Xu, X Dai, Optimum beamforming for MIMO multicasting. EURASIP J. Adv. Signal Process. **2013**(121), 1–15 (2013)
37. M Grant, S Boyd, CVX: Matlab software for disciplined convex programming (2011). http://cvxr.com/cvx. Accessed Apr 2011
38. S Boyd, L Vandenberghe, *Convex optimization*. (Cambridge university press, Cambridge, 2009)
39. RT Marler, JS Arora, Survey of multi-objective optimization methods for engineering. Struct. Multidiscip. Optim. **26**(6), 369–395 (2004)
40. Z-Q Luo, W-K Ma, AM-C So, Y Ye, S Zhang, Semidefinite relaxation of quadratic optimization problems. IEEE Signal Process. Mag. **27**(3), 20–34 (2010)
41. K Cumanan, Z Ding, B Sharif, GY Tian, KK Leung, Secrecy rate optimizations for a MIMO secrecy channel with a multiple-antenna eavesdropper. IEEE Trans. Veh. Technol. **63**(4), 1678–1690 (2014)
42. A Ben-Tal, A Nemirovski, *Lectures on modern convex optimization: analysis, algorithms, and engineering applications*. vol. 2. (SIAM, Philadelphia, 2001)
43. H Weingarten, Y Steinberg, S Shamai (Shitz), The capacity region of the Gaussian multiple-input multiple-output broadcast channel. IEEE Trans. Inf. Theory. **52**(9), 3936–3964 (2006)
44. D Bertsekas, *Nonlinear programming*, 2nd edn. (Athena Scientific, Belmont, 1999)

# Decoupled signal detection for the uplink of massive MIMO in 5G heterogeneous networks

Leonel Arévalo[1*], Rodrigo C. de Lamare[1], Martin Haardt[2] and Raimundo Sampaio-Neto[1]

**Abstract**

Massive multiple-input multiple-output (MIMO) systems are strong candidates for future fifth-generation (5G) heterogeneous cellular networks. For 5G, a network densification with a high number of different classes of users and data service requirements is expected. Such a large number of connected devices needs to be separated in order to allow the detection of the transmitted signals according to different data requirements. In this paper, a decoupled signal detection (DSD) technique which allows the separation of the uplink signals, for each user class, at the base station (BS) is proposed for massive MIMO systems. A mathematical signal model for massive MIMO systems with centralized and distributed antennas in heterogeneous networks is also developed. The performance of the proposed algorithm is evaluated and compared with existing detection schemes in a realistic scenario with distributed antennas. A sum-rate analysis and a computational cost study for DSD are also presented. Simulation results show an excellent performance of the proposed algorithm when combined with linear and successive interference cancellation detection techniques.

**Keywords:** Massive multiple-input multiple-output (MIMO) systems, Heterogeneous networks, Fifth-generation (5G) cellular networks, Decoupled signal detection (DSD)

## 1 Introduction

Large-scale multiple-input multiple-output (MIMO) systems, also known as massive MIMO, are a promising technology which use a large number of antennas to serve a high number of user terminals at the same time without requiring extra bandwidth resources [1–4]. This new greater scale version of traditional MIMO systems, where a restricted number of antennas is used, is designed to exploit the benefits of extra degrees of freedom obtained by the use of more antennas [5]. Massive MIMO can increase the spectral efficiency 10 times or more when compared with its predecessor [6]. In the Long Term Evolution (LTE) standard [7], which uses traditional MIMO systems, with as many as eight antenna ports at the base station (BS), and operates in the frequency-division duplex (FDD) mode, the users estimate the channel response and feed it back to the BS. For massive MIMO with hundreds of antenna elements, this might not be feasible, due to the large number of channel coefficients that each user needs to estimate being proportional to the number of the antennas at the BS. In this paper we focus on the uplink, the reason is that the most natural transmission mode to operate in massive MIMO is the time-division duplexing (TDD) mode, where a reciprocity between the uplink and downlink channels can be obtained, if we use appropriate calibration techniques to combat the distortions induced by hardware imperfections, since the base station can offer more processing resources aimed at estimating the channel between users' terminals and the BS. On the downlink, it is possible to use different precoding schemes to mitigate the interference for the received signals at the mobile users. Such precoding schemes rely on the channels estimated on the uplink.

*Correspondence: leonel_arevalo@cetuc.puc-rio.br
[1]Center for Studies in Telecommunications (CETUC), Pontifical Catholic University of Rio de Janeiro (PUC-Rio), Rio de Janeiro, Brazil
Full list of author information is available at the end of the article

One of the main research challenges of massive MIMO is to develop computationally simple ways to process the large number of signals received at the BS. The interference between antennas and users, propagation effects such as correlation, path loss and shadowing, thermal noise and signal degradation due to the hardware imperfections need to be suppressed. Linear detection techniques such as maximum radio combining (MRC) and zero forcing (ZF) are a good option in terms of computational complexity; however, their performance is not compatible with the growing demand for high data rates. The performance of linear detectors can be improved using some nonlinear sub-optimal detector based on successive interference cancellation (SIC) [8], e.g., multibranch SIC (MB-SIC) [9, 10] and multifeedback SIC (MF-SIC) [11]. However, SIC-based detectors have a considerable computational cost for high-dimensional systems. If all signals from all active users are coupled in the detection process, the BS could spend unnecessary processing resources since some types of users might not require a very high performance.

In the next generation of wireless communication systems [12], it is expected that a large number of users with different configurations and requirements are connected to the network. Therefore, it is necessary to design heterogeneous networks capable of interconnecting the different user types with each other [13]. The received signals from this large number of connected devices such as metering equipment, sensors, environmental monitoring devices, health care gadgets, security management products, smart grid components, smart phones, and tablets need to be separated in order to detect the transmitted information according to their different data requirements. In this context, distributed antenna systems (DAS) with massive MIMO are a promising alternative for the 5G cellular architecture [14], where the BS will be equipped with a large number of antennas and some remote antenna arrays or radio heads will be distributed around the cell and connected to the BS via optical fiber. The signals associated with different remote antenna arrays are processed at the BS. DAS have low path loss effects and improve the coverage and the spectral efficiency [15, 16]. The energy consumption of users is reduced and the transmission quality is improved due to the shorter distances between users and some remote antenna arrays. For this vision of 5G wireless networks, which includes a combination of massive MIMO, heterogeneous networks, and distributed antenna systems, efficient signal processing techniques at the BS are necessary.

In this paper, we propose an algorithm for the uplink of massive MIMO systems to separate the combined received signal of all users at the BS into independent signals for each user class. The proposed decoupled signal detection (DSD) applies a decomposition into multiple independent single user class signals, where all users in a class have the same data requirements and a common complex modulation. Assuming that the channel state information (CSI) was previously estimated, DSD employs a common channel inversion and QR decomposition to decouple the received signal. Applying the proposed algorithm, the computational cost of the signal processing is reduced and it is possible to have flexibility on the detection procedures at the BS. A signal model for heterogeneous networks with different classes of users and an arbitrary configuration of centralized antenna systems (CAS) and DAS is also introduced in this paper. A sum rate analysis and a computational complexity study for the proposed DSD are presented. The performance of the proposed scheme is evaluated in a realistic scenario with propagation effects and compared with existing approaches.

The remainder of this paper is organized as follows. In Section 2 the proposed massive MIMO signal model is presented. The proposed DSD scheme is presented in Section 3. The sum rate analysis for the DSD scheme is described in Section 4. Section 5 presents simulation results. Section 6 gives the conclusions.

## 2 Massive MIMO signal model

5G cellular networks will be able to support a large number and diverse classes of devices, i.e., they will be heterogeneous, where a single macro cell may need to support 10,000 or more low-rate devices, along with its traditional high-rate mobile users [17]. This will require joint user classification according to their data rate requirements. Thus, a user in a high definition video call may have higher data rate requirements compared to a mobile user transmitting voice. Then we can classify them in different classes of users. In this section, a signal model for heterogeneous networks with different classes of users, as depicted in Fig. 1, and an arbitrary configuration of CAS and DAS is presented. We consider the uplink channel scenario of a massive MIMO system with $N$ different classes of active users transmitting simultaneously signals to one base station (BS) equipped with $D$ remote antenna arrays distributed around the cell and $N_B$ receive antennas at the BS. Each remote array of antennas has $Q$ antennas linked to the BS via wired links. Therefore, the total number of receive antennas is $N_r = N_B + DQ$. The choice of $N_B$, $D$, and $Q$ is made based on the features of the network and type of application scenario. For example, suppose that we have a city with a high density of users or devices in the center of a cell and sparsely distributed users or devices in the remaining part of the cell. In this case, we could use a number of centralized antennas to deal with the high density of users and distributed antennas to serve the remaining devices. The cardinality of the $n$th user class $\mid C_n \mid$ represents the number of users of

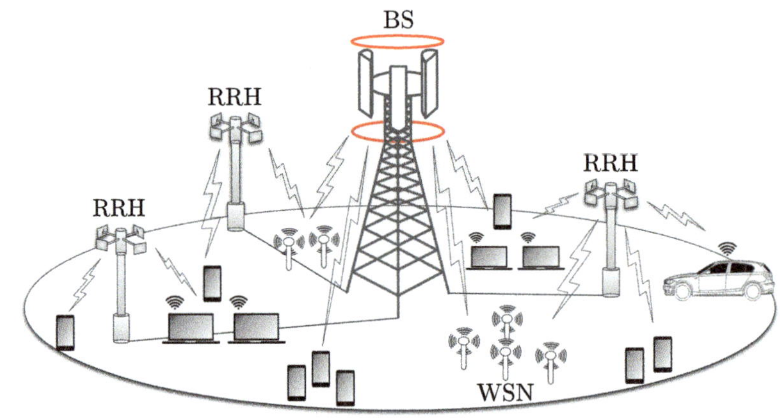


**Fig. 1** Heterogeneous wireless network: base station (BS), remote radio head (RRH), and wireless sensor networks (WSN)

the class $n$. The total number of active users is given by $K = \sum_{n=1}^{N} \mid C_n \mid$. The $k$th user in the $n$th user class transmits data divided into $N_{t_{k,n}}$ sub-streams through $N_{t_{k,n}}$ antennas, where $N_r \geqslant N_t = \sum_{n=1}^{N} \sum_{k=1}^{|C_n|} N_{t_{k,n}}$ and $N_t$ is the total number of transmit antennas. The received signal vector at the BS from all active users in all user classes is given by

$$\mathbf{y} = \sum_{n=1}^{N} \sum_{k=1}^{|C_n|} \mathbf{\Upsilon}_{k,n} \bar{\mathbf{H}}_{k,n} \mathbf{s}_{k,n} + \mathbf{n}, \tag{1}$$

where $\mathbf{s}_{k,n}$ is the $N_{t_{k,n}} \times 1$ transmitted signal vector, by the $k$th user of the $n$th user class, at one time slot taken from a complex constellation, denoted by $\mathcal{A} = \{a_1, a_2, \ldots, a_O\}$. Each symbol has $M$ bits and $O = 2^M$. The vector $\mathbf{n}$ is an $N_r \times 1$ zero mean complex circular symmetric Gaussian noise vector with covariance matrix $\mathbf{K_n} = \mathbb{E}\left[\mathbf{n}\mathbf{n}^{\mathcal{H}}\right] = \sigma_n^2 \mathbf{I}$. Moreover, $\bar{\mathbf{H}}_{k,n}$ is the $N_r \times N_{t_{k,n}}$ channel matrix of the $k$th user in the class $n$ with elements $\bar{h}_{i,j}^{(k,n)}$ corresponding to the complex channel gain from the $j$th transmit antenna of the $k$th user to the $i$th receive antenna. For the antenna elements located at the BS and at each remote radio head, the $D+1$ sub-matrices of $\bar{\mathbf{H}}_{k,n} = \left[\left(\bar{\mathbf{H}}_{k,n}^{(1)}\right)^T, \left(\bar{\mathbf{H}}_{k,n}^{(2)}\right)^T, \ldots, \left(\bar{\mathbf{H}}_{k,n}^{(D+1)}\right)^T\right]^T$ can be modeled using the Kronecker channel model [18], expressed by

$$\bar{\mathbf{H}}_{k,n}^{(j)} = \left(\mathbf{R}_r^{(j)}\right)^{1/2} \mathbf{G}_{k,n}^{(j)} \left(\mathbf{R}_{t_{k,n}}\right)^{1/2} \tag{2}$$

where $\mathbf{G}_{k,n}^{(j)}$ has complex channel gains between the $k$th user and the $j$th radio head, obtained from an independent and identically distributed random fading model whose coefficients are complex Gaussian random variables with zero mean and unit variance. $\mathbf{R}_r^{(j)}$ and $\mathbf{R}_{t_{k,n}}$ denote the receive correlation matrix of the $j$th radio head and the transmit correlation matrix, respectively. The components of the correlation matrices $\mathbf{R}_r^{(j)}$ and $\mathbf{R}_{t_{k,n}}$ are modeled as a variation of the model described in equations (3)-(5) of reference [18]:

$$\mathbf{R} = \begin{bmatrix} 1 & \rho & \rho^4 & \ldots & \rho^{(N_a-1)^2} \\ \rho & 1 & \rho & \ldots & \vdots \\ \rho^4 & \rho & 1 & \vdots & \rho^4 \\ \vdots & \vdots & \vdots & \vdots & \vdots \\ \rho^{(N_a-1)^2} & \ldots & \rho^4 & \rho & 1 \end{bmatrix}, \tag{3}$$

where $N_a$ is the number of antennas and $\rho$ is the correlation coefficient of neighboring antennas ($\rho = \rho_{t_x}$ for the transmit antennas and $\rho = \rho_{r_x}$ for the receive antennas), i.e., a decaying of the correlation index with antenna separation faster than exponential was adopted. Note that $\rho = 0$ represents an uncorrelated scenario and $\rho = 1$ implies a fully correlated scenario. The $N_r \times N_r$ diagonal matrix $\mathbf{\Upsilon}_{k,n}$ represents the large-scale propagation effects for the user $k$ of the user class $n$, such as path loss and shadowing, given by

$$\mathbf{\Upsilon}_{k,n} = \mathrm{diag}\Big(\underbrace{\gamma_{k,n}^1 \cdots \gamma_{k,n}^1}_{N_B} \underbrace{\gamma_{k,n}^2 \cdots \gamma_{k,n}^2}_{Q} \cdots \underbrace{\gamma_{k,n}^{D+1} \cdots \gamma_{k,n}^{D+1}}_{Q}\Big), \tag{4}$$

where the parameters $\gamma_{k,n}^j$ denote the large-scale propagation effects from the $k$th user to the $j$th radio head described by

$$\gamma_{k,n}^j = \alpha_{k,n}^j \beta_{k,n}^j, \qquad j = 1, \ldots, D+1. \tag{5}$$

Here $\alpha_{k,n}^j$ is the distance based path-loss between each user and the radio heads which is calculated by

$$\alpha_{k,n}^j = \sqrt{\frac{L_{k,n}^j}{\left(d_{k,n}^j\right)^{\tau}}}, \tag{6}$$

where $L^j_{k,n}$ is the power path loss of the link associated with the user and the $j$th radio head, $d^j_{k,n}$ is the relative distance between this user and the $j$th radio head, $\tau$ is the path loss exponent chosen between 2 and 4 depending on the environment. The log normal random variable $\beta^j_{k,n}$ which represents the shadowing between user $k$ and the receiver is given by

$$\beta^j_{k,n} = 10^{\frac{\mu_{k,s}\vartheta^j_{k,s}}{10}}, \tag{7}$$

where $\mu_{k,s}$ is the shadowing spread in decibels and $\vartheta^j_{k,s}$ corresponds to a real-valued Gaussian random variable with zero mean and unit variance. Since the $N_r \times N_{t_{k,n}}$ composite channel matrix includes large-scale and small-scale fading effects, it can be denoted as $\mathbf{H}_{k,n} = \boldsymbol{\Upsilon}_{k,n}\bar{\mathbf{H}}_{k,n}$, and the expression in (1) can be written as

$$\mathbf{y} = \sum_{n=1}^{N} \mathbf{H}_n\mathbf{s}_n + \mathbf{n}, \tag{8}$$

where $\mathbf{H}_n = [\mathbf{H}_{1,n} \quad \mathbf{H}_{2,n} \ldots \mathbf{H}_{|C_n|,n}]$ and $\mathbf{s}_n = \left[\mathbf{s}^T_{1,n} \; \mathbf{s}^T_{2,n} \; \ldots \mathbf{s}^T_{|C_n|,n}\right]^T$ represent the channel matrix and the transmitted symbol vector of all users in the class $n$, respectively. The received signal vector can be expressed more conveniently as

$$\mathbf{y} = \mathbf{Hs} + \mathbf{n}, \tag{9}$$

where $\mathbf{H} = [\mathbf{H}_1 \;\; \mathbf{H}_2 \ldots \mathbf{H}_N]$ and $\mathbf{s} = \left[\mathbf{s}^T_1 \;\; \mathbf{s}^T_2 \ldots \mathbf{s}^T_N\right]^T$. The symbol vector $\mathbf{s}$ of all $N$ user classes has zero mean and a covariance matrix $\mathbf{K_s} = \mathbb{E}\left[\mathbf{ss}^{\mathcal{H}}\right] = \text{diag}(\mathbf{p})$, where the elements of the vector $\mathbf{p}$ are the signal power of each transmit antenna. To maintain a notational simplicity in the subsequent analysis, we assume that all antenna elements at the users transmit with the same average transmitted power $\sigma^2_s$, i.e., $\mathbf{K_s} = \sigma^2_s\mathbf{I}$. We assume that the channel matrix $\mathbf{H}$ was previously estimated at the BS. From (9) we can see that the signals arrive coupled at the BS. If we want to use different detection procedures for each user class according to its data requirements, we have to separate the received signal vector $\mathbf{y}$ into independent received signals for each user class. For the system model presented in this work, when the number of remote radio heads is set to zero, i.e., $D = 0$, the DAS architecture is reduced to the CAS scheme with $N_r = N_B$.

## 3 Decoupled signal detection

As presented in Section 2, in heterogeneous networks different classes of users send parallel data streams, through the massive MIMO channel operating with distributed antennas, which arrive superposed at the BS. In this 5G context, we need to separate the data streams for each category of users efficiently. In this section, we describe the proposed decoupled signal detection (DSD) which allows us to separate the received signal of the $n$th user class from the others. To this end, we consider that the process of authentication, identification, and channel estimation was already made, i.e, the BS is able to identify the users by classes according to their data requirements. Similar approaches to the proposed algorithm have been proposed for the downlink, such as block diagonalization (BD)-based techniques [19–22]. However, unlike prior work in which downlink BD is used, for the proposed DSD scheme, it is not necessary to use any precoding at the transmit side. The receiver only needs to know the channels between users and receive antennas. Moreover, the concept of separating the users with respect to the classes in heterogeneous networks according to its requirements is a new approach. The first steps to construct the concept proposed in this paper were presented in [23, 24].

The received signal vector (8) can be written as:

$$\mathbf{y} = \mathbf{H}_n\mathbf{s}_n + \sum_{m=1,m\neq n}^{N} \mathbf{H}_m\mathbf{s}_m + \mathbf{n}, \tag{10}$$

where $\mathbf{H}_n$ and $\mathbf{s}_n$ are the channel matrix and the transmitted symbol vector for the $n$th user class, respectively. From (10), we can see that the $n$th user class has inter-user class interference.

### 3.1 Proposed decoupling strategy

To remove the presence of the other classes of users in the detection procedure for the $n$th user class, we can employ a linear operation to project the received signal vector $\mathbf{y}$ onto the subspace orthogonal or almost orthogonal to the subspace generated by the signals of the interfering classes. In DSD, a matrix $\mathbf{A}_n$ is calculated employing a channel inversion method and a QR decomposition [25, 26], in order to decouple the $n$th user's class received signal from other user's class signals. To compute $\mathbf{A}_n$, we construct the matrix $\tilde{\mathbf{H}}_n$ excluding the channel matrix of the $n$th user class in the following form:

$$\tilde{\mathbf{H}}_n = [\mathbf{H}_1 \ldots \mathbf{H}_{n-1} \;\; \mathbf{H}_{n+1} \ldots \mathbf{H}_N], \tag{11}$$

where $\tilde{\mathbf{H}}_n \in \mathbb{C}^{N_r\times(N_t-N_{t_n})}$ and $N_{t_n} = \sum_{k=1}^{|C_n|} N_{t_{k,n}}$ is the number of transmit antennas in the $n$th user class. After that, the objective is to obtain a matrix $\mathbf{A}_n$ that satisfies the following condition:

$$\mathbf{A}_n\tilde{\mathbf{H}}_n = \mathbf{0}, \quad \forall n \in (1\ldots N). \tag{12}$$

To compute $\mathbf{A}_n$, DSD first computes the MMSE channel inversion of the combined channel matrix $\mathbf{H}$ given by

$$\begin{aligned}\mathbf{H}^\dagger &= \mathbf{H}^{\mathcal{H}}\left(\mathbf{HH}^{\mathcal{H}} + \sigma^2\mathbf{I}\right)^{-1}\\ &= \left[(\ddot{\mathbf{H}}_1)^T \;\; (\ddot{\mathbf{H}}_2)^T \;\; \ldots \;\; (\ddot{\mathbf{H}}_N)^T\right]^T\end{aligned} \tag{13}$$

where $\sigma^2 = \sigma_n^2/\sigma_s^2$, $\mathbf{H}^\dagger \in \mathbb{C}^{N_t \times N_r}$ and $\ddot{\mathbf{H}}_n \in \mathbb{C}^{N_{t_n} \times N_r}$. Then, the matrix $\ddot{\mathbf{H}}_n$ is approximately in the null space of $\tilde{\mathbf{H}}_n$, that is

$$\ddot{\mathbf{H}}_n \tilde{\mathbf{H}}_n \approx \mathbf{0}, \quad \forall n \in (1 \ldots N). \tag{14}$$

To decouple the $n$th user group from the other user groups, we employ a QR decomposition as described by

$$\ddot{\mathbf{H}}_n = \mathbf{R}_n \mathbf{Q}_n, \quad \forall n \in (1 \ldots N), \tag{15}$$

where $\mathbf{R}_n \in \mathbb{C}^{N_{t_n} \times N_{t_n}}$ is an upper triangular matrix and $\mathbf{Q}_n \in \mathbb{C}^{N_{t_n} \times N_r}$ is a matrix with orthogonal rows and composed by approximately orthonormal basis vectors of the left null space of $\tilde{\mathbf{H}}_n$. Then we have

$$\mathbf{Q}_n \tilde{\mathbf{H}}_n \approx \mathbf{0}. \tag{16}$$

From (16), we can see that $\mathbf{Q}_n$ is a good approximation for $\mathbf{A}_n$ in (12). Using $\mathbf{A}_n = \mathbf{Q}_n$ as a linear combination with the received signal vector in (10), we have

$$\begin{aligned} \mathbf{y}_n &= \mathbf{A}_n \mathbf{y} \\ &= \mathbf{Q}_n \mathbf{H}_n \mathbf{s}_n + \mathbf{Q}_n \sum_{\substack{m=1 \\ m \neq n}}^{N} \mathbf{H}_m \mathbf{s}_m + \mathbf{Q}_n \mathbf{n}, \end{aligned} \tag{17}$$

where $\mathbf{y}_n \in \mathbb{C}^{N_{t_n} \times 1}$ is the equivalent received signal vector for the user class $n$ and the term $\mathbf{Q}_n \sum_{\substack{m=1 \\ m \neq n}}^{N} \mathbf{H}_m \mathbf{s}_m \approx \mathbf{0}$ represents the residual inter-user class interference. Then, we can transform the received signal vector into parallel single-user class signals as described by

$$\mathbf{y}_n = \check{\mathbf{H}}_n \mathbf{s}_n + \mathbf{n}_n, \quad \forall n \in (1 \ldots N), \tag{18}$$

where $\check{\mathbf{H}}_n = \mathbf{Q}_n \mathbf{H}_n \in \mathbb{C}^{N_{t_n} \times N_{t_n}}$ is the equivalent channel matrix of the $n$th user class after DSD and $\mathbf{n}_n = \mathbf{Q}_n \sum_{\substack{m=1 \\ m \neq n}}^{N} \mathbf{H}_m \mathbf{s}_m + \mathbf{Q}_n \mathbf{n} \in \mathbb{C}^{N_{t_n} \times 1}$ is the equivalent noise vector. Note that $\mathbf{H}^\dagger$ in (13) could have been provided by the pseudo-inverse of $\mathbf{H}$. This option satisfies the zero interference constraint in (12); however, it does result in a noise enhancement effect and has a restriction in terms of the dimension, i.e., $N_r \geq N_t$. The proposed strategy can be used even when $N_r < N_t$ and provides a balance between the inter-user class interference and the noise effects since the noise is taken into account in the computation of (13).

Another option to compute a basis for the left null space of $\tilde{\mathbf{H}}_n$ is performing the SVD transformation $\tilde{\mathbf{H}}_n = \tilde{\mathbf{U}}_n \tilde{\Sigma}_n \tilde{\mathbf{V}}_n^{\mathcal{H}}$, where $\tilde{\Sigma}_n \in \mathbb{C}^{N_r \times (N_t - N_{t_n})}$ is a rectangular diagonal matrix with the singular values of $\tilde{\mathbf{H}}_n$ on the diagonal, $\tilde{\mathbf{U}}_n \in \mathbb{C}^{N_r \times N_r}$ and $\tilde{\mathbf{V}}_n^{\mathcal{H}} \in \mathbb{C}^{(N_t - N_{t_n}) \times (N_t - N_{t_n})}$ are unitary matrices. If $r_n$ is the rank of $\tilde{\mathbf{H}}_n$ that corresponds to the number of non-zero singular values, i.e., $r_n = \mathrm{rank}(\tilde{\mathbf{H}}_n) \leq N_t - N_{t_n}$, the SVD can be expressed equivalently as:

$$\tilde{\mathbf{H}}_n = \left[\tilde{\mathbf{U}}_{1,n} \ \tilde{\mathbf{U}}_{0,n}\right] \ \tilde{\Sigma}_n \ \left[\tilde{\mathbf{V}}_{1,n} \ \tilde{\mathbf{V}}_{0,n}\right]^{\mathcal{H}}, \tag{19}$$

where $\tilde{\mathbf{U}}_{1,n} \in \mathbb{C}^{N_r \times r_n}$ consists of the first $r_n$ left singular vectors, $\tilde{\mathbf{U}}_{0,n} \in \mathbb{C}^{N_r \times (N_r - r_n)}$ holds the last $N_r - r_n$ left singular vectors, $\tilde{\mathbf{V}}_{1,n} \in \mathbb{C}^{(N_t - N_{t_n}) \times r_n}$ consists of the first $r_n$ singular vectors and $\tilde{\mathbf{V}}_{0,n} \in \mathbb{C}^{(N_t - N_{t_n}) \times (N_t - N_{t_n} - r_n)}$ hold the last $N_t - N_{t_n} - r_n$ singular vectors. Thus $\tilde{\mathbf{U}}_{0,n}$ and $\tilde{\mathbf{V}}_{0,n}^{\mathcal{H}}$ form an orthogonal basis for the left null space and the null space of $\tilde{\mathbf{H}}_n$, respectively. Then, the alternative solution for (12) could be:

$$\mathbf{A}_n = \tilde{\mathbf{U}}_{0,n}^{\mathcal{H}}. \tag{20}$$

Although the matrix $\tilde{\mathbf{U}}_{0,n}^{\mathcal{H}}$ eliminates the inter-user class interference effectively, i.e., $\tilde{\mathbf{U}}_{0,n}^{\mathcal{H}} \sum_{\substack{m=1 \\ m \neq n}}^{N} \mathbf{H}_m \mathbf{s}_m = \mathbf{0}$, when we use $\mathbf{y}_n = \mathbf{Q}_n \mathbf{y}$ in the first approach the noise effects in the detection procedure are mitigated due to the fact that the computation of $\mathbf{Q}_n$ takes the noise into account. Thus, noise effects are reduced which improves the performance, even in the presence of residual interference. In addition, the equivalent channel matrix $\check{\mathbf{H}}_n^1 = \tilde{\mathbf{U}}_{0,n}^{\mathcal{H}} \mathbf{H}_n$ has dimensions $(N_r - r) \times N_t$ as opposed to the matrix $\check{\mathbf{H}}_n^2 = \mathbf{Q}_n \mathbf{H}_n$ which has dimensions $N_{t_n} \times N_{t_n}$. For this reason the computational complexity of the detector is lower if we use the matrix $\mathbf{Q}_n$ to decouple the received signal vector.

The fact that we obtain a square equivalent channel matrix also allows the possibility of using lattice reduction (LR)-based detectors which have a better performance for square channel matrices [27]. Further, the computational complexity to compute the channel inversion (13) and $N$ QR decompositions (15) of matrices with dimensions $N_{t_n} \times N_r$ is much lower than the computational cost of computing $N$ SVD transformations (19) of matrices with dimensions $N_r \times (N_t - N_{t_n})$. For these reasons, in this paper we will focus on the first alternative.

As it will be presented in the next section, the equivalent received signal vector in (18) shows that the process in (11)-(17) is an effective algorithm to separate the user classes at the BS and we can consider the data stream of the $n$th user class as independent of the received signals of the other user classes. In practice, this allows the possibility of using different transmission and reception schemes for each user class. We can now implement the traditional detectors for each class of users separately which also allows the possibility of using more complex detection schemes due to the reduction of the dimensions of the matrices that need to be processed. The description of the proposed algorithm is presented in Algorithm 1.

### 3.2 Detection algorithms

In this subsection, we examine signal detection algorithms for massive MIMO in heterogeneous networks. To detect the data stream for each class of users independently, we assume that the DSD algorithm described in Algorithm 1 was previously employed.

**Algorithm 1 :** The DSD algorithm

*Channel and noise variance estimated with training sequence*

(1) Initialization: $\mathbf{H}$, $\sigma$

*Get the received signal*

(2) $\mathbf{y} = \sum_{n=1}^{N} \mathbf{H}_n \mathbf{s}_n + \mathbf{n} = \mathbf{H}\mathbf{s} + \mathbf{n}$

*Compute the combined channel inversion*

(3) $\mathbf{H}^{\dagger} = \mathbf{H}^{\mathcal{H}} \left(\mathbf{H}\mathbf{H}^{\mathcal{H}} + \sigma^2 \mathbf{I}\right)^{-1} = \left[\ddot{\mathbf{H}}_1 \; \ddot{\mathbf{H}}_2 \; \ldots \; \ddot{\mathbf{H}}_N\right]$

(4) **Do for** $n = 1$ **to** $N$

*Applying the QR decomposition*

(5) $[\mathbf{R}_n \; \mathbf{Q}_n] = \mathrm{RQ}(\ddot{\mathbf{H}}_n)$

(6) $\mathbf{A}_n = \mathbf{Q}_n$

*Compute the equivalent received signal vector*

(7) $\mathbf{y}_n = \mathbf{A}_n \mathbf{y} = \check{\mathbf{H}}_n \mathbf{s}_n + \mathbf{n}_n$

*Applying the detection procedures*

(8) $\hat{\mathbf{s}}_n = \text{Detector}\left(\mathbf{y}_n, \check{\mathbf{H}}_n = \mathbf{A}_n \mathbf{H}_n\right)$

(9) **End**

*Compute the overall estimated signal vector*

(10) $\hat{\mathbf{s}} = \left[\hat{\mathbf{s}}_1^T \; \hat{\mathbf{s}}_2^T \; \ldots \; \hat{\mathbf{s}}_N^T\right]^T$

### 3.2.1 Linear detectors

In linear detectors, the equivalent received signal vector for the $n$th user class $\mathbf{y}_n \in \mathbb{C}^{N_{t_n} \times 1}$ is processed by a linear filter to eliminate the channel effects [28]. The two linear detectors considered here are given by

$$\mathbf{W}_n^{\chi} = \begin{cases} \left(\check{\mathbf{H}}_n^{\mathcal{H}} \check{\mathbf{H}}_n\right)^{-1} \check{\mathbf{H}}_n^{\mathcal{H}}, & \chi = \text{Zero Forcing} \\ \left(\check{\mathbf{H}}_n^{\mathcal{H}} \check{\mathbf{H}}_n + \sigma^2 \mathbf{I}\right)^{-1} \check{\mathbf{H}}_n^{\mathcal{H}}, & \chi = \text{MMSE} \end{cases} \tag{21}$$

where $\sigma^2 = \sigma_n^2 / \sigma_s^2$ and $\check{\mathbf{H}}_n \in \mathbb{C}^{N_{t_n} \times N_{t_n}}$ is the equivalent channel matrix of the $n$th user class. Note that for the MMSE detector, we consider the autocorrelation matrix of the equivalent noise vector as $\mathbf{K}_{\mathbf{n}_n} \approx \sigma_n^2 \mathbf{I}$. As the residual interference is very small, an excellent performance can be obtained with this approximation. The linear hard decision of $\mathbf{s}_n$ is carried out as follows:

$$\hat{\mathbf{s}}_n = \mathbb{C}\left(\mathbf{W}_n^{\chi} \mathbf{y}_n\right), \tag{22}$$

where the function $\mathbb{C}(x)$ returns the point of the complex signal constellation closest to $x$. The linear detectors have a lower computational complexity when compared with the non-linear detectors. However, due to the impact of interference and noise, linear detectors offer a limited performance.

### 3.2.2 Successive interference cancellation

The successive interference cancellation (SIC) detector for the $n$th user class in (18) consists of a bank of linear detectors, each detects a selected component $s_{n,i}$ of $\mathbf{s}_n$. The component obtained by the first detector is used to reconstruct the corresponding signal vector which is then subtracted from the equivalent received signal to further reduce the interference in the input to the next linear receive filter. The successively cancelled received data vector that follows a chosen ordering in the $i$th stage is given by

$$\mathbf{y}_{n,i} = \mathbf{y}_n - \sum_{j=1}^{i-1} \check{\mathbf{h}}_{n,j} \hat{s}_{n,j}, \tag{23}$$

where $\check{\mathbf{h}}_{n,j}$ correspond to the columns of the channel matrix $\check{\mathbf{H}}_n$ and $\hat{s}_{n,j}$ is the estimated symbol obtained at the output of the $j$th linear detector.

### 3.2.3 Multiple-branch SIC detection

In the multi-branch scheme [10] for the $n$th user class, different orderings are explored for SIC, each ordering is referred to as a branch, so that a detector with $L$ branches produces a set of $L$ estimated vectors. Each branch uses a column permutation matrix $\mathbf{P}_n$. The estimate of the signal vector of branch $l$, $\hat{\mathbf{x}}_n^{(l)}$, is obtained using a SIC receiver based on a new channel matrix $\check{\mathbf{H}}_n^{(l)} = \check{\mathbf{H}}_n \mathbf{P}_n^{(l)}$. The order of the estimated symbols is rearranged to the original order by

$$\hat{\mathbf{s}}_n^{(l)} = \mathbf{P}_n^{(l)} \hat{\mathbf{x}}_n^{(l)}, \qquad l = 1, \ldots, L. \tag{24}$$

A higher detection diversity can be obtained by selecting the most likely symbol vector based on the ML selection rule, that is

$$\hat{\mathbf{s}}_n = \arg\min \left\| \mathbf{y}_n - \check{\mathbf{H}}_n \hat{\mathbf{s}}_n^l \right\|^2, l = 1, \ldots, L. \tag{25}$$

Other detectors could be used with the proposed DSD technique and this is up to the designer to choose the detector.

## 4 Sum-rate analysis

In this section, a performance analysis for the proposed scheme is presented in terms of the sum rate. We consider that the channel matrix $\mathbf{H}$ was previously estimated at the BS, assume Gaussian signalling and that the received signal vector was decoupled for each user class. Considering the received signal vector as presented in (18), the sum rate [29] that DSD can offer is defined as

$$R = \sum_{n=1}^{N} \log_2 \frac{\det(\mathbf{K}_{\mathbf{y}_n})}{\det(\mathbf{K}_{\mathbf{n}_n})}, \tag{26}$$

where $\mathbf{K}_{\mathbf{y}_n}$ and $\mathbf{K}_{\mathbf{n}_n}$ are the autocorrelation matrix of the equivalent received signal vector and the equivalent noise vector of the $n$th user class, respectively. It is easy to show that (26), can be expressed as

$$R = \sum_{n=1}^{N} \log_2 \det\left(\mathbf{I} + \mathbf{B}_n\mathbf{B}_n^{\mathcal{H}}\right), \tag{27}$$

where $\mathbf{B}_n = \mathbf{K}_{\mathbf{n}_n}^{-1/2}\check{\mathbf{H}}_n\mathbf{K}_{\mathbf{s}_n}^{1/2} \in \mathbb{C}^{N_{t_n}\times N_{t_n}}$. As $\mathbf{B}_n\mathbf{B}_n^{\mathcal{H}}$ is a Hermitian symmetric positive-definite square matrix, we have

$$\mathbf{B}_n\mathbf{B}_n^{\mathcal{H}} = \bar{\mathbf{Q}}_n\mathbf{\Lambda}_n\bar{\mathbf{Q}}_n^{\mathcal{H}}, \tag{28}$$

where $\bar{\mathbf{Q}}_n$ is a square unitary matrix, $\bar{\mathbf{Q}}_n\bar{\mathbf{Q}}_n^{\mathcal{H}} = \mathbf{I}$, and $\mathbf{\Lambda}_n$ is a diagonal matrix whose diagonal elements are the eigenvalues of the matrix $\mathbf{B}_n\mathbf{B}_n^{\mathcal{H}}$. Then, the reliable sum rate that the system can offer is

$$R = \sum_{n=1}^{N}\sum_{i=1}^{N_{t_n}} \log_2\left(1 + \lambda_{i,n}\right). \tag{29}$$

Note that the eigenvalues $\lambda_{i,n}$ in (29) can be obtained computing the eigenvalues of $\mathbf{B}_n^{\mathcal{H}}\mathbf{B}_n$. As mentioned before, for notational simplicity we assume that $\mathbf{K}_{\mathbf{s}_n} = \sigma_s^2\mathbf{I}$. When the DSD algorithm is applied, the equivalent noise vector for the $n$th user class $\mathbf{n}_n = \mathbf{Q}_n\sum_{\substack{m=1\\m\neq n}}^{N}\mathbf{H}_m\mathbf{s}_m + \mathbf{Q}_n\mathbf{n} \in \mathbb{C}^{N_{t_n}\times 1}$ is not white due to the residual inter-user class interference. Then its autocorrelation matrix is given by

$$\mathbf{K}_{\mathbf{n}_n} = \mathbb{E}[\,\mathbf{n}_n\mathbf{n}_n^{\mathcal{H}}] = \sigma_s^2\mathbf{Q}_n\left(\sum_{\substack{m=1\\m\neq n}}^{N}\mathbf{H}_m\mathbf{H}_m^{\mathcal{H}}\right)\mathbf{Q}_n^{\mathcal{H}} + \sigma_n^2\mathbf{I}. \tag{30}$$

Finally, the eigenvalues $\lambda_{i,n}$ in (29) are obtained from the eigenvalues of matrix $\mathbf{B}_n^{\mathcal{H}}\mathbf{B}_n$ given by

$$\mathbf{B}_n^{\mathcal{H}}\mathbf{B}_n = \frac{\sigma_s^2}{\sigma_n^2}\check{\mathbf{H}}_n^{\mathcal{H}}\left[\frac{\sigma_s^2}{\sigma_n^2}\mathbf{Q}_n\left(\sum_{\substack{m=1\\m\neq n}}^{N}\mathbf{H}_m\mathbf{H}_m^{\mathcal{H}}\right)\mathbf{Q}_n^{\mathcal{H}} + \mathbf{I}\right]^{-1}\check{\mathbf{H}}_n. \tag{31}$$

To compute the sum rate for the received signal vector in (9) when the detection is perform for all user classes together, we suppress the index $n$ from the above analysis and considering that $\mathbf{K_s} = \sigma_s^2\mathbf{I}$ and $\mathbf{K_n} = \sigma_n^2\mathbf{I}$, we get the well-known expression:

$$\bar{R} = \sum_{i=1}^{N_t} \log_2(1 + \lambda_i), \tag{32}$$

where the values $\lambda_i$ are the eigenvalues of the matrix $\mathbf{B}^{\mathcal{H}}\mathbf{B} = \frac{\sigma_s^2}{\sigma_n^2}\mathbf{H}^{\mathcal{H}}\mathbf{H}$ [30]. In Appendix 1, we show that, as well as the sum rate in (32) when all user classes are detected together, the sum rate in (29) for the proposed algorithm is independent of the detection procedure. However, the lower bound on the achievable uplink sum rate obtained by using linear detectors is different for each detector [31]. In order to analyze the behavior of the lower bound on the sum rate for the proposed scheme, we consider that a linear detector according to (21) is applied to the equivalent received signal vector (18) to detect the transmitted symbol vector of the user class $n$, then we have

$$\tilde{\mathbf{y}}_n = \mathbf{W}_n\mathbf{y}_n = \mathbf{W}_n\check{\mathbf{H}}_n\mathbf{s}_n + \mathbf{W}_n\mathbf{n}_n. \tag{33}$$

Taking the $k$th element of $\tilde{\mathbf{y}}_n$ we have

$$\tilde{y}_{k,n} = \mathbf{w}_{k,n}\check{\mathbf{h}}_{k,n}s_{k,n} + \sum_{\substack{j=1\\j\neq k}}^{|C_n|}\mathbf{w}_{k,n}\check{\mathbf{h}}_{j,n}s_{j,n} + \mathbf{w}_{k,n}\mathbf{n}_n, \tag{34}$$

where $\mathbf{w}_{k,n}$ is the $k$th row of $\mathbf{W}_n$. Modeling the noise interference, the inter-user class interference and the inter-user interference in the user class $n$ in (34) as additive Gaussian noise independent of $s_{k,n}$, considering (30) and that the channel is ergodic so that each codeword spans over a large number of realizations, we obtain the lower bound on the achievable rate for the DSD algorithm with linear detectors as described in (35).

Then, the lower bound on the achievable rate for the entire system is given by

$$R_{k,n} = \mathbb{E}\left[\log_2\left(1 + \frac{\sigma_s^2\mid\mathbf{w}_{k,n}\check{\mathbf{h}}_{k,n}\mid^2}{\sigma_s^2\sum_{\substack{j=1\\j\neq k}}^{|C_n|}\left|\mathbf{w}_{k,n}\check{\mathbf{h}}_{j,n}\right|^2 + \left|\mathbf{w}_{k,n}\left[\sigma_s^2\mathbf{Q}_n\left(\sum_{\substack{m=1\\m\neq n}}^{N}\mathbf{H}_m\mathbf{H}_m^{\mathcal{H}}\right)\mathbf{Q}_n^{\mathcal{H}} + \sigma_n^2\mathbf{I}\right]\mathbf{w}_{k,n}^{\mathcal{H}}\right|}\right)\right], \tag{35}$$

$$R = \sum_{n=1}^{N} \sum_{k=1}^{|C_n|} R_{k,n}. \tag{36}$$

For the SIC receiver, each stream is filtered by a linear detector and then, its contribution is subtracted from the received signal to improve the subsequent detection. For each layer the linear filter is recalculated. The performance of SIC detectors can be improved if we choose the cancellation order as a function of the SINR at the output of the linear detector in each layer. The lower bound for the sum rate of the proposed algorithm when a SIC detector is used for each user class, could be calculated updating the expression (35) in each layer, i.e., the values of $\mathbf{w}_{k,n}$ are recalculated for each detected stream.

## 5 Numerical results

In this section, we evaluate the performance of the proposed algorithm with different detectors in terms of the sum rate and the BER via simulations. Moreover, the computational complexity of the proposed and existing algorithms is also evaluated.

### 5.1 Sum rate

To evaluate the analytic results obtained in Section 4, the sum rate and the lower bounds for the proposed algorithm with different detection schemes will be evaluated considering CAS and DAS configuration assuming perfect CSI. For the CAS configuration, we employ $L_k = 0.7$, $\tau = 2$, the distance $d_k$ to the BS is obtained from a uniform discrete random variable distributed between 0.1 and 0.99, the shadowing spread is $\sigma_k = 3$ dB and the transmit and receive correlation coefficients are $\rho_{rx} = 0.2$ and $\rho_{tx} = 0.4$ (when $N_{t_{k,n}} > 1$), respectively. For DAS configurations, we consider a densely populated cell, where a fraction of the active users are in the center of the cell and the remaining users are in other locations of the cell. We explore different particular values for the fraction of users in the center and around the cell. Based on that, we choose specific values for $N_B$, $D$, and $Q$. For the DAS configuration, we also consider $L_{k,j}$ taken from a uniform random variable distributed between 0.7 and 1, $\tau = 2$, the distance $d_{k,j}$ for each link to an antenna is obtained from a uniform discrete random variable distributed between 0.1 and 0.5, the shadowing spread is $\sigma_{k,j} = 3$ dB and the transmit and receive correlation coefficients are $\rho_{rx} = 0.2$ and $\rho_{tx} = 0.4$, respectively.

In Fig. 2 we evaluate the sum rate in two different scenarios for the user requirements. In both cases, we fix the SNR = 10 dB and increase the number of receive antennas. For the DAS configuration, we consider $N_B = 1/2N_r$ antennas at the BS. We also consider $D = 4$ arrays of antennas distributed around the cell, each equipped

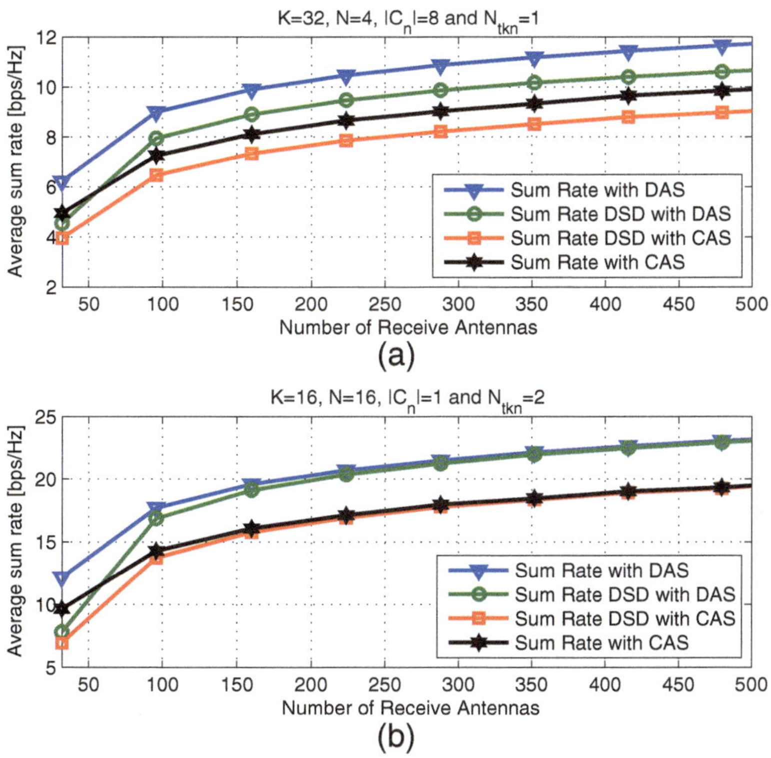


**Fig. 2 a** Sum rate vs number of receive antennas. $N = 4$, $|C_n| = 8$ users per class, $N_{t_{k,n}} = 1$ antenna per user. **b** Sum rate vs number of receive antennas. $N = 16$, $|C_n| = 1$ user per class, $N_{t_{k,n}} = 2$ antennas per user

with $Q = 1/8N_r$ antennas. For Fig. 2a, we consider $N = 4$ classes of users with $| C_n |= 8$ users each and $N_{t_{k,n}} = 1$ antenna per user. We can see that the sum rate of the proposed DSD algorithm is close to the sum rate of the traditional MIMO system and with a low computational complexity on the detection procedures as will be shown in the next subsection. For Fig. 2b, we consider 16 active users in the system and that we need to detect each user independently, i.e., $N = 16$ classes of users with $| C_n |= 1$ user at each class and $N_{t_{k,n}} = 2$ antennas per user. Under these conditions, the sum rate of the proposed scheme reaches the sum rate of the traditional MIMO system, especially for a large number of receive antennas. From the plot in Fig. 2, we can see that the sum rate for DAS is higher than that for the CAS configuration.

In Fig. 3, we compare the lower bound on the sum rate for different detectors such as, ZF, MMSE, and SIC. We consider the DAS configuration under the same conditions as in the previous experiment. For Fig. 3a, we consider $N = 3$ classes of users with $| C_n |= 10$ users at each class and $N_{t_{k,n}} = 1$ antenna per user. We can see from the plot that, similarly to the traditional MIMO systems, the lower bound on the sum rate for ZF and MMSE achieves the sum rate when $N_r$ grows. For Fig. 3b, we consider $N = 2$ classes of users with $| C_n |= 16$ users at each class and $N_{t_{k,n}} = 1$ antennas per user. We can see that the SIC-MMSE achieves the sum rate and it could be considered optimal in terms of sum rate.

In Fig. 4, we compare the lower bound sum rate versus SNR. We consider $N = 8$ classes of users with $| C_n |= 1$ user in each class and $N_{t_{k,n}} = 8$ antennas per user transmitting with high correlation between antennas $\rho_{tx} = 0.85$. We consider the DAS configuration with $N_B = 96$, $D = 4$, and $Q = 8$. We can see from the plot that the lower bounds for the proposed algorithms are very close to the lower bounds when the detection procedure is carried out together for all users. Since ZF DSD separates the user classes by computing a MMSE matrix which takes the noise component into account, it will be the only detector that outperform its coupled counter part.

Except for the ZF-DSD case, there is a small loss in the performance for DSD schemes. However, with the current processor technologies, to use a SIC MMSE in a $256 \times 256$ massive MIMO system could be infeasible. By dividing the users into classes of users, the computational complexity in the detection procedure decreases significantly and makes the use of the SIC-MMSE and more complex detectors feasible.

### 5.2 Computational complexity analysis

In this subsection, the computational complexity of the proposed algorithm is evaluated and compared with the

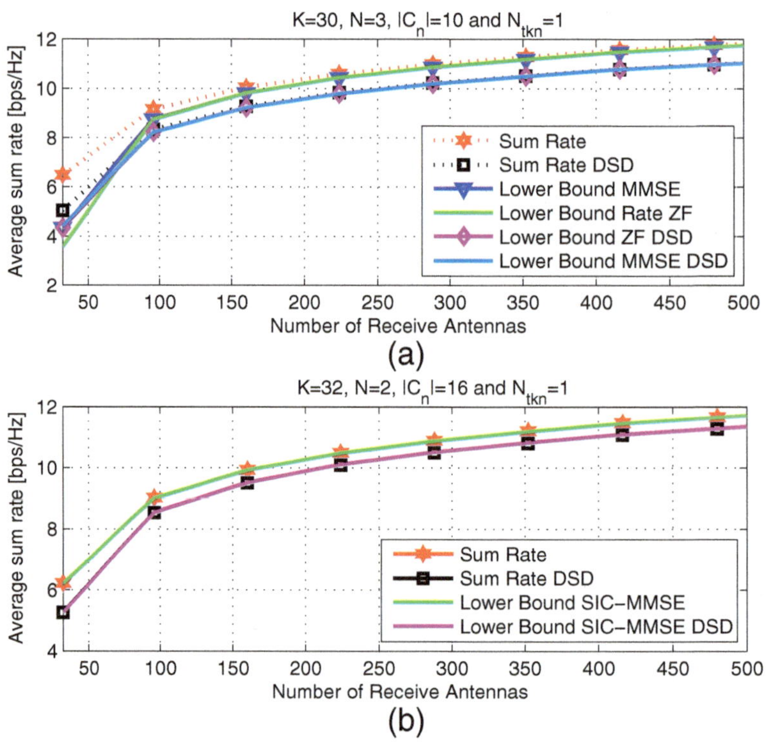


**Fig. 3 a** Sum rate vs number of receive antennas. $N = 3$, $| C_n |= 10$ users per class, $N_{t_{k,n}} = 1$ antenna per user. **b** Sum rate vs number of receive antennas. $N = 2$, $| C_n |= 16$ users per class, $N_{t_{k,n}} = 1$ antenna per user

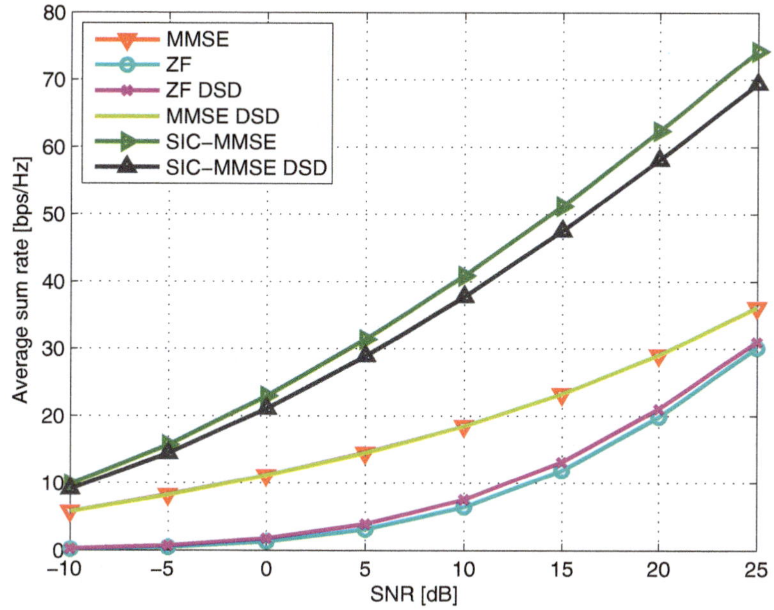


**Fig. 4** Sum rate lower bound vs SNR. $N = 8$, $| C_n | = 1$ user per class, $N_{t_{k,n}} = 8$ antennas per user. We consider DAS configuration with $N_B = 96$, $D = 4$, and $Q = 8$

traditional coupled detection schemes, when all user classes are detected together, by counting the number of floating point operations (FLOPs) per received vector **y**. Different detection schemes are considered such as MMSE, SIC, and MB-SIC. The SIC-based receivers all use MMSE detection. Furthermore, the single-branch SIC and the first branch of the MB-SIC employ norm-based ordering. We consider QPSK modulation; however, the computational cost in these detectors does not change significantly with the modulation order. The number of FLOPs for the complex QR decomposition of an $N_{t_n} \times N_r$ matrix is given in [32] as $16\left(N_r^2 N_{t_n} - N_{t_n}^2 N_r + 1/3 N_{t_n}^3\right)$. To compute the number of FLOPs required for the remaining operations, we use the Light-speed Matlab toolbox [33].

Figure 5 shows the computational complexity versus the number of user classes for different detection algorithms. We consider $K = 100$ active users, $N_{t_{k,n}} = 2$

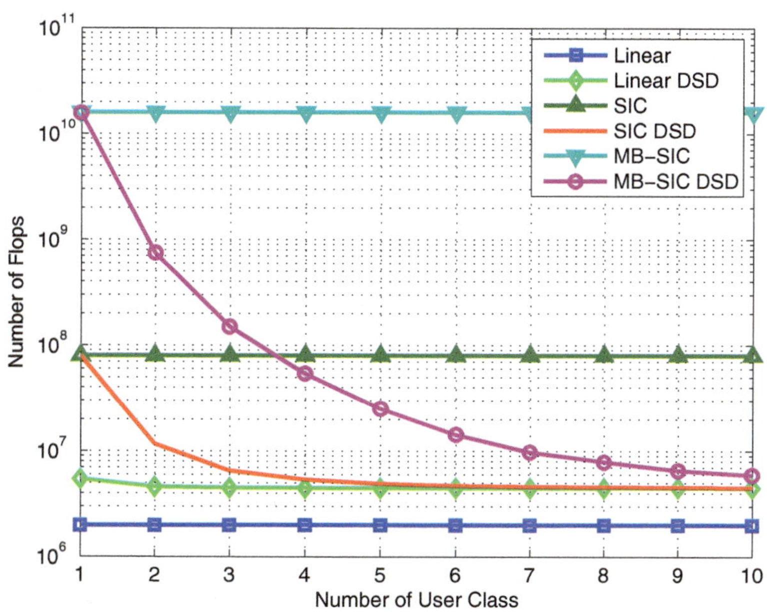


**Fig. 5** Computational complexity versus number of user class, $K = 100$ active users, $N_{t_{k,n}} = 2$ antennas per user, $N_r = 200$ receive antennas

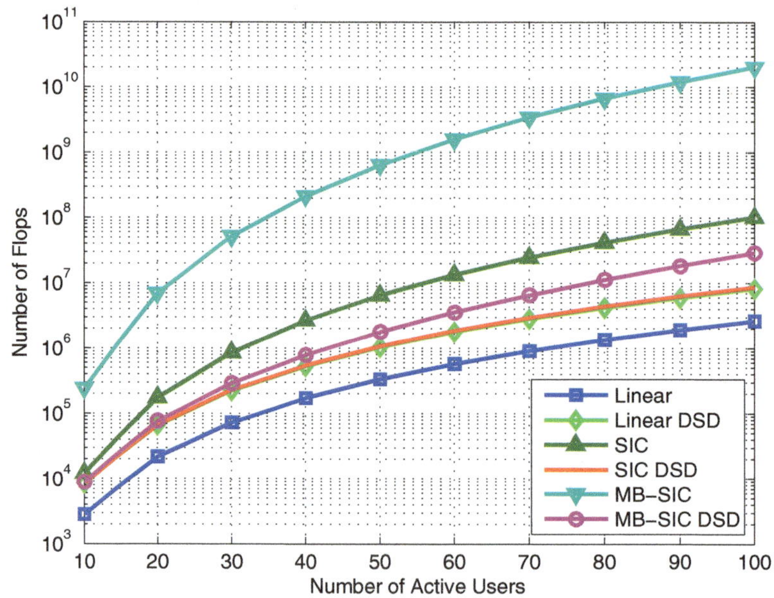


**Fig. 6** Computational complexity versus number of active user, $N = 5$ classes of user, $N_{t_{k,n}} = 2$ antennas per user, $N_t = KN_{t_{k,n}}$ transmit antennas and $N_r = 3N_t$ receive antennas

transmit antennas per user and $N_r = 200$ receive antennas distributed around the cell. We consider an increasing number of classes of users, when $K$ is not divisible by the number of classes, the number of active users is set to a smaller value so as to allow the division in $N$ classes. We can see from the figure that the complexity of the SIC and the MB-SIC detectors with the DSD algorithm is lower than the SIC and the MB-SIC coupled detectors, respectively. Furthermore, for receivers with DSD, the computational complexity is reduced as the number of user classes is increased. This fact represents an important advantage for receivers with DSD, due to the fact that it allows the use of more complex detectors for each user class according to its data requirements.

In Figs. 6 and 7, we plot the required number of FLOPs versus the number of active users and versus the number

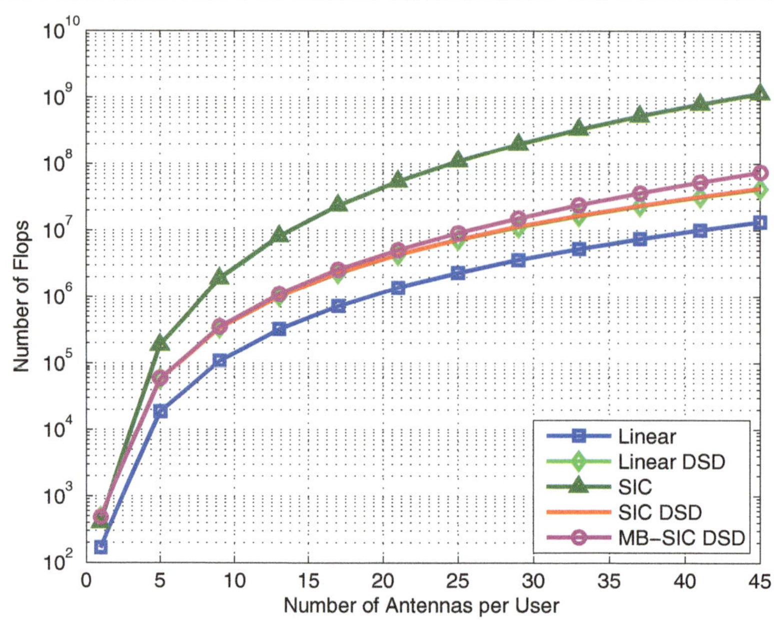


**Fig. 7** Computational complexity versus number of antennas per user, $K = 10$ active user, $N = 10$ classes of users, $N_t = KN_{t_{k,n}}$ transmit antennas and $N_r = 2N_t$ receive antennas

of antennas per user, respectively. For Fig. 6 we consider $N = 5$ classes of users, $N_{t_{k,n}} = 2$ transmit antennas per user and $N_r = 3KN_{t_{k,n}}$ receive antennas. The MB-SIC and SIC detectors with DSD have a lower computational cost than the coupled SIC detector. For this 5G context, with a high number of antennas, efficient coupled detectors are not feasible to be implemented; however, if a specific user class requires the benefits of complex detectors, the DSD algorithm reduces the cost so that more complex detectors could be applied as illustrated with MB-SIC in Fig. 6. For the results in Fig. 7, we consider that we have $K = 10$ active users and that we need to detect each user independently, i.e., $N = 10$. The number of transmit antennas per user $N_{t_{k,n}}$ is increased. We also consider that the number of receive antennas distributed around the cell is given by $N_r = 2KN_{t_{k,n}}$. The MB-SIC and SIC detectors with DSD have a significantly lower complexity when compared with the SIC detector where all users are coupled.

It is worth noting that the curves displayed in Figs. 5, 6, and 7 will have a substantial decrease if the channel does not change over a time period due to quasi static channels. In this case, the equivalent channel matrices for each user class are stored for subsequent use. It would increase the gap, in terms of the computational cost, for the detection schemes using the DSD algorithm.

### 5.3 BER performance

In this subsection, the BER performance of the proposed algorithm is evaluated using different detectors which includes linear, SIC, and MB-SIC with linear MMSE receive filters. The SIC detector of [8] uses a norm-based cancellation ordering, the MB-SIC of [10] employs a fixed number of branches, equal to the total number of transmit antennas per user class $N_{t_n}$ for the DSD schemes, and norm-based ordering in its first branch. A massive MIMO system operating in heterogeneous networks with $K$ active users is considered. We also consider the DAS configuration where the $N_r = N_B + DQ$ receive antennas are distributed around the cell in $D$ radio heads with $Q$ antennas each and the remaining $N_B$ antennas are located at the BS. We consider QPSK modulation. The SNR per transmitted information bit is defined as

$$\text{SNR} = 10 \log_{10} \frac{N_t \sigma_s^2}{M \sigma_n^2}, \tag{37}$$

where $\sigma_s^2$ is the common variance of the transmitted symbols, $\sigma_n^2$ is the noise variance at the receiver and $M$ is the number of transmitted bits per symbol. The numerical results correspond to an average of 3000 simulations runs, with $500N_t$ symbols transmitted per run. For the $N_B$ antennas at the BS, we employ $L_k = 0.3$, $\tau = 2$, the distance to the users is obtained from a uniform discrete random variable distributed between 0.4 and 0.7, the shadowing spread is $\sigma_k = 1$ dB and the transmit and receive correlation coefficients are equal to $\rho_{rx} = 0.4$. For the $R$ remote arrays of antennas, we use $L_{k,j}$ taken from a uniform random variable distributed between 0.3 and 0.5, the shadowing spread $\sigma_{k,j} = 1$ dB and the receive correlation coefficients are equal to $\rho_{rx} = 0.5$. When the number of transmit antennas at the users is $N_{t_{k,n}} > 1$, the transmit correlation coefficient is equal to $\rho_{tx} = 0.55$.

For the experiment in Fig. 8, we consider $K = 12$ user devices, where each user is equipped with $N_{t_{k,n}} = 3$ transmit antennas and $N = 3$ classes of users. The system

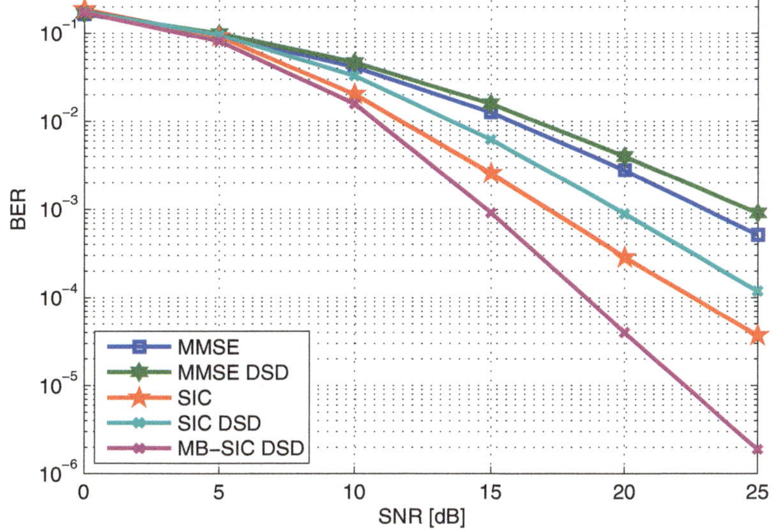


**Fig. 8** BER versus SNR with $K = 12$ active users, $N = 3$ classes of users, $N_{t_{k,n}} = 3$ transmit antennas per user and $N_r = 36$ receive antennas

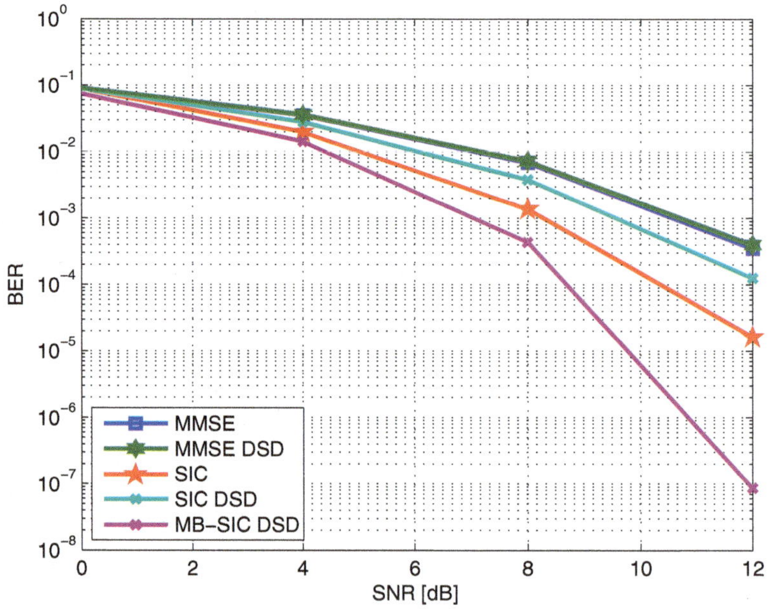


**Fig. 9** BER versus SNR with $K = 8$ active users, $N_{t_{k,n}} = 8$ transmit antennas per user, $N = 8$ classes of users, and $N_r = 128$ receive antennas

employs perfect channel state information and QPSK modulation. For the DAS configuration, we consider $N_B = 8$ receive antennas at the BS, $D = 4$ remote radio heads, and $Q = 7$ receive antennas per remote radio head. We can see from the figure that the decoupled SIC detection presents a performance close to the coupled SIC detector with a difference around 2 dB in the high SNR region. In addition the decoupled SIC detector has a drastic reduction in the computational cost when compared with its coupled version. The result also indicates a remarkable superiority in the performance for the MB-SIC receiver with the DSD scheme over the coupled SIC detector which also has a lower computational complexity than the coupled SIC detector.

In the next example, we evaluate the performance of the proposed scheme considering $K = 8$ active users,

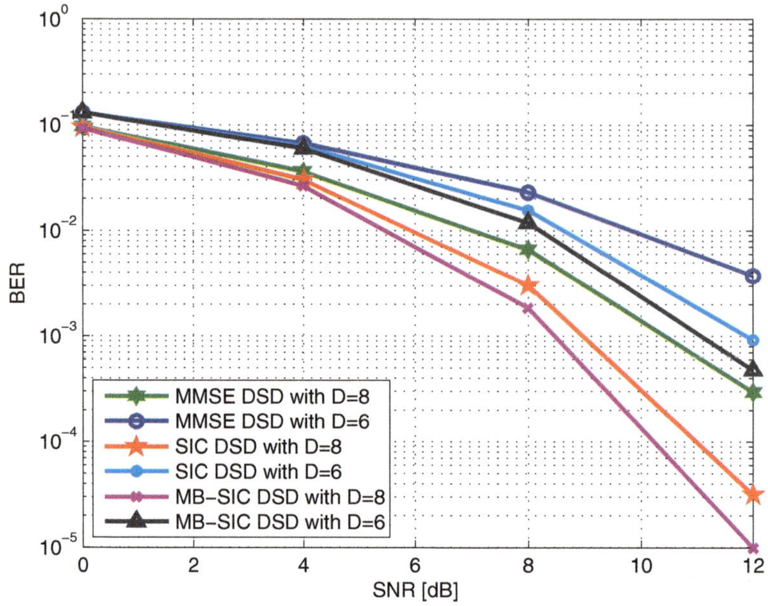


**Fig. 10** BER versus SNR with $K = 64$ active users, $N_{t_{k,n}} = 1$ transmit antenna per user, $N = 4$ classes of users, $N_B = 34$, $D$ remote antennas arrays with $Q = 7$ receive antennas per array

each user transmitting with $N_{t_{k,n}} = 8$ antennas. We also consider that we need to detect each user independently of each other, i.e., $N = 8$ classes of users with $| C_n |= 1$ user per class. For the DAS configuration, we consider $N_B = 64$ receive antennas at the BS, $D = 8$ remote radio heads, and $Q = 8$ receive antennas per remote radio head. The Fig. 9 indicates that the performance of the SIC detector with DSD is close to the SIC detector with a lower computational complexity. Note that the results for MB-SIC with DSD show a very good performance with a computational complexity much lower than the SIC detector without DSD.

In Fig. 10 we evaluate the performance of the proposed scheme with MMSE, SIC, and MB-SIC detection. We consider $K = 64$ active users, $N = 4$ classes of users, $| C_n |= 16$ users per class, and $N_{t_{k,n}} = 1$ transmit antenna per user. For the DAS configuration, we consider $N_B = 34$ receive antennas at the BS and $Q = 7$ receive antennas per remote radio head. To show the behavior of the BER performance with different numbers of distributed antennas, we consider two configurations of remote radios heads, $D = 8$ and $D = 6$. As expected, the results shows that when the number of RRHs is increased, the BER performance is improved due to the low propagation effects caused by the short distances between users and some remote antenna arrays. We also can see from the figure that the SIC and the MB-SIC detector with DSD offer an excellent BER performance with a low computational cost.

## 6 Conclusions

In this paper, a mathematical signal model and the DSD algorithm for the uplink of massive MIMO systems operating in heterogeneous cellular networks with different classes of users using CAS and DAS configurations have been presented. The proposed algorithm allows one to separate the received signals for each category of users efficiently into independent parallel single user class signals at the receiver side, applying a common channel inversion and QR decomposition and assuming that the channel matrix was previously estimated. With the proposed scheme, it is possible to handle different classes of users in heterogeneous networks and to use different modulation and/or detection schemes for each class according to its data service requirements. The main advantage of DSD is the reduction in the computational cost of efficient detection schemes that, for its high computational complexity, are not feasible for implementation when the signals received from all active users are coupled.

## Appendix 1

Let us consider that a linear detector according to (21) is applied to the equivalent received signal vector (18) to detect the transmitted symbol vector of the $n$th user class as in (33). If we define the matrix $\bar{\mathbf{A}}_n = \mathbf{W}_n\check{\mathbf{H}}_n$ and the vector $\bar{\mathbf{n}}_n = \mathbf{W}_n\mathbf{n}_n$ we can rewrite (33) as

$$\tilde{\mathbf{y}}_n = \bar{\mathbf{A}}_n\mathbf{s}_n + \bar{\mathbf{n}}_n, \tag{38}$$

then, in analogy with the analysis in Section 4, the sum rate for the $n$th user class after the detection is given by

$$R_n = \sum_{i=1}^{N_{t_n}} \log_2(1 + \bar{\lambda}_i), \tag{39}$$

where the values $\bar{\lambda}_i$ are the eigenvalues of the matrix $\bar{\mathbf{B}}_n^{\mathcal{H}}\bar{\mathbf{B}}_n$ with $\bar{\mathbf{B}}_n = \mathbf{K}_{\bar{\mathbf{n}}_n}^{-1/2}\bar{\mathbf{A}}_n\mathbf{K}_{\mathbf{s}_n}^{1/2}$. Then

$$\bar{\mathbf{B}}_n^{\mathcal{H}}\bar{\mathbf{B}}_n = \mathbf{K}_{\mathbf{s}_n}^{1/2}\bar{\mathbf{A}}_n^{\mathcal{H}}\mathbf{K}_{\bar{\mathbf{n}}_n}^{-1}\bar{\mathbf{A}}_n\mathbf{K}_{\mathbf{s}_n}^{1/2}, \tag{40}$$

where $\mathbf{K}_{\mathbf{s}_n} = \sigma_s^2\mathbf{I}$ and $\mathbf{K}_{\bar{\mathbf{n}}_n} = \mathbf{W}_n\mathbf{K}_{\mathbf{n}_n}\mathbf{W}_n^{\mathcal{H}}$. Assuming that $\mathbf{K}_{\bar{\mathbf{n}}_n}$ has an inverse, we finally obtain

$$\begin{aligned} \bar{\mathbf{B}}_n^{\mathcal{H}}\bar{\mathbf{B}}_n &= \sigma_s^2\bar{\mathbf{A}}_n^{\mathcal{H}}\left(\mathbf{W}_n\mathbf{K}_{\mathbf{n}_n}\mathbf{W}_n^{\mathcal{H}}\right)^{-1}\bar{\mathbf{A}}_n \\ &= \sigma_s^2\check{\mathbf{H}}_n^{\mathcal{H}}\mathbf{W}_n^{\mathcal{H}}\left(\mathbf{W}_n\mathbf{K}_{\mathbf{n}_n}\mathbf{W}_n^{\mathcal{H}}\right)^{-1}\mathbf{W}_n\check{\mathbf{H}}_n \\ &= \sigma_s^2\check{\mathbf{H}}_n^{\mathcal{H}}\mathbf{K}_{\mathbf{n}_n}^{-1}\check{\mathbf{H}}_n. \end{aligned} \tag{41}$$

Note that (41) and (31) will yield the same result which proves that the sum rate for the DSD algorithm is independent of the linear detection procedure.

**Acknowledgements**
This work was supported in part by grants of CNPq, FAPERJ, and DAAD.

**Competing interests**
The authors declare that they have no competing interests.

**Author details**
[1]Center for Studies in Telecommunications (CETUC), Pontifical Catholic University of Rio de Janeiro (PUC-Rio), Rio de Janeiro, Brazil. [2]Communications Research Laboratory, Ilmenau University of Technology, Ilmenau, Germany.

**References**
1. TL Marzetta, Noncooperative cellular wireless with unlimited numbers of base station antennas. IEEE Trans. Wireless Commun. **9**(11), 3590–3600 (2010)
2. EG Larsson, O Edfors, F Tufvesson, TL Marzetta, Massive MIMO for next generation wireless systems. IEEE Commun. Mag. **52**(2), 186–195 (2014)
3. F Rusek, D Persson, B Lau, E Larsson, T Marzetta, O Edfors, F Tufvesson, Scaling up MIMO: Opportunities, and challenges with very large arrays. IEEE Signal Process. Mag. **30**(1), 40–60 (2013)
4. RC de Lamare, Massive MIMO systems: Signal processing challenges and future trends. URSI Radio Science Bulletin (2013)
5. T Marzetta, Noncooperative cellular wireless with unlimited numbers of base station antennas. IEEE Trans. Wireless Commun. **9**(11), 3590–3600 (2010)
6. EG Larsson, F Tufvesson, O Edfors, TL Marzetta, Massive MIMO for next generation wireless systems. IEEE Commun. Mag. **52**(2), 186–195 (2014)
7. 3GPP-LTE, Technical specification group radio access network: evolved universal terrestrial radio access (E-UTRA): further advancements for E-UTRA physical layer aspects (release 13). 3GPP TR136.942 (2016)

8. GD Golden, CJ Foschini, RA Valenzuela, PW Wolniansky, Detection algorithm and initial laboratory results using V-BLAST space-time communication architecture. Electron. Lett. **35**(1), 14–16 (1999). doi:10.1049/el:19990058
9. RC de Lamare, R Sampaio-Neto, Minimum mean-squared error iterative successive parallel arbitrated decision feedback detectors for DS-CDMA systems. IEEE Trans. Commun. **56**(5), 778–789 (2008)
10. RC de Lamare, Adaptive and iterative multi-branch MMSE decision feedback detection algorithms for multi-antenna systems. IEEE Trans. Wirel. Commun. **12**(10), 5294–5308 (2013). doi:10.1109/TWC.2013.092013.130233
11. P Li, RC de Lamare, R Fa, Multiple feedback successive interference cancellation detection for multiuser MIMO systems. IEEE Trans. Wirel. Commun. **10**(8), 2434–2439 (2011)
12. P Demestichas, A Georgakopoulos, D Karvounas, K Tsagkaris, V Stavroulaki, J Lu, C Xiong, J Yao, 5G on the horizon: Key challenges for the radio-access network. IEEE Veh. Technol. Mag. **8**(3), 47–53 (2013)
13. H Pervaiz, L Musavian, Q Ni, in *2015 IEEE International Conference on Communication Workshop (ICCW)*. Area energy and area spectrum efficiency trade-off in 5G heterogeneous networks, (2015), pp. 1178–1183
14. W Roh, A Paulraj, MIMO channel capacity for the distributed antenna systems. IEEE Veh. Technol. Conf. **3**, 1520–1524 (2002)
15. D Wang, J Wang, X You, Y Wang, M Chen, X Hou, Spectral efficiency of distributed MIMO systems. IEEE J. Selected Areas Commun. **31**(10), 2112–2127 (2002)
16. F Römer, M Fuchs, M Haardt. Distributed MIMO systems with spatial reuse for highspeed indoor mobile radio access. 20th Meeting of the Wireless World Research Forum (WWRF), Ottawa, Canada, (2008)
17. J Andrews, S Buzzi, W Choi, S Hanly, A Lozano, A Soong, J Zhang, What will 5g be? IEEE J. Selected Areas Commun. **32**(6), 1065–1082 (2014)
18. J Kermoal, L Schumacher, K Perdersen, et al, A stochastic MIMO radio channel model with experimental validation. IEEE J. Selected Areas Commun. **20**(6), 1211–1226 (2002)
19. QH Spencer, AL Swindlehurst, M Haardt, Zero-forcing methods for downlink spatial multiplexing in multiuser MIMO channels. IEEE Trans. Signal Process. **52**(2), 461–471 (2004)
20. LU Choi, RD Murch, A transmit preprocessing technique for multiuser MIMO systems using a decomposition approach. IEEE Trans. Wireless Commun. **3**(1), 20–24 (2004)
21. V Stankovic, M Haardt, Generalized design of multi-user MIMO precoding matrices. IEEE Trans. Wireless Commun. **7**(3), 953–961 (2008)
22. CB Chae, S Shim, RW Heath, Block diagonalized vector perturbation for multiuser MIMO systems. IEEE Trans. Wireless Commun. **7**(11), 4051–4057 (2008)
23. L Arévalo, RC de Lamare, M Haardt, R Sampaio-Neto, in *2015 IEEE International Workshop on Computational Advances in Multi-Sensor Adaptive Processing (CAMSAP)*. Uplink block diagonalization for massive MIMO-OFDM systems with distributed antennas, (2015)
24. V Stankovic, M Haardt. Improved diversity on the uplink of multi-user MIMO systems. 2005 European Conference on Wireless Technology (EuWiT), (2005), pp. 113–116
25. K Zu, RC de Lamare, M Haardt, Generalized design of low-complexity block diagonalization type precoding algorithms for multiuser MIMO systems. IEEE Trans. Commun. **61**(10), 4232–4242 (2013)
26. H Sung, S Lee, I Lee, Generalized channel inversion methods for multiuser MIMO systems. IEEE Trans. Commun. **57**(11), 3489–3499 (2009)
27. L Arévalo, RC de Lamare, K Zu, R Sampaio-Neto. Multi-branch lattice reduction successive interference cancellation detection for multiuser MIMO systems. IEEE International Symposium on Wireless Communications Systems, (2014), pp. 219–223
28. A Paulraj, R Nabar, D Gore, *Introduction to space-time wireless communications*. (Cambridge University Press, 2003)
29. CK Wen, S Jin, KK Wong, On the Sum-Rate of Multiuser MIMO Uplink Channels with Jointly-Correlated Rician Fading. IEEE Trans. Commun. **59**(10), 2883–2895 (2011)
30. DNC Tse, P Viswanath, *Fundamentals of wireless communications*. (Cambridge University Press, 2005)
31. HQ Ngo, EG Larsson, TL Marzetta, Energy and spectral efficiency of very large multiuser MIMO systems. IEEE Trans. Commun. **61**(4), 1436–1449 (2013)
32. G Golub, CV Loan, *Matrix computations*. (The Johns Hopkins University Press, 1996)
33. T Minka, *The Lightspeed Matlab toolbox, efficient operations for Matlab programming, version 2.2*. (Microsoft Corp, 2007)

# 14

# Performance analysis of dual-hop satellite relay networks with hardware impairments and co-channel interference

Guo Kefeng*, Guo Daoxing, Huang Yuzhen and Zhang Bangning

**Abstract**

In this paper, we investigate the impact of hardware impairments on the outage performance and the instantaneous capacity of dual-hop decode-and-forward (DF) satellite relay networks in the presence of co-channel interference. Specifically, the exact closed-form expression for the outage probability and the analytical expression for the instantaneous capacity of the system are derived over a shadowed-Rician fading channel, which provides a fast means to evaluate the impact of hardware impairments on the system performance. Our results demonstrate that the outage floor occurs and the capacity upper bound appears when the hardware impairments exist. In addition, simulation results were provided to verify the correctness of the analytical results.

**Keywords:** Satellite relay network, Hardware impairments, Co-channel interference, Outage probability, Instantaneous capacity

## 1 Introduction

A relay is used to improve the coverage, reliability, and quality of service in wireless systems, which has become a hot research topic over the past decade [1–4]. A satellite is used as the relay to help a source transmit a signal in a farther distance when the source itself cannot transmit directly. Satellite relay communication is capable of providing seamless connectivity and high-speed broadband access for worldwide users, especially in the fields of broadcasting, disaster relief, and navigation, where the deployment of wired and wireless terrestrial networks is not economically applicable.

The links between the ground and a satellite are usually modeled by composite fading distribution to describe more accurately the amplitude fluctuation (large scale) of the transmitted signal's envelope. To this end, the shadowed-Rician distribution has been proposed in [5]. It approaches the line-of-sight (LOS) communication using the Rician distribution, while the fluctuation of the signal's amplitude is Nakagami-m distributed. Hence, it agrees very well with experimental data, whereas it demonstrates quite a manageable mathematical tractability.

Capitalizing on the latter features of the shadowed-Rician distribution, an analytical performance evaluation of dual-hop amplify-and-forward (AF) systems with a satellite relay, was provided in [6] and [7]. In [8] and [9], a single-antenna land mobile satellite (LMS) system was investigated, where the source is a satellite and the relay/destination are terrestrial nodes. Furthermore, a similar LMS system with the source-to-relay satellite link was investigated in [10].

However, most research in the area of satellite relaying assumes that the transceiver hardware of the network node is perfect. This cannot be achieved in practical communication systems. In practice, hardware suffers from several types of impairments, e.g., phase noise, I/Q imbalance, and high power amplifier (HPA) nonlinearities [11–13]. The impact of hardware impairments on single-hop wireless communication was broadly investigated in [14–16]. In addition, as indicated in [17–20], there is a fundamental capacity ceiling due to the hardware impairments, which cannot be improved by increasing the transmission power. In [17], the authors analyzed the

*Correspondence: guokefeng.cool@163.com
College of Communications Engineering, PLA University of Science and Technology, Guanghua Road, Nanjing, China

performance of multiple-input-multiple-output (MIMO) system; the capacity and the outage probability of the system were derived. In [18], the authors analyzed the system performance of the full-duplex relaying network and found that the system capacity had an upper bound in the presence of hardware impairments. In [19], the opportunistic and partial relaying selection was used in which a new way was derived to obtain the closed-form expression for the system performance. In [20], the authors analyzed the impact of hardware impairments on the fixed-gain AF relaying Nakagami-m fading channels. Above all, the system channel they analyzed was just a Rayleigh or Nakagami-m channel; there are scarcely any papers that analyzed the effect of hardware impairments on the shadowed-Rician channel.

Although the aforementioned works have significantly improved our understanding on the performance and benefits of a satellite relay network, they only considered a noise-limited scenario and the ideal case of no co-channel interference (CCI) at the destination, which are unrealistic due to the current aggressive reuse of spectrum resources. In recent years, a great deal of research has focused on the performance of conventional terrestrial systems in the presence of CCI over Rayleigh or Nakagami-m fading channels [21–23]. Due to the reuse of spectrum resources in practice, the major challenge facing satellite-terrestrial networks is the problem of interference from inter-component and/or intra component networks [24]. The authors in [25] first investigated the performance of satellite-terrestrial networks with CCI at the destination for the single antenna scenario and analyzed the system performance.

As described above, we find that it is significant to analyze the impact of hardware impairments and CCI on the satellite relay communication network in shadowed-Rician channel. Until now, to the best of the authors' knowledge, there is scarcely any paper that analyzes the impact of hardware impairments on a satellite relay network with CCI. To fill this gap, we herein investigate the joint impact of hardware impairments and CCI on the performance of a dual-hop satellite relay network.

Our main contributions are summarized as follows:

- Firstly, the closed-form expression for the outage probability is derived which is applicable to cases involving an arbitrary number of interferers, antennas, and hardware impairments, and hence, the derived expression provides a fast means to investigate the system performance.
- Secondly, the analytical expression for the instantaneous capacity of the system, which can be used to efficiently evaluate the instantaneous capacity at high signal-to-noise-rates (SNRs), is derived.
- Thirdly, in this paper, the system performance is analyzed under two channel conditions, that is, channel parameters are integers or arbitrary values.

The rest of this paper is organized as follows. In Section 2, the model we analyzed is briefly introduced. In Section 3, the cumulative distribution function (CDF) of the end-to-end signal-to-interference-plus-noise-and-distortion-ratio (SINDR), the exact expression for the outage probability, and the analytical expression for the instantaneous capacity are derived. In addition, some system performance metrics are provided. In Section 5, the analytical results are presented. Finally, in Section 5, a brief summary of our work for this paper is presented. In the Appendix, the detailed proofs for some Lemmas and Theorems are given.

## 2 System and channel models

As shown in Fig. 1, we consider a satellite relay network, where the source ($S$) communicates with destination ($D$) with the help of the satellite relay ($R$). $S$ is equipped with $N_1$ antennas, while $D$ and $R$ are equipped with a single antenna, which is a scenario applicable to satellite relay networks. We assume that the channels of the first hop and second hop follow the shadowed-Rician fading. In addition, we assume that the signal received at D is corrupted by co-channel interference. To achieve a good trade-off between the performance and implementation cost, the transmit-antenna-selection (TAS) is used at the transmitter.

The communication between $S$ and $D$ takes place in two time slots. In the first time slot, $S$ sends its signal to $R$. Similar to [2], when the hardware of each node is ideal, the received signal at $R$ is given by $y_{1i} = h_{1i}s + v_r$, where $y_{1i}$ is the received signal transmitted by the $i$-th antenna of $S$, $h_{1i}$ is the channel coefficient between the $i$-th antenna of $S$ and $R$, which follows the shadowed-Rician random variable (RV). $s_{1i}$ is the signal from the $i$-th antenna of $S$ with average power $P_{1i} = E\left[|s_{1i}|^2\right]$. $v_r$ is the additive white Gaussian noise (AWGN) which is represented as $v_r \sim \mathcal{CN}\left(0, \delta_r^2\right)$ at $R$. However, when the node of the system has hardware impairments, the received signal changes, which is given by

$$y_{1i} = h_{1i}\left(s_{1i} + \eta_{1i}\right) + v_r \left(i = 1, \ldots, N_1\right) \tag{1}$$

where $\eta_{1i}$ denotes the distortion noise with zero mean and variance $k_{1i}^2 P_{1i}$ and $k_{1i}$ is sufficient to characterize the aggregate level of impairments of the $i$-th antenna of $S \rightarrow R$ channel [26].

In the second time slot, the signal is forwarded to $D$ by $R$. The decode-and-forward (DF) relaying protocol is

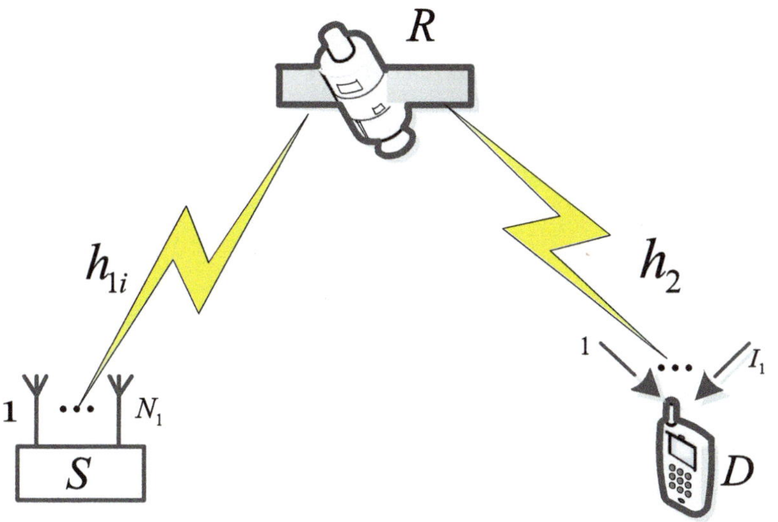


**Fig. 1** The system model

always used in the satellite relay communication. Hence, the signal received at $D$ is given by

$$y_2 = h_2 (y_r + \eta_2) + \sum_{j=1}^{I_1} g_j (s_j + \eta_j) + v_d \tag{2}$$

where $h_2$ is the independent but non-identically distributed shadowed-Rician RV between $R$ and $D$ and $g_j$ denotes the channel coefficient between the $j$-th interference ($I$) and $D$ which is an independent but non-identically distributed Rayleigh RV. $y_r$ is the signal received from $R$ with average power $P_2 = E[\,|y_r|^2]$, $s_j$ is the signal from the $j$-th interference with power $P_j = E\left[|s_j|^2\right]$, $v_d$ is the AWGN which is represented as $v_d \sim \mathcal{CN}\left(0, \delta_d^2\right)$ at $D$, $\eta_2$ denotes the distortion noise with zero mean and variance $k_2^2 P_2$, $\eta_j$ denotes the distortion noise with zero mean and variance $k_j^2 P_j$ of the $I$, and $k_2$ and $k_j$ are sufficient to characterize the aggregate level of impairments of the $R \to D$ channel and $I \to D$ channel, respectively [26]. To simplify the analysis, we assume that the hardware impairment level of the $I \to D$ channel has the same value as $k_1 = \ldots = k_j = \ldots = k_{I_1}, j = 1, \ldots, I_1$.

## 3 System performance

In this section, we present a comprehensive performance investigation of the system described in the previous section. Specifically, the description of the shadowed-Rician channel, the closed-form expression of the outage performance, and the analytical expression of the instantaneous capacity of the considered system with hardware impairments are derived.

### 3.1 The end-to-end SINDR of the system

From (1), the SNDR of $y_{1i}$ can be expressed as

$$\begin{aligned}\gamma_{1i} &= \frac{|h_{1i}|^2 P_{1i}}{|h_{1i}|^2 k_{1i}^2 P_{1i} + \delta_r^2} \\ &= \frac{\lambda_{1i}}{\lambda_{1i}(1 + \Delta_{1i}) + 1}\end{aligned} \tag{3}$$

where $\lambda_{1i} = \frac{|h_{1i}|^2 P_{1i}}{\delta_r^2}$ and $k_{1i}^2 = \Delta_{1i}$.

Because the TAS protocol is considered, the SNDR of the first hop is given by

$$\gamma_1 = \max_{i=1,\ldots,N_1} (\gamma_{1i}) \tag{4}$$

In the same way, from (2), the SINDR of the second hop is given by

$$\begin{aligned}\gamma_2 &= \frac{|h_2|^2 P_2}{|h_2|^2 k_2^2 P_2 + \sum\limits_{j=1}^{I_1} |g_j|^2 P_j \left(1 + k_j^2\right) + \delta_d^2} \\ &= \frac{\lambda_2}{\lambda_2 \Delta_2 + \sum\limits_{j=1}^{I_1} \lambda_j \left(1 + \Delta_j\right) + 1}\end{aligned} \tag{5}$$

where $\lambda_2 = \frac{|h_2|^2 P_2}{\delta_d^2}$, $\lambda_j = \frac{|g_j|^2 P_j}{\delta_d^2}$, $k_2^2 = \Delta_2$ and $k_j^2 = \Delta_j$.

According to the DF relaying protocol, the end-to-end SINDR of the system is given by

$$\gamma_e = \min(\gamma_1, \gamma_2) \tag{6}$$

### 3.2 The shadowed-Rician channel

Because the $S \rightarrow R$ and $R \rightarrow D$ channels are modeled as the shadowed-Rician fading distribution [5], the probability density function (PDF) of $\lambda_{1i}$ or $\lambda_2$ is given by

$$f_{\lambda_{lp}}(x) = \frac{\alpha_{lp}}{\bar{\lambda}_{lp}} e^{-\frac{\beta_{lp}}{\bar{\lambda}_{lp}}x} {}_1F_1\left(m_{lp}; 1; \frac{\delta_{lp}}{\bar{\lambda}_{lp}}x\right), x > 0 \quad (7)$$

where $\{l, p\} \in \{\{1, i\}, \{2\}\}$, $\bar{\lambda}_{lp}$ is the average SNR of each link, $\alpha_{lp} \triangleq \frac{\left(\frac{2b_{lp}m_{lp}}{2b_{lp}m_{lp}+\Omega_{lp}}\right)^{m_{lp}}}{2b_{lp}}$, $\beta_{lp} \triangleq \frac{1}{2b_{lp}}$, $\delta_{lp} \triangleq \frac{\Omega_{lp}}{2b_{lp}(2b_{lp}m_{lp}+\Omega_{lp})}$, $\Omega_{lp}$, $2b_{lp}$ and $m_{lp} \geq 0$ correspond to the average power of the LOS component, the average power of the multipath component, and the fading severity parameter ranging from 0 to $\infty$, respectively. In the following, we consider the two cases of $m_{lp}$ being either an integer value or an arbitrary value. In (7), ${}_1F_1(a_1; b_1; z)$ denotes the confluent hypergeometric function defined in [27].

#### 3.2.1 *Integer $m_{lp}$ value case*

If $m_{lp}$ is an integer, with the aid of [28], ${}_1F_1\left(m_{lp}; 1; \frac{\delta_{lp}}{\bar{\lambda}_{lp}}x\right)$ can be represented as (8)

$$ {}_1F_1\left(m_{lp}; 1; \frac{\delta_{lp}}{\bar{\lambda}_{lp}}x\right) = e^{\frac{\delta_{lp}}{\bar{\lambda}_{lp}}x} \times \sum_{k_{lp}=0}^{m_{lp}} \frac{(-\delta_{lp})^{k_{lp}}(1-m_{lp})_{k_{lp}} x^{k_{lp}}}{(\bar{\lambda}_{lp})^{k_{lp}}(k_{lp}!)^2} \quad (8)$$

where $(\cdot)_q$ is the Pochhammer symbol.

Substituting (8) into (7), we have

$$f_{\lambda_{lp}}(x) = \alpha_{lp} \sum_{k_{lp}=0}^{m_{lp}-1} \frac{(1-m_{lp})_{k_{lp}}(-\delta_{lp}x)^{k_{lp}}}{(k_{lp}!)^2(\bar{\lambda}_{lp})^{k_{lp}+1}} \exp\left(-\left(\frac{\beta_{lp}-\delta_{lp}}{\bar{\lambda}_{lp}}\right)x\right) \quad (9)$$

Hence, the CDF of $\lambda_{lp}$ can be given by

$$F_{\lambda_{lp}}(x_0) = \alpha_{lp} \sum_{k_{lp}=0}^{m_{lp}-1} \frac{(-\delta_{lp})^{k_{lp}}(1-m_{lp})_{k_{lp}}}{(k_{lp}!)^2(\beta_{lp}-\delta_{lp})^{k_{lp}+1}} \gamma\left(k_{lp}+1, \left(\frac{\beta_{lp}-\delta_{lp}}{\bar{\lambda}_{lp}}\right)x_0\right) \quad (10)$$

#### 3.2.2 *Arbitrary $m_{lp}$ value case*

If $m_{lp}$ is an arbitrary value, with the aid of [27], ${}_1F_1\left(m_{lp}; 1; \frac{\delta_{lp}}{\bar{\lambda}_{lp}}x\right)$ can be rewritten as

$$ {}_1F_1\left(m_{lp}; 1; \frac{\delta_{lp}}{\bar{\lambda}_{lp}}x\right) = \sum_{k_{lp}=0}^{\infty} \frac{(m_{lp})_{k_{lp}}(\delta_{lp}x)^{k_{lp}}}{(\bar{\lambda}_{lp})^{k_{lp}}(k_{lp}!)^2} \quad (11)$$

Now, to this end, by substituting (11) into (7), we have

$$f_{\lambda_{lp}}(x) = \alpha_{lp} e^{-\frac{\beta_{lp}}{\bar{\lambda}_{lp}}x} \sum_{k=0}^{\infty} \frac{(m_{lp})_{k_{lp}}(\delta_{lp}x)^{k_{lp}}}{(\bar{\lambda}_{lp})^{k_{lp}+1}(k_{lp}!)^2} \quad (12)$$

Hence, the CDF of $\lambda_{lp}$ is given by

$$F_{\lambda_{lp}}(x_0) = \alpha_{lp} \sum_{k_{lp}=0}^{\infty} \frac{(m_{lp})_{k_{lp}}}{(k_{lp}!)^2} \frac{\delta_{lp}^{k_{lp}}}{\beta_{lp}^{k_{lp}+1}} \gamma\left(k_{lp}+1, \frac{\beta_{lp}}{\bar{\lambda}_{lp}}x_0\right) \quad (13)$$

### 3.3 The outage performance

According to [1], the outage probability is defined as the probability that the instantaneous end-to-end SINDR falls below a predefined threshold $x_0$. With the help of (6), the outage probability of the system has the following expression.

$$\begin{aligned} P_{out}(x_0) &= \Pr(\gamma_e \leq x_0) \\ &= \Pr[\min(\gamma_1, \gamma_2) \leq x_0] \end{aligned} \quad (14)$$

where $P_{out}(x_0)$ is the outage probability of the system.

With the aid of [27], $P_{out}(x_0)$ can be represented as

$$P_{out}(x_0) = F_{\gamma_1}(x_0) + F_{\gamma_2}(x_0) - F_{\gamma_1}(x_0)F_{\gamma_2}(x_0) \quad (15)$$

Next, $F_{\gamma_1}(x_0)$ and $F_{\gamma_2}(x_0)$ are derived respectively.

**Lemma 1** *The CDF of $\gamma_1$ is derived as* (16) *and* (17) *respectively.*

*Proof* With the help of (4), (10), and (13), $F_{\gamma_1}(x_0)$ can be represented as (16), and (17) in both cases, which are shown at the top of this page and next page, respectively. □

$$F_{\gamma_1}(x_0) = \begin{cases} \left[\alpha_{1i}\sum_{k_{1i}=0}^{m_{1i}-1}\frac{(-\delta_{1i})^{k_{1i}}(1-m_{1i})_{k_{1i}}}{(\beta_{1i}-\delta_{1i})^{k_{1i}+1}(k_{1i}!)^2}\gamma\left(k_{1i}+1,\left(\frac{\beta_{1i}-\delta_{1i}}{\bar{\lambda}_{1i}}\right)\frac{x_0}{1-\Delta_{1i}x_0}\right)\right]^{N_1} & ,x_0 < \frac{1}{\Delta_{1i}} \\ 1, & x_0 \geq \frac{1}{\Delta_{1i}} \end{cases} \tag{16}$$

$$F_{\gamma_1}(x_0) = \begin{cases} \left[\alpha_{1i}\sum_{k_{1i}=0}^{\infty}\frac{(m_{1i})_{k_1}}{(k_{1i}!)^2}\frac{\delta_{1i}^{k_{1i}}}{\beta_{1i}^{k_{1i}+1}}\gamma\left(k_{1i}+1,\frac{\beta_{1i}}{\bar{\lambda}_{1i}}\frac{x_0}{1-\Delta_{1i}x_0}\right)\right]^{N_1} & ,x_0 < \frac{1}{\Delta_{1i}} \\ 1, & x_0 \geq \frac{1}{\Delta_{1i}} \end{cases} \tag{17}$$

**Lemma 2** *With the help of* (5), $F_{\gamma_2}(x_0)$ *is given by* (19) *and* (20), *respectively.*

*Proof* $F_{\gamma_2}(x_0)$ can be presented as

$$\begin{aligned} F_{\gamma_2}(x_0) &= F\left(\frac{\lambda_2}{\lambda_2\Delta_2+\sum_{j=1}^{I_1}\lambda_j\left(1+\Delta_j\right)+1}\leq x_0\right) \\ &= F\left(\lambda_2 \leq \frac{\sum_{j=1}^{I_1}\lambda_j\left(1+\Delta_j\right)x_0+x_0}{1-\Delta_2 x_0}\right). \end{aligned} \tag{18}$$

With the help of (10), (13), and (18), the closed-form expression for $F_{\gamma_2}(x_0)$ is derived as (19) and (20) respectively; a detailed proof is given in the Appendix. □

$$F_{\gamma_2}(x_0) = \begin{cases} \sum_{k_2=0}^{m_2-1}\sum_{i=1}^{\rho(A)}\sum_{j=1}^{\tau_i(A)}\frac{\alpha_2(-\delta_2)^{k_2}(1-m_2)_{k_2}}{\Gamma(k_2+1)(\beta_2-\delta_2)^{k_2+1}}\chi_{i,j}(A)\frac{\mu_{\langle i\rangle}^{-j}}{(j-1)!}\left\{\Gamma(j)\,\mu_{\langle i\rangle}^{j}-e^{-\frac{(\beta_2-\delta_2)(1+\Delta_j)x_0}{(1-\Delta_2x_0)\bar{\lambda}_2}}\right. \\ \left.\sum_{m=0}^{k_2}\sum_{v=0}^{m}\binom{m}{v}\frac{(1+\Delta_j)^v\Gamma(v+j)}{m!}\left[\frac{(\beta_2-\delta_2)x_0}{(1-\Delta_2x_0)\bar{\lambda}_2}\right]^m\left[\frac{(\beta_2-\delta_2)x_0}{(1-\Delta_2x_0)\bar{\lambda}_2}+\frac{1}{\mu_{\langle i\rangle}}\right]^{-(v+j)}\right\} & ,x_0 < \frac{1}{\Delta_2} \\ 1, & x_0 \geq \frac{1}{\Delta_2} \end{cases} \tag{19}$$

$$F_{\gamma_2}(x_0) = \begin{cases} \sum_{k_2=0}^{\infty}\sum_{i=1}^{\rho(A)}\sum_{j=1}^{\tau_i(A)}\frac{\alpha_2(m_2)_{k_2}}{\Gamma(k_2+1)}\frac{\delta_2^{k_2}}{\beta_2^{k_2+1}}\chi_{i,j}(A)\frac{\mu_{\langle i\rangle}^{-j}}{(j-1)!}\left\{\Gamma(j)\,\mu_{\langle i\rangle}^{j}-e^{-\frac{\beta_2x_0}{\bar{\lambda}_2(1-\Delta_2x_0)}}\right. \\ \left.\sum_{m=0}^{k_2}\sum_{v=0}^{m}\binom{m}{v}\frac{(1+\Delta_j)^v\Gamma(v+j)}{\bar{\lambda}_2 m!}\left[\frac{(\beta_2)x_0}{(1-\Delta_2x_0)\bar{\lambda}_2}\right]^m\left[\frac{\beta_2x_0}{\bar{\lambda}_2(1-\Delta_2x_0)}+\frac{1}{\mu_{\langle i\rangle}}\right]^{-(v+j)}\right\} & ,x_0 < \frac{1}{\Delta_2} \\ 1, & x_0 \geq \frac{1}{\Delta_2} \end{cases} \tag{20}$$

**Theorem 1** *When* $m_{lp}$ *is an integer,* $F_{\gamma_e}(x_0)$ *can be given by* (21).

*Proof* With the help of Lemma 1 and Lemma 2, and substituting (16) and (19) into (15), (21) is derived.

$$F_{\gamma_e}(x_0) = \begin{cases} \left[\alpha_{1i}\sum_{k_{1i}=0}^{m_{1i}-1}\frac{(-\delta_{1i})^{k_{1i}}(1-m_{1i})_{k_{1i}}}{(\beta_{1i}-\delta_{1i})^{k_{1i}+1}(k_{1i}!)^2}\gamma\left(k_{1i}+1,\left(\frac{\beta_{1i}-\delta_{1i}}{\bar\lambda_{1i}}\right)\frac{x_0}{1-\Delta_{1i}x_0}\right)\right]^{N_1} \\ +\sum_{k_2=0}^{m_2-1}\sum_{i=1}^{\rho(A)}\sum_{j=1}^{\tau_i(A)}\frac{\alpha_2(-\delta_2)^{k_2}(1-m_2)_{k_2}}{\Gamma(k_2+1)(\beta_2-\delta_2)^{k_2+1}}\chi_{i,j}(A)\frac{\mu_{\langle i\rangle}^{-j}}{(j-1)!}\times\left\{\Gamma(j)\,\mu_{\langle i\rangle}^{j}\right. \\ -e^{-\frac{(\beta_2-\delta_2)(1+\Delta_j)x_0}{(1-\Delta_2x_0)\bar\lambda_2}}\sum_{m=0}^{k_2}\sum_{v=0}^{m}\binom{m}{v}\frac{(1+\Delta_j)^v\Gamma(v+j)}{m!} \\ \left.\times\left(\frac{(\beta_2-\delta_2)x_0}{(1-\Delta_2x_0)\bar\lambda_2}\right)^m\left[\frac{(\beta_2-\delta_2)x_0}{(1-\Delta_2x_0)\bar\lambda_2}+\frac{1}{\mu_{\langle i\rangle}}\right]^{-(v+j)}\right\} \\ -\left[\alpha_{1i}\sum_{k_{1i}=0}^{m_{1i}-1}\frac{(-\delta_{1i})^{k_{1i}}(1-m_{1i})_{k_{1i}}}{(\beta_{1i}-\delta_{1i})^{k_{1i}+1}(k_{1i}!)^2}\gamma\left(k_{1i}+1,\left(\frac{\beta_{1i}-\delta_{1i}}{\bar\lambda_{1i}}\right)\frac{x_0}{1-\Delta_{1i}x_0}\right)\right]^{N_1} \\ \times\left\{\sum_{k_2=0}^{m_2-1}\sum_{i=1}^{\rho(A)}\sum_{j=1}^{\tau_i(A)}\frac{\alpha_2(-\delta_2)^{k_2}(1-m_2)_{k_2}}{\Gamma(k_2+1)(\beta_2-\delta_2)^{k_2+1}}\chi_{i,j}(A)\frac{\mu_{\langle i\rangle}^{-j}}{(j-1)!}\times\left\{\Gamma(j)\,\mu_{\langle i\rangle}^{j}\right.\right. \\ -e^{-\frac{(\beta_2-\delta_2)(1+\Delta_j)x_0}{(1-\Delta_2x_0)\bar\lambda_2}}\sum_{m=0}^{k_2}\sum_{v=0}^{m}\binom{m}{v}\frac{(1+\Delta_j)^v\Gamma(v+j)}{m!} \\ \left.\left.\times\left(\frac{(\beta_2-\delta_2)x_0}{(1-\Delta_2x_0)\bar\lambda_2}\right)^m\left[\frac{(\beta_2-\delta_2)x_0}{(1-\Delta_2x_0)\bar\lambda_2}+\frac{1}{\mu_{\langle i\rangle}}\right]^{-(v+j)}\right\}\right\}, & x_0<\min\left(\frac{1}{\Delta_{1i}},\frac{1}{\Delta_2}\right) \\ 1, & x_0\geq\min\left(\frac{1}{\Delta_{1i}},\frac{1}{\Delta_2}\right) \end{cases} \tag{21}$$

□

**Theorem 2** *When $m_{lp}$ is an arbitrary value, $F_{\gamma_e}(x_0)$ can be given by* (22).

*Proof* As with Theorem 1, with the help of Lemma 1 and Lemma 2, and substituting (17) and (20) into (15), (22) is derived.

$$F_{\gamma_e}(x_0) = \begin{cases} \left[\alpha_{1i}\sum_{k_{1i}=0}^{\infty}\frac{(m_{1i})_{k_1}}{(k_{1i}!)^2}\frac{\delta_{1i}^{k_{1i}}}{\beta_{1i}^{k_{1i}+1}}\gamma\left(k_{1i}+1,\frac{\beta_{1i}}{\bar\lambda_{1i}}\frac{x_0}{1-\Delta_{1i}x_0}\right)\right]^{N_1} \\ +\alpha_2\sum_{k_2=0}^{\infty}\sum_{i=1}^{\rho(A)}\sum_{j=1}^{\tau_i(A)}\chi_{i,j}(A)\frac{(m_2)_{k_2}}{\Gamma(k_2+1)}\frac{\delta_2^{k_2}}{\beta_2^{k_2+1}}\frac{\mu_{\langle i\rangle}^{-j}}{(j-1)!} \\ \times\left\{\Gamma(j)\,\mu_{\langle i\rangle}^{j}-e^{-\frac{\beta_2x_0(1+\Delta_j)}{\bar\lambda_2(1-\Delta_2x_0)}}\sum_{m=0}^{k_2}\sum_{v=0}^{m}\binom{m}{v}\frac{(1+\Delta_j)^v\Gamma(v+j)}{m!}\right. \\ \left.\times\left[\frac{\beta_2x_0}{\bar\lambda_2(1-\Delta_2x_0)}\right]^m\left[\frac{\beta_2x_0}{\bar\lambda_2(1-\Delta_2x_0)}+\frac{1}{\mu_{\langle i\rangle}}\right]^{-(v+j)}\right\} \\ -\left[\alpha_{1i}\sum_{k_{1i}=0}^{\infty}\frac{(m_{1i})_{k_1}}{(k_{1i}!)^2}\frac{\delta_{1i}^{k_{1i}}}{\beta_{1i}^{k_{1i}+1}}\gamma\left(k_{1i}+1,\frac{\beta_{1i}}{\bar\lambda_{1i}}\frac{x_0}{1-\Delta_{1i}x_0}\right)\right]^{N_1} \\ \times\left\{\alpha_2\sum_{k_2=0}^{\infty}\sum_{i=1}^{\rho(A)}\sum_{j=1}^{\tau_i(A)}\chi_{i,j}(A)\frac{(m_2)_{k_2}}{\Gamma(k_2+1)}\frac{\delta_2^{k_2}}{\beta_2^{k_2+1}}\frac{\mu_{\langle i\rangle}^{-j}}{(j-1)!}\right. \\ \times\left\{\Gamma(j)\,\mu_{\langle i\rangle}^{j}-e^{-\frac{\beta_2x_0(1+\Delta_j)}{\bar\lambda_2(1-\Delta_2x_0)}}\sum_{m=0}^{k_2}\sum_{v=0}^{m}\binom{m}{v}\frac{(1+\Delta_j)^v\Gamma(v+j)}{m!}\right. \\ \left.\left.\times\left[\frac{\beta_2x_0}{\bar\lambda_2(1-\Delta_2x_0)}\right]^m\left[\frac{\beta_2x_0}{\bar\lambda_2(1-\Delta_2x_0)}+\frac{1}{\mu_{\langle i\rangle}}\right]^{-(v+j)}\right\}\right\}, & x_0<\min\left(\frac{1}{\Delta_{1i}},\frac{1}{\Delta_2}\right) \\ 1, & x_0\geq\min\left(\frac{1}{\Delta_{1i}},\frac{1}{\Delta_2}\right) \end{cases} \tag{22}$$

□

### 3.4 Instantaneous capacity

In this section, we analyze the instantaneous capacity of the system, and the analytical expression for the instantaneous capacity of the system is given. The approximate value of the system at high SNRs region is also derived.

In information-theoretical view, the instantaneous capacity has the following definition which refers to the value of the instantaneous maximum mutual information between the source and the destination, which can be mathematically formulated as

$$C=\frac{1}{2}\log_2(1+\gamma_e) \tag{23}$$

where $\gamma_e$ is the SINDR of the system and $C$ is the instantaneous capacity of the system.

Now, by substituting (4), (5), and (6) into (23), the instantaneous capacity can be derived as (24) which is at the top of eleventh page.

$$\begin{aligned} C &= \frac{1}{2}\log_2(1+\gamma_e) \\ &= \frac{1}{2}\log_2\left\{1+\min\left[\max_{i=1,\ldots,N_1}\left(\frac{\lambda_{1i}}{\lambda_{1i}\Delta_{1i}+1}\right),\right.\right. \\ &\qquad \left.\left.\frac{\lambda_2}{\lambda_2\Delta_2+\sum_{j=1}^{I_1}\lambda_j(1+\Delta_j)+1}\right]\right\} \end{aligned} \tag{24}$$

From (24), we find that, when $\lambda_{1i}$ and $\lambda_2$ become infinite, (24) achieves a fixed value that is just a function of $\Delta_{1i}$ and $\Delta_2$ as in (25).

$$C=\frac{1}{2}\log_2\left[1+\min\left(\frac{1}{\Delta_{1i}},\frac{1}{\Delta_2}\right)\right] \tag{25}$$

## 4 Numerical results

In this section, Monte Carlo simulations are provided to verify the analytical results. Some representative numerical results are provided to evaluate the impact of hardware impairments of the considered systems. Here, the analytical curves of arbitrary the $m_{lp}$ value case are obtained by using the first 30 terms of the infinite series. In this section, we assume that $m_2 = 5, b_2 = 0.251, \Omega_2 = 0.279$ for the channel parameters between $R$ and $D$, $k_j = 0.1$, $P_{1i} = P_r$ and $\frac{P_j}{\delta_d^2} = 1dB$. We also assume that $N_1 = 3$ and $I_1 = 3$ for the simulations. As shown, the analytical results agree well with the Monte Carlo simulations, which imply the correctness of the derived theoretical formula.

Figure 2 plots the outage performance of the satellite relay system with ideal hardware and impaired hardware. As shown in Fig. 2, we can see that the theory results are sufficiently tight across the entire SNR with the simulation results, which verifies the correctness of our analysis. From the results, we can also see that when the channel is under heavy fading, the system has worse performance at low SNRs. At high SNRs, the outage probability for the two conditions is the same only having a relationship with the impairment level. Moreover, from Fig. 2, we know that the lower the hardware impairments' level is, the lower the outage probabilities are.

Figure 3 examines the impact of the outage threshold on the outage performance of the considered system. It can be clearly seen that when the system has hardware impairments, the outage probability increases to 1 as the outage threshold increases to a certain constant. The larger the $k_{1i}$ and $k_2$ are, the lower the outage threshold is. In contrast, when the system has ideal hardware, there exists no outage threshold for the considered system. From the results, we can also see that when the channel is under heavy fading, the system has worse performance.

Figure 4 depicts that that hardware impairments have a small impact on the instantaneous capacity at low SNRs, but are very influential at high SNRs. It also depicts that when the system has hardware impairments, the instantaneous capacity has an upper bound at high SNRs which is proved by (25). However, the instantaneous capacity at high SNRs of ideal hardware has no upper bound.

## 5 Conclusions

In this paper, we investigated dual-hop satellite relay networks with hardware impairments and co-channel interference. Specifically, the exact closed-form expression for the outage probability of the considered system was derived, which made it easy to evaluate the impact of the impairing parameters on the system performance. The results revealed that an outage floor appeared when the hardware impairments existed. In addition, the analytical expression for the instantaneous capacity was also derived, which revealed that the instantaneous capacity had an upper bound when the system had hardware impairments. However, when the channel was under heavy fading, the performance of the system was worse.

## Appendix

In this appendix, the derivations of (19) and (20) are presented. First, with the help of (10) and (12), (18) can be further represented as (26) and (27), respectively.

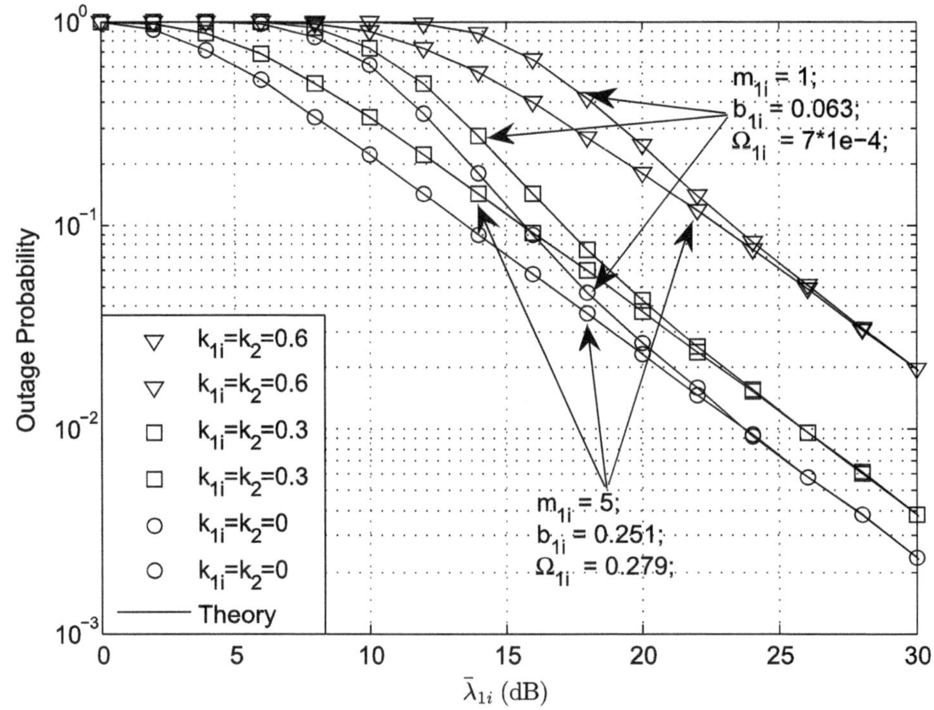


**Fig. 2** Outage probability of the system with ideal hardware and hardware impairments

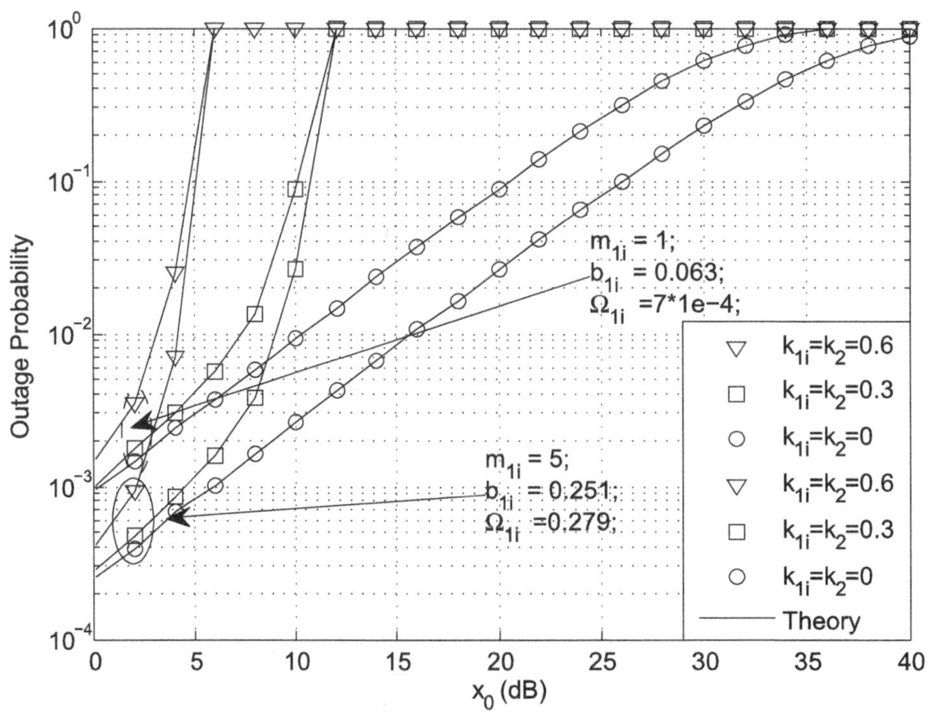


**Fig. 3** Outage probability for different threshold $x_0$

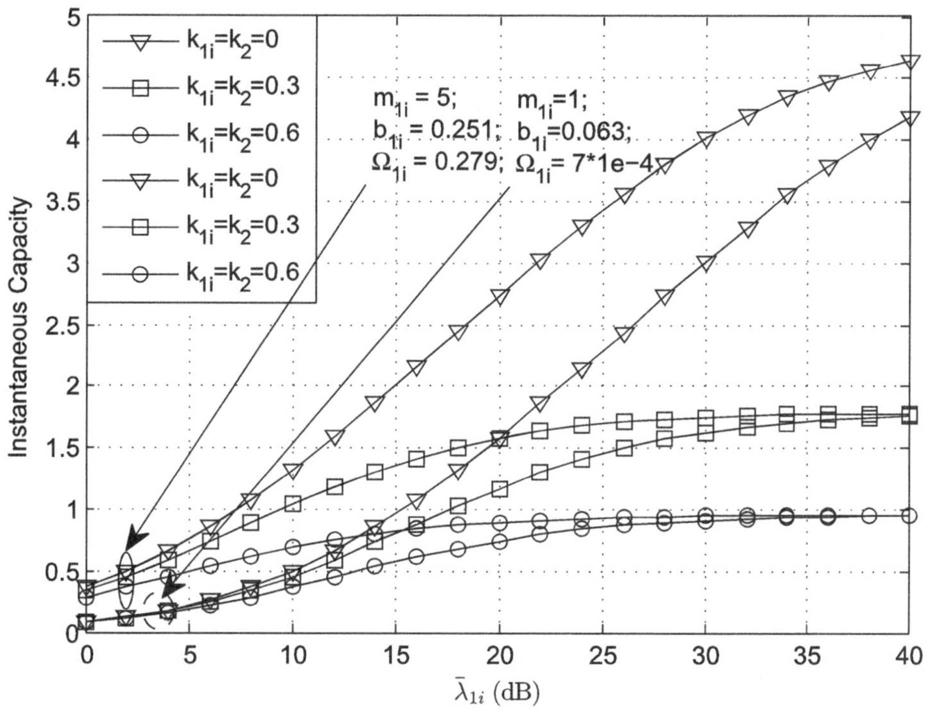


**Fig. 4** The instantaneous capacity of the system

$$F_{\gamma_2}(x_0) = \alpha_2 \sum_{k_2=0}^{m_2-1} \frac{(-\delta_2)^{k_2}(1-m_2)_{k_2}}{(k_2!)^2(\beta_2-\delta_2)^{k_2+1}} \gamma\left(k_2+1, \left(\frac{\beta_2-\delta_2}{\bar{\lambda}_2}\right)\left[\frac{\sum_{j=1}^{I_1}\lambda_j\left(1+\Delta_j\right)x_0+x_0}{1-\Delta_2 x_0}\right]\right) \tag{26}$$

$$F_{\gamma_2}(x_0) = \alpha_2 \sum_{k_2=0}^{\infty} \frac{(m_2)_{k_2}}{(k_2!)^2}\frac{\delta_2^{k_2}}{\beta_2^{k_2+1}} \gamma\left(k_2+1, \frac{\beta_2}{\bar{\lambda}_2}\left[\frac{\sum_{j=1}^{I_1}\lambda_j\left(1+\Delta_j\right)x_0+x_0}{1-\Delta_2 x_0}\right]\right) \tag{27}$$

With the help of [27. section 8.252.1] and [27. section 1.111], (26) and (27) can be expanded to (28) and (29), respectively.

$$F_{\gamma_2}(x_0) = \begin{cases} \alpha_2 \sum\limits_{k_2=0}^{m_2-1} \frac{(-\delta_2)^{k_2}(1-m_2)_{k_2}}{\Gamma(k_2+1)(\beta_2-\delta_2)^{k_2+1}} \left\{1-\left(\frac{\beta_2-\delta_2}{\bar{\lambda}_2}\right)^m e^{-\left(\frac{\beta_2-\delta_2}{\bar{\lambda}_2}\right)\left(\frac{x_0}{1-\Delta_2 x_0}\right)\left\{1-\left[\sum_{j=1}^{I_1}\lambda_j(1+\Delta_j)\right]\right\}}\right. \\ \left.\times\left\{\sum\limits_{m=0}^{k_2}\sum\limits_{\nu=0}^{m}\frac{1}{m!}\binom{m}{\nu}\left(\frac{x_0}{1-\Delta_2 x_0}\right)^m\left[\sum\limits_{j=1}^{I_1}\lambda_j\left(1+\Delta_j\right)\right]^\nu\right\}\right\}, & x_0 < \frac{1}{\Delta_2} \\ 1, & x_0 \geq \frac{1}{\Delta_2} \end{cases} \tag{28}$$

$$F_{\gamma_2}(x_0) = \begin{cases} \alpha_2 \sum_{k_2=0}^{\infty} \frac{(m_2)_{k_2}}{\Gamma(k_2+1)} \frac{\delta_2^{k_2}}{\beta_2^{k_2+1}} \left\{ 1 - \left(\frac{\beta_2}{\lambda_2}\right)^m e^{-\left(\frac{\beta_2}{\lambda_2}\right)\left(\frac{x_0}{1-\Delta_2 x_0}\right)\left\{1-\left[\sum_{j=1}^{I_1} \lambda_j (1+\Delta_j)\right]\right\}} \right. \\ \left. \times \left\{ \sum_{m=0}^{k_2} \sum_{\nu=0}^{m} \binom{m}{\nu} \frac{1}{m!} \left(\frac{x_0}{1-\Delta_2 x_0}\right)^m \left[\sum_{j=1}^{I_1} \lambda_j (1+\Delta_j)\right]^{ii} \right\} \right\} & , x_0 < \frac{1}{\Delta_2} \\ 1, & x_0 \geq \frac{1}{\Delta_2} \end{cases} \tag{29}$$

We first define $X = \sum_{j=1}^{I_1} \lambda_j$, and the PDF of $X$ can be given by [29]

$$p_X(x) = \sum_{i=1}^{\rho(A)} \sum_{j=1}^{\tau_i(A)} \chi_{i,j}(A) \frac{\mu_{\langle i \rangle}^{-j}}{(j-1)!} x^{j-1} e^{-\frac{x}{\mu_{\langle i \rangle}}}, x \geq 0 \tag{30}$$

where $A = diag\left(\mu_1, \mu_2, \cdots \mu_{I_1}\right)$, $\rho(A)$ is the number of district diagonal elements of $A$, $\mu_{\langle 1 \rangle} > \mu_{\langle 2 \rangle} > \cdots > \mu_{\langle \rho(A) \rangle}$ are the district diagonal elements in decreasing order, $\tau_i(A)$ is the multiplicity of $A$, and $\chi_{i,j}(A)$ is the $(i,j)$-th *characteristic coefficient* of $A$ [30].

From the definition of the conditional probability density function (CPDF), (18) can also be expressed as

$$F_{\gamma_2}(x_0) = \int_0^{\infty} F_{\gamma_2}(x_0|x) p_X(x)\, dx \tag{31}$$

Substituting (28), (29) and (30) into (31), and after performing some simplifications, (31) can be represented as (32) and (33).

$$F_{\gamma_2}(x_0) = \begin{cases} \int_0^{\infty} \alpha_2 \sum_{k_2=0}^{m_2-1} \sum_{i=1}^{\rho(A)} \sum_{j=1}^{\tau_i(A)} \chi_{i,j}(A) \frac{\mu_{\langle i \rangle}^{-j} (-\delta_2)^{k_2} (1-m_2)_{k_2}}{\Gamma(j)\Gamma(k_2+1)(\beta_2-\delta_2)^{k_2+1}} x^{j-1} e^{-\frac{x}{\mu_{\langle i \rangle}}} dx \\ - \int_0^{\infty} \alpha_2 \sum_{k_2=0}^{m_2-1} \sum_{i=1}^{\rho(A)} \sum_{j=1}^{\tau_i(A)} \frac{\chi_{i,j}(A) \mu_{\langle i \rangle}^{-j} (-\delta_2)^{k_2} (1-m_2)_{k_2}}{\Gamma(j)\Gamma(k_2+1)(\beta_2-\delta_2)^{k_2+1}} \left(\frac{\beta_2-\delta_2}{\lambda_2}\right)^m \\ \times \sum_{m=0}^{k_2} \sum_{\nu=0}^{m} \binom{m}{\nu} \frac{1}{m!} \left(\frac{x_0}{1-\Delta_2 x_0}\right)^m (1+\Delta_j)^{\nu} \\ \times e^{-\left(\frac{\beta_2-\delta_2}{\lambda_2}\right)\left(\frac{x_0}{1-\Delta_2 x_0}\right)(1+\Delta_j)} e^{-\left[\left(\frac{\beta_2-\delta_2}{\lambda_2}\right)\left(\frac{x_0}{1-\Delta_2 x_0}\right) - \frac{1}{\mu_{\langle i \rangle}}\right] x} x^{\nu+j-1} dx, & x_0 < \frac{1}{\Delta_2} \\ 1, & x_0 \geq \frac{1}{\Delta_2} \end{cases} \tag{32}$$

$$F_{\gamma_2}(x_0) = \begin{cases} \int_0^{\infty} \alpha_2 \sum_{k_2=0}^{\infty} \sum_{i=1}^{\rho(A)} \sum_{j=1}^{\tau_i(A)} \frac{\chi_{i,j}(A) \mu_{\langle i \rangle}^{-j} (m_2)_{k_2} \delta_2^{k_2}}{\Gamma(j)\Gamma(k_2+1)\beta_2^{k_2+1}} x^{j-1} e^{-\frac{x}{\mu_{\langle i \rangle}}} dx \\ - \int_0^{\infty} \alpha_2 \sum_{k_2=0}^{\infty} \sum_{i=1}^{\rho(A)} \sum_{j=1}^{\tau_i(A)} \frac{\chi_{i,j}(A) \mu_{\langle i \rangle}^{-j} (m_2)_{k_2} \delta_2^{k_2}}{\Gamma(j)\Gamma(k_2+1)\beta_2^{k_2+1}} \left(\frac{\beta_2}{\lambda_2}\right)^m \\ \times \sum_{m=0}^{k_2} \sum_{\nu=0}^{m} \binom{m}{\nu} \frac{1}{m!} \left(\frac{x_0}{1-\Delta_2 x_0}\right)^m (1+\Delta_j)^{\nu} \\ \times e^{-\left(\frac{\beta_2}{\lambda_2}\right)\left(\frac{x_0}{1-\Delta_2 x_0}\right)(1+\Delta_j)} e^{-\left[\left(\frac{\beta_2}{\lambda_2}\right)\left(\frac{x_0}{1-\Delta_2 x_0}\right) - \frac{1}{\mu_{\langle i \rangle}}\right] x} x^{\nu+j-1} dx, & x_0 < \frac{1}{\Delta_2} \\ 1, & x_0 \geq \frac{1}{\Delta_2} \end{cases} \tag{33}$$

With the help of [27. section 3.351.3], and after simplifications, (19) and (20) are presented.

**Acknowledgements**
This work is supported by the National Science Foundation of China (No.61501507) and the Jiangsu Provincial Natural Science Foundation of China (No.20150719).

**Competing interests**
The authors declare that they have no competing interests.

**References**
1. JN Laneman, DNC Tse, GW Wornell, Cooperative diversity in wireless networks: efficient protocols and outage behavior. IEEE Trans. Inf. Theroy. **50**(4), 1963–1968 (2004)
2. B Wang, Z Jun, Z Li, Achieveable rates and scaling laws of power-constrained wireless sensory relay networks. IEEE Commun.on Infor Theory. **52**(9), 4084–4104 (2006)
3. KS Gomadam, SA Jafar, The effect of noise correlation in amplify-and-forward relay networks. IEEE Commun.on Infor Theory. **55**(2), 731–745 (2009)
4. Y Huang, J Wang, Q Wu, C Zhong, C Li, Outage performance of spectrum sharing systems with MRC diversity under multiple primary user's interference. IEEE Commun. Lett. **16**(4), 576–579 (2014)
5. A Abdi, WC Lau, MS Alouini, M Kaveh, A new simple model for land mobile satellite channels: first-and second-order statistics. IEEE Trans. Wireless Commun. **2**(3), 519–528 (2003)
6. NI Miridakis, DD Vergados, A Michalas, Dual-hop communication over a satellite relay and shadowed Rician channels. IEEE Trans. Vehi. Tech. **64**(9), 4031–4040 (2015)
7. S Wang, M Yang, Q Guo, in *The Fourth Int. Conf. Adv. Sat. Space Commun, Chamonix*. Outage probability evaluation of land mobile satellite cooperative diversity communication system (IEEE, MontBlanc, 2012), pp. 26–30
8. A Iqbal, KM Ahmed, A hybrid satellite-terrestrial cooperative network over non identically distributed fading channels. J. Commun. **6**(7), 581–589 (2011)
9. MR Bhatnagar, MK Arti, Performance analysis of AF based hybrid satellite-terrestrial cooperative network over generalized fading channels. IEEE Commun. Lett. **17**(10), 1912–1915 (2013)
10. MR Bhatnagar, MK Arti, Performance analysis of hybrid satellite-terrestrial FSO cooperative system. IEEE Photon. Technol. Lett. **25**(22), 2197–2200 (2013)
11. E Costa, S Pupolin, m-QAM-OFDM system performance in the presence of a nonlinear amplifier and phase noise. IEEE Trans.Commun. **50**(3), 462–472 (2002)
12. T Schenk. RF imperfections in high-Rate wireless systems: impact and digital compensation (Springer, Netherlands, 2008), pp. 2033–2043
13. C Studer, M Wenk, A Burg, in *Proc. 2010 ITG Work. Smart Ant.* MIMO transmission with residual transmit-RF impairments, (Bremen, 2010), pp. 189–196
14. J Qi, S Assa, MS Alouini, in *Proc. 2012 IEEE Wireless Commun. Netw. Conf.* Analysis and compensation of I/Q imbalance in amplify-and-forward cooperative systems ((WCNC) Pairs, France, 2012), pp. 1–4
15. E Bjornson, P Zetterberg, M Bengtsson, B Ottersten, Capacity limits and multiplexing gains of MIMO channels with transceiver impairments. IEEE Commun. Lett. **17**(1), 91–94 (2013)
16. Z Xinlin, M Matthaiou, E Bjornson, M Coldrey, M Debbah, *On the MIMO capacity with residual transceiver hardware impairments*. (IEEE ICC, Jun 2014 Sydney, Australia, 2014), pp. 10–14
17. E Bjornson, J Hoydis, M Kountouris, M Debbah, in *2013 8th International Conference on Digital Signal Processing*. Hardware impairments in large-scale MISO systems: energy efficiency, estimation, and capacity limits (Fira, Greece, 2013), pp. 18–20
18. X Xia, D Zhang, K Xu, W Ma, Y Xu, Hardware impairments aware transceiver for full-duplex massive MIMO relaying. IEEE Trans. Sign. Proc. **63**(24), 6565–6580 (2015)
19. TT Duy, TQ D, DB Costa, VNQ Bao, M Elkashlan, Proactive relay selection with joint impact of hardware impairment and co-channel interference. IEEE Trans. Commun. **63**(5), 1594–1606 (2015)
20. M Maletic, M Cabarkapa, N Neskovic, D Budimir, Hardware impairments impact on fixed-gain AF relaying performance in Nakagami-m fading. Electronics Lett. **52**(2), 121–122 (2016)
21. K An, M Lin, J Ouyang, H Wei, Performance analysis of beamforming in two-hop AF relay networks with antenna correlation and interference. AEU-Int. J. Electron. Commun. **68**(7), 587–594 (2014)
22. C Zhong, S Jin, KK Wong, Dual-hop systems with noisy relay and interference-limited destination. IEEE Trans. Commun. **58**(3), 764–768 (2010)
23. YZ Huang, FA Qahtani, C Zhong, Q Wu, J Wang, H Alnuweiri, Performance analysis of multiuser multiple antenna relaying networks with co-channel interference and feedback delay. IEEE Trans. Commun. **62**(1), 59–73 (2014)
24. M Sadek, S Aissa, Personal satellite communication: technologies and challenges. IEEE Wireless Commun. **12**(5), 28–35 (2012)
25. K An, M Lin, J Ouyang, Y Huang, G Zheng, Symbol error analysis of hybrid satellite-terrestrial cooperative networks with cochannel interference. IEEE Commun. Lett. **18**(11), 1947–1950 (2014)
26. 8 hints for making and interpreting EVM measurements (2005). Tech. Rep. Agilent Technologies
27. IS Gradshteyn, IM Ryzhik, *Table of Intergals Series and Products*, 7th ed. (Elsevier, Amsterdam, 2005), pp. 5989–3144EN
28. The Wolfram Function Site. [Online], Avilable: http://functions.wolfram.com. Accessed 1998
29. A Bletsas, H Shin, MZ Win, Cooperative communication with outage-optimal opportunistic relaying. IEEE Trans. Wire. Commun. **6**(9), 3450–3460 (2007)
30. H Shin, MZ Win, MIMO diversity in the presence of double scattering. IEEE Trans. Inform. Theory. **54**(7), 2976–2996 (2008)

15

# On the security of cooperative cognitive radio networks with distributed beamforming

Weifeng Mou, Weiwei Yang* , Yuzhen Huang, Xiaoming Xu, Yueming Cai and Kaihua Wang

**Abstract**

This paper investigates the secrecy performance of amplify-and-forward (AF)-relaying cooperative cognitive radio networks (CCRNs) over Rayleigh-fading channels. Specifically, we consider practical passive eavesdropping scenarios, where the channel state information of the eavesdropper's link is not available at the secondary transmitter. In order to avoid interfering with the primary receiver and enhance the secrecy performance, collaborative distributed beamforming is adopted at the secondary relays. Closed-form and asymptotic expressions for the secrecy outage probability of CCRNs in the presence of single and multiple non-colluding eavesdroppers are derived. The asymptotic analysis reveals that the achievable secrecy diversity order of collaborative distributed beamforming with $M$ AF relays is $M-1$ regardless of the number of eavesdroppers. In addition, simulations are conducted to validate the accuracy of our analytical results.

**Keywords:** Cooperative cognitive radio networks (CCRNs), Physical layer security, Distributed beamforming

## 1 Introduction

Cognitive radio has been regarded as a potential means to improve spectral efficiency by allowing secondary users (SUs) to share the spectrum originally allocated to the primary users (PUs), as long as the generated interference aggregated at the primary receivers is below acceptable levels [1]. In order to enhance system performance and extend the coverage of secondary transmission, cooperative relay techniques have been further introduced into cognitive radio networks (CRNs), and thus, a novel network model, cooperative cognitive radio networks (CCRNs), has attracted significant interests in the research community [2]. On the other hand, due to the broadcast nature of wireless medium and the openness of cognitive radio architecture, CRNs face a more serious challenge of security, and higher layer cryptographic authentication and identification have become expensive and vulnerable to attacks [3]. Recently, a promising approach towards achieving secure communications has been developed by Wyner in [4] termed as physical layer security, the key idea of which lies in exploiting the randomness of wireless channels to ensure the security of confidential information.

*Correspondence: wwyang1981@163.com
College of Communications Engineering, PLA University of Science and Technology, Nanjing 210007, China

Due to the advantages of physical layer security, many researchers have devoted efforts to investigate the physical layer security issues in CRNs [5–10]. In [5], the secrecy performance of CRNs in terms of the secrecy outage probability and the probability of non-zero secrecy capacity was analyzed. In [6], the authors investigated the achievable secrecy rates of multiple-input single-output (MISO) CRNs with different beamforming schemes. Later, the secrecy outage performance of single-input multiple-output (SIMO) CRNs using selection combining (SC) and generalized selection combining (GSC) was investigated in [7] and [8], respectively. The authors in [9] investigated the secrecy outage and diversity performance for multi-user multi-eavesdropper CRNs. The exact and asymptotic expressions of the secrecy outage probability in multiple-input multiple-output (MIMO) CRNs with transmit antenna selection (TAS)/maximal ratio combining (MRC) were derived in [10].

In addition to the multi-antenna diversity, user cooperation can also be exploited to enhance the security of wireless transmission. The authors in [11] proposed an opportunistic decode-and-forward (DF)-relaying scheme for CCRNs against eavesdropping. In [12], we investigated the secrecy outage performance for DF-relaying CCRNs with outdated channel state information. Different joint relay and jammer selection policies were developed to enhance the secrecy outage performance for DF-relaying CCRNs in [13]. The security-reliability trade-off performance of single-relay and multi-relay selection strategies for DF-relaying CCRNs was investigated in [14]. However, very few research has considered the secrecy performance of amplify-and-forward (AF)-relaying CCRNs except [15], in which the authors investigated the security and the reliability performance of CCRNs with single AF relay in terms of the intercept probability. It is worth noting that the intercept probability is a special case of secrecy outage probability when the target secrecy data rate is set to zero. To the best of the authors' knowledge, no works have considered the secrecy outage probability of CCRNs with multiple AF relays. Besides, various advanced techniques have been employed to enhance the physical layer security of CRNs, such as multiple antenna diversity in [7–10], relay selection in [11–14], artificial noise in [13], and secrecy beamforming/precoding in [6]. However, collaborative distributed beamforming-based security enhancement in CCRNs has not been investigated yet.

Different from the aforementioned works, in this paper, we investigate the secrecy outage performance of multiple AF relay-assisted CCRNs with distributed beamforming against passive eavesdropping attacks. The main contributions of our work are summarized as follows:

1) Compared with [11–14] that only considered the relay selection for DF-relaying CCRNs and [15] that only considered the intercept probability for single AF-relaying CCRNs, we investigate the physical layer security in terms of the probability of non-zero secrecy capacity, the secrecy outage probability, the secrecy array gain, and the secrecy diversity order for multiple AF-relaying CCRNs with collaborative distributed beamforming in the presence of single and multiple non-colluding eavesdroppers, respectively, where distributed zero-forcing beamforming (D-ZFB) is employed at the relays without interfering with the primary users.
2) We derive the closed-form expressions of the probability of non-zero secrecy capacity and the secrecy outage probability as well as the asymptotic expression at high SNR regimes. Our asymptotic results accurately predict the secrecy diversity order of $M$ AF relay CCRNs with collaborative distributed beamforming, i.e., $M-1$, which is different from the results obtained in [11] and [13]. This is due to the fact that the proposed collaborative distributed beamforming scheme is designed at relays to avoid the interference at PUs at the expense of one spatial degree. In addition, numerical and simulation results are provided to verify the correctness of the proposed scheme.

## 2 System model

Let us consider a spectrum-sharing cognitive relay network [16], which consists of one secondary source ($S$), $M$ AF relays ($R_1, \cdots, R_M$), one secondary destination ($D$), one primary receiver ($P$), and one passive eavesdropper ($E$), as shown in Fig. 1. All the nodes are equipped with a single antenna and operate in half-duplex mode. We assume that the direct links from $S$ to $D$ and from $S$ to $E$ do not exist due to severe shadowing environment. The data transmission from $S$ to $D$ can only be done with the help of relays, with the possible wiretap from the eavesdropper. We assume that all channels experience slow block fading and additive white Gaussian noise (AWGN) with variance of $\sigma_0^2$, where the fading coefficients are invariant during one fading block and the corresponding channel gains follow independent Rayleigh distribution. Let $h_{ab}$ denote the channel coefficient between $a$ and $b$. The channel power gains $\left|h_{SR_i}\right|^2$ , $\left|h_{R_iD}\right|^2$ , $\left|h_{R_iE}\right|^2$, $\left|h_{R_iP}\right|^2$, and $|h_{SP}|^2$ are independent and exponentially distributed with parameters $\lambda_{SR_i}$ , $\lambda_{R_iD}$ , $\lambda_{R_iE}$, $\lambda_{R_iP}$, and $\lambda_{SP}$, respectively. Without loss of generation, all the secondary relays are close to each other and forming a cluster, i.e., $\lambda_{SR_i} = \lambda_{SR}$ , $\lambda_{R_iD} = \lambda_{RD}$ , $\lambda_{R_iE} = \lambda_{RE}$, and $\lambda_{R_iP} = \lambda_{RP}$ for all $i$. $\mathbf{h}_{SR} = \left[h_{SR_1}, \cdots, h_{SR_M}\right]^T$, $\mathbf{h}_{RD} = \left[h_{R_1D}, \cdots, h_{R_MD}\right]^T$, $\mathbf{h}_{RE} = \left[h_{R_1E}, \cdots, h_{R_ME}\right]^T$, and $\mathbf{h}_{RP} = \left[h_{R_1P}, \cdots, h_{R_MP}\right]^T$ represent the channel vectors between the secondary source and the relays, between the relays and the secondary destination, between the relays and the eavesdropper, and between the relays and the primary receiver, respectively. It is worth mentioning that we focus on a passive eavesdropper scenario, where the instantaneous channel state information (CSI) of the eavesdropper's link $\mathbf{h}_{RE}$ is not known at the secondary source. In addition, the CSI of interference links, e.g., $\left|h_{R_iP}\right|^2$ and $|h_{SP}|^2$, can be acquired through a spectrum-band manager that mediates between the primary and secondary networks. However, for those

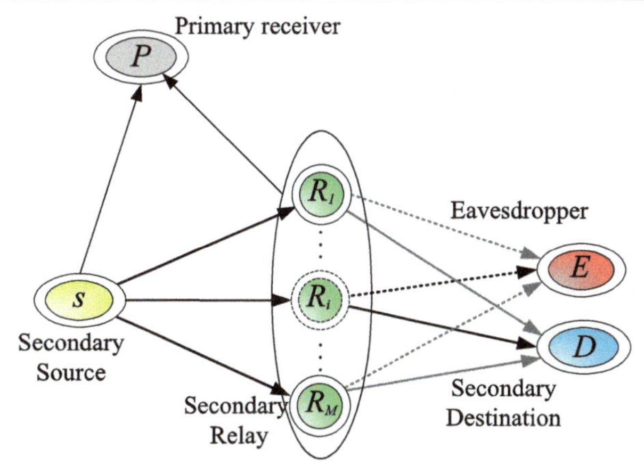

**Fig. 1** Cooperative cognitive radio network with single eavesdropper

cases where the secondary network has no knowledge of the interference channel, our results serve as performance bounds for the considered CCRNs and represent efficient system design tools [17].

The transmission is divided into two slots. In the first phase, the secondary source broadcasts the encoded confidential data. According to the underlaying spectrum-sharing approach, the secondary source adjusts its transmit power $P_S$ under a predefined threshold $I$ to guarantee the quality of service of the primary user. Hence, the transmit power at the secondary source is constrained as $P_S = I\big/|h_{sp}|^2$ [7]. In the second phase, the relays amplify and forward the received signals to the destination with the transmit power $P_R$. In order to avoid interfering with the primary transmission and maximize the mutual information of the secondary system, the D-ZFB scheme is applied to null the interference to the primary receiver, and the $M \times 1$ D-ZFB weight vector is the solution of the following problem, mathematically,
$$\begin{aligned} &\max_{\mathbf{w}_{ZF}} \left|\mathbf{h}_{RD}^H \mathbf{w}_{ZF}\right| \\ &s.t. \left|\mathbf{h}_{RP}^H \mathbf{w}_{ZF}\right| = 0 \\ &\|\mathbf{w}_{ZF}\|^2 = 1 \end{aligned}$$

By applying a standard Lagrangian multiplier method, the optimal weight vector is given as $\mathbf{w}_{ZF} = \mathbf{T}^{\perp}\mathbf{h}_{RD}\big/\left\|\mathbf{T}^{\perp}\mathbf{h}_{RD}\right\|$, where $\mathbf{T}^{\perp} = \mathbf{I} - \mathbf{h}_{RP}\left(\mathbf{h}_{RP}^H\mathbf{h}_{RP}\right)\mathbf{h}_{RP}^H$ is the projection idempotent matrix with rank $(M-1)$.

Thus, the equivalent received signal-to-noise ratio (SNR) at $D$ and $E$ is given as

$$\gamma_D^{AF} = \frac{\gamma_1\gamma_2}{\gamma_1 + \gamma_2 + 1} \tag{1}$$

$$\gamma_E^{AF} = \frac{\gamma_1\gamma_E}{\gamma_1 + \gamma_E + 1} \tag{2}$$

where $\gamma_1 = \left(I/\sigma_0^2\right)\|\mathbf{h}_{SR}\|^2/|h_{SP}|^2$, $\gamma_2 = \left(P_R/\sigma_0^2\right)\left|\mathbf{T}^{\perp}\mathbf{h}_{RD}\right|^2$, and $\gamma_E = \left(P_R/\sigma_0^2\right)\left|\mathbf{h}_{RE}^H\mathbf{w}_{ZF}\right|^2$ denote the equivalent instantaneous SNRs at the $S$-$R$, $R$-$D$, and $R$-$E$ links, respectively.

Considering that $\|\mathbf{h}_{SR}\|^2$ is a chi-square random variable with $2M$ degrees of freedom with parameter $\lambda_{SR}$ due to $h_{SR_i}, i = 1, \cdots, M$ being i.i.d exponential distribution variables with parameter $\lambda_{SR}$, and $|h_{SP}|^2$ is an exponential random variable with parameter $\lambda_{SP}$, then the cumulative distribution function (CDF) of $\gamma_1$ is given by

$$F_{\gamma_1}(\gamma) = \left(\frac{\gamma}{\bar{\gamma}_1 + \gamma}\right)^M \tag{3}$$

where $\bar{\gamma}_1 = (I\lambda_{SR})\big/\left(\lambda_{SP}\sigma_0^2\right)$.

Let $\hbar_{RD} = \mathbf{h}_{RD}\big/\sqrt{\lambda_{RD}}$, then each entry of $\hbar_{RD} = \mathbf{h}_{RD}\big/\sqrt{\lambda_{RD}}$ is i.i.d. $\mathcal{CN}(0,1)$. Considering $\mathbf{T}^{\perp} = \mathbf{I} - \mathbf{h}_{RP}\left(\mathbf{h}_{RP}^H\mathbf{h}_{RP}\right)\mathbf{h}_{RP}^H$ is the projection idempotent matrix with rank $(M-1)$, $\left(\mathbf{T}^{\perp}\right)^H\mathbf{T}^{\perp}$ is a Hermitian matrix with rank $(M-1)$. Based on [18, Theorem 2], $\hbar_{RD}^H\left(\mathbf{T}^{\perp}\right)^H\mathbf{T}^{\perp}\hbar_{RD}$ is a chi-square random variable with $2(M-1)$ degrees of freedom. Then, the CDF of $\gamma_2 = \left(P_R/\sigma_D^2\right)\left|\mathbf{T}^{\perp}\mathbf{h}_{RD}\right|^2 = \left(P_R\lambda_{RD}/\sigma_D^2\right)\hbar_{RD}^H\left(\mathbf{T}^{\perp}\right)^H\mathbf{T}^{\perp}\hbar_{RD}$ is given by

$$F_{\gamma_2}(\gamma) = 1 - \frac{1}{\Gamma(M-1)}\Gamma\left(M-1, \frac{\gamma}{\bar{\gamma}_2}\right) \tag{4}$$

where $\bar{\gamma}_2 = P_R\lambda_{RD}/\sigma_0^2$, $\Gamma(\cdot)$ is the gamma function [19, (8.310.1)], and $\Gamma(\cdot,\cdot)$ is the incomplete gamma function [19, (8.350.2)].

## 3 Secrecy performance analysis with single eavesdropper

In this section, we characterize secrecy performance of the multiple AF-relaying cooperative cognitive radio network with collaborative distributed beamforming in terms of the probability of non-zero secrecy capacity, the secrecy outage probability, the secrecy array gain, and the secrecy diversity order.

Based on (1) and (2), the instantaneous secrecy capacity of the considered network is given by [8]

$$C_S = \max\left\{\frac{1}{2}\log_2\left(1+\gamma_D^{AF}\right) - \frac{1}{2}\log_2\left(1+\gamma_E^{AF}\right), 0\right\} \tag{5}$$

Throughout this work, we focus on the passive eavesdropping scenario, where the secondary source does not have the channel information of the equivalent $S$-$R$-$E$ link and has no choice but to encode the confidential data with a constant codeword rate $R_S > 0$, which is the target secrecy data rate for the considered system. If $R_S$ is less than secrecy capacity, perfect secrecy can be guaranteed. Otherwise, $E$ can eavesdrop on the confidential data; therefore, perfect secrecy is compromised.

### 3.1 The probability of non-zero secrecy capacity

In this subsection, we examine the condition for the existence of non-zero secrecy capacity. According to (5), the probability of non-zero secrecy capacity is formulated as

$$\Pr\{C_S > 0\} = \Pr\{\gamma_2 > \gamma_E\} = 1 - \int_0^\infty F_{\gamma_2}(\gamma) f_{\gamma_E}(\gamma)\, d\gamma \tag{6}$$

where the probability density function (PDF) of $\gamma_E$ is $f_{\gamma_E}(\gamma) = \exp\left(-\gamma/\bar{\gamma}_E\right)/\bar{\gamma}_E$ and $\bar{\gamma}_E = P_R\lambda_{RE}/\sigma_0^2$.

Substituting (4) into (6) and using $\Gamma(n+1,x) = \Gamma(n+1)\exp(-x)\sum_{m=0}^{n}\frac{x^m}{\Gamma(m+1)}$, we can derive the closed-form expression of the probability of non-zero secrecy capacity with the aid of [19, 3.351.3] as

$$\Pr\{C_S>0\}=\frac{1}{\Gamma(M-1)}\int_0^\infty\Gamma\left(M-1,\frac{\gamma}{\bar{\gamma}_2}\right)\exp\left(-\gamma/\bar{\gamma}_E\right)/\bar{\gamma}_E d\gamma$$
$$=\sum_{m=0}^{M-2}\frac{1}{\Gamma(m+1)\bar{\gamma}_E}\left(\frac{1}{\bar{\gamma}_2}\right)^m\int_0^\infty\gamma^m\exp\left(-\frac{\bar{\gamma}_2+\bar{\gamma}_E}{\bar{\gamma}_2\bar{\gamma}_E}\gamma\right)d\gamma$$
$$=\sum_{m=0}^{M-2}\frac{\bar{\gamma}_2}{\bar{\gamma}_E}\left(\frac{\bar{\gamma}_E}{\bar{\gamma}_2+\bar{\gamma}_E}\right)^{m+1} \quad (7)$$

It is obvious that the probability of non-zero secrecy capacity is independent of the tolerable interference level $I$ in the primary network, which is due to the fact that D-ZFB is employed to null the interference to PUs in the second phase.

### 3.2 Secrecy outage probability

The secrecy outage probability is defined as the probability of $C_S<R_S$ and can be expressed as

$$P_{\text{out}}=\Pr\left\{\frac{1}{2}\log_2\left(1+\gamma_D^{AF}\right)-\frac{1}{2}\log_2\left(1+\gamma_E^{AF}\right)<R_S\right\} \quad (8)$$

Substituting (1) and (2) into (8), the secrecy outage probability is rewritten as

$$P_{\text{out}}=\Pr\left\{\frac{1+\frac{\gamma_1\gamma_2}{\gamma_1+\gamma_2+1}}{1+\frac{\gamma_1\gamma_E}{\gamma_1+\gamma_E+1}}<\gamma_{th}\right\} \quad (9)$$

where $\gamma_{th}=2^{2R_S}$ denotes the secrecy SNR threshold.

Furthermore, in order to be mathematically tractable, we apply some approximations into (10) as follows

$$P_{\text{out}}\overset{a}{\approx}\Pr\left\{\frac{\frac{\gamma_1\gamma_2}{\gamma_1+\gamma_2+1}}{\frac{\gamma_1\gamma_E}{\gamma_1+\gamma_E+1}}<\gamma_{th}\right\}$$
$$\overset{b}{\approx}\Pr\left\{\frac{(\gamma_1+\gamma_E)\gamma_2}{(\gamma_1+\gamma_2)\gamma_E}<\gamma_{th}\right\}$$
$$\overset{c}{\approx}\Pr\left\{\min\left(\frac{\gamma_1}{(\gamma_{th}-1)},\frac{\gamma_2}{\gamma_{th}}\right)<\gamma_E\right\} \quad (10)$$

where we employ the approximations of $(1+x)/(1+y)\approx x/y$ in (a), $xy/(1+x+y)\approx xy/(x+y)$ in (b), and $\frac{xy}{x+y}\approx\min(x,y)$ in (c). These approximations have been used in [20, 21], and the effect of the approximation error can be neglected in the medium and high SNR regimes.

Let $Z=\min\left(\frac{\gamma_1}{(\gamma_{th}-1)},\frac{\gamma_2}{\gamma_{th}}\right)$, and the secrecy outage probability is further expressed as

$$P_{\text{out}}\approx\int_0^\infty\Pr\{Z<\gamma\}f_{\gamma_E}(\gamma)\,d\gamma \quad (11)$$

To solve the above integral, we first give the CDF of $Z$ as follows:

$$F_Z(\gamma)=\Pr\{Z<\gamma\}$$
$$=1-\left[1-F_{\gamma_1/(\gamma_{th}-1)}(\gamma)\right]\left[1-F_{\gamma_2/\gamma_{th}}(\gamma)\right]$$
$$=\sum_{k=1}^{M}\sum_{m=0}^{M-2}\binom{M}{k}\frac{(-1)^k}{\Gamma(m+1)}\left(\frac{\gamma_{th}\gamma}{\bar{\gamma}_2}\right)^m\left[1+\frac{(\gamma_{th}-1)\gamma}{\bar{\gamma}_1}\right]^{-k}\exp\left(-\frac{\gamma_{th}\gamma}{\bar{\gamma}_2}\right) \quad (12)$$

Now, substituting (12) into (11), and using $\left(\frac{x}{\alpha+x}\right)^M=\sum_{k=0}^{M}\binom{M}{k}(-1)^k\left(1+\frac{x}{\alpha}\right)^{-k}$, the secrecy outage probability is given by

$$P_{\text{out}}\approx 1+\frac{1}{\bar{\gamma}_E}\sum_{k=1}^{M}\sum_{m=0}^{M-2}\binom{M}{k}\frac{(-1)^k}{\Gamma(m+1)}\left(\frac{\gamma_{th}}{\bar{\gamma}_2}\right)^m$$
$$\times\int_0^\infty\left(1+\frac{(\gamma_{th}-1)\gamma}{\bar{\gamma}_1}\right)^{-k}\gamma^m\exp\left(-\left(\frac{\gamma_{th}}{\bar{\gamma}_2}+\frac{1}{\bar{\gamma}_E}\right)\gamma\right)d\gamma \quad (13)$$

Then, with the help of [19, (9.222.1)], the approximated expressions of the secrecy outage probability can be derived as

$$P_{\text{out}}\approx 1+\frac{1}{\bar{\gamma}_E}\sum_{k=1}^{M}\sum_{m=0}^{M-2}\binom{M}{k}(-1)^k\left(\frac{\gamma_{th}}{\bar{\gamma}_2}\right)^m$$
$$\times\frac{\alpha^{(m+k)/2}}{\beta^{(m-k)/2+1}}\exp\left(\alpha\beta/2\right)W_{-(k+m)/2,(m-k)/2+0.5}(\alpha\beta) \quad (14)$$

where $\alpha=\bar{\gamma}_1/(\gamma_{th}-1)$, $\beta=\gamma_{th}/\bar{\gamma}_2+1/\bar{\gamma}_E$ and $W_{m,n}(x)$ is the Whittaker function [19, (9.222.1)]. It is noted from (14) that the secrecy outage probability expression is derived in closed form and can be easily applied to evaluate the secrecy performance with arbitrary numbers of relays, arbitrary average SNRs, and arbitrary interference constraint level.

### 3.3 Asymptotic secrecy outage probability

In this subsection, in order to exploit a novel insight of design, the asymptotic analysis for the secrecy outage probability was provided to characterize the behavior under the high SNR of legitimate links. The asymptotic result enables us to explicitly examine the impact of collaborative distributed beamforming on the secrecy performance in terms of the secrecy diversity order. Without loss of generation, let $\lambda_{SR}=\lambda_{RD}=\lambda\to\infty$, namely $\bar{\gamma}_2=\kappa\bar{\gamma}_1=\bar{\gamma}\to\infty$, where $\kappa=P_R\lambda_{SP}/I$. Here, $\bar{\gamma}\to\infty$ corresponds to the scenario where the relays have high received SNRs at the first phase and the secondary destination are located much closer to the relays than the eavesdropper, which is a practical scenario of interest. It is obvious that $\bar{\gamma}/\bar{\gamma}_E=\lambda/\lambda_{RE}\to\infty$ for a given $\bar{\gamma}_E$ when $\bar{\gamma}\to\infty$. Similar to [7], we denote $MER=\lambda_{SD}/\lambda_{RE}$ as the ratio of average channel gain from the relays to the secondary destination to that from the relays to the eavesdropper.

With the aid of the Maclaurin series expansion [19, (1.211.1)], we can rewrite (12) as

$$F_Z^{\infty}(\gamma) \overset{\bar{\gamma}\to\infty}{=} \frac{1}{\Gamma(M)}\left(\frac{\gamma_{th}\gamma}{\bar{\gamma}}\right)^{M-1} + o\left(\frac{\gamma}{\bar{\gamma}}\right) \tag{15}$$

where $o(\cdot)$ denotes higher order terms.

Substituting (15) and the PDF of $\gamma_E$ $f_{\gamma_E}$ into (11), and using [19, (3.351.3)], the asymptotic secrecy outage probability is calculated as

$$\begin{aligned} P_{\text{out}}^{\infty} &= \int_0^{\infty} \frac{1}{\Gamma(M)}\left(\frac{\gamma_{th}\gamma}{\bar{\gamma}}\right)^{M-1} \frac{1}{\bar{\gamma}_E}\exp\left(-\frac{\gamma}{\bar{\gamma}_E}\right)d\gamma \\ &= \gamma_{th}^{M-1} MER^{-(M-1)} \end{aligned} \tag{16}$$

Obviously, one can observe from (16) that the secrecy outage probability behaves as $MER^{-(M-1)}$ for $MER \to \infty$. Thus, the achievable secrecy diversity order of the proposed collaborative distributed beamforming yields to

$$d_{\text{single}} = M - 1, \tag{17}$$

and the achievable secrecy array gain is given by

$$\Theta_{\text{single}} = \gamma_{th}^{-1} \tag{18}$$

which implies that as the number of relays increases, the slope of the secrecy outage probability curve becomes steeper when $MER \to \infty$ in the geometric sense. Therefore, collaborative distributed beamforming can effectively improve the physical layer security of cognitive transmission to defend against eavesdropping attacks. However, comparing with the relay selection scheme in [11] and [13], the proposed collaborative distributed beamforming scheme cannot achieve the full diversity order, i.e., $d = M$. This is because the proposed collaborative distributed beamforming scheme allocates the power into the projected channels to enforce zero interference to the PUs, which reduces the achievable secrecy diversity order to $M - 1$.

## 4 Secrecy performance with multiple eavesdroppers

In this section, we extend our cooperative cognitive system model to $K$ non-colluding eavesdroppers, as shown in Fig. 2. We assume that all the eavesdroppers are located close to each other, which implies the same average received SNR at eavesdroppers. We denote $\mathbf{h}_{RE_k} = \left[h_{R_1E_k}, \cdots, h_{R_ME_k}\right]^T$ as the channel vector between the relays and the $k$th eavesdropper. Then, the equivalent instantaneous SNR between the relays and the $k$th eavesdropper can be denoted by $\gamma_{E_k} = \left(P_R/\sigma_0^2\right)\left|\mathbf{h}_{RE_k}^H \mathbf{w}_{ZF}\right|^2$, which is exponentially distributed with the same parameter $\bar{\gamma}_E$.

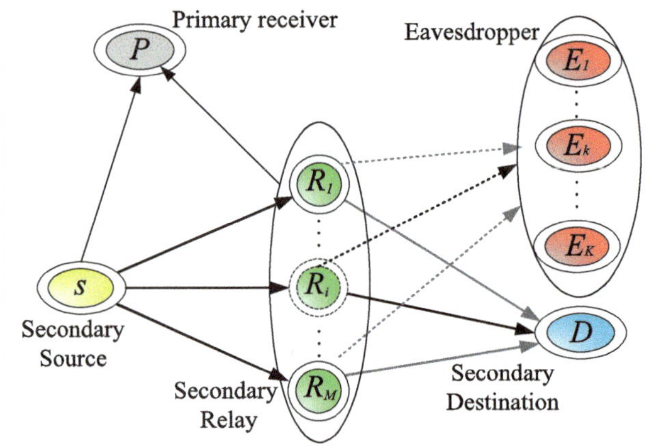


**Fig. 2** Cooperative cognitive radio network with $K$ eavesdroppers

Under this scenario, the instantaneous secrecy capacity should be rewritten as

$$\begin{aligned} C_S^K = \max\Bigg\{ &\frac{1}{2}\log_2\left(1 + \frac{\gamma_1\gamma_2}{\gamma_1+\gamma_2+1}\right) \\ &- \max_{k=1,\cdots,K}\left(\frac{1}{2}\log_2\left(1+\frac{\gamma_1\gamma_{E_k}}{\gamma_1+\gamma_{E_k}+1}\right)\right), 0\Bigg\} \end{aligned} \tag{19}$$

Similar to (6), the probability of non-zero secrecy capacity with $K$ non-colluding eavesdroppers can be formulated as

$$\begin{aligned} \Pr\left\{C_S^K > 0\right\} &= \Pr\left\{\gamma_2 > \max_{k=1,\cdots,K}\gamma_{E_k}\right\} \\ &= 1 - \int_0^{\infty} F_{\gamma_2}(\gamma) f_{\gamma_E^K}(\gamma)\,d\gamma \end{aligned} \tag{20}$$

where $f_{\gamma_E^K}(\gamma)$ denotes the PDF of $\gamma_E^K = \max_{k=1,\cdots,K}\gamma_{E_k}$.

Considering that $\gamma_{E_k}, k = 1,\cdots,K$ are i.i.d exponential distribution variables, the CDF and PDF of $\gamma_E^K$ can be expressed as

$$F_{\gamma_E^K}(\gamma) = \sum_{K=0}^{K}\binom{K}{k}(-1)^k \exp\left(-\frac{k\gamma}{\bar{\gamma}_E}\right) \tag{21}$$

and

$$f_{\gamma_E^K}(\gamma) = \sum_{K=0}^{K-1}\binom{K-1}{k}\frac{K}{\bar{\gamma}_E}(-1)^k \exp\left(-\frac{(k+1)\gamma}{\bar{\gamma}_E}\right) \tag{22}$$

Substituting (4) and (22) into (20), the probability of non-zero secrecy capacity with $K$ non-colluding eavesdroppers can be derived as

$$\Pr\{C_S > 0\} = \sum_{m=0}^{M-2}\sum_{k=0}^{K-1}\binom{K-1}{k}(-1)^k\frac{K\bar{\gamma}_2}{\bar{\gamma}_E}\left[\frac{\bar{\gamma}_E}{(k+1)\bar{\gamma}_2+\bar{\gamma}_E}\right]^{m+1} \tag{23}$$

Similarly, the secrecy outage probability with $K$ non-colluding eavesdroppers is further expressed as

$$\begin{aligned} P_{\text{out}}^{K} &\approx \int_0^\infty \Pr\{Z<\gamma\} f_{\gamma_E^K}(\gamma)\, d\gamma \\ &= 1+\sum_{l=1}^{M}\sum_{m=0}^{M-2}\sum_{k=0}^{K-1}\binom{K-1}{k}\binom{M}{1}\frac{K(-1)^{k+l}}{\Gamma(m+1)\bar{\gamma}_E}\left(\frac{\gamma_{th}}{\bar{\gamma}_2}\right)^m \\ &\quad\times \int_0^\infty \left(1+\frac{\gamma}{\alpha}\right)^{-l}\gamma^m \exp\left(-\left(\frac{\gamma_{th}}{\bar{\gamma}_2}+\frac{(k+1)}{\bar{\gamma}_E}\right)\gamma\right)d\gamma \\ &\overset{t=\gamma/\alpha}{=} 1+\sum_{l=1}^{M}\sum_{m=0}^{M-2}\sum_{k=0}^{K-1}\binom{K-1}{k}\binom{M}{1}\frac{K(-1)^{k+l}\alpha^{m+1}}{\Gamma(m+1)\bar{\gamma}_E}\left(\frac{\gamma_{th}}{\bar{\gamma}_2}\right)^m \\ &\quad\times \int_0^\infty (1+t)^{-l}t^m \exp(-\eta\alpha t)\, dt \\ &= 1+\sum_{l=1}^{M}\sum_{m=0}^{M-2}\sum_{k=0}^{K-1}\binom{K-1}{k}\binom{M}{1}\frac{K(-1)^{k+l}}{\bar{\gamma}_E}\left(\frac{\gamma_{th}}{\bar{\gamma}_2}\right)^m \\ &\quad\times \frac{\alpha^{(m+l)/2}}{\eta^{(m-l)/2+1}}\exp(\alpha\eta/2)\, W_{-(m+l)/2,(m-l)/2+0.5}(\alpha\eta) \end{aligned} \tag{24}$$

where $\eta=\gamma_{th}/\bar{\gamma}_2+(k+1)/\bar{\gamma}_E$.

The asymptotic secrecy outage probability with $K$ non-colluding eavesdroppers can be derived as

$$\begin{aligned} P_{\text{out}}^{K,\infty} &\approx \sum_{K=0}^{K-1}\binom{K-1}{k}\frac{K}{\bar{\gamma}_E}(-1)^k\frac{1}{\Gamma(M)}\left(\frac{\gamma_{th}}{\bar{\gamma}_2}\right)^{M-1} \\ &\quad\times\int_0^\infty \gamma^{M-1}\exp\left(-\frac{(k+1)\gamma}{\bar{\gamma}_E}\right)d\gamma \\ &= \left[\sum_{k=0}^{K-1}\binom{K-1}{k}(-1)^k K(k+1)^{-M}\gamma_{th}^{M-1}\right] MER^{-(M-1)} \end{aligned} \tag{25}$$

It is clear that the achievable secrecy diversity order in the presence of $K$ non-colluding eavesdroppers is still $M-1$, which means that the secrecy diversity performance is independent of the number of eavesdroppers. However, the achievable secrecy array gain is different due to the presence of multiple non-colluding eavesdroppers, which can be yielded as

$$\Theta_{\text{multiple}}=\left[\sum_{k=0}^{K-1}\binom{K-1}{k}(-1)^k K(k+1)^{-M}\right]^{-1/(M-1)}\gamma_{th}^{-1} \tag{26}$$

It is indicated from (18) and (26) that the achievable secrecy array gain decreases with increasing $K$, which results in a decrease in the secrecy outage performance.

## 5 Numerical and simulation results

In this section, Monte Carlo simulations are provided to validate our analytical expressions over Rayleigh-fading channels. We assume that $R_s = 0.1$ bit/s/Hz, $P_R = 20$ dB, $\sigma_0^2 = 1$, $\lambda_{SR} = \lambda_{RD} = 1$, and $\lambda_{SP} = \lambda_{RP} = 0.3$. It is worth noting that the primary and secondary users are spatially separated in two different wireless networks. As such, an average channel gain ($\lambda_{SP}$ and $\lambda_{RP}$) between two heterogeneous users from the different networks is set to be smaller than that ($\lambda_{SR}$ and $\lambda_{RD}$) between two heterogeneous users from the same networks.

Figure 3 plots the secrecy outage probability versus $I$ for various $M$ over Rayleigh-fading channels. The analytical results based on (14) are in precise agreement with the Monte Carlo simulations. It is shown that for a fixed $M$, the secrecy outage probability decreases with increasing $I$, which increases the transmitting power at the secondary source and improves the transmission reliability of S-R links. We can also see that there exists a secrecy outage performance floor in the high $I$ region. It is because the secrecy outage performance is determined by second hop links when $I \to \infty$. On the other hand, the secrecy outage performance is improved with larger $M$, as more relays do not only help improve the communication quality of S-R links but also enhance security of R-D links. As shown in Fig. 3, as the number of relays increases from $M = 3$ to 5, the secrecy outage performance floor is significantly reduced.

Figures 4 and 5 plot the secrecy outage probability and the probability of non-zero secrecy capacity versus $MER$ for various $M$ in the presence of a single eavesdropper. The exact curves precisely agree with the Monte Carlo simulation results, which validates the correctness of our analysis. As shown in the figures, as the number of relays increases from $M = 2$ to 5, the secrecy outage probability and the probability of non-zero secrecy capacity are significantly improved. The reason is that more relays do not only help improve the communication quality of S-R links but also enhance the security of the R-D links. In order to predict the secrecy diversity order and the secrecy array gain, the asymptotic secrecy outage probability curves based on (16) are also drawn in Fig. 5. We can see that the asymptotic curves well approximate the analytical curves in the high $MER$ region, which verifies

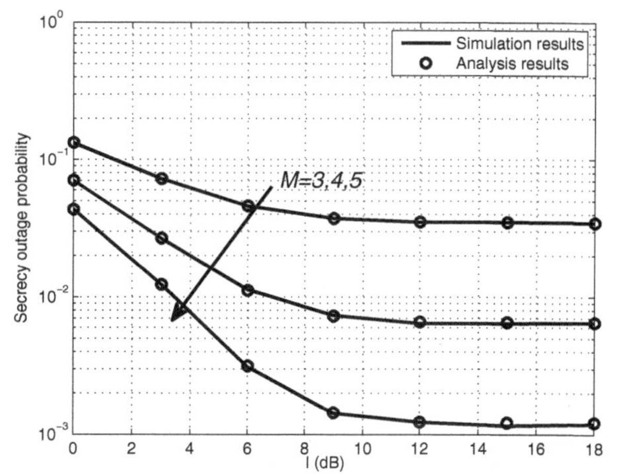


**Fig. 3** Secrecy outage probability versus $MER$ for various $I$ ($\lambda_{RE} = 0.2$)

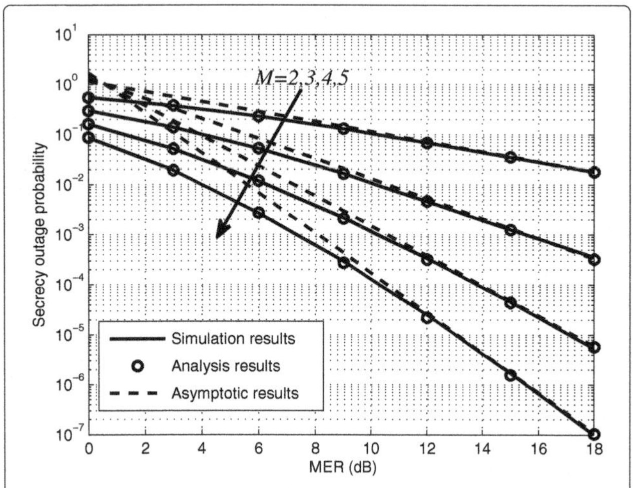


**Fig. 4** Secrecy outage probability versus *MER* for various *M* ($K = 1$ and $I = 10$ dB)

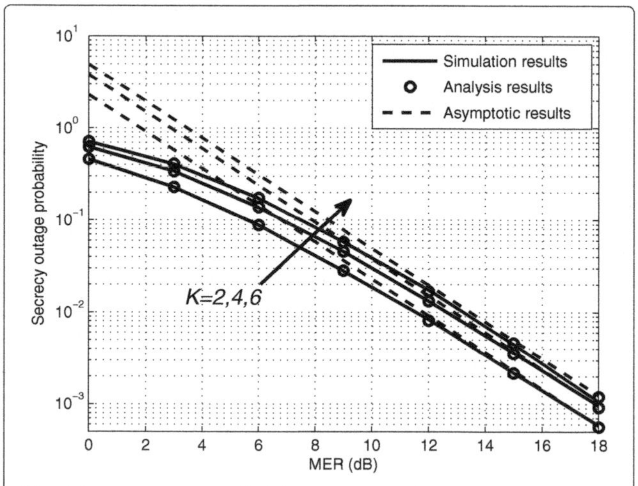


**Fig. 6** Secrecy outage probability versus *MER* for various *K* ($M = 3$ and $I = 10$ dB)

the derived asymptotic expressions. As $MER \to \infty$, the secrecy outage probability decreases at a faster speed with an increasing number of relays $M$. This is because the achievable secrecy diversity order is $M - 1$.

Figures 6 and 7 plot the secrecy outage probability and the probability of non-zero secrecy capacity versus *MER* in the presence of $K$ non-colluding eavesdroppers ($M = 3$ and $I = 10$ dB). As observed from the figure, we can find that for different values of *MER* and $K$, the analytical results match well the simulation. In addition, the asymptotic secrecy outage probability curves based on (25) converge exactly in the high *MER* region in Fig. 6. Moreover, the slopes of the curves of the secrecy outage probability are in parallel, which verifies the secrecy diversity order of $M-1$ regardless of the number of eavesdroppers $K$. From Fig. 7, it is evident that the probability of non-zero secrecy capacity improves with increasing *MER*. In addition, the probability of non-zero secrecy capacity improves at a faster speed with smaller $K$.

## 6 Conclusions

In this paper, we have investigated the security performance of multiple AF-relaying CCRNs with collaborative distributed beamforming scheme against passive eavesdropping attacks. To show the advantages of the collaborative distributed beamforming scheme, we have derived closed-form expressions of the probability of non-zero secrecy capacity and secrecy outage probability as well as the asymptotic expression over Rayleigh-fading channels, which provide an efficient means to evaluate the impact of key parameters on the security performance. Moreover,

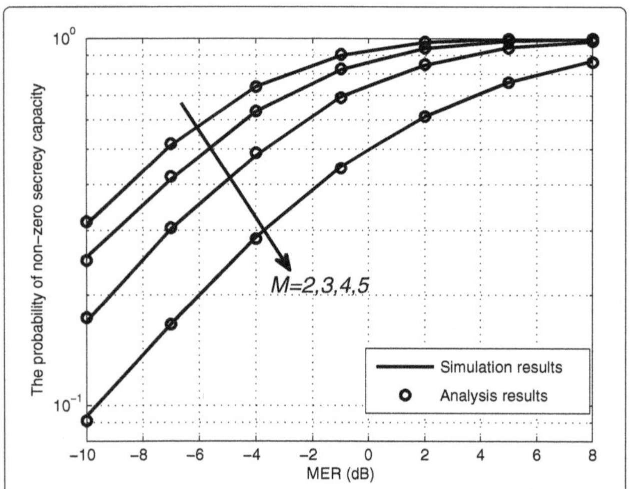


**Fig. 5** Probability of non-zero secrecy capacity versus *MER* for various *M* ($K = 1$ and $I = 10$ dB)

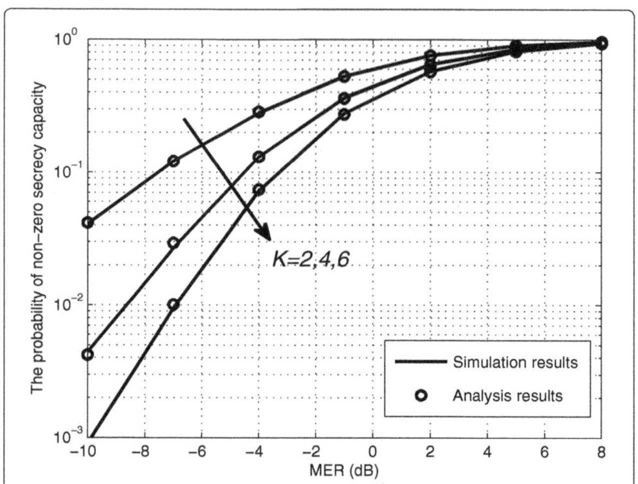


**Fig. 7** Probability of non-zero secrecy capacity versus *MER* for various *K* ($M = 3$ and $I = 10$ dB)

our asymptotic results demonstrated that the achievable secrecy diversity order of collaborative distributed beamforming with $M$ AF relays is $M - 1$ regardless of the number of eavesdroppers.

**Acknowledgements**
This work is supported by the National Natural Science Foundation of China (No. 61471393 and 61501507) and the Jiangsu Provincial Natural Science Foundation of China (No. BK20150719) and the China Postdoctoral Science Foundation under a Special Financial (No. 2013T60912).

**Competing interests**
The authors declare that they have no competing interests.

**References**

1. D Wyner, The wire-tap channel. Bell Syst. Tech. J. **54**(8), 1355–1367 (1975)
2. A Jovicic, P Viswanath, Cognitive radio: an information-theoretic perspective. IEEE Trans. Inf. Theory. **55**(9), 3945–3958 (2009)
3. X Chen, H-H Chen, W Meng, Cooperative communication for cognitive radio networks—from theory to application. IEEE Commun. Surv. Tutor. **16**(3), 1180–1192 (2014)
4. Z Shu, Y Qian, S Ci, On physical layer security for cognitive radio networks. IEEE Netw. **27**(3), 28–33 (2013)
5. C Tang, G Pan, T Li, Secrecy outage analysis of underlay cognitive radio unit over Nakagami-m fading channels. IEEE Wirel. Commun. Lett. **3**(6), 609–612 (2014)
6. Y Pei, Y-C Liang, L Zhang, KC Teh, KH Li, Secure communication over MISO cognitive radio channels. IEEE Trans. Wirel. Commun. **9**(4), 1494–1502 (2010)
7. Y Zou, X Li, Y-C Liang, Secrecy outage and diversity analysis of cognitive radio systems. IEEE J. Sel. Areas Commun. **32**(11), 2222–2236 (2014)
8. M Elkashlan, L Wang, TQ Duong, GK Karagiannidis, A Nallanathan, On the security of cognitive radio networks. IEEE Trans. Veh. Techol. **64**(8), 3790–3795 (2015)
9. H Lei, H Zhang, IS Ansari, C Gao, Y Guo, G Pan, KA Qaraqe, Secrecy outage performance for SIMO underlay cognitive radio systems with generalized selection combining over Nakagami-$m$ channels. IEEE Trans. Veh. Techol. **65**(12), 10126–10132 (2016)
10. H Zhao, Y Tan, G Pan, Y Chen, N Yang, Secrecy outage on transmit antenna selection/maximal ratio combining in MIMO cognitive radio networks. IEEE Trans. Veh. Techol. **65**(12), 10236–10242 (2016)
11. Y Zou, J Zhu, L Yang, Y-C Liang, Y-D Yao, Securing physical layer communications for cognitive radio networks. IEEE Commun. Mag. **53**(9), 48–54 (2015)
12. W Yang, X Xu, Y Cai, in *IEEE WCNC 2014*. Secrecy outage analysis for DF underlay CRNs with outdate CSI (Wireless Communications and Networking Conference (WCNC), 2014 IEEE, Istanbul, 2014)
13. Y Liu, L Wang, TT Duy, M Elkashlan, TQ Duong, Relay selection for security enhancement in cognitive relay networks. IEEE Wirel. Commun. Lett. **4**(1), 46–49 (2015)
14. Y Zou, B Champagne, W-P Zhu, L Hanzo, Relay-selection improves the security-reliability trade-off in cognitive radio systems. IEEE Trans. Commun. **63**(1), 215–228 (2015)
15. Q Gu, G Wang, L Gao, M Peng, Security-reliability performance of cognitive AF relay-based wireless communication system with channel estimation error. EURASIP J. Adv. Signal Process. **2014**, 1–11 (2015)
16. W Yang, K Wang, X Xu, J Zhou, in *WOCC 2016*. Secure transmission for AF relaying spectrum-sharing systems with collaborative distributed beamforming (Wireless and Optical Communication Conference (WOCC), 2016 25th, Chengdu, 2016)
17. A Afana, V Asghari, A Ghrayeb, S Affes, On the performance of cooperative relaying spectrum-sharing system with collaborative distributed beamforming. IEEE Trans. Commun. **62**(3), 857–871 (2014)
18. RJ Pavur, Quadratic forms involving the complex Gaussian. M.Sc. dissertation, Math. Dept., Texas Tech Univ.Lubbock, TX, (1980)
19. IS Gradshteyn, IM Ryzhik, *Table of integrals, series, and products*, sixth ed. (Academic Press, London, 2000)
20. I Krikidis, Opportunistic relay selection for cooperative networks with secrecy constraints. IET Commun. **4**(15), 1787–1791 (2010)
21. A Behnad, X Wang, Accuracy of harmonic mean approximation in performance analysis of multihop amplify-and-forward relaying. IEEE Wirel. Commun. Lett. **3**(2), 125–128 (2014)

16

# Modelling and implementation of virtual radio resources management for 5G Cloud RAN

Sina Khatibi[1,2], Luísa Caeiro[3*], Lúcio S. Ferreira[4], Luis M. Correia[1] and Navid Nikaein[5]

**Abstract**

The virtualisation of Radio Access Networks (RANs) is one of the goals in designing 5G mobile networks. This paper aims at presenting a proof of concept for the virtualisation of radio resources using Open Air Interface (OAI), a software-based Long-Term Evolution (LTE) eNodeB physical emulator. OAI was extended to support multi-tenancy, representing diverse Virtual mobile Network Operators (VNOs) with different Service Level Agreements (SLAs). A comprehensive analytical model for managing the virtual radio resources has been proposed, with two key parts: estimation of available radio resources and their allocation to different VNOs. The estimation is performed by the model, and the allocation is managed by OAI scheduling. Various scenarios and use cases are studied in this virtual RAN environment, network performance being evaluated for different situations, by varying guaranteed levels, serving weights, and used services. Results show that the proposed approach offers almost the same capacity to guaranteed VNOs regardless of other existing VNOs, experiencing at worst a degradation of 32% of its initial allocated data rate, without violation of the guaranteed data rate. The data rate allocated to best effort VNOs may decrease up to 7% of its initial value, which is acceptable, to guarantee other more demanding SLAs.

**Keywords:** Virtualisation, Radio resources, Cloud, C-RAN, 5G

## 1 Introduction

Future mobile networks will face an even faster growth of mobile data demand [1]. Nevertheless, drastic temporal and geographical variations of traffic, in addition to the shortage of network capacity, put a very challenging situation to operators [2]. The common provisioning approach used in Radio Access Networks (RANs) of considering only busy hours is no longer acceptable. It leads to an inefficient resource usage with high CAPital and OPerational Expenditures (CAPEX and OPEX, respectively) costs. Thus, operators have to find a practical, flexible, and cost-efficient solution for their networks, in order to address this issue, as they have scarce radio resources available. On the other hand, the appearance of Virtual Network Operators (VNOs), which aim to provide specific services without owning an infrastructure can be seen as a definitive trend. The combination of these two challenges lead to the motivation of the current paper, addressing the need of an efficient management of varying radio resources shared among multiple VNOs, guaranteeing agreed Service Level Agreements (SLAs).

The design of a future Internet architecture favours multiple coexisting solutions, designed and customised to satisfy specific network requirements, rather than trying to achieve a global architecture that fits all [3]. A network environment with multiple virtual networks coexisting on the same physical infrastructure is supported by the concept of network virtualisation [4]. Regardless of the wired or wireless nature of networks, virtualisation can be considered as the process of splitting the entire network into multiple concurrent ones, non-transparent to each other [5]. In this sense, the virtualisation of wireless networks is considered as a technology in which the physical network infrastructure and its radio resources are abstracted and sliced into virtual wireless network resources. These can then be shared among multiple parties, while holding isolation in addition to the other functionalities [6, 7].

The virtualisation of RAN has been addressed in recent studies [8]. In particular, virtualisation strategies for

* Correspondence: luisa.caeiro@estsetubal.ips.pt
[3]IPS-EST Setúbal, Setúbal and INESC-ID, Lisbon, Portugal
Full list of author information is available at the end of the article

5G in advanced computational platforms, such as cloud computing, have been presented [9]. The advantages of these so-called Virtual RANs (V-RANs) can be summarised as follows:

- Elimination of cell boundaries [10, 11]: in contrast to traditional solutions, the network determines which Base Stations (BSs) the Mobile Terminal (MT) is associated with, in order to provide a centre-cell experience throughout the entire network, so that the MT does not experience a degradation of link performance as it moves away from the BS.
- Isolation among virtual instances [7, 12, 13]: the isolation among VNO instances ensures that traffic and mobility, in addition to the fluctuations in the channel conditions of MTs, do not affect the Quality of Service (QoS) offered to other VNOs. It also enables multiple VNOs, with different objectives and requirements, to operate in the same physical infrastructure. Achieving a lower downtime is another advantage of having isolation among VNOs.
- Flexibility [14], programmability [15], and customisation of virtual networks [16]: they enable an easier and faster evaluation of the new proposed networking technologies, without considering the complicated interfaces and characteristics of physical infrastructures.
- Multi-Radio Access Technologies (Multi-RAT) support [15, 17]: offloading of traffic (e.g., to Wi-Fi) cannot be realised for multiple VNOs operating over the same physical infrastructure, since there may not be enough channels to separately allocate a set to each VNO [18]. The solution is to virtualise radio resources to implement a Wi-Fi virtual link for each VNO.

The virtualisation of wireless networks has proven to be an active research topic. The management of resources (e.g., radio, fibre, and processing resources) is the focus of many works [19], since the performance of virtual wireless networks is strongly affected by the way resources are used [20, 21]. Aiming at achieving an efficient mapping from virtual radio resources onto available physical ones, a negotiation-based approach focusing on Long-Term Evolution (LTE)-Advanced is proposed in [22]. The authors in [23] propose a technique that uses MT-centric metrics to provide multiple partitions of a wireless network into virtual transmission point sets. A dynamic cloud resource sharing scheme of processing resources, based on the Indian buffet game, is presented in [24]. A service-aware user-centric resource allocation scheme is presented in [25], which achieves multi-point fairness in downlink. Regarding radio resources, various studies considered various forms of spectrum sharing [8, 26, 27]. The authors in [28] presented a solution, in addition to a demo, for the virtualisation of wireless local networks. The modelling of the Virtual Radio Resource Management (VRRM) for cellular heterogeneous access networks is presented in [29–31]. The key idea is to aggregate and manage all the physical radio resources, to create virtual wireless links, and to offer Capacity-as-a-Service.

The novelty of this paper is to present, for 5G C-RAN, a proof of concept of the VRRM model using the Open Air Interface (OAI) [32], an open-source software-based implementation of LTE, spanning the full protocol stack both in E-UTRAN and EPC [33]. To achieve it, OAI is extended to support multi-tenancy, representing diverse VNOs with different SLAs. The estimation is performed by the model, and the allocation is managed by OAI scheduling. A specific scenario of usage is considered to demonstrate the performance of the VRRM model. The key principle that is demonstrated is the capability of the VRRM algorithm to interact with the scheduling algorithm in order to guarantee the specific SLAs of different VNOs, which is independent of the selected frequency bands and service profiles. In this sense, this concept can be applied to other frequency bands and types of services.

The rest of this paper is organised as follows. Section 2 presents the architecture of V-RAN. The concept of virtualisation of radio resources, and the used analytical model, is briefly addressed in Section 3. Section 4 introduces the architecture used for integration, followed by the algorithms for integration in Section 5. The reference scenario, use cases, and scenario variations, based on which numeric results were obtained, are introduced in Section 6. The analysis of results is discussed in Section 7. Finally, the conclusions are drawn in Section 8.

## 2 Architecture

The V-RAN architecture has been considered as the reference architecture for the integration of VRRM in OAI, encompassing the virtualisation of radio resources based on C-RAN [2]. Although the V-RAN architecture envisages the existence of diverse RATs, in order to meet the main objective of integration and demonstration, OAI has been chosen to build and customise an LTE BS and connected Users Equipment (UEs), in order to test different configurations and network setups and to monitor the network and MTs in real time.

The reference architecture is depicted in Fig. 1, its key architectural elements being:

- VNOs: network operators that do not own a RAN infrastructure, asking the V-RAN enabler for connectivity in terms of capacity to carry various

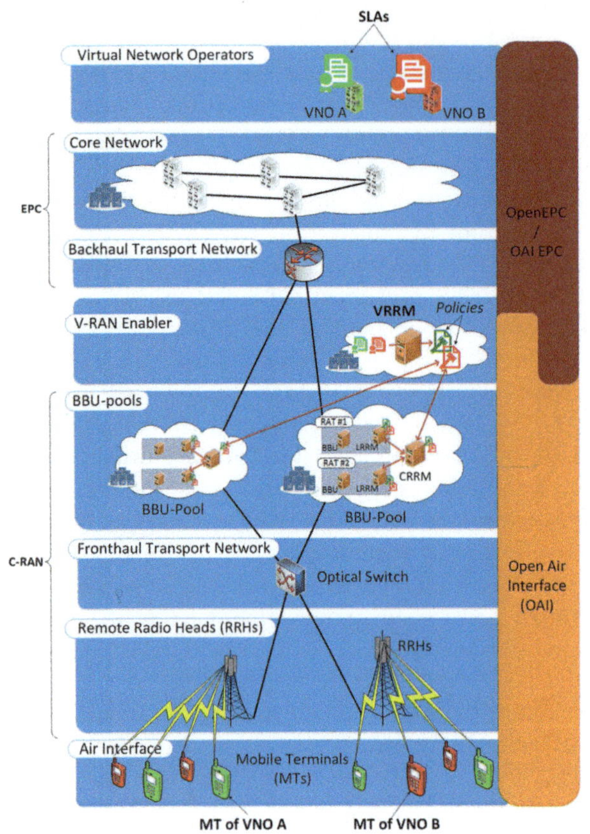

**Fig. 1** Architecture of a V-RAN

services traffic with various QoS requirements to/from their subscribers.

- Backhaul transport network: a low latency optical transport network, which connects the operators' cores to the physical infrastructure of a RAN.
- Virtualisation platform: the key difference between a C-RAN and a V-RAN. On the one hand, it is in charge of abstracting the physical infrastructure for the VNOs, while on the other hand, it handles the request of VNOs through the available physical resources. The most important functionality of the virtualisation platform is the VRRM, the highest manager, which is in charge of translating VNO requirements and SLAs through sets of polices onto the lower levels. It optimises the usage of virtual radio resources, without dealing with the management of the physical ones. Reports and monitoring information (e.g., estimated remained capacity) received from lower levels enables it to improve policies.
- Base Band Unit pools (BBU-pools) data centre: a set of Virtual Machines used for baseband processing of traffic among UEs and network cores.
- Fronthaul transport network: it carries signals between BBU pools and Remote Radio Heads (RRHs), using Common Public Radio Interfaces, with high data rates over optical fibres. The optical equipment needs to have the lowest delay possible, since the maximum round-trip delay must be below 150 μs (i.e. a maximum of 15 km of BBU-RRH distance) [34]. The optical switch must enable the scaling or migration of BBU pools in multiple data centres.
- RRHs: transceivers in charge of exchanging data and control traffic to/from UEs through the air interface, supporting multiple RATs.

By comparing C-RANs with current mobile networks, it can be seen that the BS has been divided into RRHs and BBU pools (and the transmission in between them). The virtualisation platform, which offers isolation, element abstraction, and multi-tenancy, does not exist in current networks. The changes in the architecture and the dedicated hardware replacement by Virtual Machines in data centres provide high flexibility, resource efficiency, and cost reduction.

By crossing the V-RAN architecture with the OAI wireless technology platform, one can identify the last three elements as possibly being implemented by the OAI BSs. VNOs, on the other hand, can deploy their own Open EPCs, which can be implemented by OAI EPCs.

## 3 Virtualisation of radio resources

### 3.1 Initial considerations

The concept of virtualisation of radio resources to achieve virtual wireless links, as the milestone in the realisation of end-to-end virtual networks, is presented in [29]. The proposed concept is based on the aggregation of all physical radio resources from different RATs, in order to offer VNOs on-demand, flexible, and efficient wireless connectivity. In this novel methodology, VNOs ask for wireless capacity from a set of RAN Providers to serve their subscribers, without having to deal with the physical infrastructure at all. RAN Providers, owning the physical infrastructure, are capable of offering Capacity-as-a-Service to VNOs. The advantages of RAN virtualisation compared to RAN sharing (where each operator is allocated a portion of spectrum) come from network element abstraction, isolation among virtual instances, and the ability to support multi-RATs and new business models.

The management of virtual radio resources is a complex optimisation problem, since network status and constraints vary in time. Among various possible techniques and approaches for solving this problem, partial VRRM seems to be the simplest one [30], where the main optimisation problem is broken into multiple sub-problems; in other words, the time axis is divided into decision windows, and VRRM maximises the objective function in each of these intervals, independently. However, it is worthwhile noticing that decisions in each interval affect the network state directly, and the outcome at a certain interval depends on the states in previous ones, the optimal solution needing to consider this dependency. Therefore, the output of partial VRRM is a local minimum and not the global one; nevertheless, partial VRRM is a simple solution, which can be used as the starting step and a reference point.

### 3.2 Estimation of available resources

In general, the data rate of a Radio Resource Unit (RRU) (e.g. timeslot in GSM, code in UMTS, and resource block in LTE), assigned to a UE, varies between 0 and a maximum value. This depends on various parameters, such as the RAT, modulation, and coding schemes; thus, it can be given as a function of channel quality, i.e. Signal to Interference plus Noise Ratio (SINR).

The estimation of available resources is made by VRRM, based on the model presented in [29] and the approaches presented in [35]. In this model, the network is modelled as a *K*-tier network, where each tier models the BSs of a particular class. It is assumed that the BSs in a given tier are spatially distributed as a Poisson Point Process with a given density and transmission power. The received power is assumed to be exponentially distributed (i.e. Rayleigh fading is assumed for the signal magnitude). Based on the data rates of different access technologies as a function of SINR, and the distribution of SINR, the Probability Density Function (PDF) of a single RRU of each RAT is calculated.

The PDF of the total date of a single RAT (assuming that channels are independent [36]), is equal to the convolution of all RRUs' PDFs. Likewise, the PDF of the total network data rate is computed by convolving all RATs' PDFs. By obtaining the total network PDF and Cumulative Density Function (CDF), an estimation of the available network capacity is in hand to be used in the allocation procedure.

### 3.3 Allocation of radio resources

After estimating the total network capacity, the VRRM has to distribute it among the various network services. The key objective, in the allocation of resources, is to maximise the usage efficiency, by taking restrictions into account, e.g. priority of services to different VNOs based on their SLAs.

SLAs can generally be categorised into three types:

- Guaranteed Bitrate (GB) guarantees minimum and maximum data rates, regardless of the network status. The total satisfaction of the VNO is achieved when the maximum guaranteed data rate is allocated to it. The upper boundary enables VNOs to have full control of their networks, e.g. a VNO may decide to offer VoIP to only 30% of its subscribers simultaneously, which can put into practice by choosing a guaranteed SLA for its VoIP service. It is expected that subscribers always experience a good QoS in return of relatively more expensive services.
- Best effort with minimum Guaranteed (BG) guarantees a minimum level of service (i.e. data rate), but the request for data rates higher than the guaranteed one is served in a best effort manner. In this case, although VNOs do not invest as much as former ones, they can still guarantee the minimum QoS to their subscribers. From the subscribers' viewpoint, the acceptable service (not as good as the previous ones) is offered with a relatively lower cost.
- Best Effort (BE), in which the VNO is served in a pure best effort way. Operators, and consequently their subscribers, in return, may suffer from low QoS and resource starvation during busy hours.

The allocation of radio resources is based on one objective function, presented in [30], where a set of weights are used to prioritise the allocation of data rates to the different services of the different VNOs: services with higher weights are served with higher data rates. The choice of these weights is based on the SLAs between the VNOs and VRRM; hence, the objective function for VRRM can be written as follows:

$$f^{v}_{\mathbf{R_b}}\left(\mathbf{R^{Srv}_b}\right) = f^{cell}_{\mathbf{R_b}}\left(\mathbf{R^{cell}_b}\right) + f^{WLAN}_{\mathbf{R_b}}\left(\mathbf{R^{WLAN}_b}\right) - f^{f}_{\mathbf{R_b}}\left(\mathbf{R^{f}_b}\right) \quad (1)$$

where

- $f^{cell}_{\mathbf{R_b}}$: objective function for cellular RATs
- $\mathbf{R^{cell}_b}$: vector of serving data rates from cellular networks,
- $f^{WLAN}_{\mathbf{R_b}}$: objective function for WLANs
- $\mathbf{R^{WLAN}_b}$: vector of serving data rates from WLANs
- $f^{f}_{\mathbf{R_b}}$: fairness function
- $\mathbf{R^{f}_b}$: vector of intermediate fairness variables
- $\mathbf{R_{b^{Srv}}}$: vector of serving data rates

The objective function for cellular RATs in (1) is given by

$$f^{cell}_{R_b}\left(\boldsymbol{R^{cell}_b}\right) = \sum_{i=1}^{N_{VNO}} \sum_{j=1}^{N_{srv}} W^{Srv}_{ji} R^{cell}_{b_{ji}} [Mbps] \quad (2)$$

where

- $N_{VNO}$: number of served VNOs in the VRRM
- $N_{srv}$: number of services for each VNO
- $W^{Srv}_{ji}$: weight of serving unit of data rate for service $j$ of VNO $i$, where $W^{Srv}_{ji} \in [0, 1]$

The objective function for WLANs in (1), where the goal is to minimise the collision rates, is given by

$$f_{R_b}^{WLAN}\left(R_{b[Mbps]}^{WLAN}\right) = \sum_{i=1}^{N_{VNO}}\sum_{j=1}^{N_{srv}} \left(W_{ji}^{Srv} R_{b_{ji}}^{WLAN} + W^{SRb} \frac{\overline{R_{b_j}}}{\overline{R_b^{max}}} R_{b_{ji}}^{WLAN}\right) \quad (3)$$

where

- $W^{SRb}$: weight for session average data rate, where $W^{SRb} \in [0, 1]$
- $\overline{R_b^{max}}$: maximum average data rate among all network services
- $\overline{R_{b_j}}$: average data rate for service $j$

There are also constraints in the allocation of data rates that should not be violated. The fundamental constraint is the estimated total network capacity, in addition to minimum and maximum guaranteed data rates.

In the allocation process, there may be situations where resources are not enough to meet all guaranteed capacity, the optimised allocation being no longer feasible. In these cases, constraints are relaxed by the introduction of a violation (also known as slack) parameter and weights, with which the former infeasible optimisation problem turns into a feasible one, where the optimal solution maximises the objective function and minimises the weighted average constraints violations.

The received VRRM policies and associated data rates must be translated into a specific number of RRUs for each VNO. This mapping occurs at the Local RRM level, being based on the average Modulation and Coding Scheme (MCS) received in the monitoring reports. The translation function pre-allocates the required RRUs, from the available ones of each VNO, to achieve the data rates dictated by the policies.

## 4 Integration overview

The infrastructure to implement the concept of virtualisation of radio resources is a cloud host environment of servers.

OAI [32] is an open-source software-based LTE eNodeB, enabling the integration and implementation of these concepts. The transceiver functionality is realised by a software radio front end, connected to a host computer for processing. OAI enables to build an open cellular ecosystem for flexible and low-cost 5G deployment and experimentations with the following characteristics:

- Open and integrated development environment under the control of experimenters.
- Fully software-based network functions offering flexibility to architect, instantiate, and reconfigure network components (at the edge, core, or cloud, using the same or different addressing space).
- Playground for commercial handsets, as well as application, service, and content providers.
- Rapid prototyping of 3GPP compliant and non-compliant use cases, and new concepts towards 5G, from Machine-to-Machine/Internet of Things (M2M/IoT) and Software-Defined Network (SDN) to Cloud-RAN (C-RAN) and massive Multiple Input Multiple Output (MIMO).

OAI is implemented in Linux, using a cellular network emulator [32], while VRRM is implemented in a Windows-based server, deployed on the same network domain, as depicted in Fig. 1. The VRRM server issues policies based on information and statistics received from the infrastructure emulator. The servers are connected through an internal network, managed by the cloud provider, the links carrying information and control signals among servers. A bidirectional interface between OAI and VRRM using Transmission Control Protocol (TCP) sockets was defined to:

- Transfer scenarios and configurations from OAI to VRRM.
- Send policies from VRRM to OAI.
- Retrieve real-time statistics and information of VNOs from OAI for VRRM.

In order to support VRRM policies, the OAI scheduler required modifications to differentiate groups of subscribers, one per VNO. With this approach, it is possible to impose different policies per VNO/group, offering different levels of QoS to UEs according to the VNO/group they belong to. By collecting operation statistics separately per VNO/group, the scheduler may adapt to meet the diverse requirements. The following modifications on OAI were implemented:

- Adding group-based statistics to an eNodeB
- Introducing new sets of long-time statistics, using the already implemented listing concept
- Changing the codes to initialise, fill, and use the aforementioned statistics
- Changing the MAC scheduler algorithm in order to support the groups' policies
- Adding support for bidirectional connections and the required protocols
- Changing the codes to add the groups' information into the XML file.

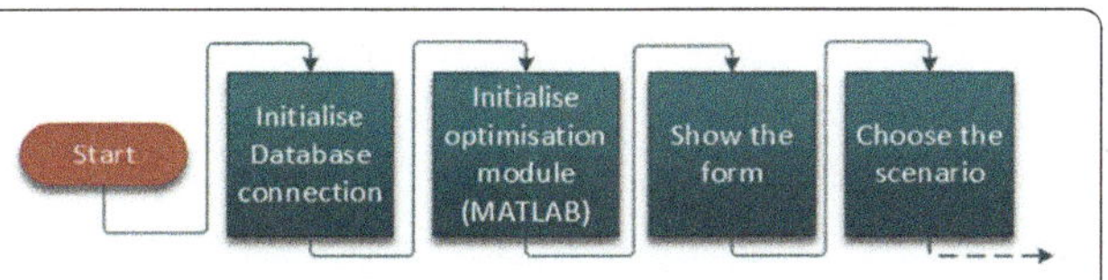

**Fig. 2** VRRM module flowchart

## 5 Algorithms for integration

### 5.1 VRRM server

#### *5.1.1 Module structure*

Figure 2 presents the flowchart of the VRRM server. Before the appearance of the Graphical User Interface on screen, there are multiple initialisations to be done. At first, the server makes a connection to the SQL Server [6]. The database is used to speed up the procedure by not re-calculating intermediate steps or results. Considering the VRRM procedure in specific, calculating the PDFs of RRUs, RATs, and network is the most time-consuming tasks during evaluations and simulations. By saving the intermediate results during the calculation of probability functions, it is possible to significantly decrease the processing time.

Afterwards, the optimisation module is initialised. The optimisation function was developed in MATLAB [37] and converted to a managed Dynamic Library Link (DLL); the optimisation is based on the interior-point method, using MATLAB's *linprog* [29]. The initialisation of MATLAB-based modules is more time-consuming than other elements; hence, the initialisation is done once at the beginning and used multiple times later on. After having all the pre-requisite modules loaded, the control panel appears.

#### *5.1.2 Communication with OAI*

In order to enable the integration of VRRM with OAI, a pre-defined scenario, designated by *OAI Integration*, has been included (Fig. 3).

After loading the real-time charts, the *TCPLink* module starts, and the user is asked to choose the right interface for communicating with the OAI server. A waiting form is shown, informing the VRRM server is waiting for receiving the packet event from OAI, meaning the connection has been established. VRRM listens over these links for scenario description and reports, which also uses to send policies to OAI.

Upon reception of a new message by VRRM, it checks for the "Scenario Description" message; then, the message is parsed and the scenario details are extracted. In the next step, the RAT module that contains the properties of the radio resources is created and initialised. After initialising the CRRM and VRRM modules, the VNOs based on the received scenario are formed. Finally, the services are added to the VNOs.

In the next stage, the VRRM model optimisation is solved, results are produced, and the new policy flag is set.

### 5.2 OAI extensions to support multiple VNO

#### *5.2.1 Module structure*

In addition to the development of the VRRM server, OAI also needs some adaptation to support multiple VNOs or groups operating in the same infrastructure. A key requirement is to gather the statistics per VNO/group, instead of all UEs.

OAI stores UEs' key performance indicators, e.g. total data rates offered to the UEs in downlink. The following group-based statistics have been added to OAI, per sub-

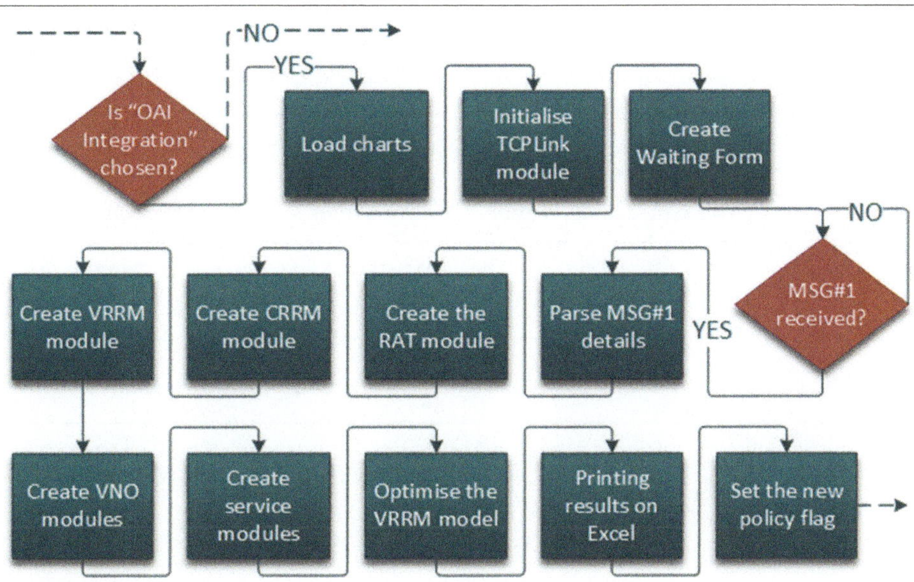

**Fig. 3** VRRM module flowchart for OAI integration

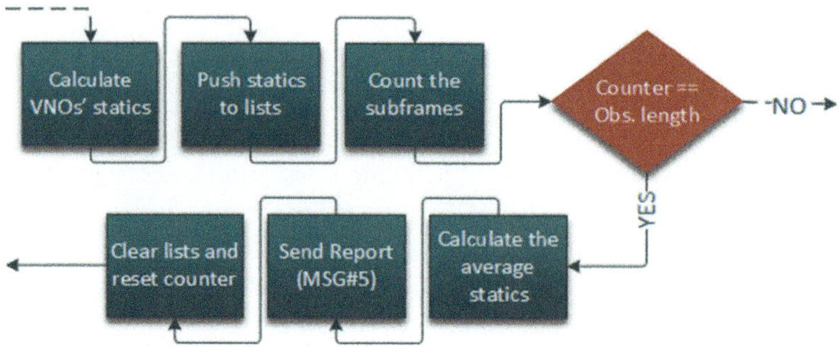


**Fig. 4** Flowchart of calculating and reporting long-term statistics

frame: MAC Protocol Data Units (PDUs), transferred size, download data rate, MCS, and active/scheduled UEs.

These statistics are used in the scheduler for enforcing the VRRM's generated policies and management of radio resources. The reports for the VRRM server contain statistics over an observation window, which may be as long as the decision window discussed in VRRM modelling [30]. In order to calculate the long-term behaviour, the statistics obtained in each sub-frame are collected by using the already implemented listing concept in OAI. The following group of variables' list was added: bitrate, MCS, and active/scheduled UEs.

These variables have to be initialised during the eNodeB MAC initialisation phase. Figure 4 shows the flowchart for collecting, calculating, and reporting the long-term statistics. As the VNO statistics become available in each sub-frame, they are pushed into the corresponding lists. At the end of each sub-frame, a counter is placed to acknowledge the number of sub-frames. When the counter's value becomes equal to the length of the observation window, the average of the aforementioned statistics pushed into the lists is calculated for the observation interval, and reported to the VRRM server, using the *Real-Time Report* message. Finally, lists are cleared, and the sub-frame counter is set to 0.

#### *5.2.2 Changing the algorithm of MAC scheduler in order to support the groups' policies*

The key methods implemented in OAI's downlink scheduler are pre-allocation and allocation (Fig. 5). The former aims at storing the UEs' Downlink–Shared Chanel (DL-SCH) buffer, and calculating the number of required RBs per UE, while in the latter, RBs are effectively allocated to UEs.

In the following, the changes introduced to this function to support multiple VNOs/groups are described. Some modifications were introduced in the UEs' sorting function. Originally, UEs were sorted based on the following aspects:

- *HARQ round*: users that are in their second round of HARQ are given higher priority.
- *Bytes in the buffer*: users that have more bytes to receive are moved to the beginning of the user list.
- *Maximum time of SDU creation*: users with more delay tolerance are pushed to the back of list, opening up space for other users.
- *UE ID*: in the case all the other criteria match, the UE ID is the last sorting criterion.

This function was changed in order to put users of higher priority VNOs on the top of the list, before checking the other criteria. Based on the policies received from VRRM, each VNO receives a portion of available RBs, the goal being to start the user list with the users of VNOs with the biggest share of RBs. The flowchart for sorting UEs is depicted in Fig. 6.

The pre-allocation sequence is modified to support the possibility of having multiple VNOs/groups with different priorities. The pre-allocation was implemented initially as it is shown in Fig. 7, without the blue boxes. When there are scheduled UEs, the RBs required to fully serve each of them are calculated; then, the possibility of allocating at least a minimum RB units (i.e. two RBs in the implementation of

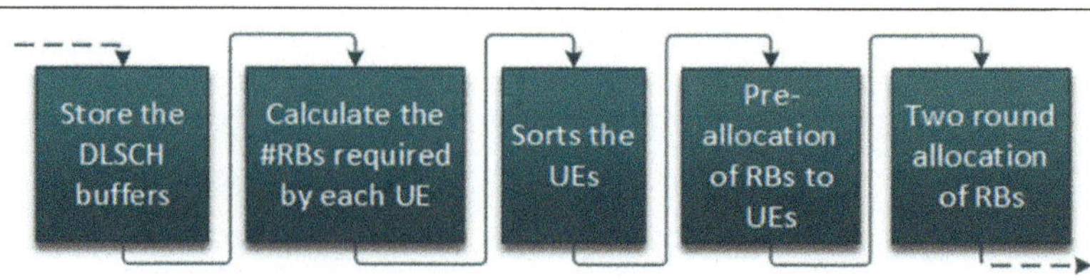


**Fig. 5** Flowchart of downlink scheduler

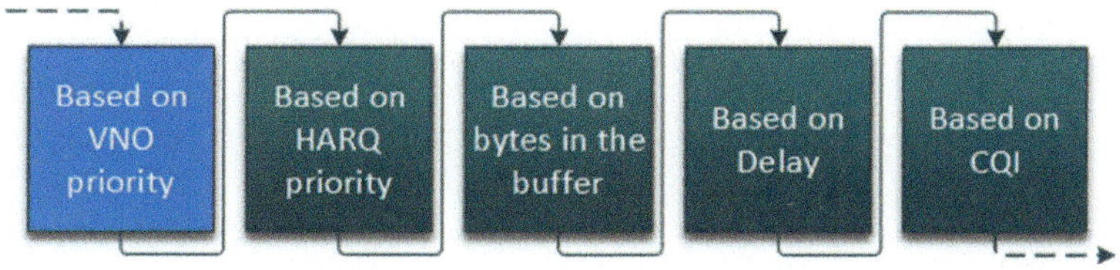


**Fig. 6** UE sorting algorithms

OAI) to all the scheduled UEs is checked. The minimum RB units are pre-allocated to all scheduled UEs when there are not enough resources; otherwise, the average RB per UE obtained in the last step is pre-allocated.

Then, based on the sorted UE list, each terminal that has to receive the control information is granted with the required RBs. Regarding data traffic, all UEs are pre-allocated the average RBs, unless the required RBs for a user is smaller than the average RBs, in which case the UE is pre-assigned only the required RBs.

In order to add multi-tenancy, the pre-allocation procedure is changed by calling a new function that maps the assigned data rate of each VNO onto its share of available resources, the blue boxes in Fig. 7. For each frame, the sharing of available resources is updated, based on the policies received from VRRM, minimum data rate for each VNO, and key performance indicators, namely, the average MCS per VNO.

The number of RBs per VNO is calculated by dividing the assigned minimum data rate by the average achieved MCS. The adaptation to the maximum number of RBs available for the eNodeB is considered in a second step, in order to avoid overbooking.

Then, the same pre-allocation procedure is done, but this time only per VNO, i.e. the average RBs per VNO is calculated, and used instead of the average RB in the previous approach. The flowchart of the pre-allocation procedure with the new algorithm is illustrated in Fig. 7. It is worthwhile noting that the primary goal was to implement the support for virtualisation of radio resources with minimum changes in OAI.

#### *5.2.3 Changing the codes to add the group information into the input XML file*

In order to input the group information, a new set of parameters identified by <PROTOCOL><MAC> tags were added to the input XML file. In the <VRRM> section, the parameters for the TCP/IP communication are defined, namely, the IP address of the machine in which VRRM is running and the TCP port. The VRRM_LINK parameter allows running OAI without the connection

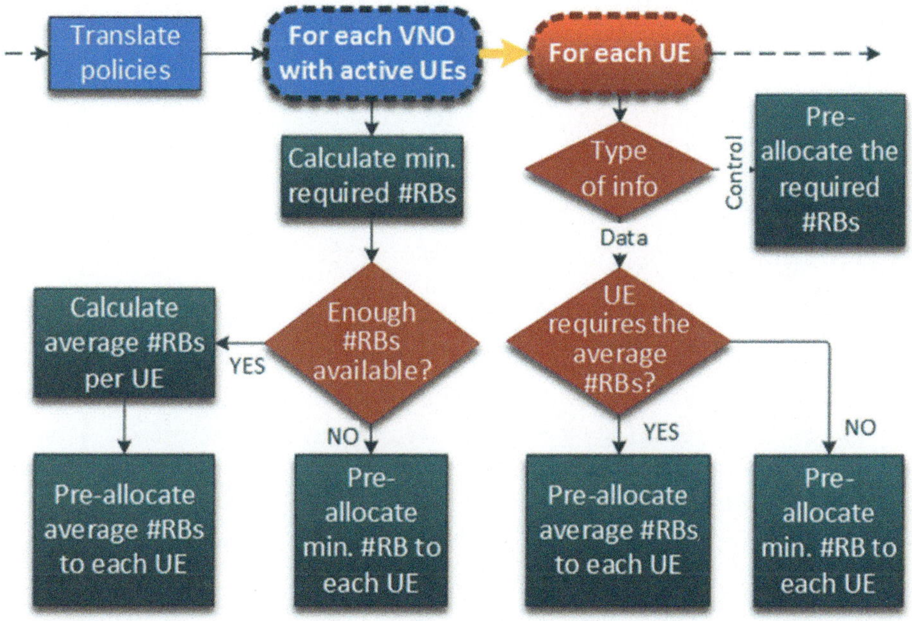


**Fig. 7** Pre-allocation procedure

**Table 1** Group type settings

| Group type | GROUP_MIN_GB | GROUP_MAX_GB |
|---|---|---|
| GB | >0 | >0 |
| BG | >0 | 0 |
| BE | 0 | 0 |

to VRRM, but maintaining the virtualisation of radio resources for serving different VNOs.

In the following lines, the number of VNOs/groups and the quantity of frames that are considered for each report sent to VRRM are set. Finally, in the <GROUPS> section, the statistic information for each group is stated. The three previously defined SLAs are considered: BE, BG, and GB; the differentiation among these groups is based on the data rate assigned to GROUP_MIN_GB and GROUP_MAX_GB (Table 1). The SERVICE_WEIGHT and VIOLATION_WEIGHT are parameters used for the optimisation performed by VRRM module, allowing distinguishing among VNOs of the same type.

#### *5.2.4 Adding support for bidirectional connection and protocols*

The communication protocol implemented between the OAI and the VRRM modules follows the sequence represented in Fig. 8.

First, the scenario description message is sent from OAI to VRRM; then, OAI requests for the policies update, and VRRM replies with a No Update Available or a Policies Update message, according to the state of policies computation, while OAI sends real-time reports and periodically requests for policies updates.

## 6 Scenario

### 6.1 Reference scenario

An urban reference scenario was defined, details being provided through the Open Air Interface Scenario Descriptor (OSD) XML file. Each OSD XML file is composed of five key parts, each of them defines a subset of components:

- *Environment/system*, including fading and antenna, as specified in Table 2. The 5 MHz bandwidth corresponds to 25 RBs. From these, 11 RBs are allocated for VRRM, the remaining data rate being reserved for the physical operator. In this sense, performance results are according to the usage of 11 RBs.
- *Topology*, including area, UE/eNodeB distribution and mobility, as specified in Table 3; a single cell with an eNodeB is located in its centre.
- *Application*, including predefined and customised traffic profiles for downlink, a Constant Bit Rate profile being used with 32 byte packets being sent continuously.
- *Emulation*, including emulation time and performance metrics.
- *VRRM*, including the number of VNOs and their requirements.

The reference scenario considers two VNOs, a BG and a BE, with four UEs each; BG has 4 Mbps of guaranteed data rate. VNO BG has higher serving priority than BE, by means of having higher serving ($W^{Srv}$) and violation ($W^{v}$) weights (0.6 vs. 0.4, and 0.64 vs. 0.36, respectively).

Two situations are taken:

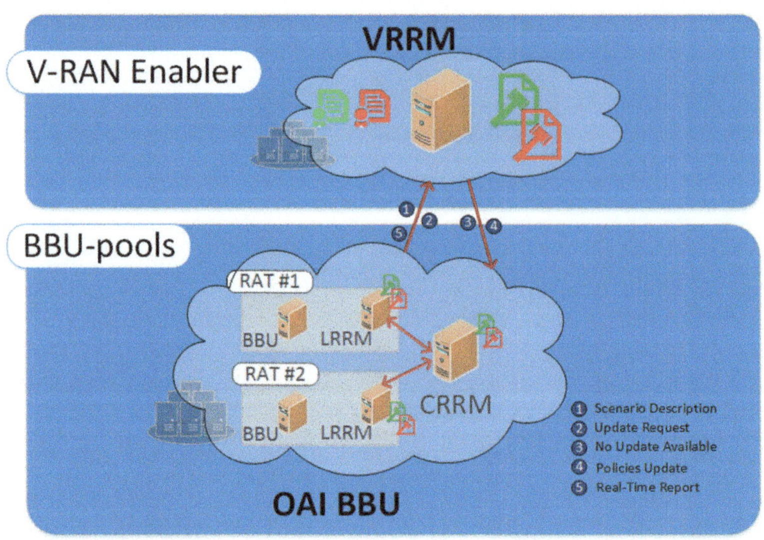


**Fig. 8** Messages exchange between VRRM and OAI

**Table 2** OAI environment/system configuration parameters

| Parameter | | | Value |
|---|---|---|---|
| SYSTEM_BANDWIDTH_MHz | | | 5 |
| SYSTEM_FREQUENCY_GHz | | | 1.9 |
| Antenna | eNB_ANTENNA | RX_NOISE_LEVEL_dB | 5 |
| | | NUMBER_OF_SECTORS | 1 |
| | | BEAM_WIDTH_dB | 1.13 |
| | | ANTENNA_GAIN_dBi | 16 |
| | | TX_POWER_dBm | 40 |
| | UE_ANTENNA | RX_NOISE_LEVEL_dB | 1 |
| | | ANTENNA_GAIN_dBi | 5 |
| | | TX_POWER_dBm | 20 |
| Fading | LARGE_SCALE | | Urban |
| | FREE_SPACE_MODEL | PATHLOSS_EXPONENT | 2 |
| | | PATHLOSS_0_dB | -100 |
| | SMALL_SCALE | | AWGN |
| WALL_PENETRATION_LOSS_dB | | | 5 |

- *Delayed VNO BG*: VNO BE subscribers join the network at the beginning of the simulation, while VNO BG ones join the network later.
- *Delayed VNO BE*: the opposite of the former situation, VNO BE's UEs being delayed.

### 6.2 Use cases

The use cases aim to demonstrate the operation of groups of users belonging to different VNOs in the same infrastructure, differentiated by different SLAs. Around the reference scenario, three use cases are defined to study the impact of several parameters (Table 4):

- *Different serving weights*: the impact of different serving weights in SLAs, considering a scenario with two VNO BE, with 0.4 and 0.6 serving weights
- *Guaranteed data rate variation*: the impact of increasing the minimum guaranteed data rate from 0 up to the total network capacity of a VNO BG
- *Different SLAs*: the combined impact of serving weight and minimum guaranteed data rate variations, two pairs of BE-BG VNOs being considered, one with serving weight 0.6 and the other with 0.4, the guaranteed data rate being gradually decreased from 4 Mbps down to 0 Mbps

## 7 Analysis of results

### 7.1 Performance of the scenario without VRRM

To demonstrate the performance of VRRM, a specific scenario was defined, consisting of a partial capacity of the 5 MHz bandwidth in the 1.9 GHz band of an LTE eNB, emulated by OAI [32]. To have an accurate knowledge of the emulator behaviour, the performance of the scenario without VRRM additional features is assessed. In fact, due to the needed computational resources of an LTE BS running on a general purpose platform, some options were taken, which affect the achieved throughputs, but do not affect the study and performance of the implemented VRRM policies that aim at differentiating users.

For the used traffic profile, a single UE achieves a maximum download data rate of 2.28 Mbps using 16QAM. Figure 9 presents the impact of the increase of the number of UEs on the network, showing the number of scheduled UEs per sub-frame, their data rate, and the total monitored network data rate, which are the key metrics to analyse the performance of VRRM.

It can be observed that by increasing the number of UEs from one to three results in an increase of the average network data rate, and a constant one for the UEs, but beyond that, the UEs' data rate decreases, as the network data rate is bounded by 6.9 Mbps, decreasing due to the rise of signalling. Figure 9 evidences also that the average number of scheduled UEs is limited to 4.7, since the number of scheduled UEs per sub-frame is restricted (in LTE) by the Number of Control Channel Elements (NCCE). For a system bandwidth of 5 MHz, NCCE allows to schedule five UEs per sub-frame; still, this number is reduced to four in some sub-frames that are used to send system signalling and control. As a result from

**Table 3** OAI topology configuration parameters

| Parameter | | | | Value |
|---|---|---|---|---|
| Area | X_m | | | 800 |
| | Y_m | | | 800 |
| Mobility | UE_MOBILITY | RANDOM_UE_DISTRIBUTION | NUMBER_OF_NODES | 10 |
| | | UE_MOBILITY_TYPE | | Static |
| | eNB_MOBILITY | eNB_INITIAL DISTRIBUTION_ | | Fixed |
| | | eNB_COORDINATES | POS_x | 400 |
| | | | POS_y | 400 |
| | | RANDOM_eNB_ DISTRIBUTION | NUMBER_ OF_CELLS | 1 |
| | | eNB_MOBILITY_TYPE | | Static |

**Table 4** Use case summary

| Use case | VNOs | Target |
|---|---|---|
| Different serving weights | 2 VNO BE | Isolated effect of serving weights |
| Guaranteed data rate variation | 1 VNO BG<br>1 VNO BE | Isolated effect of guaranteed data rate |
| Combined impact | 2 VNO BG<br>2 VNO BE | VNOs of each type with different weights. Increase of VNO BG data rate |

this limitation, it can be also observed in Fig. 9 that when the number of active UEs is increased, the average data rate per user decreases, as only a sub-set of them is fairly scheduled on each sub-frame.

### 7.2 Reference scenario

The impact of demand changes by VNOs with different SLAs (BG and BE) is studied. This is achieved by delaying the usage of one VNO, its UEs entering the network only after a pre-defined time of 4 s.

The network capacity available for VNOs is estimated theoretically by VRRM as being 6.9 Mbps, which is very close to the maximum achieved in simulations with OAI, of 7.2 Mbps (an error smaller than 4%). Based on this estimation, VRRM optimises the allocation of resources to the two VNOs, according to their SLAs (as described in Section 3.3); a resulting policy is issued to LRRM (Local RRM, implemented by OAI) to share the available resources: 5.7 Mbps is allocated to VNO BG, and 1.2 Mbps to VNO BE, corresponding to 82.6 and 17.4% of capacity, respectively. These results evidence a good match between theoretical estimations and achieved performance by simulation.

Figure 10 illustrates the data rate achieved in OAI by each of the VNOs over time. In the beginning, all network capacity is assigned to the active VNO, since the UEs from the other VNO are not requesting any. As soon as the UEs of the other VNO become active, the data rate reduces: for the *Delayed VNO BG*, Fig. 10a, it can be seen that, as soon as the VNO BG starts, VNO BE reduces from 6.72 to 0.52 Mbps (a reduction of 92.3%), VNO BG reaching 4.62 Mbps, greater than the minimum guaranteed data rate; for *Delayed VNO BE*, Fig. 10b, the data rate of VNO BG only reduces 32.5% of the initial value. As expected, when the second VNO starts, VNO BG suffers a smaller impact on the achieved data rate than VNO BE, as it has a minimum guaranteed data rate. In interval A, with four UEs, both cases achieve 6.72 Mbps, similar to the value measured without VRRM, presented in Fig. 9, while in interval B, with eight UEs, the total network data rate of the VNOs is 5.14 Mbps, also in agreement with Fig. 9. On the other hand, the VNOs' measured network data rate (4.6 and 0.52 Mbps) is lower than the one estimated by VRRM (5.7 and 1.2 Mbps), due to the fact that the VRRM model considers a statistical distribution of modulations and resulting RB capacity, and neglects signalling, while OAI always considers 16QAM modulation and signalling.

The number of scheduled UEs and the data rate per scheduled UE for each of the VNOs is presented in Table 5. On average, 3.8 UEs are served when only one VNO is active, which is not affected for VNO BG when VNO BE becomes active. In contrast, for VNO BE, this value decreases to 0.95, 25.4% of its original value. Regarding the data rate, for VNO BE, each scheduled UE receives 1.81 Mbps when VNO BG is not active, decreasing to 0.55 Mbps (31.7% of its initial value) as soon as it starts its activity, due to its "pure" best effort nature. On the other hand, for VNO BG, it only reduces to 1.22 Mbps, 66.7% of its

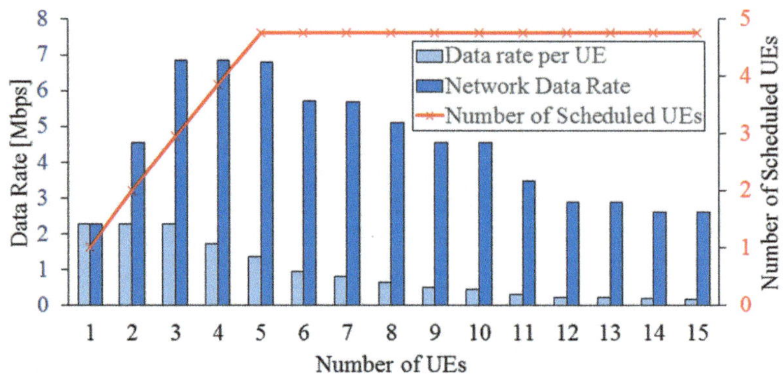


**Fig. 9** Average number of scheduled UEs and network data rate for different number of UEs, achieved in OAI

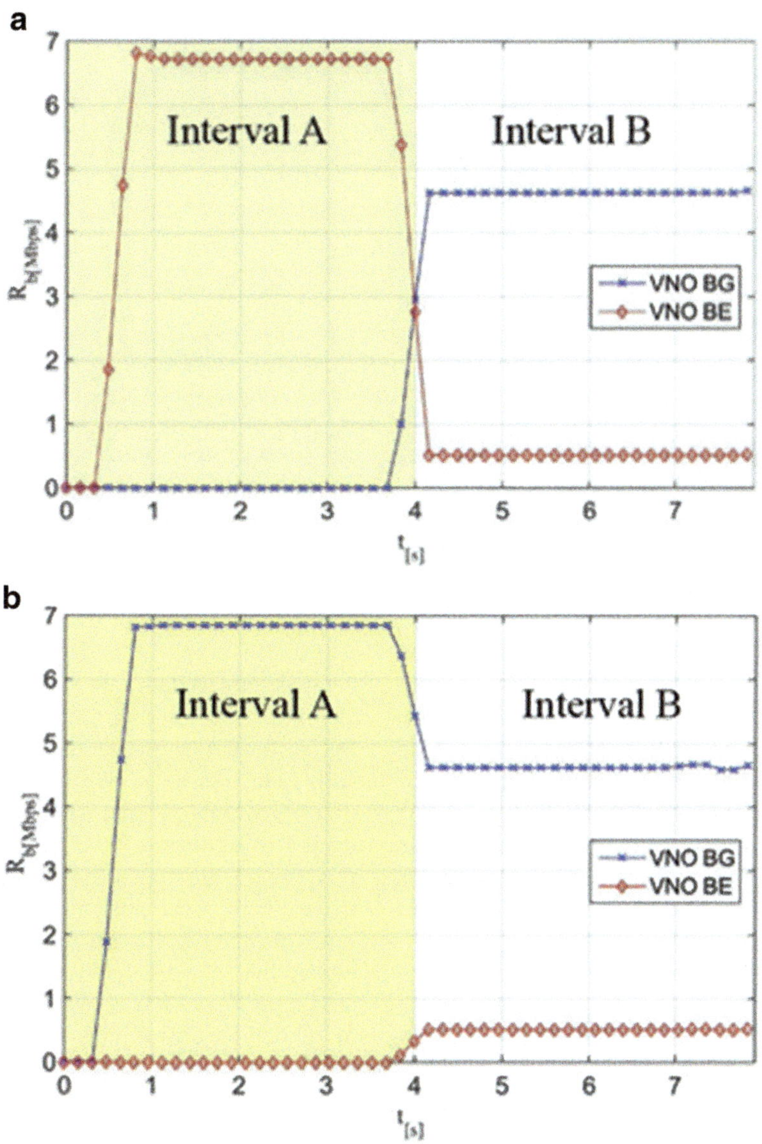


**Fig. 10** Allocated network data rates to the VNOs

**Table 5** Average number of scheduled UEs and data rate per UE

| Metric | Use case | VNO | Time interval [s] | |
|---|---|---|---|---|
| | | | A | B |
| Average number of scheduled UEs | VNO BG delayed | BG | – | 3.90 |
| | | BE | 3.80 | 0.95 |
| | VNO BE delayed | BG | 3.80 | 3.80 |
| | | BE | – | 0.95 |
| Data rate per UE [Mbps] | VNO BG delayed | BG | – | 1.22 |
| | | BE | 1.81 | 0.55 |
| | VNO BE delayed | BG | 1.81 | 1.22 |
| | | BE | – | 0.55 |

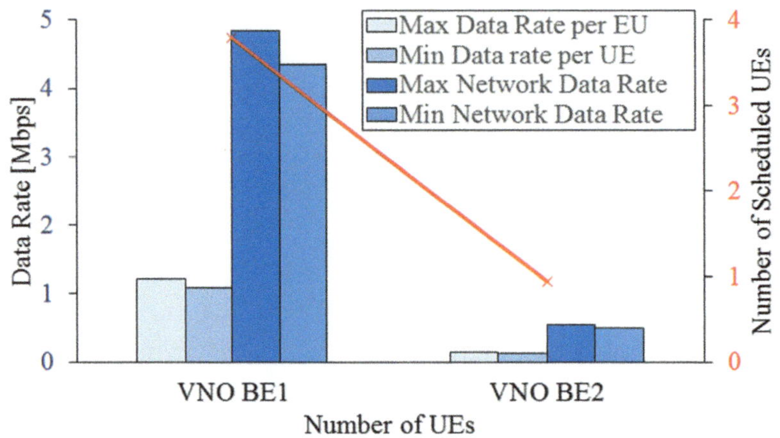


**Fig. 11** Performance of VNOs with different serving weights

original value, as expected, due to the SLA that guarantees a minimum data rate.

### 7.3 Different serving weights

The impact of different serving weights is evaluated for two VNOs BE (designated by BE1 and BE2), with four UEs each, to show the effect of prioritising one VNO relative to the other, as shown in Fig. 11. For the VNO BE1 with serving weight of 0.6, the VRRM model allocated 4.15 Mbps of network capacity, while for VNO BE2 of serving weight 0.4, this value is 2.75 Mbps. These policies result in an allocation of resources in LRRM where, in the OAI emulation environment, VNO BE1 achieves a maximum of 4.84 Mbps (89.8% of the network capacity), while VNO BE2 only 0.55 Mbps.

The number of scheduled UEs per VNO for this use case is shown in Fig. 11. Within a total of five scheduled UEs, four belong to VNO BE1, with a higher serving weight, while one UE belongs to VNO BE2. The average data rate of scheduled UEs in VNO BE1 is 1.2 Mbps, about 2.4 times higher than VNO BE2.

### 7.4 Guaranteed data rate variation

The impact of varying the guaranteed data rate for VNO BG is studied in this section. Two VNOs are considered with four UEs each, one BG and another BE, the minimum guaranteed data rate of VNO BG being varied from 1 to 4 Mbps.

Figure 12 illustrates the data rates achieved by each of these VNOs, over time. As expected, the measured data rates of VNO BG (with the highest priority) are always higher than the ones of VNO BE. For minimum guaranteed data rates of 1, 2, and 3 Mbps, the average number of scheduled UEs is very similar (2.4 and 2.3, for VNO BG and BE, respectively); however, for a minimum guaranteed data rate of 4 Mbps, almost all resources are assigned to VNO BG. In fact, the total network capacity,

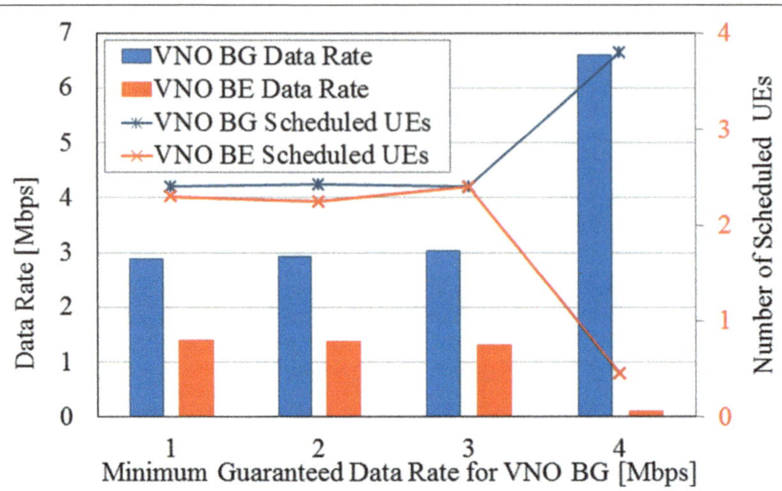


**Fig. 12** Achieved data rates per VNO and number of scheduled UEs, for different guaranteed data rates

**Table 6** VNOs' allocation policies issued by VRRM

| $R_b^{min}$ [Mbps] | VNO BG ($W^{Srv}=0.6$) | VNO BG ($W^{Srv}=0.4$) | VNO BE ($W^{Srv}=0.6$) | VNO BE ($W^{Srv}=0.4$) |
|---|---|---|---|---|
| 0 | 2.08 | 1.39 | 2.08 | 1.39 |
| 1 | 2.48 | 1.99 | 1.48 | 0.99 |
| 2 | 2.88 | 2.59 | 0.88 | 0.59 |
| 3 | 3.28 | 3.19 | 0.28 | 0.19 |
| 4 | 4.00 | 2.93 | 0.00 | 0.00 |

according to Fig. 9, for eight UEs, is 5.2 Mbps; thus, at least 76.9% of the network capacity should be assigned to VNO BG. From the VRRM side, its policy dictates an allocation of 83.1% of the network capacity to VNO BG (5.78 Mbps), and in order to achieve it, the OAI scheduler allocates almost all of the capacity to VNO BG in its first allocation round, as shown in Fig. 12. However, OAI restricts to two RBs as the minimum resources to be allocated per UE, which limits the average number of scheduled UEs to 4.7, as shown in Fig. 9.

As shown in Fig. 12, for 4 Mbps, the average number of scheduled UEs for VNO BG increases from 2.4 to 3.8 (with 83.1% of capacity), while for VNO BE, it decreases from 2.4 to 0.45. This results in a strong decrease of capacity for VNO BE, where at least two RBs must be used by each scheduled UE, leading to a large capacity made available for VNO BG, compared to VNO BE (6.6 and 0.11 Mbps, respectively).

### 7.5 Different SLAs

The goal in this section is to study the combined effect of the serving weight and the SLAs (expressed by the minimum guaranteed data rate, $R_b^{min}$). The policies calculated by the VRRM server for each SLA and VNO are presented in Table 6. According to these policies, the VNO with the higher serving weight and the higher guaranteed data rate has the highest allocated data rate.

The aforementioned policies are applied by the OAI scheduler and numeric results are generated. Figure 13 presents the maximum data rates achieved by each of the VNOs, being apparent that the scheduler manages to follow the provided policies, while the maximum number of scheduled UEs per VNO is presented in Fig. 14.

From Fig. 13, one can see that when $R_b^{min}$ for VNO BG is set to 0, in practice, all are VNO BE, only differentiated by the serving weights, in which case, the VNOs with the higher serving weights ($W^{Srv}=0.6$) receive a data rate of 2.42 Mbps; the other two VNOs receive the remaining capacity, which is less than 0.5 Mbps. By increasing VNO BGs' $R_b^{min}$ to 1 Mbps, the VNO BE with $W^{Srv}=0.4$ has no capacity left to be allocated, while for $R_b^{min}$ >3 Mbps, both VNO BE stop receiving capacity and, for $R_b^{min}=3$ Mbps, the network capacity for VNO BG is at least 6 Mbps. This means that all the available resources have to be assigned to VNO BG to meet the guaranteed data rates.

As it is apparent in Fig. 13, these two VNOs are always receiving the minimum guaranteed data rates, but the other two VNOs are not receiving any resources. When $R_b^{min}=4$ Mbps for each VNO BG, there is not enough capacity to satisfy it. VRRM detects the situation and tries to minimise the SLAs violation. As shown in Table 6, the VRRM model allocates 4 Mbps to VNO BG with $W^{Srv}=0.6$, the remaining resources (2.93 Mbps) being allocated to VNO BG with $W^{Srv}=0.4$. The results in Fig. 13 confirm that the VRRM policy was put into practice by the scheduler in OAI.

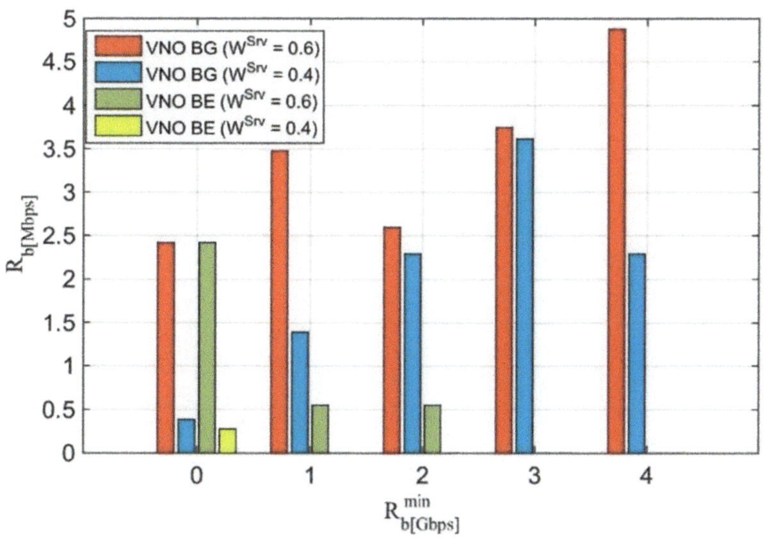


**Fig. 13** The maximum allocated data rates to the VNOs

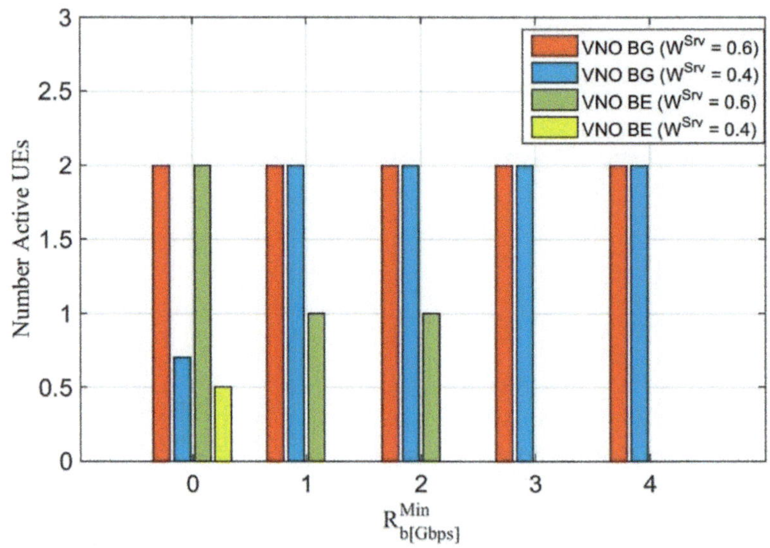


**Fig. 14** Maximum scheduled UEs

Finally, the maximum number of scheduled UEs per VNO is presented in Fig. 14. The VNOs with guaranteed SLAs have a maximum of two scheduled UEs; the maximum number of scheduled UEs for VNO BE with $W^{Srv}$ = 0.6 reaches 1, while for VNO BE with $W^{Srv}$ = 0.4 the average number of scheduled UEs does not surpass 0.5. The increase of $R_b^{min}$ results in a reduction of BE scheduled UEs, first the ones with lower serving weight, then both.

## 8 Conclusions

This paper addresses the virtualisation of radio resources as part of RAN virtualisation, a candidate solution for 5G networks, in order to achieve flexibility and efficiency. A virtual RAN operates over cloud data centres (known as C-RAN) as the physical infrastructure. A V-RAN serves multiple VNOs, which do not own access infrastructure and demand for connectivity to serve their subscribers. Virtualisation of radio resources proposes to aggregate all the radio resources from different RATs and provides virtual links for each of the VNOs based on their SLAs and requirements. The architecture for the V-RAN is illustrated and briefly discussed.

In addition, a comprehensive analytical model for managing virtual radio resources is proposed, which has two key components: the estimation of available radio resources and their allocation to different VNOs. The estimation of available resources is made by VRRM. Based on the data rates of different access technologies as a function of SINR, and the distribution of the SINR, the PDF of a single RRU of each RAT is calculated. After estimating the total network capacity, VRRM allocates it to the various services of the network, in order to maximise the usage efficiency, considering the priority of the services of different VNOs, based on their SLAs.

Furthermore, the aforementioned concept and model were implemented in OAI, a software-based LTE eNodeB physical emulator. The implementation guideline was to create multiple groups of subscribers, add support for different policies for each group in the scheduler, and create the communication links to/from the VRRM server. Various changes have been introduced to OAI, including new sets of variables.

The performance of the network in different situations, such as different minimum guaranteed data rate levels and different serving weights, is studied. The numeric results show that the proposed approach guarantees the target capacity to guaranteed VNO, regardless of the others. At worst, it experiences a reduction of 32.5% of its maximum achieved data rate, never going beyond the agreed minimum guaranteed data rate. For best effort VNOs, the allocated data rate may decrease 93% of its initial value, which is acceptable due to its nature, to guarantee other more demanding SLAs.

In conclusion, the proposed approach is of value, and its implementation in a software-based emulator shows its validity.

**Abbreviations**

BBU: Base band unit; BE: Best effort; BG: Best effort with minimum guaranteed; BS: Base station; CAPEX: CAPital and oPerational eXpenditures; CDF: Cumulative density function; C-RAN: Cloud-RAN; DLL: Dynamic library link; GB: Guaranteed bitrate; IoT: Internet of things; LTE: Long-term evolution; M2M: Machine-to-machine; MIMO: Multiple input multiple output; MT: Mobile terminal; Multi-RAT: Multi-radio access technologies; OAI: Open air interface; OPEX: OPerational expenditures; PC: Personal computer; PDF: Probability density function; PDU: Protocol data unit; QoS: Quality of service; RAN: Radio access network; RRH: Remote radio heads; RRU: Radio resource unit; SDN: Software-defined network; SINR: Signal to interference plus noise ratio; SLAs: Service level agreements; TCP: Transmission control protocol; UE: User equipment; VNO: Virtual mobile network operator; V-RAN: Virtual rAN; VRRM: Virtual radio resource management


#### Acknowledgements
The research leading to these results was partially funded by EU's Seventh Framework Programme Mobile Cloud Networking project (FP7-ICT-318109).

#### Funding
The research leading to these results was partially funded by the European Union's Seventh Framework Programme Mobile Cloud Networking project (FP7-ICT-318109); the remaining funding was obtained via internal projects from the authors' institutions. The sole responsibility on the paper and its findings is of the authors, and the funding institutions had no role in it.


#### Authors' contributions
The authors have contributed jointly to all parts on the preparation of this manuscript, and all authors read and approved the final manuscript.

#### Competing interests
The authors declare that they have no competing interests.


#### Author details
[1]IST/INESC-ID, University of Lisbon, Lisbon, Portugal. [2]Nomor Research GmbH, Munich, Germany. [3]IPS-EST Setúbal, Setúbal and INESC-ID, Lisbon, Portugal. [4]Lisbon Lusíada University and INESC-ID, Lisbon, Portugal. [5]Eurecom, Biot, France.


#### References

1. Cisco Systems, Global Mobile Data Traffic Forecast Update, 2012 – 2017, *Visual network index (VNI) white paper* (Cisco Systems, San José, 2013)
2. H Guan, T Kolding, P Merz, Discovery of cloud-RAN, in *Proc. of NSN Cloud-RAN Workshop* (Beijing, 2010). http://www.thecom.co.il/files/wordocs/articledownload.pdf
3. Y Zaki, Z Liang, C Goerg, A Timm-Giel, LTE wireless virtualization and spectrum management, in *Proc. of 3rd Joint IFIP Wireless and Mobile Networking Conference* (IEEE, Budapest, 2010)
4. NMMK Chowdhury, R Boutaba, A survey of network virtualization. Comput Netw **54**(5), 862–876 (2010)
5. W Xin, P Krishnamurthy, D Tipper, Wireless network virtualization, in *Proc. of International Conference on Computing, Networking and Communications* (IEEE, San Diego, 2013)
6. Microsoft, *SQL server*, 2015 [Online]
7. L Chengchao, FR Yu, Wireless network virtualization: a survey, some research issues and challenges. IEEE Commun Surv Tutorials **17**(1), 358–380 (2015)
8. XC Perez, J Swetina, T Guo, R Mahindra, S Rangarajan, Radio access network virtualization and sharing for future mobile carrier networks. IEEE Commun Mag **51**(7), 27–35 (2013)
9. Huawei Technology, *5G: new air interface and radio access virtualisation*. Huawei White Paper, 2015
10. China Mobile Research Institute, *C-RAN—road towards green radio access network* (Shanghai, China Mobile Research Institute, 2011)
11. Z Heli, W Weidong, L Xi, J Hong, User association scheme in cloud-RAN based small cell network with wireless virtualization, in *Proc. of IEEE Conference on Computer Communications* (IEEE, Hong Kong, China, 2015)
12. L Chengchao, FR Yu, Wireless virtualization for next generation mobile cellular networks. IEEE Wirel Commun **22**(1), 61–69 (2015)
13. Z Zhu, P Gupta, Q Wang, S Kalyanaraman, Y Lin, H Franke, S Sarangi, Virtual base station pool: towards a wireless network cloud for radio access networks, in *Proc. of 8th ACM International Conference on Computing Frontiers* (ACM Digital Library, Ischia, 2011)
14. J Monserrat, G Mange, V Braun, H Tullberg, G Zimmermann, O Bulakci, METIS research advances towards the 5G mobile and wireless system definition. EURASIP J Wirel Commun Netw **53**(1), 1–16 (2015)
15. I Chih-Lin, S Han, Z Xu, Q Sun, Z Pan, 5G: rethink mobile communications for 2020+. Philos Trans R Soc A Math Phys Eng Sci **374**, No. 2062 (2016)
16. I Chih-Lin, C Rowell, H Shuangfeng, X Zhikun, L Gang, P Zhengang, Toward green and soft: a 5G perspective. IEEE Commun Mag **52**(2), 66–73 (2014)
17. V Borges, K Cardoso, E Cerqueira, M Nogueira, A Santos, Aspirations, challenges, and open issues for software-based 5G networks in extremely dense and heterogeneous scenarios. EURASIP J Wirel Commun Netw **164**(1), 1–13 (2015)
18. O Bejarano, EW Knightly, P Minyoung, IEEE 802.11ac: from channelization to multi-user MIMO. IEEE Commun Mag **51**(10), 84–90 (2013)
19. D Pompili, A Hajisami, TX Tran, Elastic resource utilization framework for high capacity and energy efficiency in cloud RAN. IEEE Commun Mag **54**(1), 26–32 (2016)
20. X Hongyu, P Mugen, C Yuanyuan, C Hsiao-Hwa, Joint mode selection and resource allocation for downlink fog radio access networks supported D2D, in *Proc. of 11th International Conference on Heterogeneous Networking for Quality, Reliability, Security and Robustness* (IEEE, Taipei, 2015)
21. MI Kamel, L Long Bao, A Girard, LTE wireless network virtualization: dynamic slicing via flexible scheduling, in *Proc. of IEEE 80th Vehicular Technology Conference* (Vancouver, 2014)
22. G Tseliou, F Adelantado, C Verikoukis, Resources negotiation for network virtualization in LTE-A networks, in *Proc. of IEEE International Conference on Communications* (Sydney, 2014)
23. K Zarifi, H Baligh, M Jianglei, M Salem, A Maaref, Radio access virtualization: cell follows user, in *Proc. of 25th IEEE Annual International Symposium on Personal, Indoor, and Mobile Radio Communication* (IEEE, Washington DC, 2014)
24. S Kim, Dynamic C-RAN resource sharing scheme based on a hierarchical game approach. EURASIP J Wirel Commun Netw **3**(1), 1–12 (2016)
25. A Beylerian, T Ohtsuki, Multi-point fairness in resource allocation for C-RAN downlink CoMP transmission. EURASIP J Wirel Commun Netw **2016**(1), 12 (2016)
26. M Kalil, A Shami, Y Yinghua, Wireless resources virtualization in LTE systems, in *Proc. of IEEE Conference on Computer Communications* (IEEE, Toronto, Toronto, 2014)
27. Y Mao, L Yong, J Depeng, Y Jian, S Li, Z Lieguang, Opportunistic spectrum sharing based resource allocation for wireless virtualization, in *Proc. of 7th International Conference on Innovative Mobile and Internet Services in Ubiquitous Computing* (Taichung, 2013)
28. D Garlisi, F Giuliano, I Tinnirello, P Gallo, F Gringoli, G Bianchi, Deploying virtual MAC protocols over a shared access infrastructure using MAClets, in *Proc. of IEEE Conference on Computer Communications* (Turin, 2013)
29. S Khatibi, LM Correia, Modelling of virtual radio resource management for cellular heterogeneous access networks, in *Proc. of IEEE 25th Annual International Symposium on Personal, Indoor, and Mobile Radio Communications* (IEEE, Washington, DC, 2014)
30. S Khatibi, LM Correia, A model for virtual radio resource management in virtual RANs. EURASIP J Wirel Commun Netw **2015**(1), 68 (2015)
31. S Khatibi, LM Correia, Modelling virtual radio resource management with traffic offloading support, in *Proc. of IEEE 24th European Conference on Network and Communication* (IEEE, Paris, 2015)
32. Open Air Interface, 2015 (http://www.openairinterface.org). Accessed Feb 2015
33. Mobile Cloud Networking, Algorithms and mechanisms for the mobile network cloud, in *Project Report*, ed. by C Parada, 2014
34. B Haberland, F Derakhshan, H Grob-Lipski, R Klotsche, W Rehm, P Schefczik, M Soellner, Radio base stations in the cloud. Bell Labs Tech J **18**(1), 129–152 (2013)
35. S Khatibi, LM Correia, The effect of channel quality on virtual radio resource management, in *Proc. of IEEE 82nd Vehicular Technology Conference* (IEEE, Boston, 2015)
36. A Papoulis, SU Pillai, *Probability, random variables, and stochastic processes* (McGraw-Hill, 2002)
37. The MathWorks Inc, *MATLAB and statistics toolbox* (Natick, 2015). http://www.mathworks.com. Accessed Feb 2014

# Permissions

All chapters in this book were first published in EURASIP JWCN, by Springer International Publishing AG.; hereby published with permission under the Creative Commons Attribution License or equivalent. Every chapter published in this book has been scrutinized by our experts. Their significance has been extensively debated. The topics covered herein carry significant findings which will fuel the growth of the discipline. They may even be implemented as practical applications or may be referred to as a beginning point for another development.

The contributors of this book come from diverse backgrounds, making this book a truly international effort. This book will bring forth new frontiers with its revolutionizing research information and detailed analysis of the nascent developments around the world.

We would like to thank all the contributing authors for lending their expertise to make the book truly unique. They have played a crucial role in the development of this book. Without their invaluable contributions this book wouldn't have been possible. They have made vital efforts to compile up to date information on the varied aspects of this subject to make this book a valuable addition to the collection of many professionals and students.

This book was conceptualized with the vision of imparting up-to-date information and advanced data in this field. To ensure the same, a matchless editorial board was set up. Every individual on the board went through rigorous rounds of assessment to prove their worth. After which they invested a large part of their time researching and compiling the most relevant data for our readers.

The editorial board has been involved in producing this book since its inception. They have spent rigorous hours researching and exploring the diverse topics which have resulted in the successful publishing of this book. They have passed on their knowledge of decades through this book. To expedite this challenging task, the publisher supported the team at every step. A small team of assistant editors was also appointed to further simplify the editing procedure and attain best results for the readers.

Apart from the editorial board, the designing team has also invested a significant amount of their time in understanding the subject and creating the most relevant covers. They scrutinized every image to scout for the most suitable representation of the subject and create an appropriate cover for the book.

The publishing team has been an ardent support to the editorial, designing and production team. Their endless efforts to recruit the best for this project, has resulted in the accomplishment of this book. They are a veteran in the field of academics and their pool of knowledge is as vast as their experience in printing. Their expertise and guidance has proved useful at every step. Their uncompromising quality standards have made this book an exceptional effort. Their encouragement from time to time has been an inspiration for everyone.

The publisher and the editorial board hope that this book will prove to be a valuable piece of knowledge for researchers, students, practitioners and scholars across the globe.

# List of Contributors

**Caimei Fu, Youming Li, Ming Jin, Gang Wang and Peng Lei**
College of Information Science and Engineering, Ningbo University, Ningbo, China

**Yucheng He**
School of Information Science and Engineering, National Huaqiao University, Xiamen, China

**Weidong Gao, Lin Ma and Gang Chuai**
Beijing University of Posts and Telecommunications, Beijing 100876, China

**Junhai Luo and Zuoting Liu**
School of Electronic Engineering, University of Electronic Science and Technology of China, No. 2006, Xiyuan Ave, West Hi-Tech Zone, 611731 Chengdu, China

**Alireza Vahid**
Electrical Engineering Department, University of Colorado, Denver, Denver 80204, CO, USA

**Vaneet Aggarwal**
School of Industrial Engineering, Purdue University, West Lafayette 47907, IN, USA

**A. Salman Avestimehr**
Electrical Engineering Department, University of Southern California, Los Angeles 90089, CA, USA

**Ashutosh Sabharwal**
Electrical and Computer Engineering Department, Rice University, Houston 77251, TX, USA

**Ke Yan, Guangchun Luo, Ling Tian, Qi Jia and Chengzong Peng**
School of Computer Science and Engineering, University of Electronic Science and Technology of China, Chengdu 611731, China

**Oscar D. Ramos-Cantor, Ganapati Hegde and Marius Pesavento**
Communication Systems Group, Darmstadt University of Technology, 64283 Darmstadt, Germany

**Jakob Belschner**
Technology Innovation Division, Deutsche Telekom AG, 64295 Darmstadt, Germany

**Mingyu Lee, Yunmin Kim and Tae-Jin Lee**
College of Information and Communication Engineering, Sungkyunkwan University, Suwon 16419, South Korea

**Shunlan Zhang and Qing Peng**
Guangxi Key Laboratory of Wireless Wideband Communication and Signal Processing, Jinji Road, Guilin 541004, China

**Ju Ni**
Library of Guilin University of Electronic Technology, Jinji Road, Guilin 541004, China

**Dina Deif and Yasser Gadallah**
Electronics and Communications Department, School of Sciences and Engineering, The American University in Cairo, AUC Avenue, New Cairo 11835, Egypt

**Anas M. Salhab and Salam A. Zummo**
Electrical Engineering Department, King Fahd University of Petroleum & Minerals, Dhahran 31261, Saudi Arabia

**Husam Y. Alzaq and B. Berk Ustundag**
Department of Computer Engineering, Faculty of Computer Engineering, Istanbul Technical University, Maslak, Ayazaga, 34469 Istanbul, Turkey

**Weidong Mei, Zhi Chen, Jun Fang and Shaoqian Li**
National Key Laboratory of Science and Technology on Communications, University of Electronic Science and Technology of China, No. 2006, Xiyuan Ave, West Hi-Tech Zone, Chengdu 611731, China

**Leonel Arévalo, Rodrigo C. de Lamare and Raimundo Sampaio-Neto**
Center for Studies in Telecommunications (CETUC), Pontifical Catholic University of Rio de Janeiro (PUC-Rio), Rio de Janeiro, Brazil

**Martin Haardt**
Communications Research Laboratory, Ilmenau University of Technology, Ilmenau, Germany

**Guo Kefeng , Guo Daoxing, Huang Yuzhen and Zhang Bangning**
College of Communications Engineering, PLA University of Science and Technology, Guanghua Road, Nanjing, China

**Weifeng Mou, Weiwei Yang , Yuzhen Huang, Xiaoming Xu, Yueming Cai and Kaihua Wang**
College of Communications Engineering, PLA University of Science and Technology, Nanjing 210007, China

**Luis M. Correia**
IST/INESC-ID, University of Lisbon, Lisbon, Portugal

**Sina Khatibi**
IST/INESC-ID, University of Lisbon, Lisbon, Portugal Nomor Research GmbH, Munich, Germany

**Luísa Caeiro**
IPS-EST Setúbal, Setúbal and INESC-ID, Lisbon, Portugal

**Lúcio S. Ferreira**
Lisbon Lusíada University and INESC-ID, Lisbon, Portugal

**Navid Nikaein**
Eurecom, Biot, France

# Index

## P



## Q



## R



## S



## T



## U



## V



## W